F.C.

Engineering Economy

PRENTICE HALL INTERNATIONAL SERIES
IN INDUSTRIAL AND SYSTEMS ENGINEERING

W. J. Fabrycky and J. H. Mize, Editors

ALEXANDER *The Practice and Management of Industrial Ergonomics*
AMOS AND SARCHET *Management for Engineers*
ARMINE, RITCHEY, MOODIE, AND KMEC *Manufacturing Organization and Management,* 6/E
ASFAHL *Industrial Safety and Health Management,* 2/E
BABCOCK *Managing Engineering and Technology*
BADIRU *Expert Systems Analysis in Engineering and Manufacturing*
BANKS AND CARSON *Discrete-Event System Simulation*
BANKS AND FABRYCKY *Procurement and Inventory Systems Analysis*
BEIGHTLER, PHILLIPS, AND WILDE *Foundations of Optimization,* 2/E
BLANCHARD *Logistics Engineering and Management,* 4/E
BLANCHARD AND FABRYCKY *Systems Engineering and Analysis,* 2/E
BROWN *Systems Analysis and Design for Safety*
BUSSEY AND ESCHENBACH *The Economic Analysis of Industrial Projects,* 2/E
CANADA AND SULLIVAN *Economic and Multi-Attribute Evaluation of Advanced Manufacturing Systems*
CHANG AND WYSK *An Introduction to Automated Process Planning Systems*
CHANG, WYSK, AND WANG *Computer Aided Manufacturing*
CLYMER *Systems Analysis Using Simulation and Markov Models*
ELSAYED AND BOUCHER *Analysis and Control of Production Systems,* 2/E
FABRYCKY AND BLANCHARD *Life-Cycle Cost and Economic Analysis*
FABRYCKY, GHARE, AND TORGERSEN *Applied Operations Research and Management Science*
FABRYCKY AND THUESEN *Economic Decision Analysis,* 2/E
FRANCIS, MCGINNIS, AND WHITE *Facility Layout and Location: An Analytical Approach,* 2/E
GIBSON *Modern Management of the High-Technology Enterprise*
HALL *Queing Methods: For Services and Manufacturing*
HAMMER *Occupational Safety Management and Engineering,* 4/E
HUTCHINSON *An Integrated Approach to Logistics Management*
IGNIZIO *Linear Programming in Single- and Multiple-Objective Systems*
KUSIAK *Intelligent Manufacturing Systems*
MUNDEL *Improving Productivity and Effectiveness*
MUNDEL *Motion and Time Study: Improving Productivity,* 6/E
OSTWALD *Engineering Cost Estimating,* 3/E
PHILLIPS AND GARCIA-DIAZ *Fundamentals of Network Analysis*
PULAT *Fundamentals of Industrial Ergonomics*
SANDQUIST *Introduction to System Science*
SMALLEY *Hospital Management Engineering*
TAHA *Simulation Modeling and SIMNET*
THUESEN AND FABRYCKY *Engineering Economy,* 8/E
TURNER, MIZE, CASE, AND NAZEMETZ *Introduction to Industrial and Systems Engineering,* 3/E
WHITEHOUSE *Systems Analysis and Design Using Network Techniques*
WOLFF *Stochastic Modeling and the Theory of Queues*

Eighth Edition

ENGINEERING ECONOMY

G. J. Thuesen

Georgia Institute of Technology

W. J. Fabrycky

Virginia Polytechnic Institute and State University

PRENTICE HALL Englewood Cliffs, New Jersey 07632

Library of Congress Cataloging-in-Publication Data

THUESEN, G. J.
Engineering economy / G.J. Thuesen, W.J. Fabrycky. — 8th ed.
p. cm.
Includes bibliographical references and index.
ISBN 0-13-279928-6
1. Engineering economy. I. Fabrycky, W. J. (Wolter J.)
II. Title.
TA177.4.T47 1993
658.115—dc20 92-28371
CIP

Acquisitions Editor: Marcia Horton
Production Editor: Joe Scordato
Copy Editor: Bob Lentz
Designer: Meryl Poweski
Prepress Buyer: Linda Behrens
Manufacturing Buyer: Dave Dickey
Supplements Editor: Alice Dworkin
Editorial Assistant: Dolores Mars

Printed in the United States of America
10 9 8 7 6 5 4 3 2 1

ISBN 0-13-279928-6

Prentice-Hall International (UK) Limited, *London*
Prentice-Hall of Australia Pty. Limited, *Sydney*
Prentice-Hall Canada Inc., *Toronto*
Prentice-Hall Hispanoamericana, S.A., *Mexico*
Prentice-Hall of India Private Limited, *New Delhi*
Prentice-Hall of Japan, Inc., *Tokyo*
Simon & Schuster Asia Pte. Ltd., *Singapore*
Editora Prentice-Hall do Brasil, Ltda., *Rio de Janeiro*

To H. G. Thuesen—
the founding author of this text—
whose wisdom as an educator, writer,
inventor, friend, and parent has
inspired those he influenced

Contents

Preface

This Eighth Edition of *Engineering Economy* continues to emphasize the concepts and techniques of analysis useful in evaluating the worth of systems, products, and services in relation to their cost. Our objective is to help the reader appreciate the significance of the economic aspects of engineering and to become proficient in the evaluation of engineering proposals in terms of worth and cost. Economic feasibility is the essential prerequisite of successful engineering for the increasingly competitive global marketplace.

The engineering approach to problem solution has advanced and broadened to the extent that success often depends upon the ability to deal with both economic and physical factors. Being accustomed to the use of facts and being proficient in computation, engineers should accept the responsibility for providing interpretation of their work. It is easier for engineers to master the fundamental concepts of economic analysis necessary to bridge the gap between the physical and economic aspects of engineering application than for persons who are not technically trained to acquire the necessary technical background. To aid the engineer in so doing is the primary aim of this book.

A secondary aim of this book is to acquaint the engineer with operations and operational feasibility. Economic factors in the operation of systems and equipment can no longer be left to chance, but must be considered during the design process. A basic understanding of mathematical modeling of operation is becoming more important as complex operational systems require the attention of a larger number of engineers. Accordingly, the section entitled Operation Economy has been broadened to emphasize the connection between design and operations.

Those familiar with earlier editions of this text will note that we have retained the basic conceptual approach with considerable emphasis on examples. Also retained is the functional factor designation system originated by the late H. G. Thuesen. Symbols in the system have been changed as was suggested by the American National Standards Institute Committee on Industrial Engineering Terminology.

Significant format changes have been made in this edition. Most equations are now numbered and key equations are displayed. Definitions are set apart and displayed. Principles are highlighted. Also, the reader will find key points listed at the end of each chapter.

We have had no difficulty in teaching this material to engineering sophomores as well as to upper-division students in management, economics, and the physical sciences who wish to obtain an introduction to engineering from the economic point of view. Elementary calculus is the only mathematical background required.

This text contains more than enough material for a three-semester-hour course. For a course of shorter duration some material will have to be omitted. This may be easily done, since the foundation topics are concentrated in the first ten chapters.

We would like to acknowledge those who recieved earlier versions of the manuscript for this book: Theo A. DeWinter of Boston University, Lawrence M. Seiford of the University of Massachusetts at Amherst, Sheng-Hsien Teng of the University of Wisconsin at Madison, James M. Daschbach of the University of Toledo, and John Malindretos of the New Jersey Institute of Technology.

It is our pleasure to acknowledge the very useful comments offered by students and practicing professionals who used prior editions of this text over many years. Also, we acknowledge the expert editorial and word-processing assistance provided by Mrs. LaVonda Matherly and Mrs. Joene Owen.

G. J. Thuesen
W. J. Fabrycky

Part One

INTRODUCTION TO ENGINEERING ECONOMY

The first part of this text introduces engineering economy—its relationship to engineering and to the engineering process. The focus is on technical concepts to support the process for performing engineering economy studies. Selected economic concepts and cost classifications are then provided as a foundation for quantitative methods that will appear in subsequent chapters. Next, interest and interest rate are introduced as the basis for time value analysis. Finally, Part One presents the concept of the earning power of money as the primary justification for borrowing to increase productivity.

1

Engineering and Engineering Economy

Engineering activities of analysis and design are not an end in themselves. They are a means for satisfying human wants. Thus, engineering has two concerns: the materials and forces of nature, and the needs of people. Because of resource constraints, engineering must be closely associated with economics. It is essential that engineering proposals be evaluated in terms of worth and cost before they are undertaken. In this chapter and throughout the text, we emphasize that an essential prerequisite of successful engineering application is economic feasibility.

1.1 ENGINEERING AND SCIENCE

Engineering is not a science, but an application of science. It is an art composed of skill and ingenuity in adapting knowledge to the uses of humanity. The Accreditation Board for Engineering and Technology has adopted the definition

> **Engineering** is the profession in which a knowledge of the mathematical and natural sciences gained by study, experience, and practice is applied with judgment to develop ways to utilize, economically, the materials and forces of nature for the benefit of mankind.

This, like most other accepted definitions, emphasizes the applied nature of engineering.

The role of the scientist is to add to mankind's accumulated body of systematic knowledge and to discover universal laws of behavior. The role of the engineer is to apply this knowledge to particular situations to produce products and services. To the engineer, knowledge is not an end in itself but is the raw material from which he or she fashions structures, systems, products, and services. Thus, engineering involves the determination of the combination of materials, forces, information, and human factors that will yield a desired result. Engineering activities are rarely carried out for the satisfaction that may be derived from them directly. With few exceptions, their use is confined to satisfying human wants.

Modern civilization depends to a large degree upon engineering. Most products and services used to facilitate work, communication, transportation, and national defense and to furnish sustenance, shelter, and health are directly or indirectly a result of engineering activity. Engineering has also been instrumental in providing leisure time for pursuing and enjoying culture. Through the development of instant communication and rapid transportation, engineering has provided the means for both cultural and economic improvement of humanity.

Science is the foundation upon which the engineer builds toward the advancement of mankind. With the continued development of science and the worldwide application of engineering, the general standard of living may be expected to improve and further increase the demand for those things that contribute to people's love for the comfortable and beautiful. The fact that these human wants may be expected increasingly to engage the attention of engineers is, in part, the basis for the incorporation of humanistic and social considerations in engineering curricula. An understanding of these fields is essential as engineers seek solutions to the complex sociotechnological problems of today.

1.2 THE BI-ENVIRONMENTAL NATURE OF ENGINEERING

Engineers are confronted with two important interconnected environments, the *physical* and the *economic.* Their success in altering the physical environment to produce products and services depends upon a knowledge of physical laws. However, the worth of these products and services lies in their utility measured in economic terms. There are numerous examples of structures, machines, processes, and systems that exhibit excellent physical design but have little economic merit.

Want satisfaction in the economic environment and engineering proposals in the physical environment are linked by the production or the construction process. Figure 1.1 illustrates the relationship between engineering proposals, production or construction, and want satisfaction.

In dealing with the physical environment engineers have a body of physical laws upon which to base their reasoning. Such laws as Boyle's law, Ohm's law, and Newton's laws of motion were developed primarily by collecting and

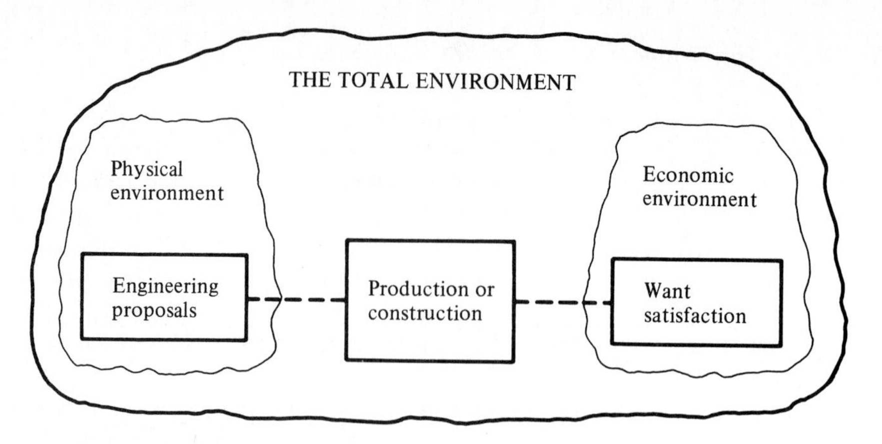

FIGURE 1.1. The physical and economic environments.

comparing numerous similar instances and by the use of an inductive process. These laws may then be applied by deduction to specific instances. They are supplemented by many formulas and known facts, all of which enable the engineer to come to conclusions that match the facts of the physical environment within narrow limits. Much is known with certainty about the physical environment.

Much less, particularly of a quantitative nature, is known about the economic environment. Since economics is involved with the actions of people, it is apparent that economic laws must be based upon their behavior. Economic laws can be no more exact than the description of the behavior of people acting singly and collectively.

The usual function of engineering is to manipulate the elements of one environment, the physical, to create value in a second environment, the economic. However, engineers sometimes have a tendency to disregard economic feasibility and are often appalled in practice by the necessity for meeting situations in which action must be based on estimates and judgment. Yet today's engineering graduates are increasingly finding themselves in positions in which their responsibility is extended to include economic considerations.

There are those, and some are engineers, who feel that engineers should restrict themselves to the consideration of physical factors and leave the economic and humanistic aspects of engineering to others; some would not even consider these aspects as coming under engineering. The reason may be that those who take pleasure in discovering and applying the well-ordered certainties of the physical environment find it difficult to adjust their thinking to consider the complexities of the economic environment.

Engineers can readily extend their inherent ability of analysis to become proficient in the analysis of the economic aspects of engineering application. Furthermore, the engineer who aspires to a creative position in engineering will find proficiency in economic analysis helpful. The large percentage of engineers who will eventually be engaged in managerial activities will find such proficiency a necessity.

Initiative for the use of engineering rests, for the most part, upon those who will concern themselves with social and economic consequences. To maintain the initiative, engineers must operate successfully in both the physical and economic sectors of the total environment. It is the objective of *engineering economy* to prepare engineers to cope effectively with the bi-environmental nature of engineering application.

1.3 PHYSICAL AND ECONOMIC EFFICIENCY

Both individuals and organizations possess limited resources. This makes it necessary to produce the greatest output for a given input—that is, to operate at high efficiency. Thus, the search is not merely for a *good* opportunity for the employment of limited resources, but for the *best* opportunity.

People are continually seeking to satisfy their wants. They give up certain utilities in order to gain others that they value more. This is essentially an economic process, in which the objective is the maximization of economic efficiency.

Engineering is primarily a producer activity that comes into being to satisfy human wants. Its objective is to get the greatest end result per unit of resource expenditure. This is essentially a physical process in which the objective is the maximization of physical efficiency, which may be stated as

$$\boxed{\text{efficiency (physical)} = \frac{\text{output}}{\text{input}}.} \tag{1.1}$$

If interpreted broadly enough, physical efficiency is a measure of the success of engineering activity in the physical environment. However, the engineer must be concerned with two levels of efficiency. On the first level is physical efficiency expressed as outputs divided by inputs of such physical units as Btu's, kilowatts, and foot-pounds. When such physical units are involved, efficiency will always be less than unity, or less than 100%.

On the second level are economic efficiencies. These are expressed in terms of economic units of output divided by economic units of input, each expressed in terms of a medium of exchange such as money. Economic efficiency may be stated as

$$\boxed{\text{efficiency (economic)} = \frac{\text{worth}}{\text{cost}}.} \tag{1.2}$$

It is well known that physical efficiencies over 100% are not possible. However, economic efficiencies can exceed 100% and must do so for economic ventures to be successful.

Physical efficiency is related to economic efficiency. For example, a power plant may be profitable in economic terms even though its physical efficiency in converting units of energy in coal to electrical energy may be relatively low. As an example, in the conversion of energy in a certain plant, assume that the physical efficiency is only 36%. Assuming that output Btu's in the form of electrical energy have an economic worth of $14.65 per million and that input Btu's in the form of coal have an economic cost of $1.80 per million, then

$$\text{efficiency (economic)} = \frac{\text{Btu output} \times \text{worth of electricity}}{\text{Btu input} \times \text{cost of coal}}$$

$$= 0.36 \times \frac{\$14.65}{\$1.80} = 293\%.$$

Since physical processes are of necessity carried out at efficiencies less than 100% and economic ventures are feasible only if they attain efficiencies greater than 100%, it is clear that in feasible economic ventures the economic worth per unit of physical output must always be greater than the economic cost per unit of physical input. Consequently, economic efficiency must depend more upon the worth and cost per unit of physical outputs and inputs than upon physical efficiency. Physical efficiency is always significant, but only to the extent that it contributes to economic efficiency.

In the final evaluation of most ventures, even those in which engineering plays a leading role, economic efficiencies must take precedence over physical efficiencies. This is because the function of engineering is to create utility in the economic environment by altering elements of the physical environment.

1.4 THE ENGINEERING PROCESS

Engineering activities dealing with elements of the physical environment take place to meet human needs that arise in an economic setting, as was illustrated by Figure 1.1. The engineering process employed from the time a particular need is recognized until it is satisfied may be divided into a number of phases. These phases are discussed in this section.

1.4.1 Determination of Objectives

One important phase of the engineering process involves the search for new objectives for engineering application—to find out what people need and want that can be supplied by engineering. In the field of invention, success is not a direct result of the fabrication of a new device; rather, it depends upon the capability of the invention to satisfy human wants. Thus, market studies seek to learn what the desires of people are. Automobile manufacturers make surveys to learn what mechanical, comfort, and style features people want in transportation. Highway commissions make traffic counts to learn what construction programs will be of greatest use. Considerations of physical and economic feasibility come only after what is wanted has been determined.

The things that people want may be the result of logical considerations, but more often they are the result of emotional drives. There appears to be no logical reason why one prefers a certain make of car, a certain type of work, or a certain style of clothes. The bare necessities needed to maintain physical existence, in terms of calories of nourishment, clothing, and shelter, are limited and may be determined with a fair degree of certainty. But the wants that stem from emotional drives seem to be unlimited.

Economic limitations are continually changing with people's needs and wants. Physical limitations are continually being pushed back through science and engineering. In consequence, new openings revealing new opportunities are continually developing. For each successful venture an opening through the barrier of economic and physical limitations has been found.

The facet of the engineering process that seeks to learn of human wants requires not only a knowledge of the limitations of engineering capability but also a general knowledge of sociology, psychology, political science, economics, literature and other fields related to the understanding of human nature. A knowledge of these fields is recognized to be useful or essential in most branches of modern engineering.

1.4.2 Identification of Strategic Factors

The factors that stand in the way of attaining objectives are known as *limiting factors.* An important element of the engineering process is the identification of the limiting factors restricting accomplishment of a desired objective. Once the limiting factors have been identified, they are examined to locate *strategic factors*—those factors which can be altered to remove limitations restricting the success of an undertaking.

The understanding that results from the delineation of limiting factors and their further consideration to arrive at the strategic factors often stimulates ideas for improvements. There is obviously no point in operating upon some factors. Consider, for example, a situation in which a truck driver is hampered because he has difficulty in loading a heavy box. Three factors are involved: the pull of gravity, the mass of the box, and the strength of the man. Not much success would be expected from an attempt to lessen the pull of gravity. Nor is it likely to be feasible to reduce the mass of the box. A stronger man might be secured, but it seems more logical to consider overcoming the need for strength by devices to supplement the strength of the man. This analysis leads to consideration of lifting devices that might circumvent the limiting factor of strength.

The identification of strategic factors is important, for it focuses concentration on those areas in which success is obtainable. This may require inventive ability, or the ability to put known things together in new combinations, and is distinctly creative in character. The means that will achieve the desired objective may consist of a procedure, a technical process, or a mechanical, organizational, or managerial change. Strategic factors limiting success may be circumvented by operating on engineering, human, and economic factors individually and jointly.

1.4.3 Determination of Means

The determination of means is subordinate to the identification of strategic factors, just as the identification of strategic factors is necessarily subordinate to the determination of objectives. Strategic factors may be altered in many different ways. Each possibility must be evaluated to determine which will be most successful in terms of overall economy. Engineers are well equipped by training and experience to determine means for altering the physical environment. If the means devised to overcome strategic factors come within the field of engineering, they may be termed *engineering proposals.*

It may be presumed that a knowledge of facts in a field is a necessity for creativeness in that field. For example, it appears that a person who is proficient in the science of combustion and machine design is more likely to contrive an energy-saving internal-combustion engine than a person who has little or no such knowledge. It also appears that knowledge of costs and people's desires as well as of engineering is necessary to conceive of opportunities for profit that involve engineering.

Engineers have the opportunity to be creative by considering human and economic factors in their work. The machine designer may design tools or machines that will require a minimum of maintenance and that can be operated with less fatigue and greater safety. The highway designer may consider durability, cost, and safety. Engineers in any capacity can see to it that both private and public projects are planned, built, and operated in accordance with good engineering practice. Engineering is an expanding profession. People have great confidence in the integrity and ability of engineers. Perhaps nothing will enhance the image of the individual engineer and the profession of engineering more than the acceptance of responsibilities that go beyond the determination of means.

1.4.4 Evaluation of Engineering Proposals

It is usually possible to accomplish a desired result by several means, each of which is feasible from the technical aspects of engineering application. The most desirable of the several proposals is the one that can be performed at the least cost. The evaluation of engineering proposals in terms of comparative cost is an important facet of the engineering process and an essential ingredient in the satisfaction of wants with maximum economic efficiency. Although engineering alternatives are most often evaluated to determine which is most desirable economically, exploratory evaluations are also made to determine if any likely engineering proposal can be formulated to reach a goal profitably.

A wide range of factors may be considered in evaluating the worth and cost of engineering proposals. When investment is required, the time value of money must be considered. Where machinery and plants are employed, depreciation becomes an important factor. Most proposals involve organized effort, thus making labor costs an important consideration. Material is an important ingredient which may lead to market analyses and a study of procurement policy. Risks of a physical and economic nature may be involved and must be evaluated. Where the accepted engineering proposal is successful, a net income will be derived, thus making the consideration of income tax necessary. This text offers specific instruction in methods of analysis pertaining to each of these factors and general instruction regarding the engineering process as a whole.

1.4.5 Assistance in Decision Making

Engineering is concerned with action to be taken in the future. Therefore, an important facet of the engineering process is to improve the certainty of decision with respect to the want-satisfying objective of engineering application. Correct decisions can offset many operating handicaps. On the other hand, incorrect decisions may and often do hamper all subsequent action. No matter how expertly a bad decision is carried out, results will be at best unimpressive and at worst disastrous.

To make a decision is to select a course of action from among several. A correct decision is the selection of that course of action which will result in an outcome more desirable than would have resulted from any other selection. Decision rests upon the possibility of choice—that is, on the fact that there are alternatives from which to choose. The engineer acting in a creative capacity proceeds on the thought that there is a most desirable solution if it can be found.

Engineers are becoming increasingly aware that many sound proposals fail because those who might have benefited from them did not understand their significance. A prospective user of a good or service is primarily interested in its worth and cost. The person who lacks an understanding of engineering may find it difficult or even impossible to grasp the technical aspects of a proposal sufficiently to arrive at a measure of its economic desirability. The uncertainty so engendered may easily cause loss of confidence and a decision to discontinue consideration of the proposal.

The logical determination and evaluation of alternatives in tangible terms has long been recognized as integral to the engineering process. The success of engineers in dealing with this element of application is responsible, in part, for the large percentage of engineers who are engaged either directly or indirectly in decision-making activities.

1.5 A PLAN FOR ENGINEERING ECONOMY STUDIES

The engineering process described in the previous section involves a creative element and includes the employment of engineering economy studies. These studies can be made either haphazardly or on the basis of a logical plan. This plan in-

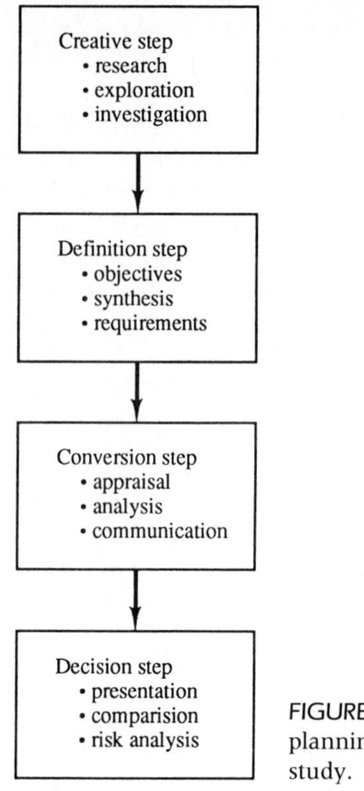

FIGURE 1.2. Steps in planning for an economy study.

volves a creative step, a definition step, a conversion step, and a decision step as illustrated in Figure 1.2.

1.5.1 The Creative Step

Engineers, whether engaged in research, design, construction, production, operations, or management activity, must be concerned with the efficient use of limited resources. When known opportunities fail to hold sufficient promise for the employment of resources, more promising opportunities are sought. People with vision accept the premise that better opportunities exist than are known to them. This view, accompanied by initiative, leads to exploratory activities aimed at finding the better opportunities. Exploration, research, investigation, and similar activities are creative. In such activities, steps are taken into the unknown to find new possibilities which may then be evaluated to determine if they are superior to those that are known.

Opportunities are not made; they are discovered. The person who concludes that there is no better way makes a self-fulfilling prophecy. When the belief is held that there is no better way, a search for one will not be made, and a better way will not be discovered. Any situation embraces groups of facts of which some may be known and others unknown. New opportunities for profit are fashioned from the facts as they exist.

Some successful ideas depend upon the discovery of new facts. New facts may become known through research effort or by accident. Research is effort consciously directed to the learning of new facts. In pure research, facts are sought without regard for their specific usefulness, on the premise that a stockpile of knowledge will in some way contribute to human welfare. Much progress rests without doubt upon facts discovered from efforts to satisfy curiosity.

Both new facts and new combinations of facts may be consciously sought. The creative aspect of engineering economy consists in finding new facts and new combinations of facts out of which may be fashioned opportunities to provide profitable service through the application of engineering.

The creative step is of major importance in economy studies. Since the mental processes involved are in large measure illogical, this step must be approached with considerable alertness and curiosity and a willingness to consider new ideas and unconventional patterns of thought.

Aside from the classical statement that "inspiration is 99% perspiration," there are few guides to creativeness. It appears that both conscious application and inspiration may contribute to creativeness. Some people seem to be endowed with marked aptitudes for conceiving new ideas.

1.5.2 The Definition Step

The definition step consists of defining alternatives that originated in the creative step, or that have been selected for comparison in some other way. A complete and all-inclusive alternative rarely emerges in its final state. It begins as a hazy but interesting idea. The attention of the individual or group is then directed to analysis and synthesis, and the result is a definite proposal. In its final form, an alternative should consist of a complete description of its objectives and its requirements in terms of inputs and outputs.

Both different ends and different methods are embraced by the term alternative. All proposed alternatives are not necessarily attainable. Some are proposed for analysis even though there seems to be little likelihood that they will prove feasible. The idea is that it is better to consider many unprofitable alternatives than to overlook one that is profitable. Alternatives that are not considered cannot be adopted, no matter how desirable they may actually be.

In addition to the alternatives formally set up for evaluation, another alternative is always present—that of making no decision. The decision not to decide may be a result of either active consideration or passive failure to act; it is usually motivated by the thought that there will be opportunities in the future that will prove more profitable than any known at present.

In the first stage of the definition step, the engineer's aim should be to delineate each alternative on the basis of its major and subordinate physical units and activities. The purpose of this stage is to ensure that all factors of each alternative and no others will be considered.

The second stage of the definition step consists of enumerating the prospective items of output and input of each alternative, in quantitative physical terms as far as possible and then in qualitative terms. Though qualitative items cannot be expressed numerically, they may often be of major importance. They should be made visible so that they may be considered in the final evaluation.

1.5.3 The Conversion Step

In order to compare alternatives properly, it is important that they be converted to a common measure. The common measure usually chosen for economic comparison is value expressed in terms of money.

The first phase of the conversion step is to convert the prospective output and input items enumerated in the definition step into receipts and disbursements at specified dates. This phase consists essentially of appraising the unit value of each item of output or input and determining their total amounts by computation. On completion, each alternative should be expressed in terms of definite cash flows occurring at specified dates in the future, plus an enumeration of qualitative considerations that have been impossible to reduce to monetary terms. For such items the term "irreducibles" is often employed.

The second phase of the conversion step consists of placing the estimated future cash flows for all alternatives on a comparable basis, considering the time value of money. This involves employment of the techniques presented in future chapters. Selection of the particular technique depends upon the situation, and its appropriateness is a matter of judgment. Consideration of inherent inaccuracies in estimates of the future outputs and inputs may be considered part of the conversion step and should not be overlooked.

The final phase of the conversion step is to communicate the essential aspects of the economy study, together with an enumeration of irreducibles, so that they may be considered by those responsible for making the decision. Responsibility for the acceptance or rejection of an engineering proposal is exercised more often than not by persons who have not been concerned with the technical aspects of the proposal. Also, the persons who control acceptance are likely to lack understanding of technical matters.

A proposal should be explained in terms that will best present its significance to those who will control its acceptance. The aim should be to take such persons on an excursion into the future to experience what will happen if the proposal is accepted or rejected. For example, suppose that a proposal for a new water pollution control system is to be presented. Since those who must decide if it should be adopted rarely have the time and background to go into and appreciate all the technical details involved, the significance of these details in terms of economic results must be made clear. Of interest to the decision makers will be such things as the present outlay required, capital-recovery period, flexibility of the system in event of regulation changes, effect upon wastewater quality, and difficulties of financing. Cost and other data should be broken down and presented so that attention may be easily focused upon pertinent aspects of the proposal. Diagrams, graphs, pictures, and even models should be used where they will contribute to understanding.

1.5.4 The Decision Step

On completion of the conversion step, quantitative and qualitative outputs and inputs for each alternative form the basis for comparison and decision. Quantitative input may be deducted from quantitative output to obtain quantitative profit, or the ratio of quantitative profit to quantitative input may be found. Each of these measures is then supplemented by the qualitative considerations enumerated.

Decisions between alternatives should be made *on the basis of their differences.* Thus, all identical factors can be canceled out for the comparison of any two or more alternatives at any step in an economy study. In this process great care must be exercised that factors canceled as being identical are actually of the same significance. Unless it is very clear that factors considered for cancellation are identical, it is best to carry them through the first stage of the decision step. This may entail a greater amount of computation and documentation, but the added complexity and loss of time is ordinarily insignificant in comparison with the value to be derived.

When a diligent search uncovers insufficient information to reason the outcome of a course of action, the problem is to render as accurate a decision as the lack of facts permits. In such situations there is a decided tendency, on the part of many, to make little logical use of the data that are available, on the thought that since some rough estimating has to be done on some elements of the situation, the estimate might as well embrace the entire situation. But an alternative may usually be subdivided into parts, and the available data are often adequate for a complete or nearly complete evaluation of several of the parts. The segregation of the known and unknown parts is in itself additional knowledge. Also, the unknown parts, when subdivided, frequently are recognized as being similar to parts previously encountered and thus become known.

After a situation has been carefully analyzed and the possible outcomes have been evaluated as accurately as possible, a decision must be made. Even after all the data that can be brought to bear on a situation have been considered, some areas of uncertainty may be expected to remain. If a decision is to be made, these areas of uncertainty must be bridged by consideration of nonquantitative data or by the evaluation of irreducibles. Some call the type of evaluation involved in the consideration of irreducibles *intuition;* others call it *hunch* or *judgment.*

When there is complete knowledge of all facts concerned and their relationships, reason can supplant judgment, and predictions become a certainty. Judgment tends to be qualitative. Reason is both qualitative and quantitative. Judgment is at best an informal consideration and weighing of facts; at its worst it is merely wishful thinking. Judgment appears to be an informal process for considering information, past experience, and feeling in relation to a problem. No matter how sketchy factual knowledge of a situation may be, some sort of a conclusion can always be drawn in regard to it by judgment.

Whatever it be called, it is inescapable that this type of thinking—or, perhaps better, this type of feeling—must always be the final part in coming to a decision about the future. There is no other way if action is to be taken. There appears to be a marked difference in people's abilities to come to sound conclusions when some facts relative to a situation are missing. Perhaps much more attention should be devoted to developing sound judgment, for those who possess it are richly rewarded. But as effective as intuition, hunch, or judgment may sometimes be, this type of thinking should be reserved for those areas where facts on which to base a decision are missing.

An important aim of engineering economy studies is to gather and analyze the facts so that reason may be used to the fullest extent in arriving at a decision. In this way judgment can be reserved for parts of the situation where factual knowledge is absent. This idea is embraced in the statement, "Figure as far as you can, then add judgment."

1.6 ENGINEERING ECONOMY AND THE ENGINEER

Economy, the attainment of an objective at low cost in terms of resource input, has always been associated with engineering. During much of history the limiting factor has been predominantly physical. Thus, a great innovation, the wheel, awaited invention, not because it was useless or costly, but because the mind of man could not synthesize it earlier. But, with the development of science, things have become physically possible that people are interested in only slightly or not at all. Thus, a new type of transportation system may be perfectly feasible from the physical standpoint but may enjoy limited use because of its first cost or cost of operation.

The engineer is often concerned by the lack of certainty associated with the economic aspect of engineering. However, it must be recognized that economic considerations embrace many of the subtleties and complexities characteristic of people. Economics deals with the behavior of people individually and collectively, particularly as their behavior relates to the satisfaction of their wants.

The wants of people are motivated largely by emotional drives and tensions and to a lesser extent by logical reasoning processes. A part of human wants can be satisfied by physical goods and services such as food, clothing, shelter, transportation, communication, entertainment, medical care, educational opportunities, and personal services; but people are rarely satisfied by physical things alone. In food, sufficient calories to meet their physical needs will rarely satisfy. People want the food they eat to satisfy their emotional needs as well as energy needs. In consequence we find people concerned with the flavor of food, its consistency, the china and silverware with which it is served, the person or persons who serve it, the people in whose company it is eaten, and the "atmosphere" in which it is served. Similarly, many desires are associated with clothing and shelter, in addition to those required merely to meet physical needs.

Those who have a part in satisfying human wants must accept the uncertain action of people as a factor with which they must deal, even though they find such action unexplainable. Much or little progress has been made in learning how to predict human behavior, depending upon one's viewpoint. The idea that human reactions will someday be well enough understood to be predictable is accepted by many people; but even though this has been the objective of the thinkers of the world since the beginning of time, it appears that progress in psychology has been meager compared to the rapid progress made in the physical sciences. In spite of the fact that human reactions can be neither predicted nor explained, they must be considered by those who are concerned with satisfying human wants.

Since economic factors are the strategic consideration in most engineering activities, engineering practice may be either responsive or creative. If the engineer takes the attitude that he should restrict himself to the physical, he is likely to find that the initiative for the application of engineering has passed on to those who will consider economic and social factors.

The engineer who acts in a *responsive* manner acts on the initiative of others. The end product of his work has been envisioned by another. Although this position leaves him relatively free from criticism, this freedom is gained at the expense of professional recognition and prestige. In many ways, he is more of a technician than a professional. Responsive engineering is, therefore, a direct hindrance to the development of the engineering profession.

The *creative* engineer, on the other hand, not only seeks to overcome physical limitations, but also initiates, proposes, and accepts responsibility for the success of projects involving human and economic factors. The general acceptance by engineers of the responsibility for seeing that engineering proposals are both technically and economically sound, and for interpreting proposals in terms of worth and cost, may be expected to promote confidence in engineering as a profession.

KEY POINTS

- Engineering activities are not an end in themselves but are a means for satisfying human wants.
- The essential prerequisite of successful engineering application is economic feasibility.
- Engineering is not a science; it is an art composed of skill and ingenuity in adapting knowledge to the uses of humanity.
- Engineering is concerned with two interconnected environments, the physical and economic.

- In the evaluation of most ventures, economic efficiency must take precedence over physical efficiency.
- Phases in the engineering process include the determination of objectives, the identification of strategic factors, the determination of means, the evaluation of engineering proposals, and assistance in decision making.
- The logical plan for engineering economy studies involves the creative step, the definition step, the conversion step, and the decision step.
- Economy in the attainment of an objective in terms of resource input has always been a characteristic of engineering.

QUESTIONS

1. Contrast the role of the engineer with the role of the scientist.
2. What is the important difference between a physical law and an economic law?
3. Why should the engineer be concerned with both the physical and the economic aspects of the total environment?
4. Give an example of a product that would be technically feasible but would have little economic merit.
5. Explain how physical efficiency below 100% may be converted to economic efficiency above 100%.
6. Why must economic efficiency take precedence over physical efficiency?
7. As compared with the economic aspect of engineering application, give reasons why the physical aspect is decreasing in relative importance.
8. What is involved in seeking new objectives for engineering application?
9. What is a limiting factor; a strategic factor? Give an example of each.
10. Give reasons why engineers are particularly well prepared to determine means for the attainment of an objective.
11. How may engineers assist in decision making?
12. Discuss the potential benefits of the plan for engineering economy studies.
13. Apply the plan for engineering economy studies to an engineering problem of your choice.
14. How is the creative step related to the satisfaction of human wants?
15. List the method for discovering means to employ resources more profitably.
16. Why is it not possible to consider all possible alternatives?
17. Discuss the nature of judgment and explain why it is applicable to more situations than reason.

18. Explain why decisions must be based on differences between alternatives occurring in the future.

19. Why are decisions relative to the future based upon estimates instead of upon the facts that will apply?

20. Why is judgment always necessary in coming to a decision relative to an outcome in the future?

21. Why should the economic interpretation of engineering proposals be made by engineers?

22. Contrast the responsive engineer with the creative engineer.

2

Some Economic and Cost Concepts

Concepts are crystallized thoughts that have withstood the test of time. They are usually qualitative in nature and not necessarily universal in application. Economic and cost concepts, if carefully related to fact, may be useful in suggesting approaches to problems in engineering economy. Those given in this chapter are not exhaustive. Others will be found throughout the text. The ability to arrive at correct decisions depends jointly upon a sound conceptual understanding and the ability to handle the quantitative aspects of the problem. In this chapter we give special attention to some basic concepts useful in economy studies.

2.1 CONCEPTS OF VALUE AND UTILITY

The term *value* has a variety of meanings. In economics,

> **Value** is a measure of the worth that a person ascribes to a good or a service.

Thus, the value of an object is inherent not in the object but in the regard that a person has for it. Value should not be confused with the cost or the price of an object. There may be little or no relation between the value a person ascribes to an article and the cost of providing it, or the price that is asked for it.

The utility that an object has for an individual is determined by him or her. In economics,

Utility is a measure of the power of a good or a service to satisfy human wants.

Thus, the utility of an object, like its value, inheres not in the object itself but in the regard that a person has for it. Utility and value, in the sense used here, are closely related. The utility that an object has for a person is the satisfaction he or she derives from it. Value is an appraisal of utility in terms of a medium of exchange.

In ordinary circumstances a large variety of goods and services is available to an individual. The utility that available items may have in the mind of a prospective user may be expected to be such that the desire for them will range from abhorrence, through indifference, to intense desire. The evaluation of the utility of various items is not ordinarily constant but may be expected to change with time. Each person also possesses either goods or services that he may render. These have the utility for the person himself that he regards them to have. These same goods and possible services may also be desired by others, who may ascribe to them very different utilities. The possibility for exchange exists when each of two persons possesses utilities desired by the other.

If the supernatural is excluded from consideration, all that has utility is physically manifested. This statement is readily accepted in regard to physical objects that have utility, such as an automobile, a house, or a steak dinner. But this statement is equally true in regard to the more intangible things. Music, which is regarded as pleasing to people, is manifested to them as air waves that strike their ears. Pictures are manifested as light waves. Even friendship is realized only through the five senses and must, therefore, have its physical aspects. It follows that utilities must be created by changing the physical environment.

For example, the consumer utility of raw steak can be increased by altering its physical condition by an appropriate application of heat. In the area of producer utilities the machining of a bar of steel to produce a drive shaft for a tractor is an example of creating utility by manipulation of the physical environment. The purpose of most engineering effort is to determine how physical factors may be altered to create the most utility for the least cost in terms of the utilities that must be given up.

2.2 CONSUMER AND PRODUCER GOODS

Two classes of goods are recognized by economists: consumer goods and producer goods.

Consumer goods are the goods and services that directly satisfy human wants.

Examples of consumer goods are television sets, houses, shoes, books, orchestras, and health services.

Producer goods are the goods and services that satisfy human wants indirectly as part of the production or construction process.

Broadly speaking, the ultimate end of all engineering activity is to supply goods and services that people may consume to satisfy their needs and desires. Producer goods are, in the long run, used as a means to an end—namely, that of producing goods and services for human consumption. Examples of this class of goods are bulldozers, machine tools, ships, and railroad cars. Producer goods are an intermediate step in people's efforts to supply their wants. They are not desired for themselves, but because they may be instrumental in producing something that can be consumed.

Once the kind and amount of consumer goods to be produced has been determined, the determination of the kinds and amounts of producer goods and facilities needed to produce them may be approached objectively. The energy, ash, and other contents of coal, for instance, can be determined very accurately and are the basis for evaluating the utility of the coal. The extent to which producer utility may be considered by logical processes is limited only by factual knowledge and the ability to reason.

2.2.1 Utility of Consumer Goods

People will consider two kinds of utility. One kind embraces the utility of goods and services that they intend to consume personally for the satisfaction they get out of them. Thus, it seems reasonable that the utility a person ascribes to goods and services that are consumed directly is in large measure a result of subjective, nonlogical mental processes. This may be inferred from the fact that sellers of consumer goods apparently find emotional appeals more effective than factual information. Early automobile advertising took the form of objective information related to design and performance, but more recent practice stresses such subjective aspects as beauty, comfort, and prestige values.

Some kinds of human wants are much more predictable than others. The demand for food, clothing, and shelter, which are needed for bare physical existence, is much more stable and predictable than the demand for those items that satisfy human emotional needs. The amount of foodstuffs needed for existence is ascertainable within reasonable limits in terms of calories of energy. Clothing and shelter requirements may be fairly accurately determined from climate data. But once humans are assured of physical existence, they reach out for satisfactions related to being a person rather than merely to being a physical organism.

An analysis of advertising and sales practices used in selling consumer goods will reveal that they appeal primarily to the senses rather than to reason, and perhaps rightly so. If the enjoyment of consumer goods stems almost exclusively from how one feels about them rather than what one reasons about them, it seems logical to make sales presentations on the basis of what customers ascribe utility to.

2.2.2 Utility of Producer Goods

The second kind of utility that an object or service may have for a person is as a means to an end. Producer goods are not consumed for direct satisfaction but as a means of producing consumer goods, usually by facilitating alteration of the physical environment.

Although the utility of consumer goods is primarily determined subjectively, the utility of producer goods as a means to an end may be, and usually is, considered objectively. In this connection, consider the satisfaction of the human want for harmonic sounds, as in a concert of recorded music. Suppose it has been decided that the desire for a certain recorded concert can be met by 100,000 pressings of a stereo disc. Then the organization of artists, technicians, and equipment necessary to produce the discs becomes predominantly objective in character. The amount of material that must be compounded and processed to form one disc is calculable to a high degree of accuracy. If a concern has been making CD's for some time, it will know the various operations that are to be performed and the unit times for performing them. From these data, the kind and amount of producer service, the amount and kind of labor, and the number of various types of machines are determinable within rather narrow limits. Whereas the determination of the kinds and amounts of consumer goods needed at any one time may depend upon the most subjective of human considerations, the problems associated with their production are quite objective by comparison.

2.3 ECONOMIC ASPECTS OF EXCHANGE

Economy of exchange occurs when utilities are exchanged by two or more people. In this connection, a utility means anything that a person may receive in an exchange that has any value whatsoever—for example, a lathe, a dozen pencils, a meal, music, or a friendly gesture.

2.3.1 Mutual Benefit in Exchange

A buyer will purchase an object when money is available and when he or she believes that the good has equal or greater utility than the amount required to purchase it. Conversely, a seller will sell an object when he or she believes that the amount of money to be received for the object has greater utility than the object. Thus, an exchange will not be effected unless at the time of exchange both parties believe that they will benefit. Exchanges are made when they are thought to result in mutual benefit. This is possible because the objects of exchange are not valued equally by the parties to the exchange.

Economy of exchange is possible because consumer utilities are evaluated by the consumer almost entirely, if not entirely, by subjective considerations. Accordingly, an exchange of consumer utilities results in a gain for both parties. People have different needs by virtue of their history and their current situation. This is the reason that people can be found who will subjectively evaluate consumer utilities so that an exchange will permit each to gain.

Assuming that each party to an exchange of producer utilities correctly evaluates the objects of exchange in relation to his situation, what makes it possible for each person to gain? The answer is that the participants are in different economic environments. For example, a merchant buys lawn mowers from a manufacturer. At a certain volume of activity the manufacturer finds that he can produce and distribute mowers at a total cost of $90 per unit, and the merchant buys a number of the mowers at a price of $110 each. The merchant then finds that by expending an average of $40 per unit in selling effort, he can sell a number of the mowers to homeowners at $195 each. Both participants profit by the exchange. The reason that the manufacturer profits is that his environment is such that he can sell to the merchant for $110 a number of mowers that he cannot sell elsewhere at a higher price, and that he can manufacture lawn mowers for $90 each. The reason that the merchant profits is that his environment is such that he can sell mowers at $195 each by applying $40 of selling effort upon a mower of certain characteristics that he can buy for $110 each from the manufacturer in question, but not for less elsewhere.

Why doesn't the manufacturer enter the merchandising field and thus increase his profit, or why doesn't the merchant enter the manufacturing field? The answer is that neither the manufacturer nor the merchant can do so unless each changes his environment. The merchant, for example, lacks physical plant equipment and an organization of engineers and workers competent to manufacture mowers. Also, he may be unable to secure credit necessary to engage in manufacturing, although he may easily secure credit in greater amounts for merchandising activities. It is quite possible that he cannot alter his environment so that he can build mowers for less than $110. Similar reasoning applies to the manufacturer. Exchange consists essentially of physical activity designed to transfer the control of things from one person to another. Thus, even in exchange, utility is created by altering the physical environment.

Each party in an exchange should seek to give something that has little utility for him but that will have great utility for the receiver. In this manner each exchange can result in the greatest gain for each party. Nearly everyone has participated in such a favorable exchange. When a car becomes stuck in snow, only a slight push may be required to dislodge it. The slight effort involved in the dislodging push may have very little utility for the person giving it, so little that he expects no more compensation than a friendly nod. On the other hand, it might have very great utility for the person whose car was dislodged, so great that he may offer a substantial tip. The aim of much sales and other research is to find products that not only will have great utility for the buyer but that can be supplied at a low cost—that is, have low utility for the seller. The difference between the utility that a specific good or service has for the buyer and the utility it has for the seller represents the profit or net benefit that is available for division between buyer and seller. This difference may be called the *range of mutual benefit* in exchange.

The factors that may determine a price within the range of mutual benefit at which exchange will take place are infinite in variety. They may be either subjective or objective. A person seeking to sell may be expected to make two evaluations: the minimum amount he will accept and the maximum amount a prospective buyer can be induced to pay by persuasion. The latter estimate may be based upon mere conjecture or upon a detailed analysis of buyers' subjective and objective situations. In bargaining, it is usually advantageous to obscure one's situation. Thus, sellers will ordinarily refrain from revealing the costs of the goods or services they are seeking to sell or from referring a buyer to a competitor who is willing to sell at a lower price.

2.3.2 Persuasion in Exchange

It is not uncommon for an equipment salesperson to call on a prospective customer, describe and explain a piece of equipment, state its price, offer it for sale, and have the offer rejected. This is concrete evidence that the equipment item does not possess sufficient utility at the moment to induce the prospective customer to buy it. In such a situation, the salesperson may be able to induce the prospect to listen to further sales talk, during which the prospect may decide to buy on the basis of the original offer. This is concrete evidence that the machine now possesses sufficient utility to induce the prospective customer to buy. Because there was no change in the equipment or the price at which it was offered, there must have been a change in the customer's attitude or regard for it. The pertinent fact is that a proposition which was at first undesirable, now has become desirable as a result of a change in the customer, not in the proposition.

What brought about the change? A number of reasons could be advanced. Usually it would be said that the salesperson persuaded the customer to buy. In other words, the salesperson induced the customer to believe something, namely, that the machine had sufficient utility to warrant its purchase. There are many aspects to persuasion. It may amount only to calling attention to the availability of an item. A person cannot purchase an item he does not know exists. A part of a salesperson's function is to call attention to the things he or she has to sell.

It is observed that persuasive ability is much in demand, is often of inestimable beneficial consequences to all concerned, and is usually richly rewarded. Persuasion as it applies to the sale of goods is of economic importance to industry. Manufacturers must dispose of the goods they produce. They can increase the salability of their products by building into them greater customer appeal in terms of greater usefulness, greater durability, or greater beauty, or they may elect to accompany their products to market with greater persuasive effort in the form of advertising and sales promotion. Either plan will require expenditure, and both are subject to the law of diminishing returns. It is an interesting study in economy to determine what levels of perfection of product and sales effort will be most profitable.

Whatever the approach, factual or emotional, persuasion consists of taking a person on an excursion into the future in an attempt to show and convince him what will happen if he acts in accordance with a proposal. The purpose of engineering economic analysis is to estimate, on as factual a basis as possible, what the economic consequences of a decision will be. It is therefore a useful technique in persuasion.

2.4 CLASSIFICATIONS OF COST

The ultimate objective of engineering application is the satisfaction of human needs and wants. But human wants are not satisfied without cost. Alternative engineering proposals will differ in the costs they involve relative to the objective of want satisfaction. The engineering proposal resulting in the least cost will be considered best, if its end result is identical to that of competing proposals.

A number of cost classifications have come into use to serve as a basis for economic analysis. As concepts, these classifications are useful in calling to mind the source and effect of costs that will have a bearing on the end result of a proposal. This section will define and discuss these cost classifications.

2.4.1 First Cost

When any activity commences, initial or immediate costs will be incurred.

First cost is the initial cost of capitalized property, including transportation, installation, and other related initial expenditures.

The chief advantage in recognizing this classification is that it calls attention to a group of costs associated with the initiation of a new activity that might not otherwise be given proper consideration. Ordinarily, this classification is limited to costs that occur only once for any given activity.

First cost is usually made up of a number of cost elements that do not recur after an activity is initiated. For purchased equipment, these include the purchase price plus shipping cost, installation cost, and training cost. For a fabricated structure, system, or item of equipment, they include engineering design and development cost, test and evaluation cost, and construction or production cost as well as shipping, installation, and training costs.

Many activities that otherwise may be profitable cannot be undertaken because their associated first cost represents too high a level of investment. Many engineering proposals that are otherwise sound are not initiated because the first cost involved is beyond the reach of the controlling organization.

2.4.2 Operation and Maintenance Cost

Whereas first cost occurs only once in getting an activity started,

> **Operation and maintenance cost** is that group of costs experienced continually over the useful life of the activity.

Included in this cost category are labor costs for operating and maintenance personnel, fuel and power costs, operating and maintenance supply costs, spare and repair part costs, costs for insurance and taxes, and a fair share of indirect costs called overhead or burden. These costs can be substantial, and often they exceed the first cost in total amount. The timing of their occurrence differs substantially, however, in that operating and maintenance costs occur over time until the structure, system, or equipment is retired from service.

Many complex structures and items of equipment require elaborate logistic support systems to sustain their operation. Operating and maintenance costs are incurred by these support systems themselves and are rightfully charged back to the primary activity or venture being supported. Included in the costs of support systems are the usual labor, energy, material, and overhead costs as well as the often overlooked cost of holding inventories of spare and repair parts, transportation and logistic costs, and the costs of communication and coordination. Only in recent years have support systems been separately considered as important adjuncts to the operation and maintenance of complex structures and systems.

2.4.3 Fixed Cost

Fixed costs arise from making preparation for the future. A robot is purchased now in order that labor costs may be reduced in the future. Materials that may never be needed are purchased in large quantities and stored at much expense and with some risk in order that idleness of production facilities and people may be avoided. Research is carried on with no immediate benefit in view in the hope that it will pay in the long run. The investments that give rise to fixed cost are made in the present in the hope that they will be recovered with a profit as a result of reductions in variable costs or of increases in income.

> **Fixed cost** is that group of costs involved in a going activity whose total will remain relatively constant throughout the range of operational activity.

Fixed costs are made up of such cost items as depreciation, maintenance, taxes, insurance, lease rentals, interest on invested capital, sales programs, certain administrative expenses, and research. These costs arise from the decisions of the past and in general are not subject to rapid change. Volume of operational activity, on the other hand, may fluctuate widely and rapidly. As a result, fixed costs per unit may easily go out of control. This is probably the major cause of business failure, for few have the foresight or luck to make commitments in the present that will fit requirements of the future even reasonably well. Since fixed costs cannot be changed readily, consideration must be focused upon maintaining a satisfactory volume and character of activity.

The concept of fixed cost has a wide application. For example, certain losses in the operation of an engine are in some measure independent of its output of power. Among its fixed costs, in terms of energy for a given speed and load, are those for the power to drive the fan, the valve mechanism, and the oil and fuel pumps. Almost any task involves preparation independent of its extent. Thus, to paint a small area may require as much effort for the cleaning of a brush as to paint a large area. Similarly, manufacturing involves fixed costs that are independent of the volume of output.

In practice, fixed costs are only relatively fixed, and their total may be expected to increase somewhat with increased activity. The increase probably will not follow a smooth curve but will vary in accordance with the characteristics of the enterprise. Consider a plant of several units that has been shut down or is operating at zero volume. Heat, light, janitor, and many other services are not required. Many of these services must be reinstated if the plant is to operate at all. If reinstated only on a minimum basis, they will probably be adequate for quite a range of activity. Further increases in activity will require expenditures for other services that cannot be provided to just the extent needed. Thus, what are termed "fixed costs" may be expected to increase in some stepped pattern with an increase in activity.

2.4.4 Variable Cost

Variable cost is related to the rule of use or to the activity level. For example, the consumption of fuel by an engine may be expected to be proportional to its output of power, and the amount of paint used may be expected to be proportional to the area painted. In manufacturing, the amount of material needed per unit of product may be expected to remain constant and, therefore, the material cost will vary directly with the number of units produced. In general, all costs such as direct labor, direct material, direct power, and the like, which can readily be allocated to each unit produced, are considered to constitute variable costs, and the balance of the costs of the enterprise are regarded as fixed.

> **Variable cost** is that group of costs which vary in some relationship to the level of operational activity.

Variable expense may be expected to increase in a stepped pattern. To increase production beyond a certain extent, another machine may be added. Even though its full capacity may not be utilized, a full crew may need to be employed to operate it. Also, an increase in productivity may be expected to result in the use of materials in greater quantities and thus in their purchase at a lower cost per unit due to quantity discounts and volume handling.

2.4.5 Incremental and Marginal Cost

The terms incremental cost and marginal cost refer to essentially the same concept. The word *increment* means increase.

> **Incremental cost** is the additional cost that will be incurred as the result of increasing output by one more unit.

Reference is usually made to an increase of cost in relation to some other factor, thus resulting in such expressions as incremental cost per ton, incremental cost per gallon, or incremental cost per unit of production. The term *marginal cost* refers specifically to an increment of output whose cost is barely covered by the monetary return derived from it.

Figure 2.1 illustrates the nature of fixed and variable cost as a function of output in units. The incremental cost of producing 10 units between outputs of 60 and 70 units per day is illustrated to be \$8. Thus, the average incremental cost of these 10 units may be computed as Δ cost/Δ output = \$8/10 = \$0.80 per unit.

In actual situations, it is ordinarily difficult to determine incremental cost. There is no general approach to the problem, but each case must be analyzed on the basis of the facts that apply to it at the time, considering the future period involved. Incremental costs can be overestimated or underestimated, and either

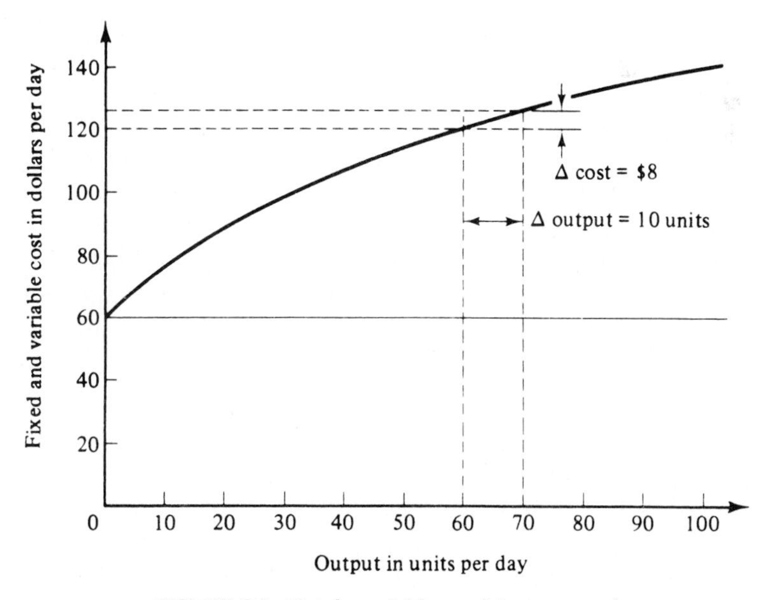

FIGURE 2.1. Fixed, variable, and incremental cost.

error may be costly. Overestimation of incremental costs may obscure a profit possibility; underestimation, on the other hand, may lead to the undertaking of an activity that will result in a loss. Thus, accurate information is necessary if sound decisions are to be made.

2.4.6 Sunk Cost

Most decision makers seek the course of action that is expected to result in the most favorable future benefits. Because only the future consequences of investment alternatives can be affected by current decisions, an important principle in economy studies is to disregard cost incurred in the past.

> A **sunk cost** is a past cost that cannot be altered by future action and is therefore irrelevant.

Although the principle that sunk costs should be ignored seems reasonable, it is quite difficult for many people to apply. For example, suppose that two years ago one purchased 1,000 shares of stock of $26 per share and now it is worth $15 per share. In all probability other stocks are available that have a better future than the stock presently possessed. Many people react to this type of situation by holding onto their present stock until they can recover their losses. Thus, they avoid an open admission of the losses and thereby of their failure in judgment. However, it should be clear that because of better opportunities elsewhere it is certainly sound decision making to acknowledge losses by selling the present stock and to use what money remains more productively from now into the future. It is the emotional involvement with past or sunk costs that makes it difficult for such costs to be ignored in practice.

2.5 LIFE-CYCLE COST

Most products, systems, and structures are brought into being and utilized over a life cycle that begins with the identification of a need and ends with phaseout and disposal. The time orientation of activities includes conceptual and preliminary design, detail design and development, production and/or construction, product utilization, and disposal. In general, the life cycle has two major phases as shown in Figure 2.2: acquisition and utilization. A strong relationship exists between the life-cycle concept as described here and the concept supporting Figure 1.1.

During the acquisition phase, nonrecurring costs are incurred, and these constitute the first cost of the structure or system. During utilization, recurring costs are experienced. Therefore,

> **Life-cycle cost** is defined as all costs, both nonrecurring and recurring, that occur over the life cycle.

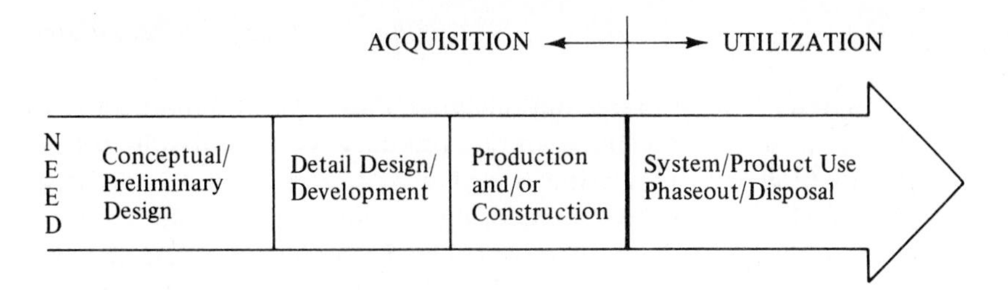

FIGURE 2.2. The system/product life cycle.

The ultimate value of products which result from engineering is measured in economic terms. However, the economic aspects of design are often not examined until detail design is almost complete. By then it is too late. Many beneficial design decisions can be made during the design phase of the life cycle that will minimize the cost of operating and maintaining the product during use. Ordinarily the objective should be to minimize the sum of all costs incurred over the life cycle. Figure 2.3 illustrates how little can be done to improve the life-cycle cost of a product once the design is complete.

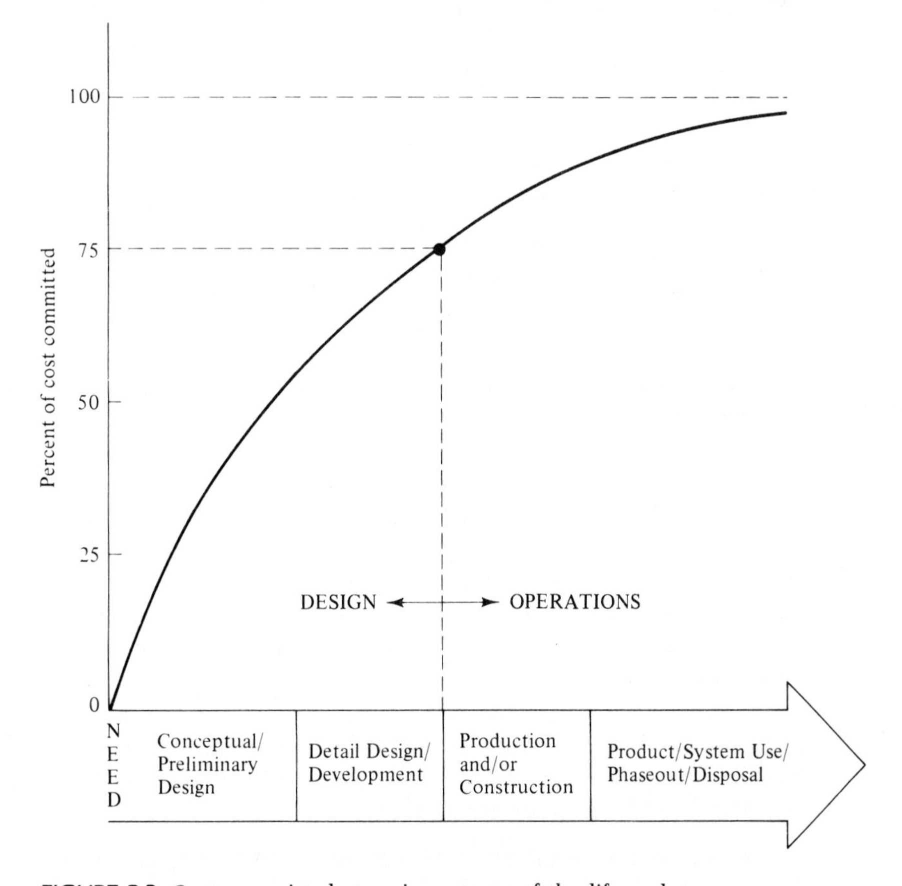

FIGURE 2.3. Cost committed at various stages of the life cycle.

Life-cycle cost and economic analysis should originate early in the product life cycle, during conceptual and preliminary design. Missing from much work in cost analysis for engineering is a view of the entire life cycle. Like many other technological functions, engineering economy often comes into play after the detailed design has been finalized. This is too late. It is estimated that about 75 percent of the life-cycle cost is committed by this point in the design process, as illustrated in Figure 2.3. Thus, it is during the early design activities of the life cycle that life-cycle cost containment can be most effective.

2.6 INTEREST AND INTEREST RATE

Charging a rental for the use of money is a practice dating back to the time of earliest recorded history. The ethics and economics of interest have been a subject of discussion for philosophers, theologians, statesmen, and economists throughout the ages.

> **Interest** is a rental amount charged by financial institutions for the use of money.

The concept of interest can be extended to capital assets, which "borrow" from their owner, repaying through the earnings generated. This economic gain from the use of money is what gives money its time value. Because engineering projects require the investment of money, it is important that the time value of the money used be properly reflected in the evaluation of these projects.

> **Interest rate,** or the rate of capital growth, is the rate of gain received from an investment.

Usually this rate of gain is stated on a per-year basis, and it represents the percentage gain realized on the money committed to the undertaking. Thus, an 11% interest rate indicates that for every dollar of money used, an additional $0.11 must be returned as payment for the use of that money. This interest rate is determined by market forces involving supply and demand. The price (interest rate) is determined by mutual agreement between the borrower and the lender and is known as the *market rate.*

In one aspect, interest is an amount of money *received* as a result of investing funds, either by lending it or by using it in the purchase of materials, labor, or facilities. Interest received in this connection is gain or profit. In another aspect, interest is an amount of money *paid out* as a result of borrowing funds. Interest paid in this connection is a *cost.*

2.6.1 Interest Rate from the Lender's Viewpoint

A person who has a sum of money is faced with several alternatives regarding its use:

1. The person may exchange the money for goods and services that will satisfy his or her personal wants. Such an exchange would involve the purchase of consumer goods.
2. The person may exchange the money for productive goods or instruments. Such an exchange would involve the purchase of producer goods.
3. The person may hoard the money, either for the satisfaction of gloating over it, or awaiting an opportunity for its subsequent use.
4. The person may lend the money, asking only that the original sum be returned at some future date.
5. The person may lend the money on the condition that the borrower will repay the initial sum plus interest at some future date.

The person anticipating an inflation rate in excess of the interest rate may be tempted to choose alternatives 1 or 2 above, knowing that such choices will cost more later.

If the decision is to lend the money with the expectation of its return plus interest, the lender must consider a number of factors in deciding on the interest rate. The following are the most important.

1. What is the probability that the borrower will not repay the loan? The answer may be derived from the integrity of the borrower, his wealth, his potential earnings, and the value of any security granted the lender. If the chances are three in a hundred that the loan will not be repaid, the lender is justified in charging 3% of the sum to compensate him for the risk of loss.
2. What expense will be incurred in investigating the borrower, drawing up the loan agreement, transferring the funds to the borrower, and collecting the loan? If the sum of the loan is $1,000 for a period of one year and the lender values his efforts at $20, then he is justified in charging 2% of the sum to compensate for the expense involved.
3. What net amount will compensate for being deprived of electing other alternatives for disposing of the money? Assume that $6 per hundred or 6% is considered as adequate return, considering the investment opportunities foregone.
4. What is the probability that the interest rate may change due to inflationary effects? If rates of inflation are expected to be higher during the term of the loan, a higher rate would be appropriate, with the opposite being true for falling rates of inflation.

On the basis of the reasoning above, with no change anticipated in inflation, the interest rate arrived at will be 3% plus 2% plus 6%, or 11%. Therefore, an interest rate may be thought of, for convenience, as being made up of percentages for (1) risk of loss, (2) administrative expenses, and (3) pure gain or profit after adjustment for the effect of inflation.

2.6.2 Interest Rate from the Borrower's Viewpoint

In most cases the alternatives open to the borrower for the use of borrowed funds are limited by the lender, who may grant the loan only on condition that it be used for a specific purpose. Except as limited by the conditions of a loan, the borrower has available essentially the same alternatives for the use of money as a person who has ownership of money. However, the borrower is faced with the necessity of repaying the amount borrowed and the interest on it in accordance with the conditions of the loan agreement or suffering the consequences. The consequences may be loss of reputation, seizure of property or of other moneys, or the placing of a lien on future earnings. Society provides many pressures, legal and social, to induce a borrower to repay a loan. Default may have serious and even disastrous consequences to the borrower.

The prospective borrower's viewpoint on the rate of interest will be influenced by the use he or she intends to make of borrowed funds. If a person borrows the funds for personal use, the interest rate paid will be a measure of the amount the person is willing to pay for the privilege of having satisfactions immediately instead of in the future.

If funds are borrowed to finance business operations expected to result in a gain, the interest to be paid must be less than the expected gain. An example of this is the common practice of banks and similar enterprises of borrowing funds to lend to others. In this case it is evident that the amount paid out as interest, plus risks incurred, plus administrative expenses must be less than the interest received on the money reloaned, if the practice is to be profitable. A borrower may be expected to seek to borrow funds at the lowest interest rate possible.

2.7 THE TIME VALUE OF MONEY

Because money can earn at a certain interest rate through its investment for a period of time, a dollar received at some future date is not worth as much as a dollar in hand at present. This relationship between interest and time leads to the concept of the *time value of money.*

A dollar in hand now is worth more than a dollar received n years from now. Why? Because having the dollar now provides the opportunity for investing that dollar for n years more than the dollar to be received n years hence. Since money has *earning power,* this opportunity will earn a return, so that after n years the original dollar plus its interest will be a larger amount than the $1 received at that time. Thus, the fact that money has a time value means that equal dollar amounts at different points in time have different value as long as the interest rate that can be earned exceeds zero. This relationship between money and time is illustrated in Figure 2.4.

It is also true that money has time value because the *purchasing power* of a dollar changes through time. During periods of inflation the amount of goods that can be bought for a particular amount of money decreases as the time of purchase occurs further out in the future. Therefore, when considering the time value of money it is important to recognize both the earning power of money and the purchasing power of money.

When inflation is present, a borrower will consider its effect on the current interest rate. In times of inflationary increases, it generally pays to be a borrower, especially if the inflation rate exceeds the interest rate. Conversely, in deflationary times, borrowed funds are an extra burden, for the interest rate is likely to exceed the deflation rate. The key to these relationships is the purchasing power of future money in relation to its time value as determined by the interest rate.

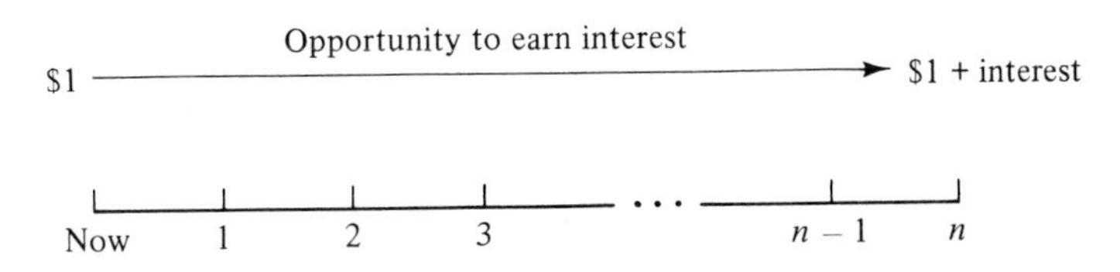

FIGURE 2.4. Illustration of the time value of money.

2.8 THE EARNING POWER OF MONEY

Funds borrowed for the prospect of gain are commonly exchanged for goods, services, or instruments of production. This leads to consideration of the *earning power of money* that may make it profitable to borrow.

Consider the example of Mr. Digg, who manually digs ditches for underground cable. For this he is paid $0.40 per linear foot and averages 200 linear feet per day. Weather conditions limit this kind of work to 180 days per year. Thus, he has an income of $80 per day worked, or $14,400 per year.

An advertisement brings to his attention a power ditcher that can be purchased for $8,000. Mr. Digg buys the ditcher after borrowing $8,000 at 14% interest. The machine will dig an average of 800 linear feet per day. By reducing the price to $0.30 per linear foot he can get sufficient work to keep the machine busy when the weather will permit.

Estimated operating and maintenance costs for the ditching machine are $40 per working day. At the end of the year the machine is worthless because it is worn out. A summary of the venture follows:

Receipts		
Amount of loan	$ 8,000	
Payment for ditches dug, 180 days × 800 ft × $0.30	43,200	$51,200
Disbursements		
Purchase of ditcher	$ 8,000	
Operating and maintenance, 180 days × $40	7,200	
Interest on loan, $8,000 × 0.14	1,120	
Repayment of loan	8,000	$24,320
Receipts Less Disbursements		$26,880

An increase in net earnings for the year over the previous year of $26,880 − $14,400 = $12,480 is enjoyed by Mr. Digg.

This example is an illustration of what is known as the "earning power of money." It was an instrument of production, the power ditcher, that enabled Mr. Digg to increase his earnings. Borrowed money made it possible for the instrument of production to be employed.

Others also gain when producer goods are profitably employed. The public gains by having ditches dug for $0.30 instead of $0.40 per foot. Also, tax revenue will increase, owing to the greater net earnings enjoyed by Mr. Digg. Increasing productivity through the employment of equipment makes these gains possible.

2.9 THE PURCHASING POWER OF MONEY

Prices for goods and services are driven upward or downward because of numerous factors at work within the economy. The cumulative effect of these factors determines the amount of price change. For example, increases in productivity and in the availability of goods tend to reduce prices, while government policies such as price supports and deficit financing tend to increase prices. When all such effects are taken together, the most common result has been that prices increase.

Inflation and deflation are terms that describe changes in price levels in an economy. Without addressing the causes of the changes in price levels, the focus will be on the methods needed to determine the rate of change of price levels. Because inflation has been a much more common occurrence than deflation, the material presented in subsequent chapters will deal primarily with inflation. However, the methods are general and will accommodate situations where economy is experiencing general price reductions.

To properly consider the time value of money in engineering economy studies, it is essential that both the earning power of money and its purchasing power be reflected properly. The concept of the purchasing power of money, along with the analytical techniques needed to incorporate this concept into engineering economy studies, is treated separately in subsequent chapters.

KEY POINTS

- The utility of an object, like its value, inheres not in the object itself but in the regard a person has for it.
- All that has utility is physically manifested.
- Two classes of goods are recognized by economists: consumer goods and producer goods.
- Economy of exchange is possible because both parties may benefit, owing to subjective considerations.
- Alternative engineering proposals will differ in the costs they involve relative to the objective of want satisfaction.
- Life-cycle cost considers all costs over the life cycle and is therefore a complete statement of the cost of a product, system, or structure.
- Interest is an amount charged for the use of money by those who have saved.
- The relationship between interest and time leads to the concept of the time value of money.
- Investment of money into producer goods which increase productivity leads to the concept of the earning power of money.
- Inflation and deflation act to alter the purchasing power of money.

QUESTIONS

1. Define value; utility.
2. Explain how utilities are created.
3. Describe the two classes of goods recognized by economists.
4. Contrast the utility of consumer goods with the utility of producer goods.
5. Why is it that the utility of consumer goods is determined subjectively, whereas the utility of producer goods is usually determined objectively?
6. Explain the economy of exchange.

7. Why is it possible for both parties to profit by an exchange?

8. What change might be brought about by persuasion in an exchange situation?

9. What elements combine to make up first cost?

10. How may first cost be a limiting factor in successful engineering activity?

11. What elements combine to make up operation and maintenance cost?

12. Contrast nonrecurring and recurring costs associated with a venture.

13. Give an example of a situation where one should evaluate both the acquisition cost and the cost of operation.

14. Discuss the difference between fixed cost and variable cost.

15. List some difficulties associated with classifying a cost as either fixed or variable.

16. What is an incremental cost?

17. Define sunk cost and explain why it should not be considered in engineering economy studies.

18. Define and explain what is meant by life-cycle cost.

19. Explain the difference in life-cycle cost committed and cost incurred.

20. Why is it so important to address early life-cycle activities carefully?

21. What is the difference between interest and an interest rate?

22. What is the market rate for interest and how is it determined?

23. What factors are considered by a lender in determining the interest rate to be sought?

24. How does inflation affect the behavior of a potential borrower?

25. What is meant by the time value of money?

26. Why does money have earning power?

27. High interest rates can nullify the beneficial effect of the earning power of money. Explain.

Part Two

INTEREST FORMULAS AND EQUIVALENCE

Economic equivalence is the fundamental operative principle for evaluating engineering alternatives in terms of worth and cost. In this part of the text, mathematical and computational tools and methods for determining economic equivalence are derived and explained. This includes the derivation of interest formulas for various cash flows and compounding assumptions, methods for equivalence calculation applicable to different economic situations, and approaches to incorporating inflationary effects in the determination of economic equivalence. Thus, Part Two provides the quantitative foundations for engineering economy studies.

3 Interest Formula Derivations

Engineering economy is concerned with the evaluation of engineering alternatives. These alternatives are usually described by estimating the amount and timing of future receipts and disbursements. Since the time value of money is concerned with the effect of time and interest rate on monetary amounts, this effect must be given primary consideration in engineering economy studies. This chapter presents the mathematical bases for considering the time value of money by contrasting simple and compound interest, describing cash flows, and deriving several categories of interest formulas.

3.1 SIMPLE AND COMPOUND INTEREST

The rental rate for a sum of money is usually expressed as the percent of the sum that is to be paid for its use for a period of one year. Interest rates are also quoted for periods other than one year, known as *interest periods*. This section compares simple and compound interest approaches for determining the effect of the time value of money.

3.1.1 Simple Interest

Under simple interest, the interest owed upon repayment of a loan is proportional to the length of time the principal sum has been borrowed. The interest earned may be found in the following manner. Let I represent the interest earned, P the principal amount, n the interest period, and i the interest rate. Then,

$$\boxed{I = Pni.} \tag{3.1}$$

Suppose that $1,000 is borrowed at a simple interest rate of 18% per annum. At the end of one year, the interest owed would be

$$I = \$1{,}000(1)(0.18) = \$180.$$

The principal plus interest would be $1,180 and would be due at the end of the year.

A simple-interest loan may be made for any period of time. Interest and principal become due only at the end of the time period. When it is necessary to calculate the interest due for a fraction of a year, it is common to consider the year as composed of 12 months of 30 days each, or 360 days. For example, on a loan of $100 at an interest rate of 18% per annum, for the period February 1 to April 20, the interest due on April 20 along with the principal sum of $100 would be 0.18($100)(80 ÷ 360) = $4.

3.1.2 Compound Interest

When a loan is made for several interest periods, interest is calculated and payable at the *end* of each interest period. There are a number of loan repayment plans. These range from paying the interest when it is due to accumulating the interest until the loan is due. For example, the payments on a 4-year loan of $1,000 at 16% interest per annum, payable when due, would be calculated as shown in Table 3.1.

If the borrower does not pay the interest earned at the end of each period and is charged interest on the *total* amount owed (principal plus interest), the interest is said to be *compounded.* The interest owed in the previous year becomes part of the total amount owed for this year. This year's interest charge includes

TABLE 3.1.
CALCULATION OF COMPOUND INTEREST WHEN INTEREST IS PAID ANNUALLY

Year	Amount Owed at Beginning of Year	Interest to Be Paid at End of Year	Amount Owed at End of Year	Amount to Be Paid by Borrower at End of Year
1	$1,000.00	$160.00	$1,160.00	$ 160.00
2	1,000.00	160.00	1,160.00	160.00
3	1,000.00	160.00	1,160.00	160.00
4	1,000.00	160.00	1,160.00	1,160.00

TABLE 3.2.
CALCULATION OF COMPOUND INTEREST WHEN INTEREST IS PERMITTED TO COMPOUND

Year	Amount Owed at Beginning of Year (A)	Interest to Be Added to Loan at End of Year (B)	Amount Owed at End of Year (A + B)	Amount Paid by Borrower at End of Year
1	$1,000.00	$1,000.00 × 0.16 = $160.00	$1,000(1.16) = $1,160.00	$ 00.00
2	1,160.00	1,160.00 × 0.16 = 185.60	$1,000(1.16)^2$ = 1,345.60	00.00
3	1,345.60	1,345.60 × 0.16 = 215.30	$1,000(1.16)^3$ = 1,560.90	00.00
4	1,560.90	1,560.90 × 0.16 = 249.75	$1,000(1.16)^4$ = 1,810.64	1,810.64

interest that has been earned on previous interest charges. For example, a loan of $1,000 at 16% interest compounded annually for a 4-year period will produce the results shown in Table 3.2.

Although the financial arrangements shown in Tables 3.1 and 3.2 require that the interest be calculated on the unpaid balance, the two cases produce different results because of the way payments are made. In the first case, payment of interest at the time it is due avoids the payment of interest on interest. The reverse is true in the second payment scheme. Thus, the effect of compound interest depends upon the payment amounts and when they are made.

3.2 DESCRIBING CASH FLOWS OVER TIME

In most engineering economy studies, only small elements of an enterprise are considered. For example, studies are often made to evaluate the consequences of the purchase of a single equipment item in a complex of many facilities. In such cases, it would be desirable to isolate the individual item from the whole by some means analogous to the "free-body" diagram in mechanics. Thus it would be necessary to itemize all receipts and all disbursements that would arise from the acquisition and operation of the equipment being considered. Then the disbursements could be subtracted from the receipts. This difference would represent profit or gain, from which the investment's return could be calculated.

To aid in identifying and recording the economic effects of investment alternatives, a graphical description of each alternative's cash transactions may be used. This graphical descriptor, referred to as a *cash flow diagram,* will provide the information necessary for analyzing an investment proposal. A cash flow diagram represents receipts received during a period of time by an upward arrow (an increase in cash) located at the period's end. The arrow's height may be proportional to the magnitude of the receipts during that period. Similarly, disbursements during a period are represented by a downward arrow (a decrease in cash). These arrows are then placed on a time scale that spans all time periods covered by the proposal.

As an example, consider the cash flow diagrams in Figure 3.1, which pertain to the simple loan transaction described in Table 3.1. In this example the borrower receives $1,000, and this amount appears as a positive cash flow on the borrower's cash flow diagram. Each year the borrower pays $160 in interest; these amounts, plus repayment of the $1,000 borrowed, appear as negative cash flows. Also shown in Figure 3.1 is the lender's cash flow diagram. The lender experiences a negative cash flow of $1,000, followed by positive cash flows for interest received and also for repayment of the original amount loaned.

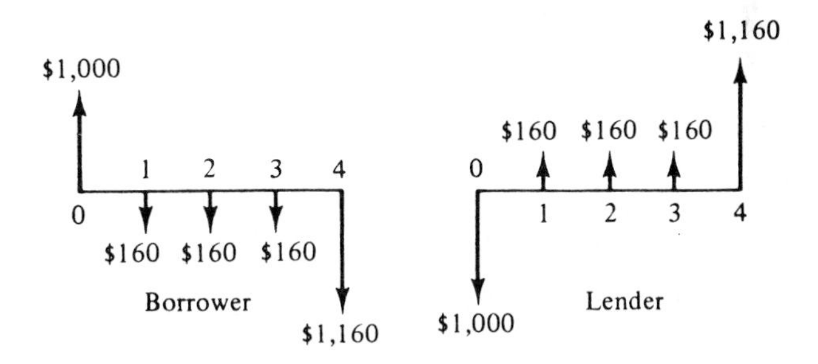

FIGURE 3.1. Cash flow diagrams.

Since there are two parties to every transaction, it is important to note that the cash flow directions in cash flow diagrams depend upon the point of view taken.

When an investment alternative has both receipts and disbursements occurring simultaneously, a net cash flow may be calculated.

> **Net cash flow** is the arithmetic sum of the receipts (+) and the disbursements (−) that occur at the same point in time.

The utilization of net cash flow implies that the net dollars received or disbursed have the same effect on an investment decision as do an investment's total receipts and disbursements considered separately.

To facilitate describing investment cash flows, the following notation will be adopted. Let

F_t = net cash flow at time t

where

$F_t < 0$ represents a net cash disbursement;

$F_t > 0$ represents a net cash receipt.

In engineering economy studies, disbursements made to implement an alternative are considered to take place at the beginning of the period embraced by the alternative. Receipts and disbursements occurring during the life of the alternative are usually assumed to occur at the end of the year or interest period in which they occur. This "year-end" convention is adopted for describing cash flows over time and for developing applicable cash flow diagrams. The remainder of this chapter derives a number of interest formulas for dealing with cash flows over time where the "year-end" convention applies.

3.3 INTEREST FORMULAS (DISCRETE COMPOUNDING, DISCRETE PAYMENTS)

The interest formulas derived in this section apply to the common situation of annual compounding interest and annual payments. The following symbols will be used. Let

i = the annual interest rate;

n = the number of annual interest periods;

P = a present principal sum;

A = a single payment, in a series of n equal payments, made at the end of each annual interest period;

F = a future sum, n annual interest periods hence.

Four important points apply in the derivation and use of interest factors for annual payments:

1. The end of one year is the beginning of the next year.
2. P is at the beginning of a year at a time regarded as being the present.
3. F is at the end of the nth year from a time regarded as being the present.
4. An A occurs at the *end* of each year of the period under consideration.

When P and A are involved, the first A of the series occurs one year after P. When F and A are involved, the last A of the series occurs simultaneously with F.

3.3.1 Single-Payment Compound-Amount Factor

If an amount P is invested now and earns at the rate i per year, how much principal and interest are accumulated after n years? The cash flow diagram for this situation is shown in Figure 3.2.

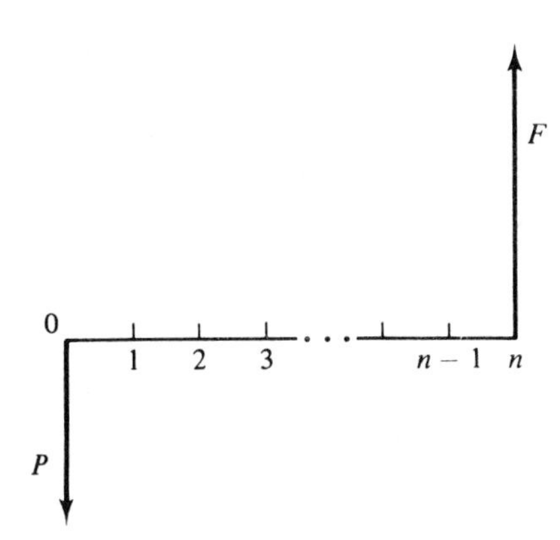

FIGURE 3.2. Single present amount and single future amount.

Since this transaction does not provide any payments until the investment is terminated, interest is compounded as shown in Table 3.2. There the interest earned is added to the principal at the end of each annual interest period. By substituting general terms in place of numerical values in Table 3.2, the results shown in Table 3.3 are obtained. The resulting factor, $(1 + i)^n$, is known as the *single-payment compound-amount factor*[1] and is designated[2]

$$\left(\overset{F/P,i,n}{\quad}\right).$$

[1] *Values for interest factors for discrete compounding interest-discrete payments are given in Appendix A.*

[2] *There are two important advantages of using the functional factor designations in place of algebraic expressions: (1) the equations for solving problems may be set up prior to looking up values of factors from the tables and inserting them in the parentheses, and (2) the source and the identity of values taken from the tables are maintained during the solution. This functional factor designation system is used throughout the text.*

TABLE 3.3.
DERIVATION OF SINGLE-PAYMENT COMPOUND-AMOUNT FACTOR

Year	Amount at Beginning of Year	Interest Earned During Year	Compound Amount at End of Year
1	P	Pi	$P + Pi = P(1+i)^1$
2	$P(1+i)$	$P(1+i)i$	$P(1+i) + P(1+i)i = P(1+i)^2$
3	$P(1+i)^2$	$P(1+i)^2 i$	$P(1+i)^2 + P(1+i)^2 i = P(1+i)^3$
n	$P(1+i)^{n-1}$	$P(1+i)^{n-1} i$	$P(1+i)^{n-1} + P(1+i)^{n-1} i = P(1+i)^n = F$

This factor may be used to find the future amount, F, of a present principal amount, P. The relationship is

$$\boxed{F = P(1+i)^n} \tag{3.2}$$

or

$$F = P(\overset{F/P,i,n}{\quad\quad}).$$

The designator used to identify the single-payment compound-amount factor is $F/P,i,n$. It appears over the parentheses where the value of the factor is to be entered. The first element in the designator, F/P, represents a ratio that identifies what the factor must be multiplied by, P, in order to find F. The i represents the interest rate per period and the n the number of periods between the occurrence of P and F.

Referring to the example of Table 3.2, if \$1,000 is invested at 16% interest compounded annually at the beginning of year one, the compound amount at the end of the fourth year will be

$$F = \$1{,}000(1 + 0.16)^4 = \$1{,}000(1.811)$$

$$= \$1{,}811.$$

Or, by use of the factor designation and its associated tabular value,

$$F = \$1{,}000(\overset{F/P,16,4}{1.811}) = \$1{,}811.$$

3.3.2 Single-Payment Present-Worth Factor

The single-payment compound-amount relationship of Equation (3.2) may be solved for P as follows:

$$\boxed{P = F\left[\frac{1}{(1+i)^n}\right].} \tag{3.3}$$

The resulting factor, $1/(1 + i)^n$, is known as the *single-payment present-worth factor* and is designated[3]

$$\overset{P/F,i,n}{(\quad)}.$$

This factor may be used to find the present worth, P, of a future amount, F.

For the investment described in Figure 3.2, the question is, "How much must be invested now at 16% compounded annually so that \$1,811 can be received 4 years hence?" This calculation is

$$P = \$1{,}811\left[\frac{1}{(1 + 0.16)^4}\right] = \$1{,}811(0.5523) = \$1{,}000.$$

Or, by using the factor designation and the interest tables,

$$P = \$1{,}811\overset{P/F,16,4}{(0.5523)} = \$1{,}000.$$

3.3.3 Equal-Payment-Series Compound-Amount Factor

In some engineering economy studies, it is necessary to find the single future value that would accumulate from a series of equal payments occurring at the end of succeeding interest periods. Such a series of cash flows is presented in Figure 3.3. The sum of the compound amounts of the several payments may be calculated by use of the single-payment compound-amount factor. For example, the calculation of the compound amount of a series of five \$100 payments made at the end of each year at 12% interest compounded annually is shown in Table 3.4.

It is apparent that the tabular method is cumbersome for calculating the compound amount for an extensive series. Therefore, it is desirable to derive a compact solution for this type of situation.

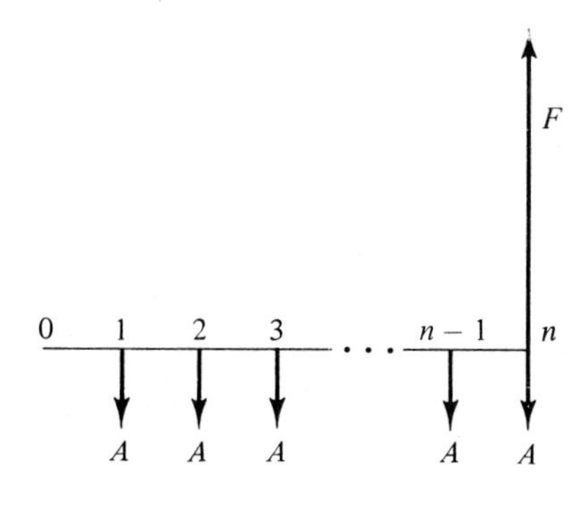

FIGURE 3.3. Equal annual series and single future amount.

[3] *Note that the single-payment compound-amount factor and the single-payment present-worth factor are reciprocals.*

TABLE 3.4.
THE COMPOUND AMOUNT OF A SERIES OF YEAR-END PAYMENTS

End of Year	Year-End Payment Times Compound-Amount Factor	Compound Amount at End of 5 years	Total Compound Amount
1	$\$100(1.12)^4$	\$157.35	
2	$100(1.12)^3$	140.49	
3	$100(1.12)^2$	125.44	
4	$100(1.12)^1$	112.00	
5	$100(1.12)^0$	100.00	\$635.28

If A represents a series of n equal payments, such as the \$100 series in Table 3.4, then

$$F = A(1) + A(1 + i) + \cdots + A(1 + i)^{n-2} + A(1 + i)^{n-1}.$$

The total future amount, F, is equal to the sum of individual future amounts calculated for each payment, A. Multiplying this equation by $(1 + i)$ results in

$$F(1 + i) = A(1 + i) + A(1 + i)^2 + \cdots + A(1 + i)^{n-1} + A(1 + i)^n.$$

Subtracting the first equation from the second gives

$$\begin{array}{rcl} F(1 + i) & = & A(1 + i) + A(1 + i)^2 + \cdots + A(1 + i)^{n-1} + A(1 + i)^n \\ -F & = & -A - A(1 + i) - A(1 + i)^2 - \cdots - A(1 + i)^{n-1} \\ \hline F(1 + i) - F & = & -A \qquad\qquad\qquad\qquad\qquad\qquad + A(1 + i)^n \end{array}$$

Solving for F gives

$$\boxed{F = A\left[\frac{(1 + i)^n - 1}{i}\right].} \qquad (3.4)$$

The resulting factor, $[(1 + i)^n - 1]/i$, is known as the *equal-payment-series compound-amount factor* and is designated

$$(\overset{F/A,i,n}{\quad}).$$

This factor may be used to find the compound amount, F, of an equal-payment series, A. For example, the future amount of a \$100 payment deposited at the end of each of the next 5 years and earning 12% per annum will be

$$F = \$100\left[\frac{(1 + 0.12)^5 - 1}{0.12}\right] = \$100(6.353) = \$635,$$

which agrees with the result found in Table 3.4. Using the factor designation and the interest tables gives

$$F = \$100\overset{F/A,\,12,5}{(\ 6.353\ )} = \$635.$$

3.3.4 Equal-Payment-Series Sinking-Fund Factor

The equal-payment-series compound-amount relationship may be solved for A as follows:

$$\boxed{A = F\left[\frac{i}{(1+i)^n - 1}\right].} \tag{3.5}$$

The resulting factor, $i/[(1+i)^n - 1]$, is known as the *equal-payment-series sinking-fund factor* and is designated[4]

$$\overset{A/F,i,n}{(\qquad)}.$$

This factor may be used to find the required end-of-year payments, A, to accumulate a future amount, F, as shown in Figure 3.3. If, for example, it is desired to accumulate \$635 by making a series of five equal annual payments at 12% interest compounded annually, the required amount of each payment will be

$$A = \$635\left[\frac{0.12}{(1+0.12)^5 - 1}\right]$$

$$= \$635(0.1574) = \$100$$

or

$$A = \$635\overset{A/F,12,5}{(\ 0.1574\ )} = \$100.$$

3.3.5 Equal-Payment-Series Capital-Recovery Factor

A deposit of amount P is made now at an annual interest rate i. The depositor wishes to withdraw the principal, plus earned interest, in a series of equal year-end amounts over the next n years. When the last withdrawal is made, there should be no funds left on deposit. The cash flow diagram for this situation is illustrated in Figure 3.4.

It has been shown previously that F is related to A by the equal-payment-series sinking-fund factor and that F and P are linked by the single-payment

[4] *The equal-payment-series compound-amount factor and the equal-payment-series sinking-fund factor are reciprocals.*

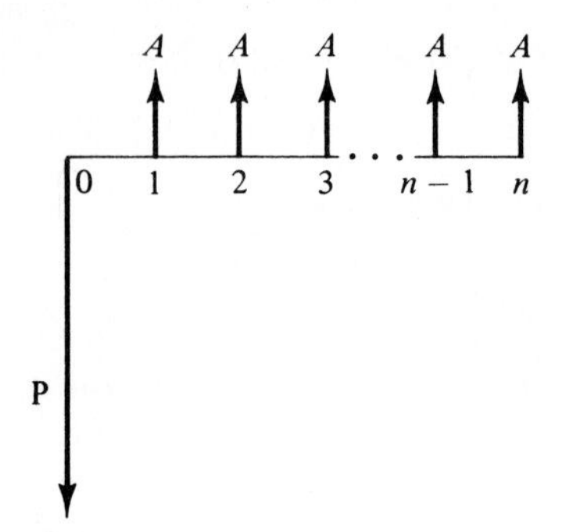

FIGURE 3.4. Equal annual series and single present amount.

compound-amount factor. The substitution of $P(1 + i)^n$ for F in the equal-payment-series sinking-fund relationship gives

$$A = P(1 + i)^n\left[\frac{i}{(1 + i)^n - 1}\right]$$

$$= P\left[\frac{i(1 + i)^n}{(1 + i)^n - 1}\right]. \quad (3.6)$$

The resulting factor, $i(1 + i)^n/[(1 + i)^n - 1]$ is known as the *equal-payment-series capital-recovery factor* and is designated

$$\left(\overset{A/P,i,n}{\quad}\right).$$

This factor may be used to find the end-of-period payments, A, that will be provided by a present amount, P. For example, \$1,000 invested at 15% interest compounded annually will provide for eight equal year-end payments of

$$A = \$1{,}000\left[\frac{0.15(1 + 0.15)^8}{(1 + 0.15)^8 - 1}\right]$$

$$= \$1{,}000(0.2229) = \$223$$

or

$$A = \$1000(\overset{A/P,15,8}{0.2229}) = \$223.$$

As each annual withdrawal is made, the amount remaining on deposit is smaller than the amount remaining after the previous withdrawal. Because the interest earned is based on the amount on deposit, the interest earned each year also diminishes. The equal-payment-series capital-recovery factor accounts for these year-by-year changes in what appears to be a complicated relationship between interest earned and amount withdrawn.

3.3.6 Equal-Payment-Series Present-Worth Factor.

To find what single amount must be deposited now so that equal end-of-period payments can be made, P must be found in terms of A. The equal-payment-series capital-recovery factor may be solved for P as follows:

$$P = A\left[\frac{(1 + i)^n - 1}{i(1 + i)^n}\right]. \tag{3.7}$$

The resulting factor, $[(1 + i)^n - 1]/i(1 + i)^n$, is known as the *equal-payment-series present-worth factor* and is designated[5]

$$\overset{P/A,i,n}{(\qquad)}.$$

This factor may be used to find the present worth, P, of a series of equal periodic payments, A, as depicted in Figure 3.4. For example, the present worth of a series of eight equal annual payments of \$223 at an interest rate of 15% compounded annually will be

$$P = \$223\left[\frac{(1 + 0.15)^8 - 1}{0.15(1 + 0.15)^8}\right]$$

$$= \$223(4.4873) = \$1{,}000$$

or

$$P = \$223(\overset{P/A,15,8}{4.4873}) = \$1{,}000.$$

3.3.7 Uniform-Gradient-Series Factor

In some cases, periodic payments do not occur in an equal series. They may increase or decrease by a constant amount. For example, a series of payments that would be uniformly increasing is \$100, \$125, \$150, and \$175 occurring at the end of the first, second, third, and fourth years. Similarly, a uniformly decreasing series would be \$100, \$90, \$80, and \$70 occurring at the end of the first, second, third, and fourth years. In each case, an equal payment series provides the base with a constant annual increase or decrease beginning at the end of the second year.

In general, a uniformly increasing series of payments for n interest periods may be expressed as $G, 2G, \ldots, (n - 1)G$, as shown in Figure 3.5, where G denotes the annual change in the magnitude of the payments. One way of evaluating such a series is to apply the interest formulas developed previously to each

[5] *The equal-payment-series capital-recovery factor and the equal-payment-series present-worth factor are reciprocals.*

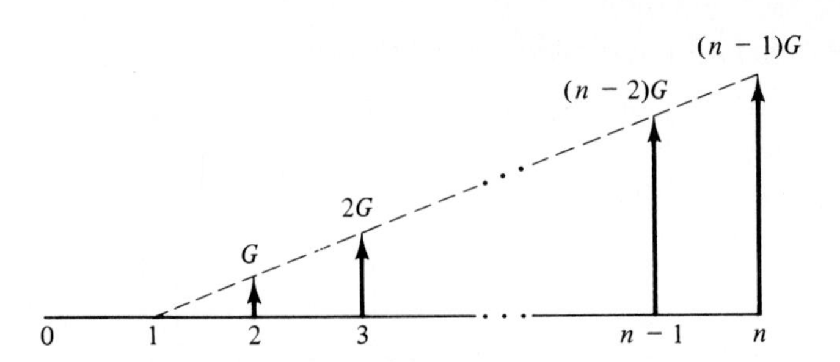

FIGURE 3.5. A uniformly increasing gradient series.

payment in the series. This method will yield good results but will be time consuming. Another approach is to reduce the uniformly increasing or decreasing series of payments to an equal-payment series so that the equal-payment-series factor can be used. Let

G = annual change or gradient;
n = the number of years;
A = the equal annual payment.

Each payment in the gradient series, $G, 2G, \ldots, (n-1)G$, can be converted to an annual amount by applying Equation 3.5 as

$$A = F(\ \overset{A/F,i,n}{}\ ) \text{ or } F\left[\frac{i}{(1+i)^n - 1}\right]$$

where F is the future amount of the gradient series. This future amount can be found by noting that the gradient series can be separated into $(n-1)$ distinct equal-payment series with annual payments of G as shown in Table 3.5. The fu-

TABLE 3.5.
GRADIENT SERIES AND AN EQUIVALENT SET OF SERIES

End of Year	Gradient Series	Set of Series Equivalent to Gradient Series
0	0	0
1	0	0
2	G	G
3	$2G$	$G + G$
4	$3G$	$G + G + G$
.	.	.
.	.	.
.	.	.
$n-1$	$(n-2)G$	$G + G + G + \cdots + G$
n	$(n-1)G$	$G + G + G + \cdots + G + G$

ture amount can be derived as follows:

$$F = G\left(\overset{F/A,i,n-1}{\quad}\right) + G\left(\overset{F/A,i,n-2}{\quad}\right) + \cdots + G\left(\overset{F/A,i,2}{\quad}\right) + G\left(\overset{F/A,i,1}{\quad}\right)$$

$$= G\left[\frac{(1+i)^{n-1}-1}{i}\right] + G\left[\frac{(1+i)^{n-2}-1}{i}\right] + \cdots + G\left[\frac{(1+i)^{2}-1}{i}\right]$$

$$+ G\left[\frac{(1+i)^{1}-1}{i}\right]$$

$$= \frac{G}{i}[(1+i)^{n-1} + (1+i)^{n-2} + \cdots + (1+i)^2 + (1+i) - (n-1)]$$

$$= \frac{G}{i}[(1+i)^{n-1} + (1+i)^{n-2} + \cdots + (1+i)^2 + (1+i) + 1] - \frac{nG}{i}.$$

The bracketed terms constitute the equal-payment-series compound-amount factor for n years. Therefore,

$$F = \frac{G}{i}\left[\frac{(1+i)^n - 1}{i}\right] - \frac{nG}{i}. \tag{3.8}$$

From Equation (3.5)

$$A = F\left[\frac{i}{(1+i)^n - 1}\right]$$

$$= \frac{G}{i}\left[\frac{(1+i)^n - 1}{i}\right]\left[\frac{i}{(1+i)^n - 1}\right] - \frac{nG}{i}\left[\frac{i}{(1+i)^n - 1}\right]$$

$$= \frac{G}{i} - \frac{nG}{i}\left[\frac{i}{(1+i)^n - 1}\right]$$

or

$$A = \frac{G}{i} - \frac{nG}{i}\left(\overset{A/F,i,n}{\quad}\right) = G\left[\frac{1}{i} - \frac{n}{i}\left(\overset{A/F,i,n}{\quad}\right)\right].$$

The resulting factor,

$$\boxed{A = G\left[\frac{1}{i} - \frac{n}{(1+i)^n - 1}\right]}, \tag{3.9}$$

is called the *uniform-gradient-series factor* and is designated[6]

$$\left(\overset{A/G,i,n}{\quad}\right).$$

As an example of the use of the gradient factor, assume that a person is planning to save $1,000 from income during this year and can increase this amount by $200 for each of the following nine years. Since the end-of-year convention is to be used unless otherwise stated, this series begins at the end of the first year, and the last amount saved occurs at the end of the tenth year. If interest is 8% compounded annually, what equal-annual series beginning at the end of year 1 and ending at year 10 would produce the same accumulation at the end of year 10 as as would be realized from the gradient series? The solution is

$$A = \$1{,}000 + \$200\left(\overset{A/G,8,10}{3.8713}\right)$$

$$= \$1{,}774 \text{ per year.}$$

The gradient factor may also be used for a uniformly decreasing gradient. Suppose that the equal-annual series equivalent to the decreasing gradient series in Figure 3.6 is desired. Visualize the cash flow in Figure 3.6 as resulting from the year-by-year subtraction of an *increasing* gradient series where $G = \$600$ from an equal-annual series of $5,000 per year. By approaching the solution in this manner no new factors are needed. The equal-annual series equivalent to this decreasing gradient series at 9% per annum is

$$A = \$5{,}000 - \$600\left(\overset{A/G,9,6}{2.2498}\right)$$

$$= \$3{,}650 \text{ per year.}$$

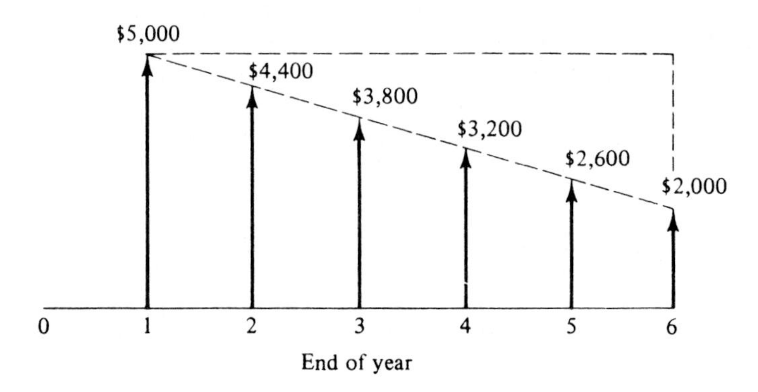

FIGURE 3.6. A uniformly decreasing gradient series.

[6] *Tabular values for the uniform-gradient-series factor are given as the last column in each interest table in Appendix A.*

3.3.8 Geometric-Gradient-Series Factor

In some situations, annual payments increase or decrease, not by a constant amount, but by a constant percentage. If g is used to designate the percentage change in the magnitude of the payment from one year to the next, the magnitude of the tth payment is related to payment F_1 as

$$F_t = F_1(1 + g)^{t-1}, \qquad t = 1, 2, \ldots, n. \tag{3.10}$$

When g is positive the series will increase, as is illustrated in Figure 3.7. When g is negative, the series will decrease.

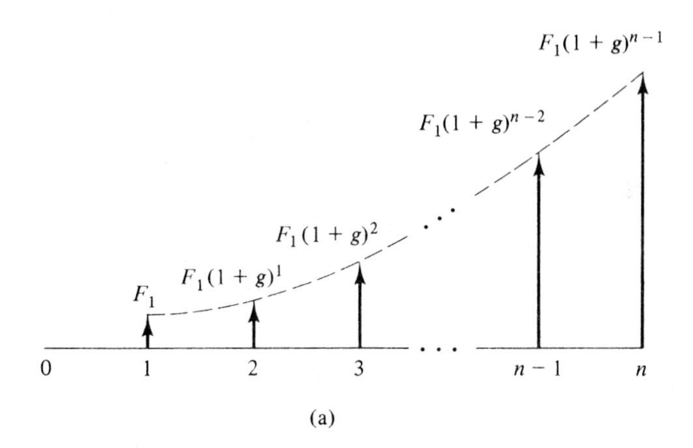

FIGURE 3.7. A geometric gradient series for $g > 0$.

To derive an expression for the present amount, P, the relationship between F_1 and F_t given by Equation (3.10) can be used, together with the single-payment present-worth factor of Equation (3.3) as

$$P = F_1\left[\frac{(1 + g)^0}{(1 + i)^1}\right] + F_1\left[\frac{(1 + g)^1}{(1 + i)^2}\right] + F_1\left[\frac{(1 + g)^2}{(1 + i)^3}\right] + \cdots + F_1\left[\frac{(1 + g)^{n-1}}{(1 + i)^n}\right].$$

Multiply each term by $(1 + g)/(1 + g)$ and simplify:

$$P = \frac{F_1}{1 + g}\left[\frac{(1 + g)^1}{(1 + i)^1} + \frac{(1 + g)^2}{(1 + i)^2} + \frac{(1 + g)^3}{(1 + i)^3} + \cdots + \frac{(1 + g)^n}{(1 + i)^n}\right].$$

Let

$$\frac{1}{(1 + g')} = \frac{1 + g}{1 + i},$$

where g' is the *growth-free rate,* and substitute for each term:

$$P = \frac{F_1}{1 + g}\left[\frac{1}{1 + g'} + \frac{1}{(1 + g')^2} + \frac{1}{(1 + g')^3} + \cdots + \frac{1}{(1 + g')^n}\right].$$

The terms within the brackets constitute the equal-payment-series present-worth factor for n years. Therefore,

$$\boxed{P = \frac{F_1}{1 + g}\left[\frac{(1 + g')^n - 1}{g'(1 + g')^n}\right]} \tag{3.11}$$

or

$$P = F_1\left[\frac{\overset{P/A,g',n}{(\qquad)}}{1 + g}\right]. \tag{3.12}$$

The factor within brackets is called the *geometric-gradient-series factor.* Its use requires finding g' from

$$\boxed{g' = \frac{1 + i}{1 + g} - 1.} \tag{3.13}$$

When g′ > 0. If $i > g$, g' will be positive, and $\overset{P/A,g',n}{(\qquad)}$ will take on its customary form. As an example, suppose that receipts from a certain venture are estimated to increase by 7% per year from a first-year base of \$360,000. The present worth of 10 years of such receipts at an interest rate of 15% may be found as follows:

$$g' = \frac{1 + 0.15}{1 + 0.07} - 1 = 7.48\%$$

$$\overset{P/A,7.48,10}{(\ 6.8704\ )} = \frac{(1 + 0.0748)^{10} - 1}{0.0748(1 + 0.0748)^{10}}$$

$$P = \$360{,}000\frac{\overset{P/A,7.48,10}{(\ 6.8704\ )}}{1.07} = \$2{,}311{,}536.$$

It may be noted that the value for $\overset{P/A,7.48,10}{(\qquad)}$ falls between the P/A values in the 7% and 8% tables. This suggests the possible use of the tables when g' is positive. However, the result will be only an approximation if linear interpolation is used.

When g′ = 0. If $i = g$, g' will be zero, and the value of $\left(\overset{P/A,g',n}{\quad}\right)$ will be n (see Table 3.6). The geometric-gradient-series factor reduces to

$$P = F_1\left[\frac{n}{1 + g}\right].$$

As an example, suppose that receipts from a certain activity are estimated to increase by 10% per year from a first-year base of $10,000. The present worth of n years of such receipts at an interest rate of 10% will be

$$P = \$10{,}000\left[\frac{n}{1.10}\right] = \$9{,}091n.$$

This is not $10,000$n$, owing to the convention of assigning the first-year payment as F_1 rather than $F_1(1 + g)$; i.e., the rate g is imposed beginning at $t = 1$, while i is in effect beginning at $t = 0$.

When g′ < 0. Finally, if $i < g$, g' will be negative, and tabular values cannot be used to evaluate the P/A factor. Equation (3.11) will have to be used directly. For example, suppose that the salary for a recent graduate is expected to increase by 12% per year from a base of $32,000 over the next five years. If the interest rate is taken to be 10% during this period, the present worth of the earnings may be found as follows:

$$g' = \frac{1 + 0.10}{1 + 0.12} - 1 = -1.79\%$$

$$\left(\overset{P/A,\,-1.79,5}{5.2801}\right) = \frac{(1 - 0.0179)^5 - 1}{-0.0179(1 - 0.0179)^5}$$

$$P = \$32{,}000\frac{\left(\overset{P/A,\,-1.79,5}{5.2801}\right)}{1.12} = \$150{,}860.$$

This amount may be compared with the present worth if there were no differential between i and g: if both were 10%. In this case, $P = \$32{,}000(5/1.10) = \$145{,}455$. The present worth differs by $5,405 because the salary is estimated to increase at a rate 2% greater than the interest rate.

The geometric-gradient-series factor may also be used for decreasing-gradient evaluations as seen in Figure 3.8. In this case, g will be negative and will result in a positive value for g' for all positive values of i. As an example, suppose that a shallow oil well is expected to produce 12,000 barrels of oil during its first year at $21 per barrel. If its yield is expected to decrease by 10% per year, the present worth of the anticipated gross revenue at an interest rate of 17% over the next seven years may be found as follows:

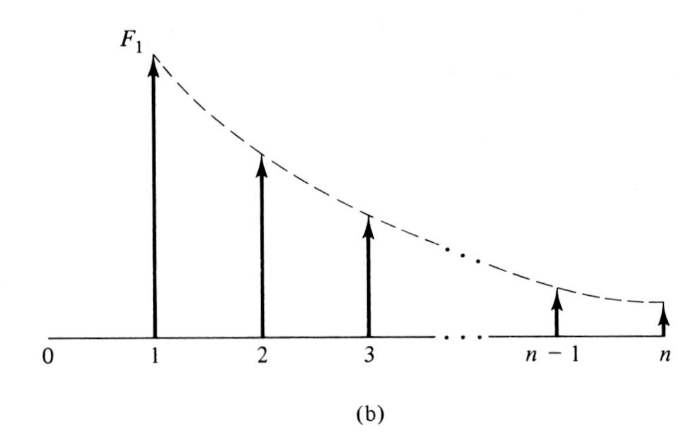

(b)

FIGURE 3.8. A geometric gradient series for $g < 0$.

$$g' = \frac{1 + 0.17}{1 - 0.10} - 1 = 0.30 \text{ or } 30\%,$$

$$P = \$21(12{,}000)\frac{\overset{P/A,30,8}{(2.9247)}}{1 - 0.10} = \$818{,}916.$$

In this example, the value for the *P/A* factor was available directly from the 30% interest table.

3.4 INTEREST-FORMULA RELATIONSHIPS

Two interest formulas derived in Section 3.3 were for the single-payment situation. Four were for equal-payment series. The last two were for gradient-payment situations incorporating the capacity to deal with increasing as well as decreasing series. When these interest formulas are used, it is essential that the cash flows conform to the format for which the factors are applicable. The schematic arrangement of the cash flows in Figure 3.9 should be helpful in this connection.

3.4.1 Interest-Factor Relationships

Numerous relationships that exist between the interest factors allow the calculation of one type factor from another. Awareness of these relationships provides the means for more effective use of the interest tables, along with a better understanding of how these factors reflect the time value of money. These relationships are as follows:

Find	Given	Formula
F	P	$F = P(F/P, i, n)$
P	F	$P = F(P/F, i, n)$
F	A	$F = A(F/A, i, n)$
A	F	$A = F(A/F, i, n)$
P	A	$P = A(P/A, i, n)$
A	P	$A = P(A/P, i, n)$
A	G	$A = G(A/G, i, n)$
P	F_1	$P = \frac{F_1(P/A, g', n)}{1 + g}$

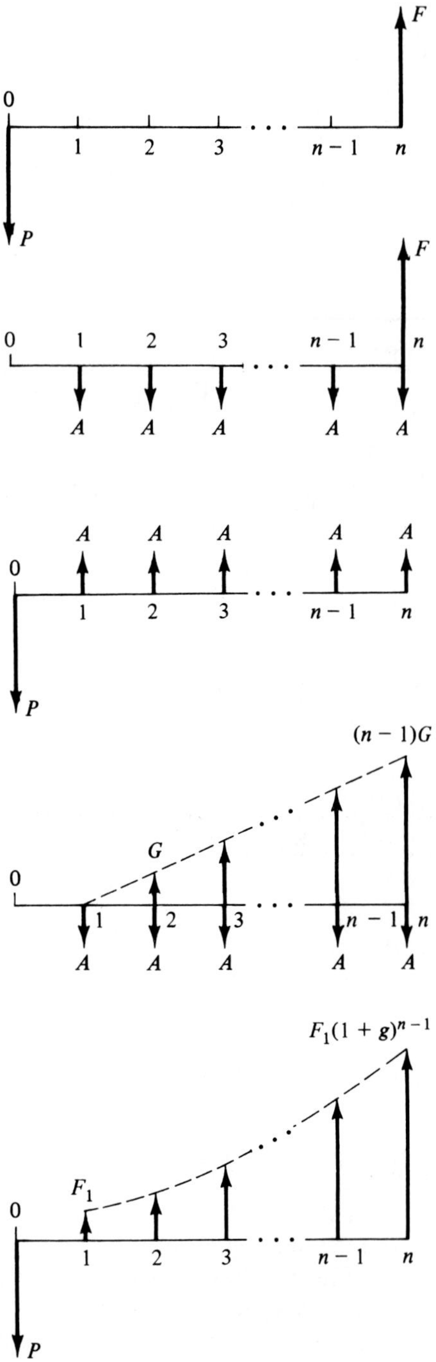

FIGURE 3.9. Schematic illustration of the use of interest factors.

1. $(F/P,i,n) = i(F/A,i,n) + 1$

2. $(P/F,i,n) = 1 - (P/A,i,n)i$

3. $(F/A,i,n) = 1 + (F/P,i,1) + (F/P,i,2) + \cdots + (F/P,i,n-1)$

4. $(A/F,i,n) = (A/P,i,n) - i$

5. $(P/A,i,n) = (P/F,i,1) + (P/F,i,2) + \cdots + (P/F,i,n)$

6. $$(A/P,i,n) = \frac{i}{1 - (P/F,i,n)}$$

3.4.2 Interest-Factor Extreme Values

In some situations it is useful to be aware of the extreme values of the common interest factors. Extreme values may help clarify situations occurring at the limit of the applicable range of the factor values. Table 3.6 gives some extreme values for $n = \infty$ and i known as well as for n known and $i = 0$.

TABLE 3.6.
EXTREME VALUES FOR INTEREST FACTORS

Interest Factor	$n = \infty$; i Known	$i = 0$; n Known
$(F/P,i,n)$	∞	1
$(P/F,i,n)$	0	1
$(F/A,i,n)$	∞	n
$(A/F,i,n)$	0	$1/n$
$(P/A,i,n)$	$1/i$	n
$(A/P,i,n)$	i	$1/n$

3.5 COMPOUNDING FREQUENCY CONSIDERATIONS

The derivations to this point have involved interest periods of only one year. In practice, however, cash flows or loan agreements may require that interest be paid more frequently, such as each half-year, each quarter, or each month. Such agreements result in interest periods of one-half year, one-quarter year, or one-twelfth year, and the compounding of interest twice, four times, or twelve times a year, respectively.

Interest rates associated with this more frequent compounding are normally quoted on an annual basis according to the following convention. When the actual or *effective* rate of interest is 3% interest compounded each six-month period, the annual or *nominal* interest is quoted as "6% per year compounded semiannually." For an effective rate of interest of 1.5% compounded at the end of each three-month period, the nominal interest is quoted as "6% per year compounded quarterly."

The nominal rate of interest is expressed on an annual basis and is determined by multiplying the actual or effective interest rate per interest period by the number of compounding periods per year.

3.5.1 Nominal and Effective Interest Rates

It is possible to establish a relationship between the effective interest rate for any time interval and the nominal interest rate per year. Let

r = nominal interest rate per year[7]
i = effective interest rate in the time interval
l = length of the time interval (in years)
m = reciprocal of the length of the compounding period (in years)

The effective interest rate for any time interval is given by[8]

$$i = \left(1 + \frac{r}{m}\right)^{l \cdot m} - 1 \tag{3.14}$$

If the interest is compounded only once in the time interval, then $l \cdot m = 1$ and

$$i = \frac{r}{m} \tag{3.15}$$

To find the applicable effective interest rate for any time interval, the following relationship may be used:

$$i = \left(1 + \frac{r}{m}\right)^{c} - 1, \qquad c \geq 1 \tag{3.16}$$

[7] *The nominal interest rate is commonly referred to in financial transactions as the annual percentage rate, APR.*

[8] *Throughout this book, i is used to designate effective interest rates and r identifies nominal interest rates. Note that the nominal rate equals the effective rate when the number of compounding periods per year is one: m = 1.*

where c is the number of compounding periods in the time interval ($c = l \times m$). When $c = 1$, Equation (3.16) reduces to Equation (3.15), and either may be used to find the effective rate per compounding period. For example, if the nominal interest rate is 9% compounded monthly, the effective rate per month is

$$i = \frac{0.09}{12} = 0.0075 \text{ or } 0.75\%.$$

If $c > 1$, the effective rate is found from Equation (3.16) for any time interval. Some examples follow:

1. Nominal rate of 12% compounded monthly with time interval of one year ($c = 12$)

$$i = \left(1 + \frac{0.12}{12}\right)^{12} - 1 = 0.1268 \text{ or } 12.68\% \text{ per year.}$$

2. Nominal rate of 18% compounded weekly with a time interval of one year ($c = 52$)

$$i = \left(1 + \frac{0.18}{52}\right)^{52} - 1 = 0.1968 \text{ or } 19.68\% \text{ per year.}$$

3. Nominal rate of 14% compounded monthly with a time interval of six months ($c = 6$)

$$i = \left(1 + \frac{0.14}{12}\right)^{6} - 1 = 0.0721 \text{ or } 7.21\% \text{ per six months.}$$

4. Nominal rate of 10% compounded weekly with a time interval of six months ($c = 26$)

$$i = \left(1 + \frac{0.10}{52}\right)^{26} - 1 = 0.0512 \text{ or } 5.12\% \text{ per six months.}$$

5. Nominal rate of 13% compounded monthly with a time interval of two years ($c = 24$)

$$i = \left(1 + \frac{0.13}{12}\right)^{24} - 1 = 0.2951 \text{ or } 29.51\% \text{ per two years.}$$

6. Nominal rate of 9% compounded semiannually with a time interval of two years ($c = 4$)

$$i = \left(1 + \frac{0.09}{2}\right)^4 - 1 = 0.1925 \text{ or } 19.25\% \text{ per two years.}$$

3.5.2 Continuous Compounding

As a limit, interest may be considered to be compounded an infinite number of times per year—that is, *continuously*. Under these conditions, the *effective* annual interest for continuous compounding is derived from Equation (3.14) with $l = 1$ as

$$i_a = \lim_{m \to \infty} \left(1 + \frac{r}{m}\right)^m - 1.$$

But since

$$\left(1 + \frac{r}{m}\right)^m = \left[\left(1 + \frac{r}{m}\right)^{m/r}\right]^r$$

and

$$\lim_{m \to \infty} \left(1 + \frac{r}{m}\right)^{m/r} = e = 2.7182,$$

then

$$i_a = \lim_{m \to \infty} \left[\left(1 + \frac{r}{m}\right)^{m/r}\right]^r - 1 = e^r - 1.$$

Therefore, when interest is compounded continuously,

$$\boxed{i_a = \text{effective annual interest rate} = e^r - 1.} \qquad (3.17)$$

3.5.3 Comparing Interest Rates

The effective interest rates corresponding to a nominal annual interest rate of 18% compounded annually, semiannually, quarterly, monthly, weekly, daily, and continuously are shown in Table 3.7.[9] Since the effective interest rate represents the actual interest earned, this rate should be used to compare the benefits of various nominal rates of interest.

For example, one might be confronted with the problem of determining whether it is more desirable to receive 16% compounded annually or 15% com-

[9] *Effective interest rates corresponding to nominal annual rates for various compounding frequencies are given in Appendix B.*

TABLE 3.7.
EFFECTIVE ANNUAL INTEREST RATES FOR VARIOUS COMPOUNDING PERIODS AT A NOMINAL RATE OF 18%

Compounding Frequency	Number of Periods per Year	Effective Interest Rate per Period	Effective Annual Interest Rate
Annually	1	18.0000%	18.0000%
Semiannually	2	9.0000	18.8100
Quarterly	4	4.5000	19.2517
Monthly	12	1.5000	19.5618
Weekly	52	0.3642	19.6843
Daily	365	0.0493	19.7142
Continuously	∞	0.0000	19.7217

pounded monthly. The effective rate of interest per year for 16% compounded annually is, of course, 16%, while for 15% compounded monthly the effective annual interest rate is

$$i_a = \left(1 + \frac{0.15}{12}\right)^{12} - 1 = 16.08\%.$$

Thus, 15% compounded monthly yields an actual rate of interest that is higher than 16% compounded annually.

3.6 INTEREST FORMULAS (CONTINUOUS COMPOUNDING, DISCRETE PAYMENTS)

In certain economic evaluations, it is reasonable to assume that continuous-compounding interest more nearly represents the true situation than does discrete compounding. Also, the assumption of continuous compounding may be more convenient from a computational standpoint in some applications. Therefore, this section presents interest formulas that may be used in those cases where discrete payments and continuous-compounding interest seem appropriate. The following symbols will be used. Let

r = the nominal annual interest rate;

n = the number of annual periods;

P = a present principal sum;

A = a single payment, in a series of n equal payments, made at the end of each annual period;

F = a future sum, n annual periods hence.

3.6.1 Single-Payment Compound-Amount Factor

The single-payment compound-amount factor may be expressed as a function of the number of compounding periods as follows:

For annual compounding: $F = P(1 + r)^n$.

For semiannual compounding: $F = P\left(1 + \frac{r}{2}\right)^{2n}$.

For monthly compounding: $F = P\left(1 + \frac{r}{12}\right)^{12n}$.

In general, if there are m compounding periods per year

$$F = P\left(1 + \frac{r}{m}\right)^{mn}.$$

When interest is assumed to compound continuously, the interest earned is instantaneously added to the principal at the end of each infinitesimal interest period. For continuous compounding, the number of compounding periods per year is considered to be infinite. Therefore,

$$F = P\left[\lim_{m \to \infty}\left(1 + \frac{r}{m}\right)^{mn}\right].$$

By rearranging terms

$$F = P\left\{\lim_{m \to \infty}\left[\left(1 + \frac{r}{m}\right)^{m/r}\right]^{rn}\right\}.$$

But,

$$\lim_{m \to \infty}\left(1 + \frac{r}{m}\right)^{m/r} = e = 2.7182.$$

Therefore,

$$\boxed{F = Pe^{rn}.} \tag{3.18}$$

The resulting factor, e^{rn}, is the *single-payment compound-amount factor* for continuous-compounding interest and is designated[10]

$$\left[\,^{F/P,r,n}\,\right]$$

[10] *Square brackets are used around the continuous-compounding factors, while parentheses enclose the discrete-compounding factors to visually distinguish these two types of factors.*

Note that any continuous-compounding, discrete-payment factor may be derived from its discrete-compounding, discrete-payment counterpart by substituting the effective continuous interest rate for i. For the factor developed in Equation (3.18) substitute

$$i = e^r - 1$$

into

$$(1 + i)^n$$

giving

$$e^{rn}.$$

3.6.2 Single-Payment Present-Worth Factor

The single-payment compound-amount relationship in Equation (3.18) may be solved for P as follows:

$$\boxed{P = F\left[\frac{1}{e^{rn}}\right].} \tag{3.19}$$

The resulting factor, e^{-rn}, is the *single-payment present-worth factor* for continuous-compounding interest and is designated

$$[\ \overset{P/F,r,n}{}\ ].$$

3.6.3 Equal-Payment-Series Present-Worth Factor

By considering each payment in a series individually, the total present worth of the series is a sum of the individual present-worth amounts as follows:

$$P = A(e^{-r}) + A(e^{-r2}) + \cdots + A(e^{-rn})$$

$$= Ae^{-r}(1 + e^{-r} + e^{-r2} + \cdots + e^{-r(n-1)})$$

which is Ae^{-r} times the geometric series $\sum_{j=0}^{n-1}\left(\frac{1}{e^r}\right)^j$. Therefore,

$$P = Ae^{-r}\left[\frac{1 - e^{-rn}}{1 - e^{-r}}\right]$$

$$\boxed{= A\left[\frac{1 - e^{-rn}}{e^r - 1}\right].} \tag{3.20}$$

The resulting factor, $(1 - e^{-rn})/(e^r - 1)$, is the *equal-payment-series present-worth factor* for continuous-compounding interest and is designated

$$\left[\ ^{P/A,r,n}\ \right].$$

3.6.4 Equal-Payment-Series Capital-Recovery Factor

The equal-payment-series present-worth relationship may be solved for A as follows:

$$\boxed{A = P\left[\frac{e^r - 1}{1 - e^{-rn}}\right]}. \tag{3.21}$$

The resulting factor, $(e^r -)/1 - e^{-rn})$, is the *equal-payment-series capital-recovery factor* for continuous-compounding interest and is designated

$$\left[\ ^{A/P,r,n}\ \right].$$

3.6.5 Equal-Payment-Series Sinking-Fund Factor

The substitution of Fe^{-rn} for P in the equal-payment-series capital-recovery relationship results in

$$A = Fe^{-rn}\left[\frac{e^r - 1}{1 - e^{-rn}}\right]$$

$$\boxed{= F\left[\frac{e^r - 1}{e^{rn} - 1}\right]}. \tag{3.22}$$

The resulting factor, $(e^r - 1)/(e^{rn} - 1)$, is the *equal-payment-series sinking-fund factor* for continuous-compounding interest and is designated

$$\left[\ ^{A/F,r,n}\ \right].$$

3.6.6 Equal-Payment-Series Compound-Amount-Factor

The equal-payment series sinking-fund relationship may be solved for F as follows:

$$\boxed{F = A\left[\frac{e^{rn} - 1}{e^r - 1}\right]}. \tag{3.23}$$

The resulting factor, $(e^{rn} - 1)/(e^r - 1)$, is the *equal-payment-series compound-amount factor* for continuous-compounding interest and is designated

$$[\;^{F/A,r,n}\;].$$

3.6.7 Uniform-Gradient-Series Factor

The equivalent annual payment, A, corresponding to a linear gradient, G, number of years, n, and interest rate, r, may be found in a similar manner as for annual compounding. It can be shown that

$$\boxed{A = G\left[\frac{1}{e^r - 1} - \frac{n-}{{}^{rn} - 1}\right].} \tag{3.24}$$

The resulting factor,

$$\left[\frac{1}{e^r - 1} - \frac{n}{e^{rn} - 1}\right],$$

is called the *uniform gradient-series factor* for continuous-compounding interest and is designated

$$[\;^{A/G,r,n}\;].$$

3.6.8 Geometric-Gradient-Series Factor

As noted earlier, substitution of the effective continuous-compounding interest rate $(e^r - 1)$ for i, the effective interest rate in the comparable discrete-compounding interest factor, will yield the continuous-compounding factor desired. Using Equation (3.13),

$$g' = \frac{(1 + i)}{1 + g} - 1$$

for the cash flow in Figure 3.7 yields the g' value neccessary for continuous compounding when the substitution is made.

$$g' = \frac{(1 + e^r - 1)}{(1 + g)} - 1 = \frac{e^r}{(1 + g)} - 1$$

resulting in

$$P = F_1\left[\frac{\left(\overset{P/A,g',n}{\qquad}\right)}{1+g}\right] \qquad (3.25)$$

as the present equivalent of a series of discrete payments growing at a rate of g percent per period when compounding is continuous.

3.7 INTEREST FORMULAS (CONTINUOUS COMPOUNDING, CONTINUOUS PAYMENTS)

In the previous derivations, payments were considered to be concentrated at discrete points in time. In many instances, however, it is reasonable to assume that cash flows occur on a relatively uniform basis throughout the year. Situations such as this involve a *funds-flow process* which may be described in terms of an annual flow rate. The following symbols will be used. Let

r = the nominal annual interest rate;

n = time expressed in years;

P = a present principal sum;

$\overline{A}$ = the uniform flow rate of money per year;

F = a future amount equal to the compound amount of a uniform flow of money at time n.

Where there is no flow of payments, as with annual payments, the compound amount and the present-worth factors are identical to those for continuous-compounding discrete payments. Thus,

$$F = Pe^{rn}$$

as was shown in Section 3.6. Its reciprocal

$$P = Fe^{-rn}$$

was also developed previously.

3.7.1 Funds-Flow Compound-Amount Factor

The following symbols will be used to develop interest formulas for the funds-flow process. Let

ΔF = a future amount equal to the compound amount ΔP. This future

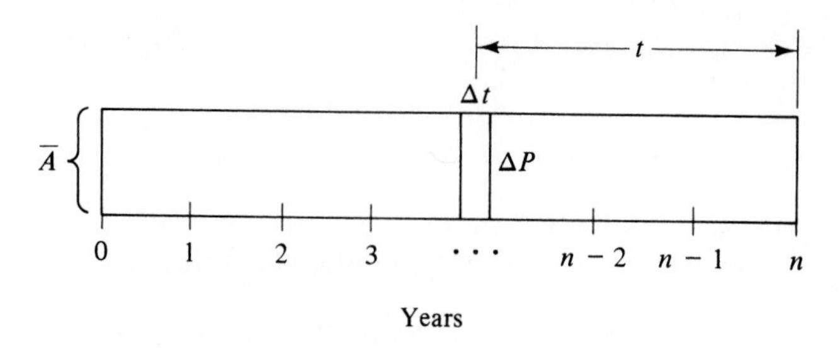

FIGURE 3.10. Uniform continuous funds flow.

amount occurs t years from time n as shown in Figure 3.10.
$\overline{A}$ = uniform rate of flow of money per year.

Since it has been shown that $F = Pe^{rn}$,

$$\Delta F = \Delta P e^{rt}.$$

But,

$$\Delta P = \overline{A}\ \Delta t$$

so that

$$\Delta F = \overline{A} e^{rt}\ \Delta t.$$

By letting Δt approach zero,

$$dF = \overline{A} e^{rt}\ dt.$$

And, for the entire interval 0 to n

$$F = \int_0^n dF = \int_0^n \overline{A} e^{rt}\ dt$$

$$F = \left[\frac{\overline{A} e^{rt}}{r}\right]_0^n = \overline{A}\left[\frac{e^{rn}}{r} - \frac{e^0}{r}\right]$$

$$\boxed{F = \overline{A}\left[\frac{e^{rn} - 1}{r}\right].} \qquad (3.26)$$

The resulting factor, $(e^{rn} - 1)/r$, is called the *funds-flow compound-amount factor* and is designated

$$\left[{}^{F/\bar{A},r,n}\right].$$

3.7.2 Funds-Flow Sinking-Fund Factor

The funds-flow compound-amount relationship may be solved for $\bar{A}$ as follows:

$$\boxed{\bar{A} = F\left[\frac{r}{e^{rn} - 1}\right].} \tag{3.27}$$

The resulting factor, $r/(e^{rn} - 1)$, is the *funds-flow sinking-fund factor* and is designated

$$\left[{}^{\bar{A}/F,r,n}\right].$$

3.7.3 Funds-Flow Capital-Recovery Factor

By using the single-payment compound-amount relationship for continuous compounding, $F = Pe^{rn}$, and the funds-flow sinking-fund relationship just derived, it is seen that

$$\bar{A} = Pe^{rn}\left[\frac{r}{e^{rn} - 1}\right]$$

$$\boxed{\bar{A} = P\left[\frac{re^{rn}}{e^{rn} - 1}\right].} \tag{3.28}$$

The resulting factor, $(re^{rn})/(e^{rn} - 1)$, is the *funds-flow capital-recovery factor* and is designated

$$\left[{}^{\bar{A}/P,r,n}\right].$$

3.7.4 Funds-Flow Present-Worth Factor

The funds-flow capital-recovery relationship may be solved for P as follows:

$$\boxed{P = \bar{A}\left[\frac{e^{rn} - 1}{re^{rn}}\right].} \tag{3.29}$$

The resulting factor, $(e^{rn} - 1)/(re^{rn})$, is the *funds-flow present-worth factor* and is designated

$$\left[\;^{P/\bar{A},r,n}\;\right].$$

3.7.5 Funds-Flow Conversion Factor

Tabulated values for the interest factors for continuous-compounding interest–discrete payments may be modified and used for the funds-flow factors. The required conversion factor may be derived by finding the year-end equivalent of a summation of an infinite number of payments occurring during the year. The equal-payment-series compound-amount factor in Equation (3.23) may be modified to reflect m interest periods per year as follows:

$$F = \frac{A}{m}\left[\frac{e^{(r/m)m} - 1}{e^{r/m} - 1}\right] = \frac{A}{m}\left[\frac{e^{r} - 1}{e^{r/m} - 1}\right].$$

But

$$\lim_{m\to\infty} \frac{A}{m}\left[\frac{e^{r} - 1}{e^{r/m} - 1}\right] = \lim_{m\to\infty} \bar{A}\left[\frac{\dfrac{e^{r} - 1}{m}}{e^{r/m} - 1}\right] = \bar{A}\left[\frac{e^{r} - 1}{r}\right]$$

$$\boxed{F = \bar{A}\left[\frac{e^{r} - 1}{r}\right].} \tag{3.30}$$

Equation (3.30) expresses the equivalence between a uniform continuous flow of funds for one year, $\bar{A}$, and a future amount at the end of the year, F. For a time span greater than one year, the same factor also provides the equivalence between a uniform flow of funds occurring at the rate of $\bar{A}$ per year and equal annual amounts A at the end of each year. Thus, for time spans greater than one year

$$\boxed{A = \bar{A}\left[\frac{e^{r} - 1}{r}\right].} \tag{3.31}$$

The resulting factor, $(e^r - 1)/r$, is called the *funds-flow conversion factor*[11] and is designated

$$\left[{}^{A/\bar{A},r}\right].$$

This conversion factor may be used with the interest factors for continuous-compounding interest-annual payments to yield values for the funds-flow factors in the following manner:

1. $\left[{}^{\bar{A}/P,r,n}\right] = \left[{}^{A/P,r,n}\right] \div \left[{}^{A/\bar{A},r}\right]$
2. $\left[{}^{P/\bar{A},r,n}\right] = \left[{}^{P/A,r,n}\right] \left[{}^{A/\bar{A},r}\right]$
3. $\left[{}^{\bar{A}/F,r,n}\right] = \left[{}^{A/F,r,n}\right] \div \left[{}^{A/\bar{A},r}\right]$
4. $\left[{}^{F/\bar{A},r,n}\right] = \left[{}^{F/A,r,n}\right] \left[{}^{A/\bar{A},r}\right].$

As an example of the use of the funds-flow conversion factor, the present amount of $8,000 per year flowing uniformly for a period of 6 years at an interest rate of 15% compounded continuously is found to be

$$P = \bar{A}\left[{}^{P/\bar{A},r,n}\right] = \bar{A}\left[{}^{P/A,r,n}\right] \left[{}^{A/\bar{A},r}\right]$$

$$= \$8{,}000\left[\overset{P/A,15,6}{3.6669}\right] \left[\overset{A/\bar{A},15}{1.078894}\right] = \$31{,}650.$$

3.8 SUMMARY OF INTEREST FORMULAS

The three groups of interest formulas derived in this chapter are summarized in Table 3.8. Each group is based on assumptions about the nature of payments and the compounding of interest. In engineering economy, the group that most accurately represents the situation under study should be used.

[11] *Values for the funds-flow conversion factor for various interest rates are given in Appendix C.*

TABLE 3.8.
SUMMARY OF INTEREST FORMULAS AND DESIGNATIONS

	Factor	Find	Given	Discrete Payments: Discrete Compouding	Discrete Payments: Continuous Compounding	Continuous Payments: Continuous Compounding
Single-Payment	Compound-Amount	F	P	$F = P(1+i)^n = P(\overset{F/P,i,n}{\quad})$	$F = Pe^{rn} = P[\overset{F/P,r,n}{\quad}]$	$F = Pe^{rn} = P[\overset{F/P,r,n}{\quad}]$
	Present-Worth	P	F	$P = F\frac{1}{(1+i)n} = F(\overset{P/F,i,n}{\quad})$	$P = F\frac{1}{e^{rn}} = F[\overset{P/F,r,n}{\quad}]$	$P = F\frac{1}{e^{rn}} = F[\overset{P/F,r,n}{\quad}]$
Equal-Payment Series	Compound-Amount	F	A	$F = A\left[\frac{(1+i)^n - 1}{i}\right] = A(\overset{F/A,i,n}{\quad})$	$F = A\left[\frac{e^{rn} - 1}{e^r - 1}\right] = A[\overset{F/A,r,n}{\quad}]$	$F = \bar{A}\left[\frac{e^{rn} - 1}{r}\right] = \bar{A}[\overset{F/\bar{A},r,n}{\quad}]$
	Sinking-Fund	A	F	$A = F\left[\frac{i}{(1+i)^n - 1}\right] = F(\overset{A/F,i,n}{\quad})$	$A = F\left[\frac{e^r - 1}{e^{rn} - 1}\right] = F[\overset{A/F,r,n}{\quad}]$	$\bar{A} = F\left[\frac{r}{e^{rn} - 1}\right] = F[\overset{\bar{A}/F,r,n}{\quad}]$
	Present-Worth	P	A	$P = A\left[\frac{(1+i)^n - 1}{i(1+i)^n}\right] = A(\overset{P/A,i,n}{\quad})$	$P = A\left[\frac{1 - e^{-rn}}{e^r - 1}\right] = A[\overset{P/A,r,n}{\quad}]$	$P = \bar{A}\left[\frac{e^{rn} - 1}{re^{rn}}\right] = \bar{A}[\overset{P/\bar{A},r,n}{\quad}]$
	Capital-Recovery	A	P	$A = P\left[\frac{i(1+i)^n}{(1+i)^n - 1}\right] = P(\overset{A/P,i,n}{\quad})$	$A = P\left[\frac{e^r - 1}{1 - e^{-rn}}\right] = P[\overset{A/P,r,n}{\quad}]$	$\bar{A} = P\left[\frac{re^{rn}}{e^{rn} - 1}\right] = P[\overset{\bar{A}/P,r,n}{\quad}]$
Gradient Series	Uniform-Gradient-Series	A	G	$A = G\left[\frac{1}{i} - \frac{n}{(1+i)^n - 1}\right] = G(\overset{A/G,i,n}{\quad})$	$A = G\left[\frac{1}{e^r - 1} - \frac{n}{e^{rn} - 1}\right] = G[\overset{A/G,r,n}{\quad}]$	
	Geometric-Gradient	P	F_1	$P = \frac{F_1}{1+g}\left[\frac{(1+g')^n - 1}{g'(1+g')^n}\right] = F_1\frac{(\overset{P/A,g',n}{\quad})}{1+g}$	$P = \frac{F_1}{1+g}\left[\frac{(1+g')^n - 1}{g'(1+g')^n}\right] = F_1\frac{(\overset{P/A,g',n}{\quad})}{1+g}$	

PROBLEMS

1. What amount will be available in 4 years if \$8,000 is invested now at 10% per year simple interest? Answer: \$11,200
2. What is the principal amount if the principal plus interest at the end of $2\frac{1}{2}$ years is \$14,000 for a simple interest rate of 18% per annum?
3. For what period of time will \$600 have to be invested to amount to \$1,500 if it earns 20% simple interest per annum? Answer: 7.5 years
4. Compare the interest earned by \$10,000 for 10 years at 12% simple interest with that earned by the same amount for 10 years at 12% compounded annually.
5. If \$160 interest is earned in two months on an investment of \$8,000, what is the annual rate of simple interest? Answer: 12%
6. A person lends \$10,000 at 9% simple interest for 4 years. At the end of this time the entire amount (principal plus interest) is invested at 11% compounded annually for 10 years. How much will accumulate at the end of the 14-year period? Answer: \$38,610
7. For an interest rate of 10% compounded annually, find:
   - **a.** How much can be loaned now if \$2,000 will be repaid at the end of 3 years?
   - **b.** How much will be required 6 years hence to repay a \$50,000 loan made now?
8. Draw a cash flow diagram for the loan and investment situation described in Problem 6 from the viewpoint of the person making the loan and the subsequent investment.
9. What will be the amount accumulated by each of the following present investments?
   - **a.** \$3,000 in 7 years at 14% compounded annually. Answer: \$7,506
   - **b.** \$1,600 in 17 years at 12% compounded annually.
   - **c.** \$20,000 in 38 years at 16% compounded annually.
   - **d.** \$3,500 in 71 years at 8% compounded annually.
   - **e.** \$5,000 in 34 years at 11.5% compounded annually.
   - **f.** \$10,000 in 150 years at 9% compounded annually.
10. What is the present value of the following future receipts?
    - **a.** \$19,000 5 years from now at 9% compounded annually. Answer: \$12,348
    - **b.** \$8,300 12 years from now at 15% compounded annually.
    - **c.** \$6,200 53 years from now at 12% compounded annually.
    - **d.** \$17,500 64 years from now at 10% compounded annually.
    - **e.** \$13,000 18 years from now at 19.2% compounded annually.

f. $5,000 10 years from now at 8% compounded annually.

11. What is the accumulated value of each of the following series of payments?
    a. $300 at the end of each year for 9 years at 12% compounded annually. Answer: $4,433
    b. $1,400 at the end of each year for 10 years at 18% compounded annually.
    c. $4,200 at the end of each year for 43 years at 11% compounded annually.
    d. $100 at the end of each year for 110 years at 10% compounded annually.
    e. $250 at the end of each year for 23 years at 9.7% compounded annually.
    f. $1,000 at the end of each year for 40 years at 7% compounded annually.
12. What equal series of payments must be put into a sinking fund to accumulate the following amounts?
    a. 62,000 in 10 years at 10% compounded annually when payments are annual. Answer: $5,763
    b. $6,500 in 8 years at 12% compounded annually when payments are annual.
    c. $18,000 in 52 years at 13% compounded annually when payments are annual.
    d. $1,000 in 3 years at 12% compounded annually when payments are annual.
    e. $5,400 in 47 years at 8% compounded annually when payments are annual.
    f. $90,000 in 72 years at 6.3% compounded annually when payments are annual.
13. What is the present value of the following series of prospective receipts?
    a. $1,500 a year for 16 years at 14% compounded annually. Answer: $9,398
    b. $230 a year for 37 years at 15% compounded annually.
    c. $1,000 a year for 9 years at 8% compounded annually.
    d. $2,500 a year for 10 years at 10% compounded annually.
    e. $900 a year for 42 years at 17.6% compounded annually.
    f. $12,000 a year for 5 years at 7.5% compounded annually.
14. What series of equal payments is necessary to repay the following present amounts?
    a. $4,000 in 5 years at 16% compounded annually with annual payments. Answer: $1,222

b. $50,000 in 10 years at 8.5% compounded annually with annual payments.

c. $9,500 in 20 years at 10% compounded annually with annual payments.

d. $37,000 in 62 years at 8% compounded annually with annual payments.

e. $10,000 in 120 years at 11% compounded annually with annual payments.

f. $100,000 in 30 years at 9.6% compounded annually with annual payments.

15. What annual equal payment series is necessary to repay the following increasing series of payments?

a. A series of 7 end-of-year payments that begins at $2,000 and increases at the rate of $100 a year with 10% interest compounded annually. Answer: $2,262

b. A series of 30 end-of-year payments that begins at $250 and increases at the rate of $50 a year with 9% interest compounded annually.

c. A series of 25 end-of-year payments that begins at $400 and increases at the rate of $200 a year with 12.5% interest compounded annually.

16. What annual equal-payment series is necessary to repay the following decreasing series of payments?

a. A series of 10 end-of-year payments that begins at $6,000 and decreases at the rate of $200 a year with 12% interest compounded annually.

b. A series of 42 end-of-year payments that begins at $10,000 and decreases at the rate of $100 a year with 8% interest compounded annually.

c. A series of 19 end-of-year payments that begins at $1,500 and decreases at the rate of $40 a year with 14.3% interest compounded annually. Answer: $1,285

17. What is the present value of the following geometrically increasing series of payments?

a. A first-year base of $2,000 increasing at 5% per year to year 10 at an interest rate of 12%. Answer: $13,585

b. A first-year base of $15,000 increasing at 10% per year to year 8 with an interest rate of 13%.

c. A first-year base of $1,000 increasing at 8% per year to year 20 with an interest rate of 8%.

18. What is the present value of the following geometrically decreasing series of payments?

a. A first-year base of $9,000 decreasing by 10% per year to year 10 with an interest rate of 17%. Answer: $30,915

b. A first-year base of $1,000,000 decreasing by 25% per year to year 4 with an interest rate of 15%.

c. A first-year base of $200,000 decreasing by 8.6% per year to year 41 with an interest rate of 13.5%.

19. What equal annual amount must be deposited for 10 years in order to provide withdrawals of $200 at the end of the second year, $400 at the end of the third year, $600 at the end of the fourth year, and so on, up to $1,800 at the end of the tenth year? The interest rate is 13% compounded annually.

20. How many years will it take for an investment to double itself if interest is compounded annually for the following interest rates?

a. 4%. Answer: 18 years

b. 5%.

c. 10%.

d. 12%.

e. 19%.

f. 40%.

21. At what rate of interest compounded annually will an investment triple itself for the following number of years?

a. 5 years.

b. 7 years.

c. 10 years.

d. 14 years

e. 28 years.

f. 35 years. Answer: 3.19%

22. Plot i as a function of n to illustrate the range of combinations of these variables that result in the doubling of an initial invested amount P.

23. What is the value of i compounded annually if $P = \$1,000$, $F = \$4,000$, and $n = 12$ years?

24. What is the value of n if $F = \$4,000$, $P = \$1,000$, and $i = 12\%$ compounded annually?

25. What rate of interest compounded annually is involved if

a. An investment of $10,000 made now will result 10 years hence in a receipt of $23,670? Answer: 9%

b. An investment of $1,000 made 18 years ago has increased in value to 4,000?

c. An investment of $2,500 made now will result 5 years hence in a receipt of $4,212?

d. An investment of $9,000 made 20 years ago has increased in value to $36,000?

26. How many years will be required for

a. An investment of \$3,000 to increase to \$6,939 if interest is 15% compounded annually? Answer: 6 years

b. An investment of \$1,000 to increase to \$7,400 if interest is 10% compounded annually?

c. An investment of \$5,000 to increase to \$302,100 if interest is 6% compounded annually?

d. An investment of \$200 to increase to \$2,824 if interest is 12.2% compounded annually?

27. Find the interest factor ($F/G,i,n$) that will convert a gradient series as defined in Table 3.5 into its future equivalent at the end of the nth year.

28. Draw the cash flow diagram for the F/A factor from the viewpoint of someone making annual deposits.

29. Draw the cash flow diagram for the A/P factor from the viewpoint of someone spending from a cash gift received at the present.

30. Find the interest factor ($P/G,i,n$) that will convert a gradient series as defined in Table 3.5 to its equivalent value at the present.

31. Rewrite the formula given for the single-payment compound-amount factor to apply to the compounding of interest at the end of each period, where p represents the number of compounding periods per year, y the number of years, and r the nominal annual rate of interest. Use P as the present sum and F as the compound amount and express F in terms of P, r, p, and y.

32. Derive a formula for finding the accumulated amount F at the end of n interest periods that will result from a series of beginning-of-period payments each equal to B if the latter are placed in a sinking fund for which the interest rate per period is i, compounded each period.

33. How would you determine a desired equal-payment-series capital-recovery factor if you only had a table of

a. Single-payment present-worth factors?

b. Equal-payment-series present-worth factors?

c. Equal-payment-series compound-amount factors?

d. Single-payment compound-amount factors?

34. How would you determine a desired equal-payment-series sinking-fund factor if you only had a table of

a. Single-payment compound-amount factors?

b. Single-payment present-worth factors?

c. Equal-payment-series compound-amount factors?

d. Equal-payment-series capital-recovery factors?

35. What effective annual interest rate corresponds to the following?

a. Nominal interest rate of 12% compounded semiannually.
b. Nominal interest rate of 12% compounded monthly.
c. Nominal interest rate of 12% compounded quarterly.
d. Nominal interest rate of 12% compounded weekly.
e. Nominal interest rate of 12% compounded daily.

36. What effective interest rate per compounding period corresponds to the following nominal interest rates?

a. r = 8% compounded quarterly. Answer: 2% per quarter.
b. r = 18% compounded monthly.
c. r = 11% compounded daily.
d. r = 14% compounded semiannually.
e. r = 9% compounded continuously.

37. Find the nominal interest rate and effective interest rate per compounding period for the following effective annual interest rates.

a. 12.36%, semiannual compounding. Answer: r = 12%, i = 6% per six months.
b. 4.06%, quarterly compounding.
c. 18.39%, monthly compounding.
d. 29.61%, weekly compounding.
e. 8.00%, daily compounding.
f. 200%, monthly compounding.

38. What nominal interest rate is paid if compounding is annual and

a. Payments of $4,500 per year for 6 years will repay an original loan of $17,000?
b. Annual deposits of $1,000 will result in $25,000 at the end of 10 years?

39. The Square Deal Loan Company offers money at 0.3% interest per week compounded weekly. What is the effective annual interest rate? What is the nominal interest rate?

40. An effective annual interest rate of 12% is desired.

a. What nominal rate should be sought if compounding is to be semiannually?
b. What nominal rate should be sought if compounding is to be quarterly?

41. How much more desirable is 10% compounded monthly than 10% compounded yearly?

42. Which of the following nominal interest rates provides the most interest earned over a year?

a. 8% compounded daily or 9% compounded annually. Answer: 9%
b. 11% compounded monthly or 12% compounded semiannually.

c. 19% compounded daily or 20% compounded annually.

d. 25% compounded weekly or 26% compounded semiannually.

e. 32% compounded monthly or 33% compounded quarterly.

f. 38% compounded monthly or 43% compounded annually.

43. Find the nominal interest rate for the following effective annual rates when compounding is continuous.

a. 10.52%.

b. 15.02%.

c. 12.00%. Answer: 11.33%

d. 31.00%.

e. 16.00%.

f. 50.00%.

44. How many years will be required for an investment to exactly double itself for the following nominal interest rates? (Note that for continuous compounding fractional years are possible.)

a. 1% compounded continuously.

b. 5% compounded continuously.

c. 10% compounded continuously. Answer: 6.93 years.

d. 11% compounded continuously.

e. 13% compounded continuously.

f. 15% compounded continuously.

45. What is the effective interest rate if a nominal rate of 12% is compounded continuously? If an effective interest rate of 8% is desired, what must the nominal rate be if compounding is continuous?

46. What is the present worth of the following prospective payments?

a. $7,500 in 30 years at an interest rate of 10% compounded continuously.

b. $9,200 in 5 years at an interest rate of 12% compounded weekly.

47. What is the accumulated value of each of the following series of payments?

a. $400 at the end of each year for 5 years at 9.8% interest compounded continuously. Answer: $2,456

b. $500 at the end of each year for 39 years at 15% interest compounded continuously.

48. What equal annual payment must be deposited into a sinking fund to accumulate $60,000 in 20 years at 13% interest compounded continuously?

49. What will be the required annual payment to repay a loan of $2,500 in 3 years if the interest rate is 8% compounded continuously? Answer: $976

50. What is the present worth of a series of equal year-end payments of $1,400 each for 10 years if the interest rate is 12.7% compounded continuously?

51. An interest rate of 10% compounded continuously is desired on an investment of $30,000. How many years will be required to recover at least the investment with the desired return if $8,000 is received each year?

52. What is the present value of the following continuous funds flows?

a. $9,000 per year for 7 years at 6% compounded continuously. Answer: $51,443

b. $700 per year for 10 years at 12% compounded continuously.

c. $400 per year for 14.3 years at 8% compounded continuously.

d. $2,300 per year for 15.8 years at 16% compounded continuously.

53. What amount will be accumulated by each of these continuous funds flow?

a. $900 per year in 6 years at 10% compounded continuously. Answer: $7,399

b. $6,000 per year in 21.2 years at 9% compounded continuously.

54. For how many years must an investment of $40,000 provide a continuous flow of funds at the rate of $8,000 per year so that an annual interest rate of 15% compounded continuously is earned?

55. How long will it take for a continuous flow of funds at the rate of $1,600 per year to accumulate to $100,000 at an interest rate of 20% compounded continuously?

56. What annual interest rate compounded continuously wll be earned on an investment of $18,300 that provides a continuous flow of funds at the rate of $4,300 annually for 7.5 years? Answer: 16.86%

57. If the interest rate is 9% compounded continuously, and funds flow at the continuous rate of $1,000 per year, find the accumulated amount at the end of 10 years.

58. Find the interest rate compounded continuously that will be earned if an investment of a continuous flow of funds at the rate of $5,600 per year for 29 years accumulates to $500,000 at the end of that time.

4 Calculating Economic Equivalence

Most computations in engineering economy require that prospective receipts and disbursements of two or more alternative proposals be placed on an equivalent basis for comparison. This calls for proper use of the interest formulas derived in the previous chapter, together with an understanding of the economic meaning of equivalence. This chapter develops the concept of economic equivalence and presents required computational methods for using interest formulas in engineering economy studies.

4.1 THE MEANING OF EQUIVALENCE

If two or more situations are to be compared, their characteristics must be placed on an equivalent basis. Which is worth more, 4 ounces of Product A or 1,800 grains of Product A? To answer this question, it is necessary to place the two amounts on an equivalent basis by use of the proper conversion factor. After conversion of ounces to grains, the question becomes: Which is worth more, 1,750 grains of Product A or 1,800 grains of Product A? The answer is now obvious.

Two things are said to be equivalent when they have the same effect.

For instance, the torques produced by applying forces of 100 pounds and 200 pounds 2 feet and 1 foot, respectively, from the fulcrum of a lever are equivalent, since each produces a torque of 200 foot-pounds.

Three elements are involved in the equivalence of sums of money. These are (1) the amounts of the sums, (2) the times of occurrence of the sums, and (3) the interest rate. Interest formulas consider time and the interest rate. Thus, they are a convenient means for calculating the equivalence of monetary amounts occurring at different points in time.

The relative merit of two or more alternatives is usually not directly apparent from a simple statement of their future receipts and disbursements. These amounts must be placed on an equivalent basis. Consider the following example: An engineer sells his patent to a corporation and is offered a choice of $12,500 now or $2,000 per year for the next 10 years, the estimated beneficial life of the patent to the corporation. The engineer is paying 12% interest on his home mortgage and will use this rate in his evaluation. The patterns of receipts are shown in Table 4.1.

A cursory examination of the receipts of the two alternatives does not reveal which is the most desirable. For instance, it is incorrect to say Alternative B is more desirable than Alternative A on the basis that the sums of receipts from those alternatives are $20,000 and $12,500, respectively. Such a statement would be correct only if the interest rate were taken to be zero.

The equivalence values for these two alternatives for an interest rate of 12% must be found by the use of interest formulas. One way to determine an

TABLE 4.1.
PATTERN OF RECEIPTS FOR TWO ALTERNATIVES

End of Year	Receipts, Alternative A	Receipts, Alternative B
0	$12,500	0
1	0	$2,000
2	0	2,000
3	0	2,000
4	0	2,000
5	0	2,000
6	0	2,000
7	0	2,000
8	0	2,000
9	0	2,000
10	0	2,000
Total Receipts	$12,500	$20,000

equivalent value for Alternative B is to calculate an amount at the present that is equivalent to 10 receipts of $2,000 each. This is

$$P = \$2{,}000(\overset{P/A,12,10}{5.6502}) = \$11{,}300.$$

This amount is equivalent to 10 future payments of $2,000 each and is directly comparable with $12,500. This is because both amounts represent money at the same point in time, the present. Thus, the engineer observes that on an equivalent basis the $12,500 lump sum is most desirable.

It should be noted that the $11,300 is only an equivalent amount determined from an anticipated series of cash receipts. An actual receipt of $11,300 would not occur, even if this alternative had been chosen. The actual receipts would be $2,000 per year for 10 years. The $11,300 is a calculated number which is directly comparable with the $12,500.

4.2 EQUIVALENCE CALCULATIONS INVOLVING A SINGLE FACTOR

The interest formulas derived in Chapter 3 express relationships that exist among the several elements making up the formulas. These formulas exhibit relationships among P, A, F, i, and n for annual compounding and among P, A, F, r, and n for continuous compounding. For the case where continuous funds flow is assumed, the formulas exhibit the relationships among P, $\bar{A}$, F, r, and n. The paragraphs that follow will illustrate methods for calculating equivalence where these interest formulas are involved.

4.2.1 Single-Payment Compound-Amount Factor Calculations

The single-payment compound-amount factors yield a sum F, at a given time in the future, that is equivalent to principal amount P for a specified interest rate i compounded annually, or r compounded continuously. For example, the compound amount on April 1, 1998, that is equivalent to a principal sum of $2,000 on April 1, 1990, for an interest rate of 9% compounded annually is

$$n = 1998 - 1990 = 8$$

$$F = P(\overset{F/P,i,n}{\quad})$$

$$= \$2{,}000(\overset{F/P,9,8}{1.993}) = \$3{,}986.$$

If interest is compounded continuously, the solution is

$$F = P\left[\overset{F/P,r,n}{\quad}\right]$$

$$= \$2{,}000\left[\overset{F/P,9,8}{2.054}\right] = \$4{,}108.$$

If the principal P, the compound amount F, and the number of years n are known, the interest rate i may be determined by interpolation in the interest tables. For example, if $P = \$300$, $F = \$525$, and $n = 9$, the solution for i is

$$F = P\left(\overset{F/P,i,n}{\quad}\right)$$

$$\$525 = \$300\left(\overset{F/P,i,9}{\quad}\right)$$

$$\left(\overset{F/P,i,9}{1.750}\right) = \frac{\$525}{\$300}$$

A search of the interest tables for annual compounding interest reveals that 1.750 falls between the single-payment compound-amount factors in the 6% and 7% tables for $n = 9$. The value from the 6% table is 1.689, and the value from the 7% table is 1.838. By linear proportion

$$i = 6 + (1)\frac{1.689 - 1.750}{1.689 - 1.838}$$

$$= 6 + \frac{0.061}{0.149} = 6.41\%.$$

The linear interpolation used for i is illustrated by Figure 4.1. Since engineering economy studies usually are based on estimates of the future, the small error introduced by interpolation will rarely be of significance.

Solution for i is possible by calculator without the use of tables as follows:[1]

$$F = P(1 + i)^n$$

$$\$525 = \$300(1 + i)^9$$

$$(1 + i)^9 = \frac{\$525}{\$300}$$

$$i = \sqrt[9]{1.750} - 1$$

$$i = 1.0642 - 1 = 0.0642, \text{ or } 6.42\%.$$

[1] *The widespread availability of electronic calculators and personal computers makes possible direct solution from the formulas without the aid of tables and possible interpolation.*

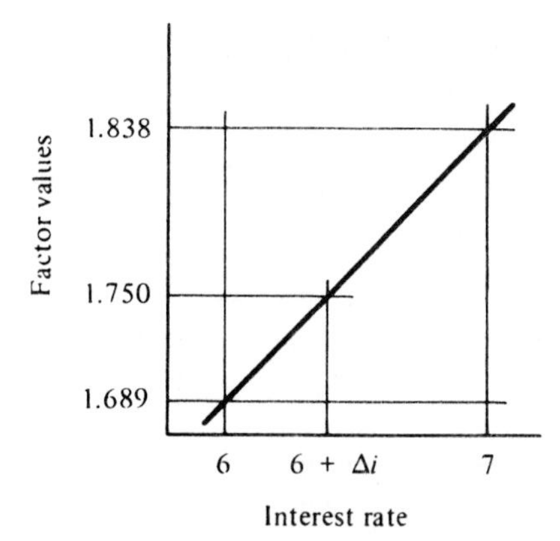

FIGURE 4.1. Interpolation for i.

If the principal sum P, its compound amount F, and the interest rate i are known, the number of years n may be determined by interpolation in the interest tables. For example, if P = \$400, F = \$800, and i = 9%, the solution for n is

$$F = P(\overset{F/P,i,n}{\quad})$$

$$\$800 = \$400(\overset{F/P,9,n}{\quad})$$

$$(\overset{F/P,9,n}{2.000}) = \frac{\$800}{\$400}.$$

A search of the 9% interest table reveals that 2.000 falls between the single-payment compound-amount factors for $n = 8$ and $n = 9$. For $n = 8$ the factor is 1.993 and for $n = 9$ it is 2.172. By linear proportion

$$n = 8 + (1)\frac{1.993 - 2.000}{1.993 - 2.172}$$

$$= 8 + \frac{0.007}{0.179} = 8.04 \text{ years.}$$

The linear interpolation used for n is illustrated by Figure 4.2.

Solution for n might have been accomplished without the use of tables as follows:

$$F = P(1 + i)^n$$

$$\$800 = \$400(1 + 0.09)^n$$

$$(1.09)^n = \frac{\$800}{\$400} = 2.000$$

$$n = 8.043 \text{ years.}$$

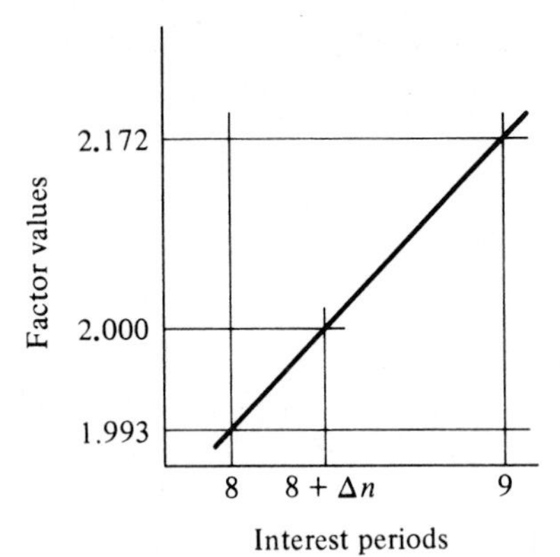

FIGURE 4.2. Interpolation for n.

The interpretation of n in this example is that 8.043 years are required for $400 to earn enough interest so that the total amount available after this time equals $800. However, when compounding occurs at the end of discrete periods, as it does in this example, the result only approximates the time required to accumulate the amount in question. The reason for this lack of accuracy is that interest is paid only at the end of each period, and at least 9 years are necessary to accumulate $800 or more. After 9 years the compound amount would actually be $868.76.

On occasion in engineering economy, the value of n cannot be found within the range of the interest table. A useful approach is based upon the fact that for this particular factor

$$\left(\overset{F/P,\,i,\,n}{\quad}\right) = \left(\overset{F/P,\,i,\,n_1}{\quad}\right)\left(\overset{F/P,\,i,\,n_2}{\quad}\right)\ldots\left(\overset{F/P,\,i,\,n_k}{\quad}\right) \tag{4.1}$$

where $n = n_1 + n_2 + \cdots + n_k$.

Suppose that the value of $\left(\overset{F/P,\,10,\,174}{\quad}\right)$ is needed. From the previous relationship the value is

$$\left(\overset{F/P,10,174}{\quad}\right) = \overset{F/P,\,10,100}{(13{,}780.612)}\ \overset{F/P,\,10,70}{(789.747)}\ \overset{F/P,\,10,4}{(1.464)}$$

$$= \$15{,}933{,}000.$$

Thus, $1 invested now will be worth $15,933,000 in 174 years if the interest is allowed to compound at 10% per year. Examples such as this dramatize the extraordinary power of compounding.

4.2.2 Single-Payment Present-Worth Factor Calculations

The single-payment present-worth factors yield a principal sum P, at a time regarded as being the present, which is equivalent to a future sum F. For example, the solution for finding the present worth of a sum equal to $400 received 12 years hence, for an interest rate of 6% compounded annually, is

$$P = F(\overset{P/F,\,i,\,n}{\quad})$$

$$= \$400\,(\overset{P/F,\,6,\,12}{0.4970}) = \$199.$$

Thus, if a person desires $400 at the *end* of 2004, $199.00 must be deposited at the *end* of 1992 (or the *beginning* of 1993) into an account paying 6% compounded annually. For continuous compounding at 6%, the same result could be obtained with less money invested now. That is,

$$P = F[\overset{P/F,\,r,\,n}{\quad}]$$

$$= \$400\,[\overset{P/F,\,6,\,12}{0.4868}] = \$195.$$

4.2.3 Equal-Payment-Series Compound-Amount Factor Calculations

The equal-payment-series compound-amount factors yield a sum F at a given time in the future, which is equivalent to a series of payments A, occurring at the end of successive years such that the last A concurs with F. The solution for finding the equivalent amount, 7 years from now, of a series of seven $40 year-end payments whose final payment occurs simultaneously with the compound amount being determined, for an interest rate of 6% is

$$F = A(\overset{F/A,i,n}{\quad})$$

$$= \$40\,(\overset{F/A,\,6,7}{8.394}) = \$336.$$

If the compound amount F, the annual payments A, and the number of years n are known, the interest rate i may be determined by interpolation in the interest tables. For example, if $F = \$441.10$, $A = \$100$, and $n = 4$, the solution for i is

$$F = A(\overset{F/A,i,n}{\quad})$$

$$\$441.10 = \$100(\overset{F/A,i,4}{\quad})$$

$$(\overset{F/A,i,4}{4.411}) = \frac{\$441.10}{\$100}.$$

This value falls between the equal-payment-series compound-amount factors in the 6% and 7% table for $n = 4$. By linear interpolation

$$i = 6 + (1)\frac{4.375 - 4.411}{4.375 - 4.440}$$

$$= 6 + \frac{0.036}{0.065} = 6.55\%.$$

Such calculations are an integral part of retirement plans in which an individual sets aside part of his earnings on a regular basis to provide for retirement. If an individual places $800 per year in a retirement plan that pays 7% compounded annually, what lump sum could be withdrawn if participation spanned 35 years?

The single amount the individual could withdraw 35 years hence is

$$F = \$800(\overset{F/A,7,35}{138.237}) = \$110{,}590.$$

Thus, the depositing of $28,000 over 35 years ($800 × 35) results in accumulated interest earnings of $82,590, yielding a total amount of $110,590. If the individual had the entire $28,000 available to invest at the beginning of the 35 years, an even greater amount would accumulate, as shown by the following calculation:

$$F = \$28{,}000(\overset{F/P,7,35}{10.677}) = \$298{,}956.$$

Suppose that an equal annual cash flow of $100 each year exists. How long will it take to accumulate $2,000 if the interest rate is 8% compounded continuously? The calculation is

$$F = A[\overset{F/A,r,n}{\quad}]$$

$$\$2{,}000 = \$100[\overset{F/A,8,n}{\quad}]$$

$$[\overset{F/A,8,n}{20.00}] = \frac{\$2{,}000}{\$100}$$

$$n = 12 + (1)\frac{19.351 - 20.000}{19.351 - 21.963}$$

$$n = 12 + \frac{0.649}{2.612} = 12.25 \text{ years}.$$

A direct approach using the interest formula requires solution for n in terms of F, A, and r as follows:

$$F = A\left[\frac{e^{rn} - 1}{e^r - 1}\right]$$

$$e^{rn} - 1 = \frac{F}{A}(e^r - 1)$$

$$rn = \log_e\left[\frac{F}{A}(e^r - 1) + 1\right]$$

$$n = \frac{\log_e\left[\frac{F}{A}(e^r - 1) + 1\right]}{r}.$$

For $F = \$2{,}000$, $A = \$100$, and $r = 8\%$,

$$n = \frac{\log_e\left[\frac{\$2{,}000}{\$100}(e^{0.08} - 1) + 1\right]}{0.08} = \frac{\log_e[20(0.0833) + 1]}{0.08}$$

$$= \frac{\log_e[2.666]}{0.08} = \frac{0.981}{0.08} = 12.26 \text{ years.}$$

When tables of interest factors are available, it is usually less time-consuming to calculate the value of the factor by interpolation from the tables, rather than to use the method illustrated.

4.2.4 Equal-Payment-Series Sinking-Fund Factor Calculations

The equal-payment-series sinking-fund factor is used to determine the amount A of a series of equal payments, occurring at the end of successive years, that are equivalent to a future sum F. The solution for finding the amount of annual sinking-fund deposits A for the period June 1, 1992, to June 1, 1999, that are equivalent to a single amount F of \$4,000 on June 1, 1999, at 12% interest is

$$A = F(\overset{A/F,i,n}{\quad})$$

$$= \$4{,}000\,(\overset{A/F,12,7}{0.0991}) = \$396.$$

Recall that all payments are end-of-period transactions, so that the first payment occurs on June 1, 1993, and the last on June 1, 1999. Solution for i and n when F, A, and n or i are known may be accomplished by interpolation in the interest tables, as was illustrated for the single-payment compound-amount factor.

Suppose a firm estimates it will require $1,000,000 six years from now for the purchase of new equipment and decides to set an amount aside each year for this purpose. If this firm is able to earn 8% compounded annually on its cash, the amount that must be deposited at the end of each of the 6 years to accumulate the $1,000,000 is

$$A = \$1{,}000{,}000\overset{A/F,8,6}{(0.1363)} = \$136{,}300.$$

The total interest earned by the firm over the 6 years is

$$\$1{,}000{,}000 - (\$136{,}300)(6) = \$182{,}200.$$

4.2.5 Equal-Payment-Series Present-Worth Factor Calculations

The equal-payment-series present-worth factors are used to find the present worth P of an equal-payment series A, occurring at the end of successive periods. For example, the present worth P, which is equivalent to a series of five $60 year-end payments beginning at the end of the first interest period after the present for an interest of 10%, is

$$P = A\overset{P/A,i,n}{(\quad\quad)}$$

$$= \$60\overset{P/A,10,5}{(3.7908)} = \$227.$$

This factor may be used to calculate the capital investment that would be justified if it would result in an annual saving each year for several years. Suppose an energy-saving device is proposed that will reduce the cost of energy by $10,000 per year for 15 years. If the interest rate is 8%, the capital investment that can be justified is any amount less than

$$P = \$10{,}000\overset{P/A,8,15}{(8.5595)} = \$85{,}595.$$

4.2.6 Equal-Payment-Series Capital-Recovery Factor Calculations

The equal-payment-series capital-recovery factors are used to determine the amount A of each payment of a series of payments occurring at the end of successive periods which is equivalent to a present sum P. For example, the solution for finding the annual year-end payment for 5 years that is equivalent to an amount P of $18,000 at the present for an interest rate of 15% is

$$A = P\overset{A/P,i,n}{(\quad\quad)}$$

$$= \$18{,}000\overset{A/P,15,5}{(0.2983)} = \$5{,}369.$$

The recovery of invested capital is a common problem in deciding to invest. For the situation above, an investment of $18,000 in an asset would have to yield an annual benefit of at least $5,369 for the venture to break even.

When funds are flowing continuously and the interest rate is compounded continuously, it is necessary to use the funds-flow factors developed in Chapter 3. To find the funds-flow equivalent of a present sum of $600 when the time period is 12 years and the interest rate is 10%, calculate

$$\bar{A} = P[\overset{\bar{A}/P,r,n}{\quad}] = P[\overset{A/P,r,n}{\quad}] \div [\overset{\bar{A}/A,r}{\quad}]$$

$$\bar{A} = \$600[\overset{\bar{A}/P,\,10,12}{\quad}] = \$600[\overset{A/P,\,10,12}{0.15050}] \div [\overset{\bar{A}/A,\,10}{1.0517}] = \$85.86 \text{ per year.}$$

Direct calculation from Equation (3.28) gives

$$\bar{A} = \$600\left[\frac{(0.10)e^{(0.10)(12)}}{e^{(0.10)(12)} - 1}\right] = \$600\left[\frac{0.332}{2.32}\right] = \$85.86 \text{ per year.}$$

4.2.7 Uniform-Gradient-Series Factor Calculations

The uniform-gradient-series factors are used to determine the amount A of each payment of a series of payments, occurring at the end of successive periods, that is equivalent to either a uniformly increasing or uniformly decreasing series of payments over the same span of time. For example, if a series of 15 payments exists, with the first payment being $100 and each subsequent payment increasing by $10, the amount A that is equivalent to this series if the interest rate is 20% is

$$A = \$100 + G(\overset{A/G,i,n}{\quad})$$

$$= \$100 + \$10(\overset{A/G,\,20,15}{3.9589}) = \$140.$$

This result indicates that $140 each year for 15 years is equivalent to the series of payments specified.

For a uniformly decreasing gradient, the same format would be used, except that G would be negative. As an example, suppose there is a series of 5 payments beginning at $800 and decreasing by $100 each year. If $r = 12\%$ compounded continuously, the amount A equivalent to this series is

$$A = \$800 - G[\overset{A/G,r,n}{\quad}]$$

$$= \$800 - \$100[\overset{A/G,\,12,5}{1.7615}] = \$624.$$

This result means that five receipts of $624 would be equivalent to the uniformly decreasing gradient series specified.

4.2.8 Geometric-Gradient-Series Factor Calculations

Equivalence calculations for the geometric gradient series result in a P that is equivalent to an increasing or a decreasing geometric gradient. The increase or decrease depends upon the sign of g, the constant percentage change from year to year. Three different situations exist for both the increasing and the decreasing gradient, depending upon the relationship of g and i: $i > g$, $i = g$, and $i < g$. Example situations for each of these were given in Chapter 3.

4.3 EQUIVALENCE CALCULATIONS INVOLVING CASH FLOWS

When a calculation of equivalence involving several interest factors is to be made, some difficulty may be encountered in organizing the cash flows and the interest factors. Also, until some experience is gained with this situation, it may be difficult to keep track of the lapse of time. For complex problems of this type, a schematic representation in tabular or graphical form may be helpful.

4.3.1 Tabular Representation of Cash Flow

As an example, suppose it is required to determine what amount at the present is equivalent to the following cash flow for an interest rate of 12%: $300 end of year 6; $60 end of years 9, 10, 11, and 12; $210 end of year 13; $80 end of years 15, 16, and 17. These payments may be represented schematically as illustrated in Figure 4.3.

The plan of attack is to determine the amount at the beginning of year 1 that is equivalent to the various payments in the cash flow described in Figure 4.3. By converting the various payments to their equivalents at the same point in time, we can then determine the total equivalent amount by direct addition.

When interest is earned, monetary amounts can be directly added only if they occur at the same point in time.

To use the interest formulas properly, recall that P occurs at the beginning of an interest period and that F and A payments occur at the end of interest periods. For instance, the group of four $60 payments are converted to a single equivalent amount of $182.24 at the end of year 8, which is one interest period before the first $60 payment. This is in accordance with the convention of the conversion formula, which requires that the amount P occur one interest period

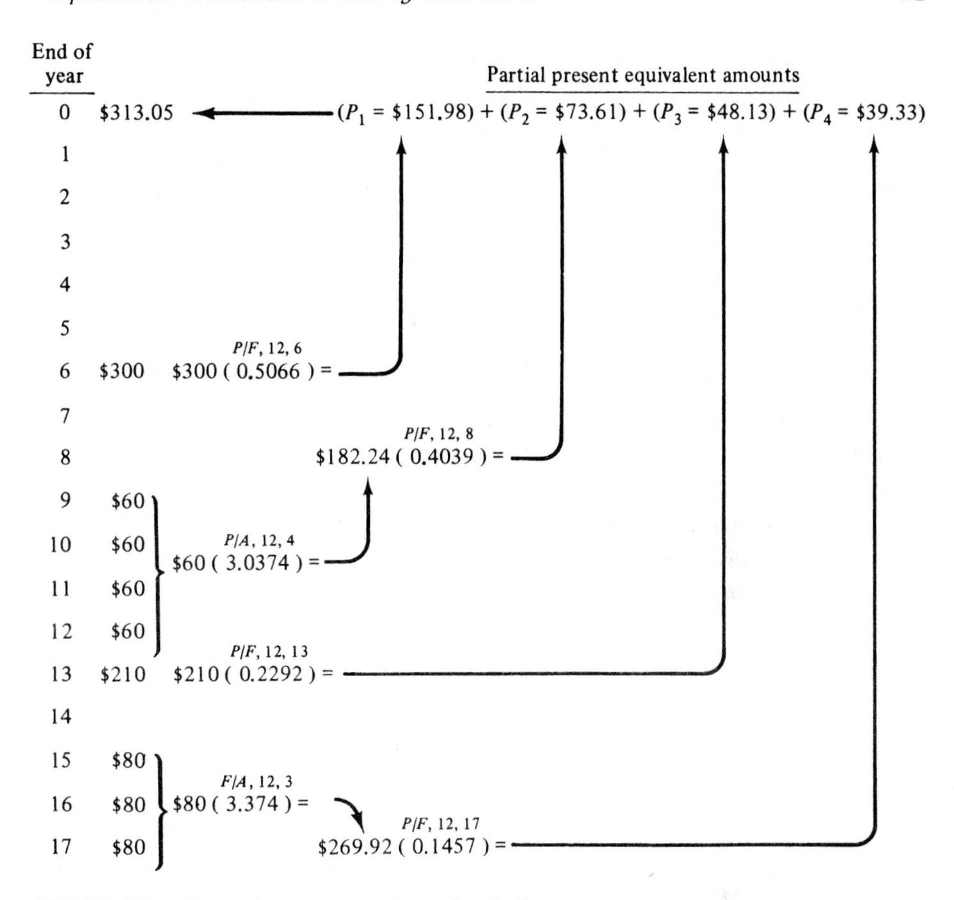

FIGURE 4.3. Schematic representation of cash flow.

prior to the first A payment. In Figure 4.3., the $269.92 as of the end of year 17 represents the equivalent future worth of the three $80 payments. Note that the $269.92 amount concurs with the last $80. This is in accordance with the convention adopted for the derivation of the factor to find F when given a series of payments A.

The sequence of calculations in the solution of this problem is clearly indicated in the table. The position of the arrowhead following each multiplication represents the position of the result with respect to time. The intermediate quantities $182.24 and $269.92 need not have been found. Much time may be saved if we indicate all calculations to be made prior to looking up factor values from the tables and making calculations. In the above example this might have been done as follows:

$$P_1 = \$300(\overset{P/F,12,6}{0.5066}) = \$151.98$$

$$P_2 = \$60(\overset{P/A,12,4}{3.0374})(\overset{P/F,12,8}{0.4039}) = 73.61$$

$$P_3 = \$210(\overset{P/F,12,13}{0.2292}) = 48.13$$

$$P_4 = \$80(\overset{F/A,12,3}{3.374})(\overset{P/F,12,17}{0.1457}) = 39.33$$

$$P = P_1 + P_2 + P_3 + P_4 = \$313.05$$

If interest had been 12% compounded continuously instead of 12% compounded annually, values would have been taken from a table for continuous compounding interest. All other calculations would have remained the same.

The amount of calculation required in a given situation can be kept to a minimum by the proper selection of interest factors. Consider the example shown in Figure 4.4. By recognizing that the cash flow beginning at the end of year 4 is a gradient series with a gradient of \$20 per year, it is possible to convert that series into an equivalent equal annual series by the following calculation:

$$A = G(\overset{A/G,i,n}{\quad})$$

$$A = \$20(\overset{A/G,10,8}{3.0045}) = \$60.09.$$

Note that the equal payments of \$60.09 begin at the end of year 3, although the gradient series begins at the end of year 4. This situation arises be-

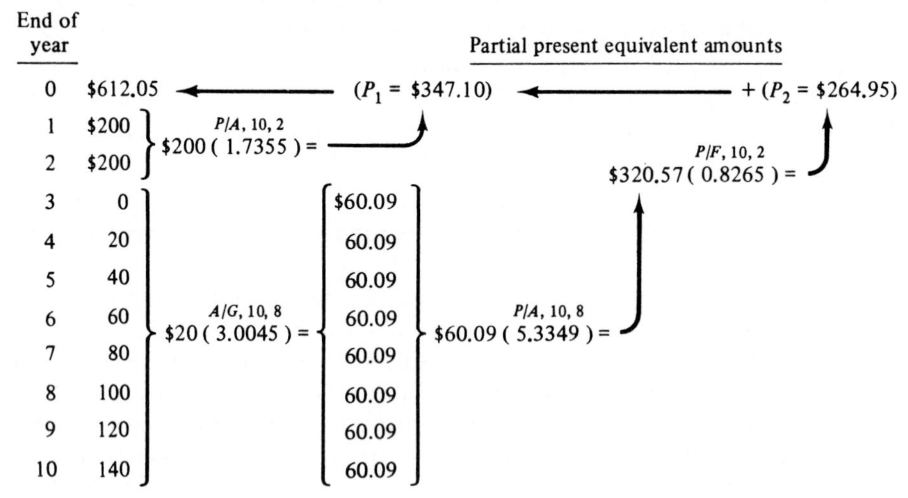

FIGURE 4.4. Schematic of cash flow with gradient.

cause the uniform-gradient-series factor assumes the gradient occurs one period following the first payment in the equal-payment series.

The equivalence at the beginning of year 1 of the cash flow shown in Figure 4.4. is calculated as follows:

$$P_1 = \$200\overset{P/A,10,2}{(1.7355)} = \$347.10$$

$$P_2 = \$20\overset{A/G,10,8}{(3.0045)}\overset{P/A,10,8}{(5.3349)}\overset{P/F,10,2}{(0.8265)} = \underline{\$264.95}$$

$$P = P_1 + P_2 = \$612.05$$

In the previous examples, the equivalences of the cash flows were found at the beginning of year 1. In many cases the equivalence amount that is desired is a single payment at some point in time other than the present, or it may be an equal-payment series over some time span. The calculations required to find various cash flow equivalents are quite similar to those shown in Figures 4.3 and 4.4. For instance, to find 10 equal annual payments that are equivalent to the cash flow in Figure 4.4., it is only necessary to convert the single-payment equivalent of \$612.05 at the beginning of year 1 into the desired equivalent using

$$A = P\overset{A/P,i,n}{(\quad)}$$

$$= \$612.05\overset{A/P,10,10}{(0.1628)} = \$99.64.$$

Since a number of different calculations will lead to the same solution for this type of problem, an effort should be made to minimize the amount of computation required. For example, in this problem each cash payment could have been converted to its equivalent at the beginning of year 1 by using the appropriate single-payment present-worth factor. This approach would have required 10 factors, while only four factors were required in the approach used above.

4.4 PRINCIPLES OF EQUIVALENCE

After observing a variety of calculations for finding equivalence between cash flows, it is evident that the use of the interest formulas is fundamental to this process. However, to this point, the discussion has been confined to examples demonstrating the manipulation of interest formula, interest factors, and interest rates. The first part of this section presents examples that illustrate one or more of the fundamental principles of equivalence. These principles are summarized at the end of the section.

4.4.1 Equivalence Between Cash Flows

In engineering economy the meaning of equivalence pertaining to value in exchange is of primary importance. For example, a present amount of $300 is equivalent to $798 if the amounts are separated by 7 years and if the interest rate is 15%. This is so because a person who considers 15% to be a satisfactory rate of interest would be *indifferent* to receiving $300 now or $798 seven years from now.

Equivalent cash flows are those that have the same value and the calculated expression of equivalence can be used as a basis for choice.

This equivalence between the two cash flows in Figure 4.5 may be illustrated by use of the single-payment formulas. A sum of $300 in the present is equivalent to

$$\$300(1 + 0.15)^7 = \$300\overset{F/P,15,7}{(\ 2.660\ )} = \$798$$

7 years from now. Similarly, $798 to be received 7 years from now is equivalent to

$$\$798\left[\frac{1}{(1 + 0.15)^7}\right] = \$798\overset{P/F,15,7}{(0.3759)} = \$300$$

at the present.

The first calculation gives the equivalent value 7 years hence of the $300 amount presently in hand. The second calculation gives the amount now that is equivalent to $798 received 7 years from now. The decision to restrict the analysis to the present, or to a time 7 years from now, is made only for computational convenience.

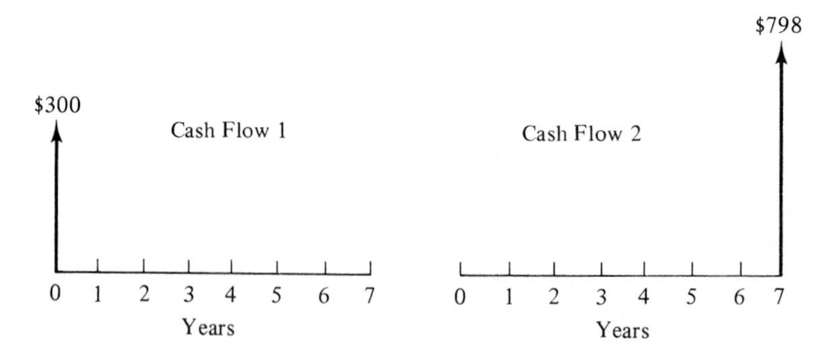

FIGURE 4.5. Equivalent cash flows at 15%.

Equivalence can be established at any point in time, since it is known that for one cash flow to be equivalent to another, their equivalent values must be equal at any point in time.

As an illustration, the equivalent worth of Cash Flow 1 at the arbitrarily chosen point in time of $n = 10$ is

$$\$300\overset{F/P,15,10}{(4.046)} = \$1{,}214.$$

And, for Cash Flow 2, the equivalent worth at $n = 10$ is

$$\$798\overset{F/P,15,3}{(1.521)} = \$1{,}214.$$

Two or more distinct cash flows are equivalent if they are equivalent to the same cash flow.

4.4.2 Equivalence for Different Interest Rates

In the examples presented to this point it has been assumed that the interest rate used in the interest formula remains fixed over the entire time span of the cash flow. In actuality interest rates frequently change over time, and it is important that cash flows be analyzed under these conditions.

As cash flows are converted to their equivalences from one time period to the next, the interest rate associated with each time period must be reflected in the calculation.

For example, the cash flow in Figure 4.6 represents three different interest rates applicable over the 5-year time span. To calculate the equivalent amount P at the present for this set of conditions, find

$$P = \left\{\$200 + \left[\$100 + \$100\overset{P/A,10,2}{(1.7355)}\right]\overset{P/F,7,1}{(0.9346)}\overset{P/F,3,4}{(0.8885)}\right\}\overset{P/F,3,4}{(0.8885)} = \$380.$$

The factor $\$100\overset{P/A,10,3}{(\quad)}$ was not used, since it would determine the equivalent value of the three \$100 payments at $t = 2$. This time span includes two different interest rates, and they cannot both be accounted for by a single factor.

Now suppose one wishes to determine the single-payment equivalent to P at $t = 5$. This future value is

$$F = \$380\overset{F/P,3,8}{(1.267)}\overset{F/P,7,1}{(1.07)}\overset{F/P,10,2}{(1.210)} = \$623.$$

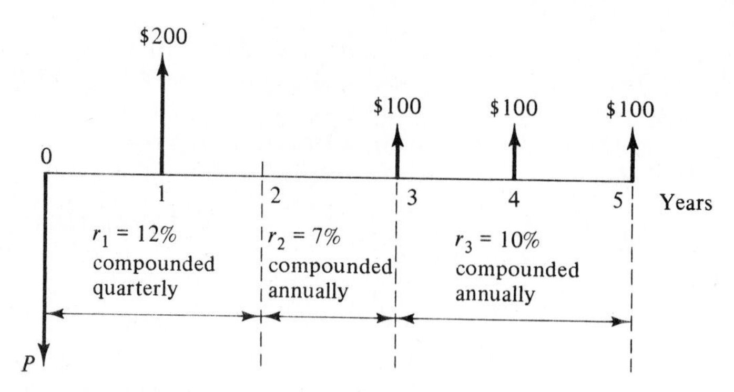

FIGURE 4.6. Equivalence for changing rates of interest.

This amount is equivalent not only to P but to the four receipts shown in Figure 4.6. This fact can be verified by moving those cash flows to $t = 5$ using the interest rates specified.

Once a single-payment equivalent is found, it is a straightforward operation to find any other cash flow equivalent to the original one. As an example, find the equal-payment-series cash flow A that runs from $t = 1$ to $t = 5$ that is equivalent to the four receipts in Figure 4.6. This is

$$\$380 = A\overset{P/F,3,4}{(0.8885)} + A\overset{P/F,3,8}{(0.7894)} + A\overset{P/F,7,1}{(0.9346)}\overset{P/F,3,8}{(0.7894)}$$

$$+ A\overset{P/F,10,1}{(0.9091)}\overset{P/F,7,1}{(0.9346)}\overset{P/F,3,8}{(0.7894)} + A\overset{P/F,10,2}{(0.8265)}\overset{P/F,7,1}{(0.9346)}\overset{P/F,3,8}{(0.7894)}$$

$$A = \$103 \text{ per year.}$$

This relationship can be restated in a more compact form using the effective rate per year for the first two terms and analyzing the remaining three cash flows together. The annual equivalent is

$$\$380 = A\overset{P/A,12.55,2}{(\ 1.6779\ )} + \left[A + A\overset{P/A,10,2}{(1.7355)}\right]\overset{P/F,7,1}{(0.9346)}\overset{P/F,3,8}{(0.7894)}$$

$$A = \$103.$$

In most of the examples presented, it is assumed that the interest rate remains constant over the entire time span of the cash flow. This assumption is commonly used in engineering economy studies.

4.4.3 Equivalence Between Receipts and Disbursements

In many applications the equivalence between known disbursements (receipts) and an unknown receipt (disbursement) is to be found for a given interest rate.

Figures 4.3. and 4.4 exhibited examples of this. Also, calculations are often required to determine an unknown interest rate that brings about equivalence between known receipts and disbursements.

A general principle of equivalence states that the actual interest rate earned on an investment is the one that sets the equivalent receipts equal to the equivalent disbursements.

Consider the cash flow diagram in Figure 4.7. The interest rate that sets the receipts equivalent to the disbursements is to be found. We can accomplish this by finding the value for i that satisfies the relationship

$$\$1{,}000 + \$500(\overset{P/F,i,1}{\quad}) + \$250(\overset{P/F,i,5}{\quad})$$
$$= \$482(\overset{P/A,i,3}{\quad})(\overset{P/F,i,1}{\quad}) + \$482(\overset{P/A,i,2}{\quad})(\overset{P/F,i,5}{\quad}).$$

By trial and error, 10% is found to the value for i that sets the receipts identified as positive values equal to the disbursements designated by negative values at $t = 0$.

One of the principles of equivalence given previously states that equivalent cash flows will exhibit their equivalence regardless of the point in time used as a basis for the equivalence calculation. This can be demonstrated for the cash flow in Figure 4.7 by arbitrarily picking $n = 5$ as the basis for calculation as follows:

$$\$1{,}000(\overset{F/P,10,5}{1.611}) + \$500(\overset{F/P,10,4}{1.464}) + \$250 = \$482(\overset{F/A,10,3}{3.310})(\overset{F/P,10,1}{1.100})$$
$$+ \$482(\overset{P/A,10,2}{1.7355})$$

$$\$2{,}593 = \$2{,}593.$$

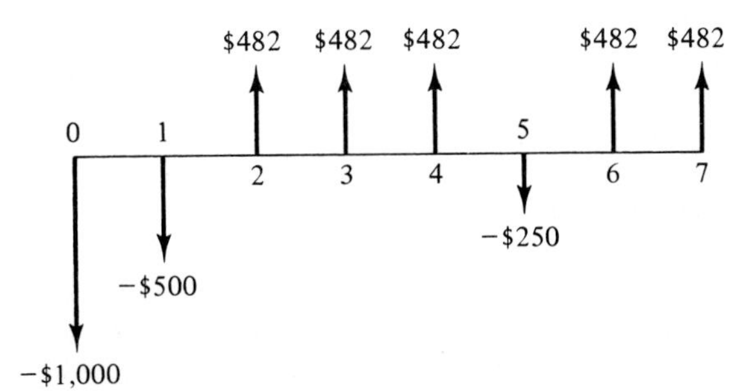

FIGURE 4.7. Cash flows for receipts and disbursements.

If the receipts and disbursements of a cash flow are equivalent for some interest rate, the cash flows of any equivalent portion of the investment are equal at that interest rate to the negative (−) of the equivalent amount of the cash flows that constitute the remaining portion of the investment.

Referring to Figure 4.8 as an example, this principle guarantees that the equivalent of Cash Flow 1 will equal the negative of Cash Flow 2 at 10%. Cash Flows 1 and 2 have been arbitrarily selected and are circled in Figure 4.8. The procedure followed distinguishes between receipts by a positive sign (+) and disbursements by a negative sign (−). The equivalence at $n = 4$ between Cash Flows 1 and 2 is calculated as follows:

$$-\$1{,}000\overset{F/P,10,4}{(1.464)} - \$500\overset{F/P,10,3}{(1.331)} + \$482\overset{F/A,10,3}{(3.310)}$$

$$= -\left[-\$250\overset{P/F,10,1}{(0.9091)} + \$482\overset{P/A,10,2}{(1.7355)}\overset{P/F,10,1}{(0.9091)}\right]$$

$$-\$534 = -\$534.$$

The arbitrarily selected portion of the cash flow could have been evaluated at any point in time, as shown earlier in this section.

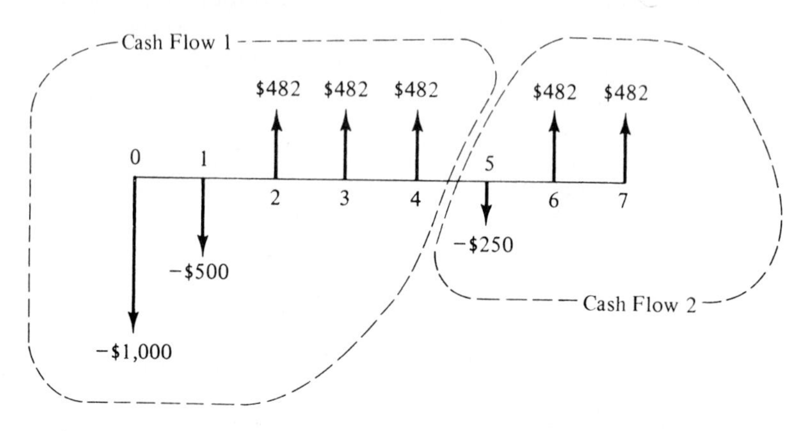

FIGURE 4.8. Figure 4.7, partitioned into two cash flows.

4.5 EQUIVALENCE CALCULATIONS WITH MORE FREQUENT COMPOUNDING

Although annual compounding of interest is the situation most often encountered, any length of time can be selected as the compounding period. Usually, these periods are discrete time intervals such as a day, week, month, 3 months,

6 months, or a year, depending on the financial institution or financial instrument involved. Even continuous compounding is sometimes adopted where the compounding periods are infinitesimal and the effective annual interest rate is the maximum that can be earned for a given nominal interest rate.

Three situations can arise with regard to the compounding frequency and the frequency of payments received. These are:

1. The compounding periods and the occurrence of payments coincide.
2. The compounding periods occur more frequently than the receipt of payments.
3. The compounding periods occur less frequently than the receipt of payments.

These three situations together with example calculations are presented in this section.

4.5.1 Compounding and Payment Periods Coincide

The interest factors of Section 3.3 require the use of the effective interest rate per period. The effective annual interest rate was used exclusively, while the payments occurred annually. The use of compounding for other than an annual period, where the compounding period and the payment periods coincide, is illustrated by the examples below.

If payments of $100 occur semiannually at the end of each 6-month period for 3 years and the nominal interest rate is 12% compounded semiannually, the present worth P is determined as follows:

$$i = \frac{12\%}{2 \text{ periods}} = 6\% \text{ per semiannual period}$$

$$n = (3 \text{ years})(2 \text{ periods per year}) = 6 \text{ periods}$$

$$P = A(\overset{P/A,i,n}{\quad}) = \$100(\overset{P/A,6,6}{4.9173}) = \$491.73.$$

As a second example, suppose that a person borrows $2,000 and is to repay this amount in 24 equal monthly installments of $99.80 over the next 2 years. Interest is compounded monthly on the unpaid balance of the loan. What are the effective interest rate per month and the nominal interest rate being paid for this loan? What is the effective annual rate of interest on the loan? The solution is found as follows:

$$\$99.80 = \$2{,}000(\overset{A/P,i,24}{\quad})$$

$$(\overset{A/P,i,24}{\quad}) = 0.0499.$$

A search of the interest tables reveals that the above factor value is found for $i = 1\frac{1}{2}\%$. Since the periods in this problem are months, the effective monthly interest rate is $1\frac{1}{2}\%$. The nominal rate of interest is

$$r = (1\tfrac{1}{2}\% \text{ per month})(12 \text{ months}) = 18\% \text{ per year compounded annually}$$

and the effective annual interest rate from Equation (3.16) with $l = 1$ is

$$i_a = \left(1 + \frac{r}{m}\right)^m - 1 = \left(1 + \frac{0.18}{12}\right)^{12} - 1 = 19.56\% \text{ per year.}$$

This result may also be found from Appendix B, Table B.1.

4.5.2 Compounding More Frequent Than Payments

When the compounding period is shorter than the payment period, the principle of matching the interest rate with the interest period should be applied. This is done by using the appropriate effective interest rate for the interest period assumed (see Section 3.5). An example will be used to illustrate this principle.

Suppose a deposit of \$100 is placed in a bank account at the end of each year for the next 3 years. The bank pays interest at the rate of 6% compounded quarterly. How much will be accumulated in this account at the end of 3 years?

The required calculation can be based on the compounding periods, which are 3 months in length, as follows:

$$i = \frac{6\%}{4 \text{ quarters}} = 1\tfrac{1}{2}\% \text{ per quarter.}$$

The amount accumulated in the account is

$$F = \$100\overset{F/P,1\frac{1}{2},8}{(\;1.127\;)} + \$100\overset{F/P,1\frac{1}{2},4}{(\;1.061\;)} + \$100 = \$318.80.$$

The first term indicates that the first \$100 deposited at the end of the first year will earn interest for the next 8 quarters. The second term indicates the second deposit will earn interest for the next 4 quarters, and the last term is the \$100 deposited at the end of the third year.

A second method is to find the effective interest rate for the payment period (1 year in this case) and then make all calculations on the basis of that period. The effective annual interest rate from Equation (3.16) with $l = 1$ is

$$i_a = \left(1 + \frac{r}{m}\right)^m - 1.$$

In this case, $m = 4$ and $r = 6\%$. Therefore,

$$i_a = \left(1 + \frac{0.06}{4}\right)^4 - 1 = 6.14\%.$$

The solution is

$$F = \$100\overset{F/A,6.14,3}{(\ 3.188\ )} = \$318.80.$$

Consider the situation of continuous compounding interest. The interest formulas for continuous interest derived in Section 3.6 can be modified to accommodate equal payments that occur more frequently than once a year. When there are c payments per year, let

$$n = c(\text{number of years})$$

and

$$r = \frac{\text{nominal interest rate per year}}{c}. \tag{4.2}$$

For example, suppose it is desired to find a future amount at the end of 5 years that would result from end-of-month deposits of $1,000 made throughout the 5-year period. Assume that the interest earned on these deposits is 15% compounded continuously. Then,

$$n = (12 \text{ periods per year})(5 \text{ years})$$

$$= 60 \text{ periods}$$

$$r = \frac{15\%}{12 \text{ periods}}$$

$$= 1.25\%.$$

Since there are no continuous compounding interest tables, it is necessary to calculate the factor value from its algebraic form as given by Equation (3.23)

$$F = A\left[\frac{e^{rn} - 1}{e^{r} - 1}\right]$$

$$= \$1,000\left[\frac{e^{(0.0125)(60)} - 1}{e^{0.0125} - 1}\right]$$

$$= \$1,000\left[\frac{1.1170}{0.0126}\right]$$

$$= \$88,650.$$

As another example, suppose that \$10,000 is placed into an account where the interest rate is 8% compounded continuously. What is the size of equal annual withdrawals that can be made over the next 5 years so that the account balance will equal zero after the last withdrawal?

For this example, the effective interest rate per year is calculated as

$$\begin{aligned} i_a &= e^r - 1 \\ &= e^{0.08} - 1 \\ &= 8.33\% \end{aligned}$$

which may also be found in Appendix B, Table B.1. The solution for the equal annual withdrawals is found from

$$A = \$10{,}000\left(\overset{A/P,\,8.33,5}{0.2526}\right) = \$2{,}526$$

where the value for $\left(\overset{A/P,8.33,5}{}\right)$ is found from Equation (3.21).

Suppose that the example above is modified so that the equal withdrawals are to be made quarterly over the 5-year time span. With quarterly payments, the calculations must be modified as follows:

$$\begin{aligned} r &= \frac{8\%}{4 \text{ quarters}} \\ &= 2\% \text{ per quarter compounded continuously} \\ i &= e^r - 1 \text{(effective rate per quarter)} \\ &= e^{0.02} - 1 \\ &= 2.02\% \text{ per quarter} \\ n &= \text{(4 periods per year)(5 years)} \\ &= 20 \text{ periods.} \end{aligned}$$

Solution for the equal quarterly withdrawals is found from

$$A = \$10{,}000\left(\overset{A/P,\,2.02,20}{0.0613}\right) = \$613$$

where $\left(\overset{A/P,2.02,20}{0.0613}\right) = \left[\overset{A/P,2,20}{0.0613}\right]$ as found from Equation (3.21).

4.5.3 Compounding Less Frequent Than Payments

Most financial institutions calculate the interest to be paid for an interest period by applying the interest rate for that period to the amount of funds on deposit for the full interest period. *Usually no interest is paid for funds deposited during an interest period.* Funds deposited during an interest period begin to earn interest for the following interest period. Similarly, funds removed during an interest period earn no interest for that period. Thus, deposits and withdrawals made during an interest period are placed at the end of the period within which the transactions occur.

As an example of this situation, consider the cash flow diagrams in Figure 4.9. The first diagram illustrates monthly deposits and monthly withdrawals involving an account with 12% compounded quarterly. The second diagram illustrates the relocation of deposits and withdrawals to conform to the end-of-period requirement. Calculations of equivalence cannot be made with the first diagram. However, when the conversion is made, the compounding periods and payment periods coincide, permitting the calculations to proceed as previously described in Section 4.5.1.

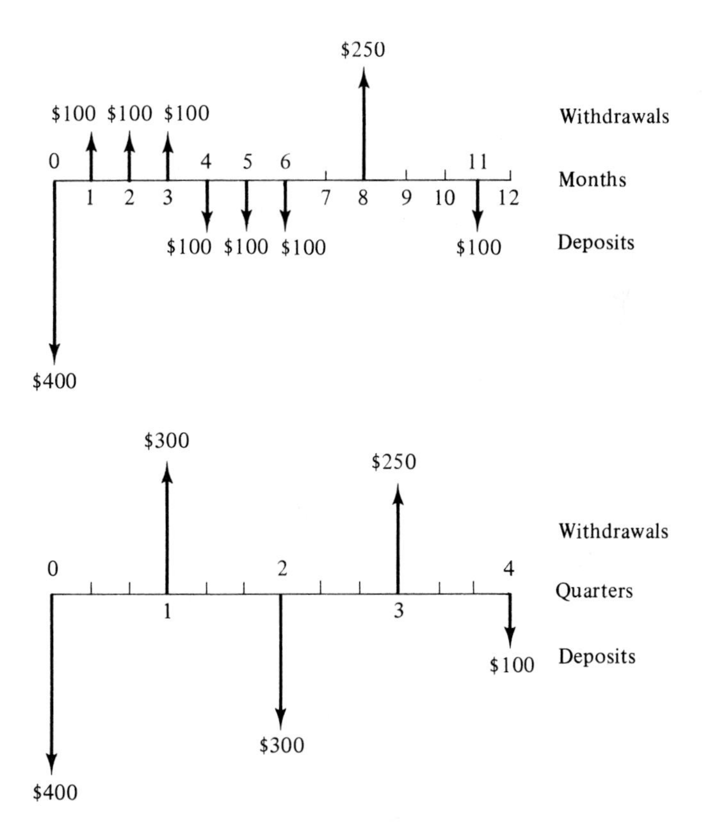

FIGURE 4.9. Cash flow convention for compounding less frequent than payments.

4.6 EQUIVALENCE CALCULATIONS INVOLVING BONDS

A *bond* is a financial instrument setting forth the conditions under which money is borrowed. It consists of a pledge by a borrower of funds to pay a stated amount or percent of interest on the par or face value at stated intervals and to repay the par value at a stated time. Bonds are commonly written with par values in multiples of $1,000. A typical $1,000 bond may, for example, embrace a promise to pay its holder $60 one year after purchase and each succeeding year until the principal amount or par value of $1,000 is repaid on a designated date. Such a bond would be referred to as a 6% bond with interest payable annually. Bonds may also provide for interest payments to be made semiannually or quarterly.

4.6.1 Bond Price and Interest

Bonds are bought and sold, since they represent pledges to pay and thus have value. The market price of a bond may range above or below its par or face value, depending upon prevailing and anticipated market conditions.

Suppose that an individual can purchase (for $900) a $1,000 municipal bond that pays 6% tax-free interest semiannually. If the bond will mature to its face value in 7 years, what will the equivalent rate of interest be? In financial terms, what is the bond's "yield to maturity"? The cash flow diagram for this situation is illustrated in Figure 4.10.

The **yield to maturity** is defined as the rate of return experienced from the bond investment from the current date until the bond matures.

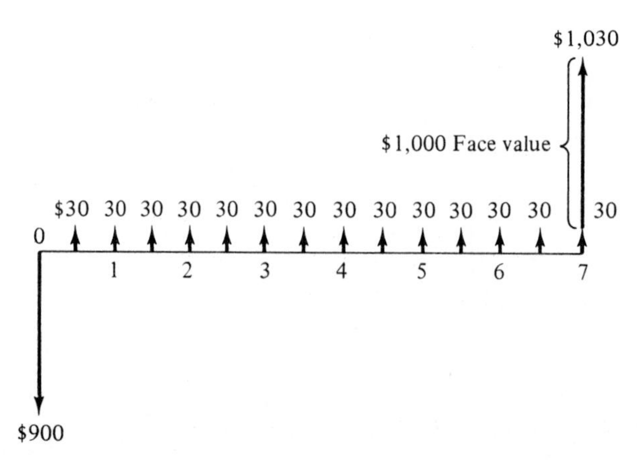

FIGURE 4.10. Cash flow diagram for bond.

The *yield to maturity* may be found by determining the interest rate that makes an expenditure of $900 in the present equivalent to the present worth of the anticipated receipts as follows:

$$\$900 = \$30(\overset{P/A,i,14}{\qquad}) + \$1{,}000(\overset{P/F,i,14}{\qquad}).$$

The solution for i must be by trial and error. At 3%, the present worth of the receipts is

$$\$30\overset{P/A,3,14}{(11.2961)} + \$1{,}000\overset{P/F,3,14}{(0.6611)} = \$1{,}000.$$

The present worth of the receipts at 4% is

$$\$30\overset{P/A,4,14}{(10.5631)} + \$1{,}000\overset{P/F,4,14}{(0.5775)} = \$894.$$

The value of i that makes the present worth of the receipts equal to $900 lies between 3% and 4%. By interpolation,

$$i = 3\% + 1\%\left[\frac{\$1{,}000 - \$900}{\$1{,}000 - \$894}\right]$$

$$= 3\% + 0.94\% = 3.94\%$$

per semiannual period. The nominal annual interest rate is

$$r = (3.94\% \text{ semiannually})(\text{two 6-month periods}) = 7.88\%$$

and the effective annual interest rate or *yield to maturity* for this bond is

$$i_a = \left(1 + \frac{0.0788}{2}\right)^2 - 1 = 1.0804 - 1 = 8.04\%.$$

Because the bond's market price is less than its par value, the real return earned (8.04%) exceeds the interest amount stated on the bond (6%). Also, this real return will not be diminished by income taxes, since muncipal bonds are tax free investments.

An investor considering purchase of a bond may require a certain minimum interest return. For example, if a nominal rate of 10% compounded semiannually is required on the municipal bond described in Figure 4.10, what is the maximum amount that can be paid now for it?

The solution requires the determination of an amount in the present, P, that is equivalent to the series of receipts in prospect as required by the bond covenant. This is stated as

$$P = \$30\overset{P/A,5,14}{(9.8987)} + \$1{,}000\overset{P/F,5,14}{(0.5051)} = \$802.06.$$

Since $802.06 is well below the current market price of the bond, it is unlikely that an exchange will take place at this price.

4.6.2 The Bond Market

Bond prices change over time, because they are influenced by the risk of nonpayment of interest or par value, supply and demand, and the future outlook regarding inflation. These factors act with the current yield and the yield to maturity to establish the price at which a bond will change hands.

The *current yield* of a bond is the interest earned each year as a percentage of the current price, often called the *coupon rate*. The current yield for the example in Figure 4.10 is $60 ÷ $900 = 6.67%. This yield provides an indication of the immediate annual return realized from the investment. The *yield to maturity* is found by solving for i, as illustrated in Section 4.6.1. If a bond is purchased at a discount (the current price being less than its face value) and held to maturity, the investor earns both the interest receipts and the difference between the purchase price and the face value. If the bond is purchased at a premium (the current price being greater than its face value) and held to maturity, the investor earns the interest receipts but gives up the difference between the purchase price and the face value. Thus, the yield to maturity reflects not only the interest receipts but also the gains or losses incurred if the bond is held to maturity.

There can be a significant difference between the yield to maturity and the current yield of a bond. Because the market price of a bond may be more or less than its face value, the current yield or the yield to maturity may differ considerably from the stated interest rate on the bond. Only when a bond is purchased at its face value will its current yield and its yield to maturity equal the coupon rate.

As the general levels of interest rates desired by investors change, the market prices of bonds change, as do bond yields. Thus, when the prevailing interest available to investors increases or decreases, the market price of bonds decreases or increases accordingly. The purchaser of a bond may be expected to take advantage of the rising and falling of bond prices. By purchasing a bond when interest rates are high and selling it when rates are lower, the investor can realize a gain due to the increase in its market value. Of course, the investor has the option of holding the bond until maturity to earn the yield to maturity regardless of changes in the general level of interest rates.

4.7 EQUIVALENCE CALCULATIONS INVOLVING LOANS

A *loan* is an agreement between a borrower and a lender stipulating the amount of money to be provided, the manner by which the money is to be repaid, the collateral to be used, and other pertinent information. Although there are classes of standard loan agreements, the variety of loan arrangements are numerous, owing to the practice of negotiation between the borrower and the lender.

The basic equivalence calculations for loans are presented in this section. Because of the wide range of loan agreements, focus will be on certain types of loans encountered most frequently by individuals and businesses. Included are examples for real estate ownership and commercial loans for the financing of automobiles, appliances, and other consumer products.

4.7.1 Effective Interest on a Loan

Borrowers should be aware of the difference between the actual interest cost of a loan and the interest rate stated by the lender. As an illustration, consider the "add-on" loan used to finance many purchases of consumer goods. In this type of loan, the total interest to be paid is precalculated and added to the principal. Principal plus this interest amount is then paid in equal monthly payments.

The effective interest rate that sets the receipts equal to the disbursements on an equivalent basis is the rate that properly reflects the true interest cost of the loan.

Suppose that an individual wishes to purchase a home appliance for $300. The salesperson indicates that the interest rate will be 20% add-on, and the payments can be made over one year. The calculation for the total amount owed is $300 + 0.20($300) = $360. With payment over 12 months, the monthly payment will be $360/12 = $30. Figure 4.11 illustrates the cash flow for this add-on loan.

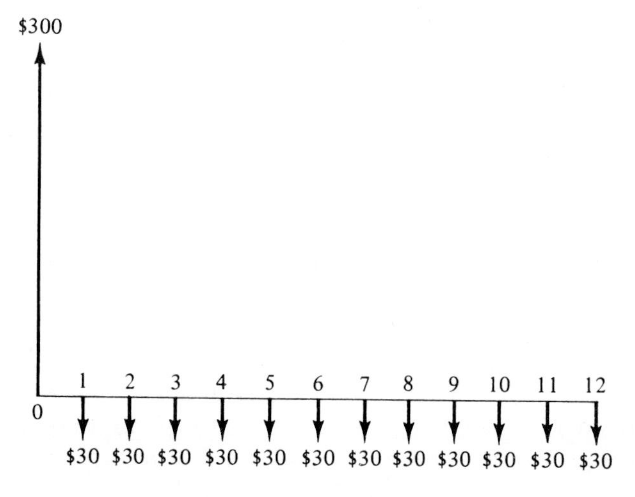

FIGURE 4.11. Cash flow for add-on loan.

The actual or effective interest rate for this loan situation is calculated by finding the value for i that sets the receipts equal to the disbursements, using Equation (3.3) as follows:

$$\$300 = \$30\overset{P/A,i,12}{(10.000)}.$$

The results are

$$i = 2.9\% \text{ per month compounded monthly}$$

$$r = 2.9\%(12) = 34.8\% \text{ per year}$$

$$i_a = (1.029)^{12} - 1 = 40.9\% \text{ per year.}$$

Although the stated interest rate was 20%, the actual or effective annual rate being paid exceeds 40%.

4.7.2 Remaining Balance of a Loan

The remaining balance of a loan is known by various names such as the amount owed, the unrecovered balance, the unpaid balance, and the principal owed. To calculate the remaining balance of a loan after a specified number of payments have been made, it is necessary to find the equivalent of the original amount borrowed less the equivalent amount repaid up to and including the last payment made.

Suppose that \$10,000 is borrowed with the understanding that it will be repaid in equal quarterly payments over five years at an interest rate of 16% per year compounded quarterly. The quarterly payments will be

$$A = \$10{,}000\overset{A/P,4,20}{(0.0736)} = \$736.$$

Immediately after the 13th payment is made, the borrower wishes to pay off the remaining balance, U_{13}, so that the obligation will terminate. The remaining balance at the *beginning* of the 14th period is found by calculating the equivalent of the original amount loaned at this point in time, less the equivalent amount repaid, as

$$U_{13} = \$10{,}000\overset{F/P,4,13}{(\;1.665\;)} - \$736\overset{F/A,4,13}{(16.627)} = \$4{,}413.$$

Alternatively, the equivalent of the payments remaining may be found at the time the remaining balance is to be paid. For this example, the remaining balance after the 13th payment, with 7 payments remaining, is

$$U_{13} = \$736\overset{P/A,4,7}{(6.0021)} = \$4{,}418$$

where the difference is due to rounding. The fact that these two amounts will always be equal was demonstrated in Section 4.4 and shown in Figure 4.8.

The interpretation of this approach is that the lender will take a lump sum after the 13th payment that is equivalent to the payments remaining at the applicable interest rate. Since the remaining balance is equivalent to the payments remaining, the lender should be indifferent to a lump sum received immediately or an equivalent series of payments into the future.

For loans in which the interest rate changes over time, the first approach presented can be applied with recognition of the changing interest rate. Suppose that an individual borrows $6,000 to be repaid in equal monthly payments of $100 with the remaining balance due at the end of 5 years. The interest rate is to be changed each year to conform to the market rate in effect. Assume that the market rate in the first year was 1.5% per month and that it is 1.0% per month in the second year.

After the 19th payment, the remaining balance on this loan, U_{19}, is

$$\$6{,}000\overset{F/P,1.5,12}{(1.196)}\overset{F/P,1,7}{(1.072)} - \left[\$100\overset{F/A,1.5,12}{(13.041)}\overset{F/P,1,7}{(1.072)} + \$100\overset{F/A,1,7}{(7.214)}\right]$$

$$= \$5{,}573.$$

When the market interest rates for the entire 5 years are known, the procedure above may be used to compute the remaining balance at any point in time.

4.7.3 Principal and Interest Payments

The repayment schedule for most loans is made up of a portion for the payment of principal and a portion for the payment of interest on the unpaid balance. The amount upon which the interest for the period is charged is the remaining balance at the beginning of the period. A loan payment received at the end of an interest period must first be applied to the interest charge. The remaining amount is then utilized to reduce the outstanding balance of the loan.

Figure 4.12 illustrates the cash flow for a fixed-rate, fixed-payment loan. An amount, P, is borrowed and equal amounts, A, are repaid periodically over n periods. Each payment is divided into an amount that is interest and a remaining amount for reduction of the principal. Define

I_t = portion of payment A at time t that is interest,

B_t = portion of payment A at time t that is used to reduce the remaining balance,

$A = I_t + B_t$, where $t = 1, 2, \ldots, n$.

The interest charged for period, t, for any loan where interest is charged on the remaining balance, is computed by multiplying the remaining balance at the beginning of period t (end of period $t - 1$) by the interest rate.

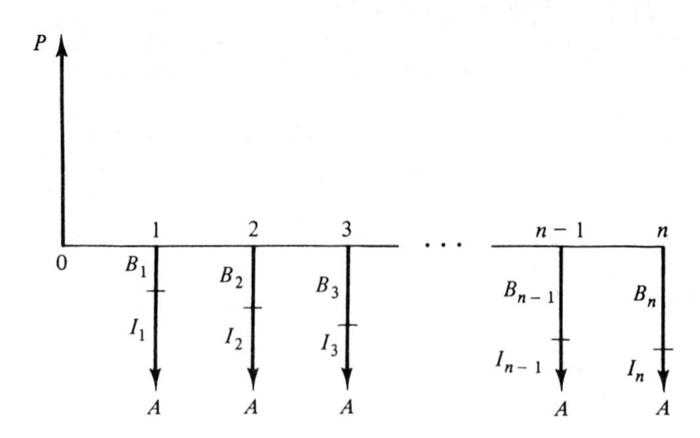

FIGURE 4.12. Cash flow for fixed-rate, fixed-payment loan.

For the fixed-rate, fixed-payment loan considered here I_t is calculated as follows:

$$I_t = A(P/A,i,n-(t-1))(i) = A(P/A,i,n-t+1)(i) \tag{4.3}$$

where $A(P/A,i,n-(t-1))$ is the balance remaining at the *end* of period $t-1$. Also,

$$B_t = A - I_t = A - A(P/A,i,n-t+1)(i)$$

$$= A\left[1 - (P/A,i,n-t+1)(i)\right].$$

From Section 3.4, $(P/F,i,n) = 1 - (P/A,i,n)(i)$. Therefore,

$$B_t = A(P/F,i,n-t+1). \tag{4.4}$$

As an example of the application of B_t and I_t to find the principal and interest payments for a fixed-rate, fixed-payment loan, let $P = \$1{,}000$ $n = 4$, and $i = 15\%$. The annual loan payment is

$$A = \$1{,}000(A/P,15,4) = \$1{,}000(0.3503) = \$350.30.$$

Table 4.2 gives the accounting for the application of this loan payment to principal and to interest for each of the four years. The total amount paid for this loan is \$350.30(4) = \$1.401.20. This is the sum of the amount paid for principal (\$1,000.14) and for interest (\$401.06) as shown in Table 4.2.

TABLE 4.2.
PRINCIPAL AND INTEREST PAYMENTS FOR A FIXED-RATE, FIXED-PAYMENT LOAN

End of Year t	Loan Payment	Payment on Principal	Interest Payment
1	$350.30	$350.30(0.5718) = $200.30 (P/F,15,4)	$150.00
2	$350.30	$350.30(0.6575) = $230.32 (P/F,15,3)	$119.98
3	$350.30	$350.30(0.7562) = $264.90 (P/F,15,2)	$ 85.40
4	$350.30	$350.30(0.8696) = $304.62 (P/F,15,1)	$ 45.68
Totals	$1,401.20	$1,000.14	$401.06

4.8 EQUIVALENCE CALCULATIONS INVOLVING WORKING CAPITAL

When investments are made in fixed assets such as a machine or production line, it is frequently necessary to have additional funds to finance any cash needs, accounts receivables, or inventories that arise from the project. This additional investment in *working capital* must not be ignored, or the actual cost of the project will be underestimated.

> **Net working capital** is defined by the accountant as a firm's short-term or current assets less its current liabilities.

The primary elements of these short-term assets include cash, customers' unpaid bills, inventories of raw materials, work-in-process, and finished goods. Each of these elements requires a commitment of funds which can either be borrowed (usually on a short-term basis) or financed from earnings. If the funds are borrowed, the cost of these working capital requirements is explicit. However, use of previous earnings for this purpose incurs an opportunity cost since income is foregone by not investing the funds in the other available opportunities.

Formulation of a cash flow description of working capital requirements is usually straightforward. Suppose a $100,000 investment in a 5-year project requires an additional $5,000 cash to cover maintenance and labor costs which may or may not materialize. With accounts receivable expected to average $8,000 over the life of the project and inventories valued at $7,000 to be carried throughout the project's life, a total of $20,000 in additional investment is required. Since it is expected that all of the investment in working capital will be recovered at the end of the project, a cash flow disbursement of $20,000 is shown at $t = 0$ along with the receipt of \$20,000 at $t = 5$. If all the other income and expenses are expected to provide a net income of \$35,000 per year and the interest rate is 20%, the annual cost that is equivalent to this cash flow is

$$A = -\$120{,}000\overset{A/P,20,5}{(0.3344)} + \$35{,}000 + \$20{,}000\overset{A/F,20,5}{(0.1344)} = -\$2{,}440 \text{ per year.}$$

Thus *with* the effects of working capital included, the equivalent disbursements exceed the equivalent receipts, and this project is *economically undesirable.*

If the cost of the working capital requirements had been omitted in this example, the annual disbursements would be reduced by

$$A = -\$20{,}000\overset{A/P,20,5}{(0.3344)} + \$20{,}000\overset{A/F,20,5}{(0.1344)} = -\$4{,}000 \text{ per year.}$$

In this case the annual equivalent of the receipts and disbursements for the project *without* the working capital consideration is $-\$2{,}440 + \$4{,}000 = \$1{,}560$, indicating that the project is *economically viable.*

This example demonstrates that the consideration of working capital requirements can have a significant impact in properly assessing a project's worth. Also the example indicates that if all the working capital is ultimately recovered at the end of the project, the cost is just the interest rate (20%) times the working capital required (\$20,000) or \$4,000 per year. This cost of \$4,000 per year represents the opportunity foregone by not having these funds available for other purposes. It is important to note that there are many circumstances where the investment in working capital is not fully recovered (inventories that deteriorate in value, loss of accounts receivable due to bad debts). Proper identification of the cash flows associated with these situations will assure the accurate assessment of the working capital costs.

KEY POINTS

- Economic equivalence is a concept central to the comparison of alternatives on a fair basis.
- Equivalence calculations involving a single factor are a step on the way to the evaluation of complex cash flows for decision situations.
- Equivalence can be maintained for cash flows regardless of the compounding frequency in force.
- Equivalent cash flows have the same economic value at the same point in time.
- Cash flows that are equivalent at one point in time are equivalent at *any* other point in time.

- If Cash Flow *A* is equivalent to Cash Flow *B* and Cash Flow *C* is equivalent to Cash Flow *B*, then Cash Flow *A must* be equivalent to Cash Flow *C*.
- The conversion of one cash flow to its equivalent at another point in time must reflect the interest rate or rates in effect for each period within the time span encompassing the equivalent cash flows.
- The actual interest rate received or paid is the interest rate that sets the equivalent receipts equal to the equivalent disbursements.
- Equivalence calculations involving bonds, loans, and working capital provide interesting application areas for this important concept.

PROBLEMS

1. What single equivalent amount will accumulate from each of the following investments?

- **a.** $4,000 in 20 years at 8% compounded semiannually. Answer: $19,204
- **b.** $16,000 in 9 years at 20% compounded quarterly.
- **c.** $1,300 in 4 years at 15% compounded monthly.
- **d.** $4,600 in 17 years at 9% compounded daily.

2. What is the present equivalent value of the following future receipts?

- **a.** $9,200 6 years from now at 9% compounded monthly. Answer: $5,372
- **b.** $1,000 10 years from now at 16% compounded quarterly.
- **c.** $23,000 7 years from now at 9% compounded semiannually.
- **d.** $104,000 80 years from now at 13% compounded weekly.

3. What series of equal payments are equivalent to the following future amounts?

- **a.** $12,000 in 8 years at 12% compounded quarterly when payments are quarterly. Answer: $229
- **b.** $3,000 in 11 years at 8% compounded semiannually when payments are annual.
- **c.** $24,000 in 12 years at 8% compounded quarterly when payments are semiannual.
- **d.** $14,000 in 5 years at 9% compounded monthly when payments are quarterly.
- **e.** $9,000 in 8 years at 14% compounded semiannually when payments are monthly.

4. What series of equal payments are equivalent to the following present amounts?
   a. $12,000 in 8 years at 6% compounded semiannually with semiannual payments. Answer: $955
   b. $40,000 in 5 years at 12% compounded monthly with monthly payments.
   c. $8,000 in 3 years at 16% compounded quarterly with annual payments.
   d. $11,000 in 5 years at 9% compounded daily with weekly payments.
   e. $1,700 in 4 years at 11% compounded semiannually with quarterly payments.
5. What is the accumulated value of each of the following series of payments?
   a. $1,400 at the end of each quarter for 10 years at 16% compounded quarterly.
   b. $500 at the end of each month for 2 years at 13% compounded semiannually.
   c. $20,000 semiannually for 4 years at 9% compounded quarterly.
   d. $5,000 at the end of each month for 5 years at 11% compounded semiannually.
6. What is the present value of the following series of prospective payments?
   a. $1,000 a month for 4 years at 10% compounded semiannually.
   b. $5,000 a year for 6 years at 12% compounded quarterly.
   c. $2,000 each quarter for 10 years at 9% compounded monthly.
   d. $12,000 each year for 4 years at 11% compounded semiannually.
7. What equal-payment series is equivalent to a payment series of $12,000 at the end of the 6th year decreasing by $500 each year to the 21st year? Interest is 10% compounded annually. Answer: $9.361
8. Find the equal-annual-payment series which would be equivalent to the following increasing series of payments if the interest rate is 10% (a) compounded annually; (b) compounded continuously. Answer: b. $1,221

   $700 at the end of the first year.
   $900 at the end of the second year.
   $1,100 at the end of the third year.
   $1,300 at the end of the fourth year.
   $1,500 at the end of the fifth year.
   $1,700 at the end of the sixth year.
   $1,900 at the end of the seventh year.

9. Find the equal-annual-payment series that would be equivalent for the fol-

lowing decreasing series if the interest rate is 8% (a) compounded annually; (b) compounded continuously.

$3,000 at the end of the first year.
$2,600 at the end of the second year.
$2,200 at the end of the third year
$1,800 at the end of the fourth year.
$1,400 at the end of the fifth year.

10. Find the present-value equivalent to the following geometrically increasing series of payments.

a. A first-year base of $4,000 increasing at 4% per year to year 15 at an interest rate of 17% compounded continuously. Answer: $23,659

b. A first-year base of $100 increasing at 10% per year to year 12 at an interest rate of 6% compounded continuously.

c. A first-year base of $7,000 increasing at 7% per year to year 20 at an interest rate of 19% compounded continuously.

11. Find the present value equivalent to the following geometrically decreasing series of payments.

a. A first-year base of $10,000 decreasing at 3% per year to year 7 at an interest rate of 9% compounded continuously. Answer: $45,879

b. A first-year base of $20,000 decreasing at the rate of 14% per year to year 30 at an interest rate of 8% compounded continuously.

c. A first-year base of $4,000 decreasing at a rate of 15% per year to year 10 at an interest of 20% compounded continuously.

12. If compounding is quarterly, what effective annual interest rate and what nominal interest rate will make the following values of P and F equivalent for the values of n shown?

a. $P = \$1{,}000$; $F = \$3{,}000$; $n = 6$ years.
Answer: 20.1% per year; 18.7% compounded quarterly

b. $P = \$3{,}300$; $F = \$9{,}000$; $n = 9$ years.

c. $P = \$1{,}700$; $F = \$12{,}000$; $n = 20$ years.

d. $P = \$2{,}500$; $F = \$1{,}200$; $n = 5$ years.

13. If compounding is semiannual, find the effective semiannual interest rate, the effective annual interest rate, and the nominal interest rate that will make the following values of P and A equivalent for the values of n shown.

a. $P = \$5{,}000$; $A = \$600$ semiannual payments; $n = 6$ years.
Answer: 6.1% per six months; 12.57% per year; 12.2% compounded semiannually

b. $P = \$10{,}000$; $A = \$3{,}000$ semiannual payments; $n = 2$ years.

c. $P = \$7{,}871$; $A = \$900$ annual payments; $n = 14$ years.

d. P = \$8,500; A = \$1,361 every two years; n = 18 years.
Answer: 1.94% per six months; 8% per two years; 3.88% compounded semiannually

14. If compounding is monthly, find the effective monthly interest rate, the effective annual interest rate, and the nominal interest rate that will make the following values of F and A equivalent for the values n shown.

a. F = \$12,000; A = \$350 monthly payments; n = 2 years.

b. F = \$4,000; A = \$53 monthly payments; n = 5 years.

c. F = \$20,000; A = \$394 semiannual payments; n = 12 years.

d. F = \$43,000; A = \$2,000 quarterly payments; n = 4 years.
Answer: 1.25% per month; 16.08% per year; 15% compounded monthly

15. A series of equal payments of \$1,600 for 25 years is equivalent to what present amount at an interest rate of 8% compounded quarterly; compounded continuously?

16. A series of equal quarterly payments of \$720 extends over a period of 10 years. What amount at the present is equivalent to this series at 18% interest compounded annually; compounded quarterly; compounded continuously?

17. A flow of funds of \$4,400 per year is deposited into a sinking fund. What amount will be accumulated at the end of 5 years if the interest rate is 12% compounded annually; compounded monthly; compounded continuously?

18. An interest rate of 10% compounded continuously is desired on an investment of \$15,000. How many years will be required to recover the capital with the desired interest if \$2,500 is received each year? Answer. 9.375 years

19. What uniform flow of payments for 12 years is equivalent to a series of equal end-of-year payments of \$750 for 10 years at 8% compounded continuously?

20. Find the following equivalent amounts if the interest rate is 9% compounded annually.

a. The present value of a series of prospective payments, \$1,500 a year for 120 years. Answer: \$16,666

b. The accumulated value of series of prospective payments, \$10 a year for 120 years.

c. The equal series of payments to be paid into a sinking fund to accumulate \$1,000,000 in 120 years.

d. The equal series of payments to be made to repay a present amount of \$85,000 in 120 years.

21. For an interest rate of 16% compounded semiannually, find

a. What payment can be made now to prevent an expense of $220 every 6 months for the next 7 years?

b. What semiannual deposit into a fund is required to total $15,000 in 8 years?

22. A mutual stock fund has grown at a rate of 16% compounded annually since its beginning. If it is anticipated that it will continue to grow at this rate, how much must be invested every year so that $60,000 will be accumulated at the end of 12 years?

23. A series of payments—$10,000, first year; $9,000, second year; $8,000, third year; $7,000, fourth year; and $6,000, fifth year—is equivalent to what present amount at 10% interest compounded annually; compounded continuously? Solve using the gradient factors and then by using only the single-payment present-worth factors.

24. A manufacturing firm is seeking a loan of $400,000 to finance production of a newly patented product line. Owing to a good reception of the product at its introductory showing, a bank had agreed to lend the firm an amount equal to 90% of the present worth of firm orders received for delivery during the next 5 years. The orders received are as follows:

50,000 during the first year.
40,000 during the second year.
30,000 during the third year.
20,000 during the fourth year.
10,000 during the fifth year.

If the product will set for $4 each, will the present worth of the orders received justify the loan required? Interest is 12% compounded annually.

25. What single amount at the end of the fifth year is equivalent to a uniform annual series of $1,000 per year for 12 years? The interest rate is 7% compounded annually. Answer: $11,144

26. A series of 10 annual payments of $7,500 is equivalent to three equal payments at the end of years 6, 10, and 15 at 15% interest compounded annually. What is the amount of these three payments?

27. A series of equal payments of $300 is received quarterly for 6 years. After the first 6 years the quarterly payments are doubled in size, and these larger payments are received for 8 more years. Find the single present amount equivalent to this series of payments received over the 14-year period if the interest rate is

a. 12% compounded quarterly. Answer: $11,098

b. 12% compounded monthly.

c. 12% compounded continuously.

28. What single payment at the end of year 5 is equivalent to a uniform flow of payments of $2,300 per year beginning at the start of year 3 and ending at the end of year 15? Interest is 8% compounded continuously. Answer: $23,629

29. A construction firm is considering the purchase of an air compressor. The compressor has the following end-of-year maintenance costs:

Year	1	2	3	4	5	6	7	8
Maintenance costs	$800	$800	$900	$1,000	$1,100	$1,200	$1,300	$1,400

What is the present equivalent maintenance cost if the interest rate is 12%?

30. A company must make license payments for a process that they have adopted for a new plant. The payments will begin at $10,000, and the first payment is expected to be made 3 years from the present when the plant is completed and in production. Payment will be made every three months thereafter, and the license payments are expected to increase by $500 each quarter. What single present amount is equivalent to the series of license payments made over an 8-year period if the interest rate is 8% compounded quarterly?

31. A manufacturer pays a patent royalty of $1.15 per unit of a product he manufactures, payable at the end of each year. The patent will be in force for an additional 5 years. For this year he manufactures 8,000 units of the product, but it is estimated that output will increase by 10% per year in the 4 succeeding years. He is considering asking the patent holder to terminate the present royalty contract in exchange for a single payment at present, or in exchange for equal annual payments to be made at the end of each of the 5 years. If 8% interest is used, what is (a) the present single payment and (b) the annual payments that are equivalent to the royalty payments in prospect under the present agreement.

32. An increasing annual uniform-gradient series begins at the end of the second year and ends after the fifteenth year. What is the value of the gradient *G* that makes the gradient series equivalent to a uniform flow of payments of $900 per month for 7 years at 10% compounded continuously? Answer: $1,415

33. An individual's salary is now $32,000 per year, and he anticipates retiring in 30 years. If his salary is increased by 10% each year and he deposits 5% of his yearly salary into a fund that earns 8% interest compounded annually, what will be the amount accumulated at the time of retirement?

34. A petroleum engineer estimates that the present production of 400,000 barrels of oil during this year from a group of 10 wells will decrease at the rate of 15% per year for years 2 through 10. Oil is estimated to be worth

$25 per barrel. If the interest rate is 10% compounded annually, what is the equivalent present amount of the prospective future receipts from the wells?

35. The operating and maintenance expenses on a machine are expected to increase $\frac{1}{2}$% per month. This month's expenses are $2,000. Find the equal *annual* series that is equivalent to the monthly expenses over 5 years for an interest rate of 21% compounded monthly. Answer: $29,978 per year

36. $14,780 was invested, from which can be withdrawn a geometric gradient series of annual payments decreasing at the rate of 10% per year. The first payment received was $4,000, and it occurred one year after the investment, followed by the remaining five payments (a total of 6 withdrawals). What is the rate of interest earned from this investment?

37. A $5,000 deposit is made at the present. Over the first 10 years the deposit will earn 8% compounded annually. For the following 5 years the rate of interest will be 12% compounded quarterly. Annual withdrawals begin exactly 16 years from the present with an initial withdrawal of $1,500. Withdrawals are to increase at the rate of 6% per year, and the interest rate beyond 15 years from the present is 10% compounded annually. For how many years can such withdrawals be made?

38. A cash flow series is increasing geometrically at the rate of 8% per year. The initial payment at $t = 1$ is $5,000, with increasing annual payments ending at $t = 20$. The interest rate in effect is 15% compounded annually for the first 7 years and 5% compounded annually for the remaining 13 years. Find the present amount that is equivalent to this cash flow.

39. A city that was planning an addition to its water supply and distribution system contracted to supply water to a large industrial user for 10 years under the following conditions: The first 5 years of service were to be paid for in advance, and the last 5 years at a rate of $45,000 a year payable at the beginning of each year. Two years after the system is in operation the city finds itself in need of funds and desires that the company pay off the entire contract so that the city can avoid a bond issue.

   a. If the city uses 9% interest compounded annually in calculating a fair receipt on the contract, what amount can they expect? Answer: $147,328

   b. If the company uses 20% interest compounded annually, how much is the difference between what the company would consider a fair value for the contract and what the city considers to be a fair value? Answer: $53,864

40. A city power plant wishes to install a feed-water heater in their steam generation system. It is estimated that the increase in efficiency will pay for the heater 1 year after it is installed, and a contractor has promised he can install the heater in 5 months. If the venture is undertaken and it is found that the heater does pay for itself in a year by saving $1,400 per month, what amount was paid to the contractor at the last of each month of construction? Assume that the saving of $1,400 occurs at the last of each month and that the money paid to the contractor could have been invested elsewhere at 12% compounded monthly.

41. A manufacturing company purchased electrical services for the next 5 years to be paid for with $70,000 now. The service after 5 years will be $15,000 per year beginning with the sixth year. After 2 years service the company, having surplus profits, requested to pay for another 5 years service in advance. If the electrical company elected to accept payment in advance, what would each company set as a fair settlement to be paid if (a) the electrical company considered 15% compounded annually as a fair return, and (b) the manufacturing company considered 12% a fair return?

42. A student borrowed $3,500 from the ABC Loan Company to buy a used car with an agreement to repay $1,500 at the end of each of the first 2 years and $2,000 at the end of the third year. What rate of interest makes the receipts and disbursements equal to each other for this loan agreement?

43. An individual is purchasing a $15,000 automobile, which is to be paid for in 48 monthly installments of $395. What nominal annual interest rate is being paid for this financing arrangement if interest is compounded monthly?

44. As usually quoted, the prepaid premium of insurance policies covering loss by fire and storm for a 3-year period is 2.5 times the premium for 1 year of coverage. What rate of interest does a purchaser receive on the additional present investment if he purchases a 3-year policy now rather than three 1-year policies at the beginning of each of the years? Answer: 21.6%

45. A woman wishes to borrow $10,000 to purchase an automobile. The lender, Friendly Bob, indicates that two charges will determine the cost of the loan, which will be repaid monthly over 5 years. The first is an initiation charge of 2% of the loan, which is payable at the time the loan is made. The second is an add-on rate of 8%, where the monthly payments are determined as follows:

$$\$10{,}000(8\%) = \text{add-on interest for 1 year}$$
$$\text{Total add-on interest for 5 years} = 5(\$800) = \$4{,}000$$

$10,000 amount borrowed
$ 4,000 total add-on interest
$14,000 ÷ 60 months = $233.33 per month

Find the effective monthly interest rate, the nominal interest rate, and the effective annual rate being charged on this loan if compounding is monthly.

46. A young couple have decided to make advance plans for financing their 3 year-old son's college education. Money can be deposited at 7% compounded annually. What annual deposit on each birthday from the fourth to the 17th inclusive must be made to provide $7,000 on each birthday from the 18th to the 21st inclusive?

47. A woman is planning to retire in 30 years. She wishes to deposit a regular amount every 3 months until she retires, so that beginning one year following retirement she will receive annual payments of $25,000 for the next 20 years. How much must be deposited if the interest rate is 8% compounded quarterly?

48. An investor desires to make an investment in bonds, provided he can realize 10% on his investment. How much can he afford to pay for a $10,000 bond that pays 7% interest annually and will mature 20 years hence? Answer: $7,447

49. How much can be paid for a $5,000, 14% bond with interest paid semiannually, if the bond matures 12 years hence? Assume the purchaser will be satisfied with 9% interest compounded semiannually, since the bonds were issued by a very stable and solvent company.

50. A man expects to be able to earn from his bank 8% compounded annually for the next 4 years and 10% compounded annually thereafter. A $10,000, 12% bond maturing in 9 years is offered for $11,000. Should the bond be purchased, with the bond's interest payments being invested at the bank interest rate, or should the $11,000 be invested in the bank for the entire 9 years?

51. A bond is offered for sale for $1,120. Its face value is $1,000 and the interest is 11% payable annually. What yield to maturity will be received if the bond matures 10 years hence? Find the bond's current yield. Answer: 9.12%; 9.82%

52. A $1,000, 8% bond is offered for sale for $900. If interest is payable annually and the bond will mature in 5 years, what is its yield to maturity and its current yield?

53. A $1,000 bond will mature in 10 years. The annual rate of interest is 6% payable semiannually. If compounding is semiannual and the bond can be purchased for $870, what is the yield to maturity in terms of the effective annual rate earned? Indicate the bond's current yield.

54. The selling price of a $10,000, 10% municipal bond is $12,000. If the bond pays interest semiannually and will mature in 20 years, find the current yield and the yield to maturity. Assume that interest is compounded annually.

55. A man deposits a sum of money in his savings account which earns interest at a rate of 6% compounded annually. He would like to pay his installment loan ($250 a year) from this bank account for 3 years.

a. How much should he deposit so that his third payment just depletes the balance?

b. Calculate and present in tabular form the remaining balance just after payment is made at each point in time, including the present.

56. A widow received $100,000 from an insurance company after her husband's death. She plans to deposit this amount in a savings account that earns interest at a rate of 7% compounded annually for 5 years.

a. If she wants to withdraw equal annual amounts from the account for 5 years, with the first withdrawal occurring one year after the deposit, how much are these disbursements?

b. Calculate and present in tabular form the remaining balance just after withdrawal is made at each point in time.

57. A contractor borrowed $10,000 with an agreement to repay the loan over 4 years in equal annual payments at an interest rate of 15%.

a. How much are these payments?

b. Calculate and present in tabular form the remaining balance just after payment is made at each point in time, including the time the loan is made.

58. A student borrowed $5,000, which she will repay in 30 equal monthly installments. After her 25th payment she desires to pay the remainder of the loan in a single payment. At 15% interest compounded monthly, what is the amount of the payment?

59. A family has borrowed $100,000 to purchase a new home. The interest rate is fixed at 12% compounded monthly over the 25-year life of the loan. Find the remaining balance on the loan immediately after the payments listed below. Also find the amount of interest that comprises the payments listed.

a. payment 1.
b. payment 10.
c. payment 25.
d. payment 30.
e. payment 40.
f. payment 60.
g. payment 75.
h. payment 90.
i. payment 100.
j. payment 200.
k. payment 250.
l. payment 275. Answer: $23,190; $240

60. An individual purchased an $80,000 town house with a down payment of 20% and a 30-year mortgage with monthly payments. Interest is 12% compounded monthly.

a. If the house is sold at the end of 5 years for $90,000, how much equity does the individual have? (Equity is the difference between the current market value and the balance owed on the loan.)

b. Of the total amount paid on the mortgage, what portion was principal and what portion interest?

61. To purchase a used automobile, $6,000 is borrowed immediately. The repayment schedule requires equal monthly payments of $264.72 to be made over the next 24 months. After the last payment is made, any remaining balance on the loan will be paid in a single lump sum. The effective annual rate of interest on this loan is 19.56% based on monthly compounding. Find the nominal interest rate and the lump sum paid at the end of the loan.

62. An individual is borrowing $100,000 at 8% compounded annually. The loan is to be repaid in equal annual payments over 30 years. However, just after the eighth payment is made, the lender allows the borrower to double the annual payment. The borrower agrees to this increased payment. If the lender is still charging 8% compounded annually on the unpaid balance of the loan, what is the balance still owed just after the 12th payment is made? Answer: $43,270

63. Four years ago $100,000 was borrowed at 14% per year compounded annually, to be repaid in equal annual payments over 20 years. After the fourth payment (the present) the borrower is offered the opportunity to pay off the existing loan by borrowing the balance remaining under the following terms. The new loan at 12% compounded annually would require annual payments of $10,000 for the first 5 years followed by payments of $25,000 per year until the final year, when the balance remaining at the end of that year is to be *completely* paid off. (Interest on each loan is being charged on its remaining balance.)

a. How many years from the present would the final payment be made if the new loan at 12% annually were undertaken?

b. How much is the interest portion of the first $25,000 payment made under the new loan? (This is the sixth payment of the new loan.)

64. A variety of mortgages are available for the purchase of residential housing. One type, the adjustable-rate mortgage (ARM), has a fixed payment for 2 years. Then, based on any change in interest rates, measured by an interest rate index for Treasury bills, a new fixed payment is calculated. Suppose you borrowed $100,000 for 30 years. Shown below are the different interest rates and their associated payments for the first 4 years of this loan (48 monthly payments).

Years	Interest Rate	Monthly Payments
1 and 2	9% compounded monthly	$ 804.62
3 and 4	12% compounded monthly	$1021.36

a. What is the balance remaining on this loan after the 45th payment has been made?

b. How much of the 46th payment is used to reduce the principal of the loan, and how much is to pay interest?

65. A person has the following outstanding debts:

1. $10,000 borrowed 4 years ago with the agreement to repay the loan in 60 equal monthly payments. There are 12 payments outstanding, and interest on the loan is 6% compounded monthly.

2. Twenty-four monthly payments of $500 owed on a loan on which interest is charged at the rate of 1% per month on the unpaid balance.

3. A bill of $3,000 due in 2 years.

A loan company has offered to consolidate these debts if the monthly amount listed is paid over the next 5 years. If interest is compounded monthly, what monthly rate of interest will be paid if the individual accepts the loan company's offer? What will be the nominal rate? What will be the annual effective interest rate?

a. $337 per month.

b. $384 per month.

c. $301 per month.

d. $372 per month. Answer: 1.75%; 21%; 23.14%

e. $313 per month.

f. $361 per month.

g. $349 per month.

h. $325 per month.

66. A financial institution is offering the following savings plan to a family to provide for the educational expenses of their oldest child. End-of-year deposits of $480 per year will be made to the plan for 10 years, with a single payment of $6,400 returned to the family at the time the last deposit is made. The child, who is presently 14 years old, will borrow $5,000 per year for 4 years, beginning 4 years from the present. This loan of $20,000 will be repaid in 10 equal annual payments of $5,773, beginning 1 year after college is completed. What is the rate of interest paid on this financing arrangement under annual compounding? (Assume college requires exactly 4 years.)

67. A company is investing \$5,000,000 in a new plant to produce a new product. It is estimated that an additional 5% of the total cost will be necessary for working capital. The life of the product is considered to be 8 years, and it is assumed that the percentage of the initial working capital recovered at the end of the product's life is as indicated below. Find the annual equivalent cost of the investment in working capital if the interest rate is 15%.

a. 100%.

b. 95%.

c. 90%.

d. 80%.

e. 50%. Answer: \$46,612.

68. A manufacturing project requires \$300,000 for inventories, \$200,000 for accounts receivable, and \$150,000 cash in order to proceed. If, at the end of the project's 10-year life, 4% of the accounts receivable are bad debts, 3% of the product in inventory has been lost or stolen, and 10% of the cash has been used, what interest rate has been earned or lost on the investment in working capital? If the interest rate for the firm is 20%, what is the annual equivalent cost of working capital?

5 Equivalence Involving Inflation

A study of the past performance of most economies throughout the world reveals that the prices for goods and services are continually fluctuating. Historically the most common movement of prices has been up (inflation), while there have been less frequent episodes of downward price movement (deflation). This change in prices affects the amount of goods and services that can be purchased for a given amount of money. As prices increase, the purchasing power of money declines, while decreasing prices enhance its purchasing power.

To properly account for the time value of money in equivalence calculations, it is important that both the earning power of money and its purchasing power be properly reflected. This chapter presents the concept of the purchasing power of money along with the analytical methods and techniques needed to incorporate this concept into engineering economy studies.

5.1 MEASURES OF INFLATION AND DEFLATION

Inflation and deflation are terms that describe changes in price levels in an economy. Without addressing the causes of the changes in price levels, the focus here will be on the methods needed to determine the rate of change of price levels and how these changes should be reflected in engineering economy studies. Because inflation has been a much more common occurrence than deflation, the examples used will deal primarily with inflation. However, the methods presented are general, and will accommodate situations where an economy is experiencing general price reductions.

The prices for goods and services are driven upward or downward because of numerous factors at work within the economy. The cumulative effect of these factors determines the amount of price change. For example, increases in productivity and in the availability of goods tend to reduce prices, while government policies such as price supports and deficit financing tend to increase prices. When all such effects are taken together, the most common result has been that prices increase.

To measure historical price-level changes for particular commodities or the general cost of living, it is necessary to calculate a price index.

> A **price index** is the a ratio of the historical price of some commodity or service at some point in time to the price at some earlier point.

The earlier point in time is usually some selected *base* year. Thus, the index in question and other indexes can be related to the same base. If, for example:

base year, 1967 (price $\text{index}_{1967} = 100$)
commodity $\text{price}_{1967} = \$1.46$ per pound
commodity $\text{price}_{1993} = \$5.74$ per pound

then

$$\text{price index}_{1993} = \frac{\$5.74/\text{lb}}{\$1.46/\text{lb}}(100) = 393.2$$

This value for the index indicates that the 1993 price is 3.932 greater than the 1967 price.

Suppose an individual can invest $100 at the present time with the expectation of earning 15% annually for the next 5 years. At the end of 5 years the accumulated amount will be

$$\$100(\overset{F/P,15,5}{2.011}) = \$201.10$$

At the present this individual can purchase one automobile tire for $100, but suppose these tires are increasing in price at an annual rate of 10%. At the end of 5 years the same tire will cost

$$\$100(\overset{F/P,10,5}{1.611}) = \$161.10$$

Under these conditions, the individual would be misled if he ignored the change in prices. He would then gain the mistaken impression that if he invested now,

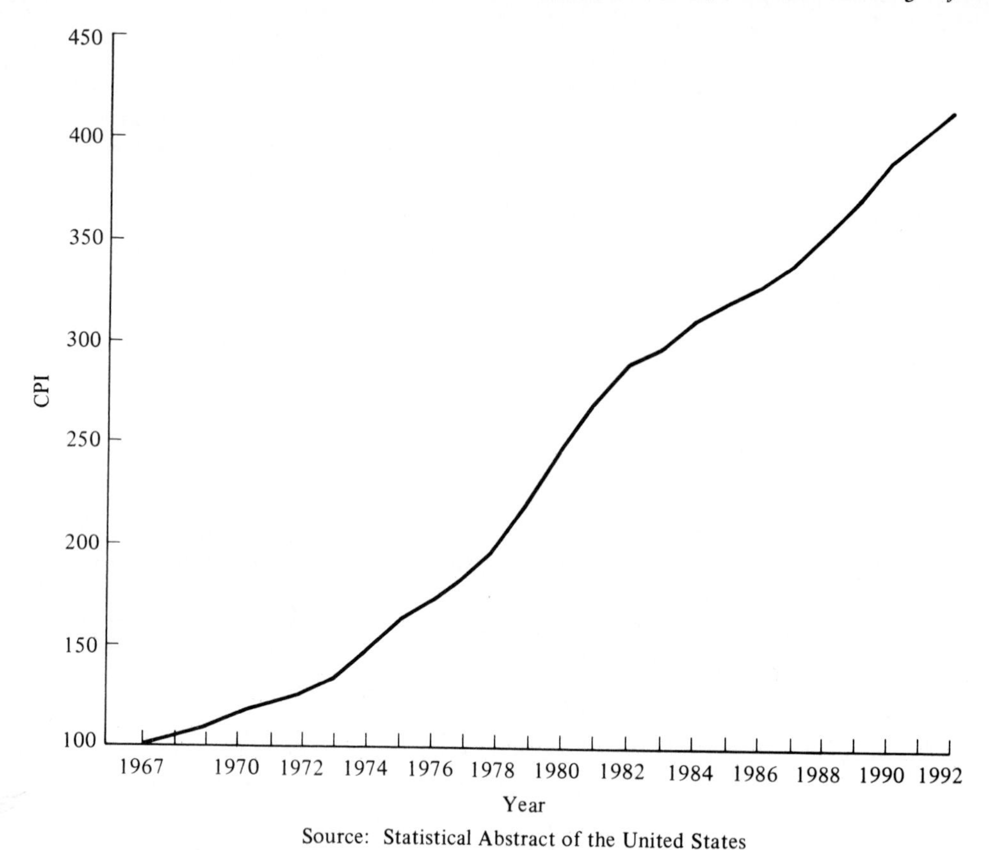

FIGURE 5.1. Consumer Price Index (1967 = 100).

the money to purchase two tires would be available at the end of five years. Actually, the money received from the investment would purchase only 1.25 tires. Thus, when considering the time value of money, one must include the impact of changes in prices (i.e., changes in the purchasing power of money) as well as the effect of its earning power.

Data for the development of price indexes are gathered and analyzed by agencies of the federal government. The Department of Commerce (Bureau of Economic Analysis) and the Department of Labor (Bureau of Labor Statistics) are the agencies primarily involved in the compilation of price indexes. Not only are price indexes prepared for individual commodities or classes of products, but composite indexes are also compiled. These composite indexes, which include the Consumer Price Index (CPI), the Producer Price Index (PPI), and the Implicit Price Index for the Gross National Product (IPI-GNP), represent different mea-

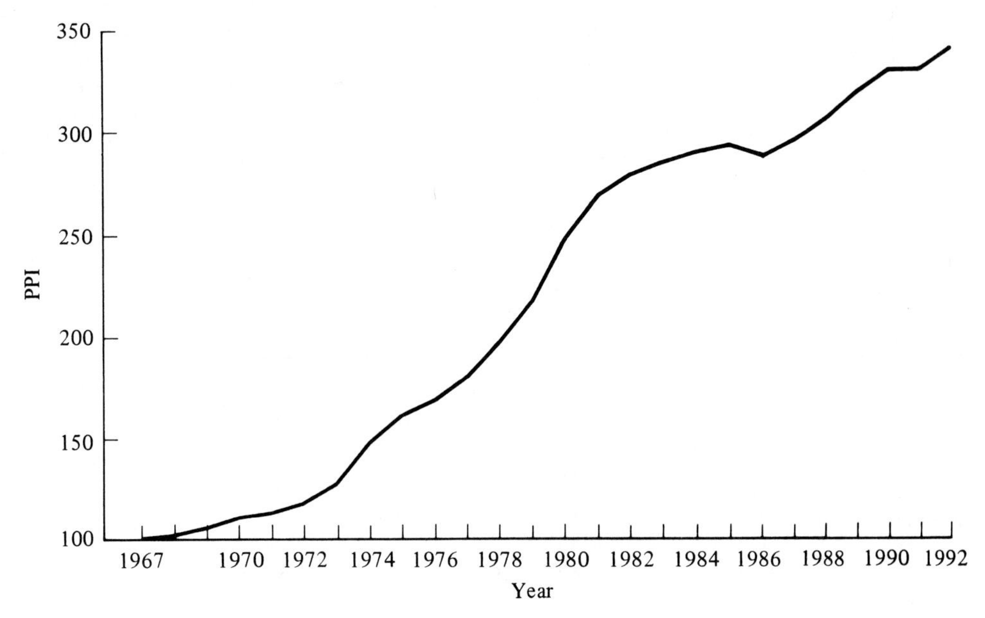

FIGURE 5.2. Producer Price Index—all commodities (1967 = 100).

sures of historical price-level changes within the economy.

Graphs for the Consumer Price Index and the Producer Price Index are presented in Figures 5.1 and 5.2. The most familiar index, the Consumer Price Index, represents the change in retail prices for a selected "market basket," including clothing, food, housing, transportation, and utilities. The purpose of this index is to measure the changes in retail prices required to maintain a fixed standard of living for the "average" consumer.[1]

The inflation rates for each of the years from 1967 to 1992 are shown in Table 5.1; note that the base year for CPI values is 1967. The variation in these annual rates from 1967 through 1992 is illustrated in Figure 5.3.

When incorporating changes of price levels in engineering economy studies, the index selected should measure those changes that are pertinent to the individual or organization undertaking the study. For example, in considering the inflationary effects on a bond purchase one might utilize the CPI if an individual was the purchaser, whereas the same purchase by an industrial organization might require a different index. All examples in this chapter will be confined to applying the CPI only. Since the methodology is general, any index can be substituted without any change in approach.

[1] *For a more detailed description of the history and concepts of the CPI, see Consumer Price Index, Report No. 517, published by the Bureau of Labor Statistics.*

TABLE 5.1.
CONSUMER PRICE INDEX AND ANNUAL INFLATION RATES, 1965–1988

Year	CPI	Annual Inflation Rate, %	Year	CPI	Annual Inflation Rate, %
1965	94.5	1.7	1979	217.4	11.3
1966	97.2	2.9	1980	246.8	13.5
1967	100.0	2.9	1981	272.4	10.4
1968	104.2	4.2	1982	289.1	6.1
1969	109.8	5.4	1983	298.4	3.2
1970	116.3	5.9	1984	311.1	4.3
1971	121.3	4.3	1985	322.2	3.6
1972	125.3	3.3	1986	328.3	1.9
1973	133.1	6.2	1987	340.4	3.7
1974	147.7	11.0	1988	354.4	4.1
1975	161.2	9.1	1989	371.4	4.8
1976	170.5	5.8	1990	391.5	5.4
1977	181.5	6.5	1991	403.6	3.1
1978	195.4	7.7	1992	415.7*	3.0*

* *Estimate*

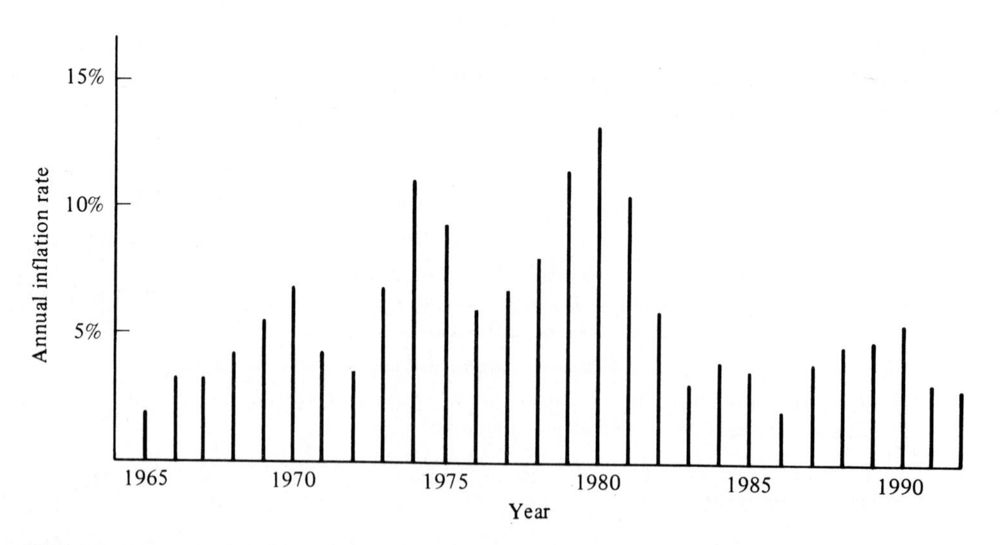

FIGURE 5.3. Annual inflation rates, 1965–1992.

5.2 THE INFLATION RATE

It is customary to utilize an annual prcentage rate representing the annual increase or decrease in price over a one-year time span. Since the rate each year is based on the previous year's prices, this rate has a compounding effect. Thus, prices that are inflating at a rate of 9% per year the first year and 8% per year the next year will have a value at the end of the second year of

$$\begin{pmatrix}\text{prices at end}\\ \text{of second year}\end{pmatrix} = (1 + 0.09)(1 + 0.08)\begin{pmatrix}\text{prices at beginning}\\ \text{of first year}\end{pmatrix}.$$

Because of the compounding nature of inflation or deflation rates, the interest formulas derived in Chapter 3 for the earning power of money may be used when computing the effects of changes in prices (i.e., changes in the purchasing power of money).

5.2.1 Computing the Inflation Rate

The historical annual rate of price increase can be computed from any of the several available indexes. Using the CPI values presented in Table 5.1, the annual inflation rate is calculated from the expression

$$\boxed{\text{annual inflation rate for year } t + 1 = \frac{\text{CPI}_{t+1} - \text{CPI}_t}{\text{CPI}_t}} \tag{5.1}$$

where CPI_t is the index of consumer prices at the end of year t. Thus, the annual inflation rate for 1980 was

$$\frac{\text{CPI}_{1980} - \text{CPI}_{1979}}{\text{CPI}_{1979}} = \frac{246.8 - 217.4}{217.4} = 0.135 \text{ or } 13.5\% \text{ per year.}$$

Most economic studies require the use of estimates that depend on expectations of *future* inflation rates. The determination of these future rates should be based on trends of historical rates, predicted economic conditions, judgment, and the other elements of economic forecasting. Like the estimation of future cash flows, the accurate prediction of future inflation rates is a difficult endeavor.

Many studies use an average annual inflation rate when the projected life of the investment is long.[2] This approach requires the estimate of a single aver-

[2] *In this chapter the average annual inflation rate is the geometric average rather than the arithmetic average.*

age rate that represents a composite of the individual yearly rates. For example, the average inflation rate from the end of 1966 to the end of 1980 (14 years) can be calculated in the following manner:

Let $\bar{f}$ = the average annual inflation rate. Then

$$97.2(1+\bar{f})^{14} = 246.8$$

$$\bar{f} = 6.9\% \text{ per year.}$$

From the end of 1972 through 1980 (8 years) the average inflation rate was

$$125.3(1+\bar{f})^{8} = 246.8$$

$$\bar{f} = 8.8\% \text{ per year.}$$

In general,

$$\boxed{\mathrm{CPI}_t(1+\bar{f})^n = \mathrm{CPI}_{t+n}.} \tag{5.2}$$

The concept of the average annual inflation rate facilitates inflation calculations. In most instances, the estimation of individual yearly inflation rates is time consuming, and using these rates usually is no more accurate than using a single composite rate. The decision whether to use individual annual inflation rates or an average composite rate should be based on judgment. One should consider the nature of the situation, its sensitivity to changes in the inflation rate, the accuracy desired, and, most of all, the information available for forecasting future inflation rates.

5.2.2 The Purchasing Power of Money

As prices increase or decrease, the amount of goods and services that can be purchased for a fixed amount of money decreases or increases accordingly. Under inflationary conditions the purchasing power of money is decreasing.

The amount of purchasing power for a dollar in a particular year relative to the goods a dollar would purchase in 1967 is presented in Table 5.2. For 1987, the purchasing power of a 1967 dollar is found using the CPI index values in Table 5.1 by calculating

$$\text{Purchasing power}_{1987} = \frac{CPI_{1967}}{CPI_{1987}} = \frac{100.0}{340.4} = 0.2938$$

TABLE 5.2.
PURCHASING POWER OF THE DOLLAR: 1940 TO 1992*

	Annual Average as Measured by			Annual Average as Measured by			Annual Average as Measured by	
Year	Producer Prices	Consumer Prices	Year	Producer Prices	Consumer Prices	Year	Producer Prices	Consumer Prices
1940	$2.469	$2.381	1962	$1.064	$1.104	1978	$0.510	$0.512
1945	1.832	1.855	1963	1.067	1.091	1979	0.459	0.461
1948	1.252	1.387	1964	1.063	1.076	1980	0.405	0.405
1949	1.289	1.401	1965	1.045	1.058	1981	0.371	0.367
1950	1.266	1.387	1966	1.012	1.029	1982	0.356	0.346
1951	1.156	1.285	1967	1.000	1.000	1983	0.351	0.335
1952	1.163	1.258	1968	0.972	0.960	1984	0.343	0.321
1953	1.175	1.248	1969	0.938	0.911	1985	0.340	0.310
1954	1.172	1.242	1970	0.907	0.860	1986	0.345	0.305
1955	1.170	1.247	1971	0.880	0.824	1987	0.338	0.294
1956	1.138	1.229	1972	0.853	0.799	1988	0.330	0.282
1957	1.098	1.186	1973	0.782	0.752	1989	0.313	0.269
1958	1.073	1.155	1974	0.678	0.678	1990	0.299	0.255
1959	1.075	1.145	1975	0.612	0.621	1991	0.299	0.248
1960	1.067	1.127	1976	0.586	0.587	1992	0.293†	0.241*
1961	1.067	1.116	1977	0.550	0.551			

SOURCE: *Statistical Abstract of the United States.*

† Estimate

*1967 = $1.00. Producer prices prior to 1961, and consumer prices prior to 1964, exclude Alaska and Hawaii. For 1940 and 1945, producer prices based on all commodities index; subsequent years based on finished goods index. Obtained by dividing the average price index for the 1967 base period (100.0) by the price index for a given period and expressing the result in dollars and cents. Annual figures are based on average of monthly data.

The average annual inflation rate from the end of 1966 through 1980 can be calculated from the following relationship, using the data in Table 5.2.

$$\text{Purchasing power}_{1980}(1 + \bar{f})^{14} = \text{Purchasing power}_{1966}$$

$$0.405(1 + \bar{f})^{14} = 1.029$$

$$\bar{f} = 6.9\%$$

Using the data in Table 5.1 for 1981, the calculation of f, the annual rate of price increase for that year, gives 10.4%.

$$246.8(1 + f)^{1} = 272.4$$

$$f = 10.4\%$$

The annual rate of decrease in purchasing power, k, for 1981 is 9.4% based on the data from Table 5.2.

$$0.405(1 - k)^{1} = 0.367$$

$$k = 9.4\%$$

Thus, the rate of price increase, f, does not usually equal k, the rate of loss of purchasing power. Because the data regarding inflation is available as either a change in prices or a loss of purchasing power, the analyst may compute f directly or find k, which can than be converted to f.

In general let,

$\bar{f}$ = the rate of increase in prices per year,
$\bar{k}$ = the rate of loss in purchasing power per year,
n = the number of years.

Then from Equation (5.2)

$$CPI_t(1 + \bar{f})^n = CPI_{t+n}$$

and by definition

$$\text{Purchasing power}_t(1 - \bar{k})^n = \text{Purchasing power}_{t+n},$$

which can be rewritten

$$\frac{CPI_{base\ year}}{CPI_t}(1 - \bar{k})^n = \frac{CPI_{base\ year}}{CPI_{t+n}}$$

giving

$$CPI_t = (1 - \bar{k})^n CPI_{t+n}.$$

Thus

$$(1 + \bar{f})^n = \frac{1}{(1 - \bar{k})^n}. \tag{5.3}$$

Based on the CPI values in Table 5.1 and the corresponding values in Table 5.2 the average annual inflation rate and the rate of loss of purchasing power from 1967 through 1981 are:

$$100(1 + \bar{f})^{13} = 246.8 \qquad 1.0(1 - \bar{k})^{13} = 0.40519$$

$$\bar{f} = 0.07196 \qquad \bar{k} = 0.06713$$

Thus, a 7.196% per-year increase in prices has the same effect as a 6.713% per-year loss in purchasing power. In this book inflation will be stated in terms of changes in price rather than changes in purchasing power.

5.3 CONSIDERING THE EFFECTS OF INFLATION

Two basic approaches are presented that allow for the simultaneous consideration of changes in money's earning power and purchasing power. These approaches are consistent and, if applied properly, will lead to identical conclusions. The first approach assumes that cash flows are measured in terms of *actual* dollars; the second uses the concept of *constant* dollars.

5.3.1 Definitions of i, i' and f

To develop the relationships between actual-dollar analysis and constant-dollar analysis, precise definitions are needed for the various interest rates to be used in the calculations. The following definitions are needed to distinguish the market interest rate, the inflation-free rate, and the inflation rate:

> The **market interest rate** represents the opportunity to earn as reflected by the actual rates of interest available in finance and business. This rate is a function of the investment activities of investors who are operating within this market. Since astute investors are well aware of the power of money to earn and the detrimental effects of inflation, the interest rates quoted in the marketplace *include* the effects of both the earning power and the purchasing power of money. When the rate of inflation increases, there is usually a corresponding upward movement in quoted interest rates.

Since the preceding and subsequent chapters generally assume that cash flows are in terms of actual dollars, the rate used in the calculations is the market interest rate. This rate has been denoted by *i*, and this designation is used consistently throughout the book to represent the interest rates available in the marketplace. When a firm states its minimum attractive rate of return (MARR) it is usually a market rate of interest.

Other names: Combined interest rate, current-dollar interest rate, actual interest rate, inflated interest rate.

The **inflation-free interest rate** represents the earning power of money with the effects of inflation *removed*. This interest rate is an abstraction. Typically it must be calculated, since it is not generally used in the transactions of the financial marketplace. The inflation-free interest rate is not quoted by bankers, stockbrokers, and other investors and is therefore not generally known to the public. If there is no inflation in an economy, then the inflation-free interest rate and the market interest rate are identical.

Other names: Real interest rate, constant-dollar interest rate.

The **inflation rate** is the annual percentage of increase in prices of goods and services. The determination of this rate is discussed in Section 5.2.

Other names: Escalation rate, rate of increase in cost of living.

When using any of the three rates just defined, we must recognize that their values will be based on estimates, reflecting future expectations.

5.3.2 Representing Cash Flows in Actual or Constant Dollars

Cash flows can be represented in terms of either actual dollars or constant dollars. These are defined as follows:

Actual dollars represent the out-of-pocket dollars received or disbursed at any point in time. This amount is measured by totaling the denominations of the currency paid or received. The cash flows presented in Chapters 3 and 4 are in terms of actual dollars.

Other names: Then-current dollars, current dollars, future dollars, escalated dollars, inflated dollars.

Constant dollars represent the hypothetical purchasing power of future receipts and disbursements in terms of the purchasing power of dollars at some base year. This base year can be arbitrarily selected, although in many analyses it is assumed to be time zero, the beginning of the investment. In subsequent analysis the base year is assumed to be time zero unless specified otherwise.

Other names: Real dollars, deflated dollars, today's dollars, zero-date dollars.

A cash flow can be expressed in terms of actual dollars either by direct assessment in actual dollars or by conversion of a constant-dollar estimate to actual dollars. Similarly, if it is desired to express the cash flow in terms of constant dollars, these dollars can be directly estimated, or the estimate can be made in actual dollars and then converted to constant dollars. The most effective approach depends upon the nature of the data regarding future cash flows and upon whether the analysis is to be done in actual or constant dollars.

The conversion of actual dollars at a particular point in time to constant dollars (based on purchasing power n years earlier) at the *same* point in time is a common requirement. When inflation has occurred at the annual percentage rate of f, this conversion is expressed as

$$\text{constant dollars} = \frac{1}{(1 + f)^n}(\text{actual dollars}). \tag{5.4}$$

Conversion of constant dollars to actual dollars for the same set of circumstances is accomplished by solving the preceding expression for actual dollars:

$$\text{actual dollars} = (1 + f)^n(\text{constant dollars}). \tag{5.5}$$

Consider the conversion of one 1985 actual dollar to 1985 constant dollars with a base year of 1967. Using Table 5.1, the CPI index gives

$$\text{constant dollars}_{1985} = \frac{100}{322.2}(\$1) = \frac{1}{3.222}(\$1) = \$0.310.$$

The constant-dollar value of \$0.310 can be verified by reference to Table 5.2, which gives the constant-dollar values for a number of years including 1985. In this example, we note the relationship $3.222 = (1 + \bar{f})^{18}$, where $\bar{f}$ is the geometric average of the inflation rate over the 18 years from 1967 to 1985. In actuality there are different inflation rates for each year, and their product over the 18 years gives

$$(1 + f_{1968})(1 + f_{1969})(1 + f_{1970}) \cdots (1 + f_{1985}) = (1 + \bar{f})^{18}.$$

If a \$5,000, 8% bond has 5 years remaining until maturity, its cash flow in actual dollars will be as shown in Table 5.3. The constant-dollar cash flow based on the purchasing power of money at the present ($t = 0$) is shown in the last column. This cash flow assumes an inflation rate of 10% per year over the next 5 years.

By examining the constant-dollar cash flow, the owner of the bond observes what the bond is providing in terms of today's purchasing power. Five years from the present the \$5,400 will purchase only what \$3,397 will purchase at present. That is, 5 years from the present a dollar received is worth only \$0.629 of its present-day purchasing power. In this example the future cash flow in actual dollars is known with certainty, since the covenant of the bond specifies the payments remaining in terms of actual dollars.

For other activities, where the level of activity remains the same over time, it may be easier to determine costs by estimating in terms of constant dollars. If

TABLE 5.3.
CONVERSION OF ACTUAL-DOLLAR CASH FLOWS TO CONSTANT-DOLLAR CASH FLOWS

Time	Cash Flow (actual dollars)	Conversion Factor	Cash Flow (constant dollars)
1	\$ 400	$\frac{1}{(1.10)^1} = \overset{P/F,10,1}{(0.9091)}$	\$ 364
2	400	$\frac{1}{(1.10)^2} = \overset{P/F,10,2}{(0.8265)}$	331
3	400	$\frac{1}{(1.10)^3} = \overset{P/F,10,3}{(0.7513)}$	301
4	400	$\frac{1}{(1.10)^4} = \overset{P/F,10,4}{(0.6330)}$	273
5	5,400	$\frac{1}{(1.10)^5} = \overset{P/F,10,5}{(0.6290)}$	3,397

an engine is to be utilized the same number of hours per year, it is reasonable to expect that the same amount of fuel will be consumed per year. Thus, this years's fuel costs will be identical to next year's fuel cost, and so on. If it is necessary to convert these constant-dollar costs to actual dollars, then Equation (5.5) can be applied.

5.3.3 Relationship Among i, i', and f

It is often desirable to compute equivalents in either the actual-dollar or the constant-dollar domain, making it important to understand the relationships between these domains. Figure 5.4 presents a single cash receipt at a point in time n years from the base year, the present. This cash receipt is shown as F in the actual-dollar domain and as F' in the constant-dollar domain.

If the expected inflation rate is f per year, it has been shown in Equation (5.4) that the factor $1/(1 + f)^n$ converts the cash flow F at n in the actual-dollar domain to a cash flow F' at n the constant-dollar domain. The factor $(1 + f)^n$ reverses this process. If the base year had not been selected at the present (say it occured 2 years prior to the present), the factor to convert actual dollars to constant dollars n years from the present would be $1/(1 + f)^{n+2}$.

The inflation rate f is required to transform dollars from one domain to the same point in time in the other domain.

To transform dollars to their equivalences at *different* points in time within the actual-dollar domain, the market interest rate (i) is used. This process is represented in Figure 5.4.

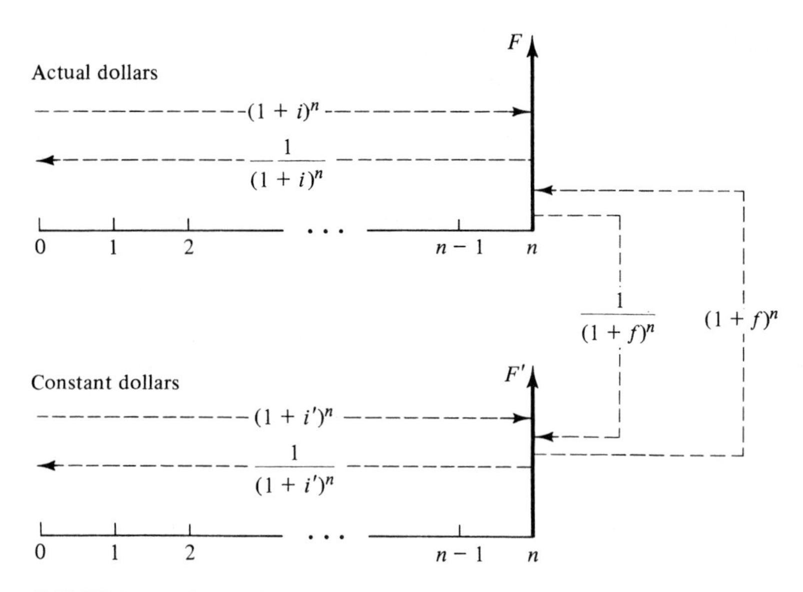

FIGURE 5.4. Relationships among i, i', and f.

In the actual-dollar domain the market interest rate, *i*, is used to calculate equivalences.

As shown in Figure 5.4., the inflation-free rate is the basis for calculating equivalences in the constant-dollar domain. The factor $1/(1 + i')^n$ finds the constant-dollar equivalence at $t = 0$ of the constant cash flow at $t = n$. Since the inflationary effects have been removed from the cash flows in the constant-dollar domain, earning-power-of-money calculations should apply an interest rate that is free of inflationary effects.

When computing equivalences in the constant-dollar domain, *i'*, the inflation-free rate must be applied.

Derivation of the relationship among i, i', and f follows when it is observed that, if the constant-dollar base year is time zero, then at time zero in Figure 5.4 actual dollars and constant dollars have identical purchasing power—that is, actual dollars at the base year will purchase the same goods or services as constant dollars. Only at the base year does this one-for-one conversion between actual and constant dollars occur.

If analysis in either the actual-dollar or the constant-dollar domain is to be consistent, the equivalent amount at the base year in either domain must be equal. Starting at $t = n$ with F in the actual-dollar domain, computing its equivalence at $t = 0$ can be accomplished in two ways. The first approach utilizes actual dollars and converts them to their equivalence at $t = 0$, the assumed base year.

$$P = F\frac{1}{(1+i)^n}$$

The second approach converts the actual dollars to constant dollars and then finds the equivalence of that constant-dollar amount at $t = 0$:

$$F' = F\frac{1}{(1+f)^n}$$

$$P = F'\frac{1}{(1+i')^n} = F\frac{1}{(1+f)^n}\frac{1}{(1+i')^n}$$

Since the P values must be equal at the base year, equating the results of the two methods of computing equivalence gives

$$F\frac{1}{(1+i)^n} = F\frac{1}{(1+f)^n}\frac{1}{(1+i')^n}$$

$$(1+i)^n = (1+f)^n(1+i')^n$$

$$1+i = (1+f)(1+i')$$

$$i = (1+f)(1+i') - 1. \tag{5.6}$$

Solving for i' yields

$$\boxed{i' = \frac{1+i}{1+f} - 1.} \tag{5.7}$$

As an example, suppose the inflation rate is 10% per year while the market interest rate is known to be 15% per year. Finding i' gives

$$i' = \frac{1.15}{1.10} - 1 = 4.55\%.$$

If a single receipt of $100 is received 12 years from the present, what is the equivalent amount of this receipt at $t = 0$?

$$P = \$100(\overset{P/F,15,12}{0.1869}) = \$18.69$$

or

$$P = \$100(\overset{P/F,10,12}{0.3186})(\overset{P/F,4.55,12}{0.5866}) = \$18.69.$$

The approach selected will usually depend on whether the result is to be presented in actual or constant dollars, whether the cash flow estimates are in actual or constant dollars, and on the ease of executing the calculations.

5.4 ANALYZING INFLATION IN INVESTMENTS

Economy studies can be done in either actual dollars or constant dollars. Although some flexibility in the method of analysis is possible, the basic principles developed in Section 5.3 must be followed. To summarize, the interest rate that is applicable depends on the domain in which the calculations are taking place.

Cash Flow Domain	*Interest Rate to Apply to Find Equivalences at Different Points in Time*
Actual dollars	Market interest rate, i
Convert between domains with inflation rate, f	
Constant dollars	Inflation-free interest rate, i'

5.4.1 Actual-Dollar vs. Constant-Dollar Analysis

As an example of actual vs. constant-dollar analysis, suppose that, if held 4 more years to maturity, a bond would have the cash flow in actual dollars shown in Figure 5.5. If the annual market interest rate is 11% and the future inflation rate is projected at 16% per year, we are to determine the present amount that is equivalent to the actual-dollar series in Figure 5.5. This example is first solved in the actual-dollar domain and then in the constant-dollar domain to show the consistency of the two methods.

Using actual-dollar analysis with $i = 11\%$,

$$P = \$1{,}500(\overset{P/A,11,4}{3.1024}) + \$10{,}000(\overset{P/F,11,4}{0.6587}) = \$11{,}241.$$

The equivalent annual series in actual dollars is computed as follows:

$$A = \$11{,}241(\overset{A/P,11,4}{0.3223}) = \$3{,}623 \text{ per year.}$$

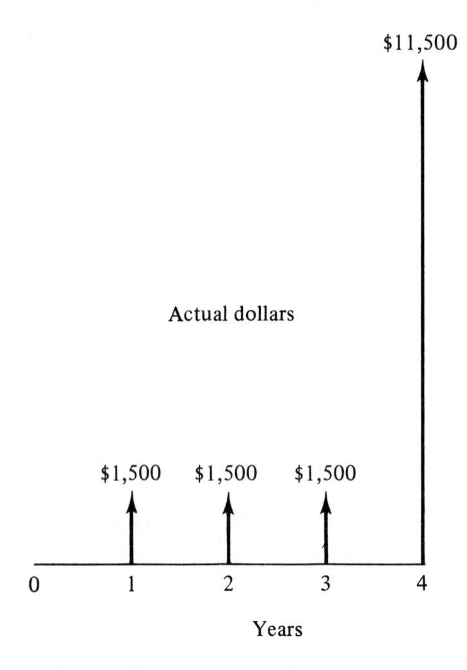

FIGURE 5.5. Remaining actual payments for a $10,000, 15% bond.

Using constant-dollar analysis (base year $t = 0$) requires the conversion of the cash flows in Figure 5.5 into constant dollars with $f = 16\%$ and gives

$$F_1' = \$1{,}500\overset{P/F,16,1}{(0.8621)} = \$1{,}293$$

$$F_2' = \$1{,}500\overset{P/F,16,2}{(0.7432)} = \$1{,}115$$

$$F_3' = \$1{,}500\overset{P/F,16,3}{(0.6407)} = \$961$$

$$F_4' = \$11{,}500\overset{P/F,16,4}{(0.5523)} = \$6{,}351.$$

The resulting constant-dollar cash flow is illustrated in Figure 5.6.

Constant-dollar analysis requires use of the inflation-free rate i', which is calculated from Equation (5.7) as

$$i' = \frac{1.11}{1.16} - 1 = 0.95690 - 1.0 = -0.0431 \text{ or } -4.31\%.$$

In this case i' is negative, since the inflation rate exceeds the market rate of interest, an unusual but possible occurrence. The factor values are determined from the algebraic form of the factors by substituting -0.0431 for the interest rate. An example of this substitution gives

$$\left(\overset{P/F,-4.31,4}{\quad}\right) = [1 + (-0.0431)]^{-4} = 1.1927.$$

Note that the *P/F* factor yields a value greater than 1 for a negative interest rate.

The present equivalent is calculated as follows:

$$P = \$1{,}293(\overset{P/F,-4.31,1}{1.0450}) + \$1{,}115(\overset{P/F,-4.31,2}{1.0921}) + \$961(\overset{P/F,-4.31,3}{1.1413})$$

$$+ \$6{,}351(\overset{P/F,-4.31,4}{1.1927}) = \$11{,}241.$$

This present equivalent amount is identical to the corresponding equivalent amount calculated using actual dollars. Thus, the consistency between the two methods is shown.

In this last example constant-dollar analysis is more time consuming than actual-dollar analysis. However, additional insight is gained by the conversion of actual dollars to constant dollars, as shown in Figure 5.6. With a 16% inflation rate the receipt of $11,500 four years from now will provide the investor with only $6,351 worth of present-day purchasing power.

The equal-payment series in constant dollars equivalent to the cash flow in Figure 5.6 is

$$A' = \$11{,}241(\overset{A/P,-4.31,4}{0.2237}) = \$2{,}515 \text{ per year.}$$

When comparing the equal-annual series in actual dollars to the corresponding series in constant dollars, the effect of inflation on the entire investment is evident. Because of the loss of purchasing power the investment will provide only $2,515 per year in present-day purchasing power. This value is contrasted to the $3,623 per year in actual dollars that would be realized on an equivalent basis from this bond.

Consider a 30-year-old woman preparing for her retirement at age 65. She estimates that she can live comfortably on $20,000 per year in terms of present-day dollars. It is estimated that the future rate of inflation will be 8% per year and that she can invest her savings at 12% compounded annually. What equal amount must this woman save each year until she retires so that she can make

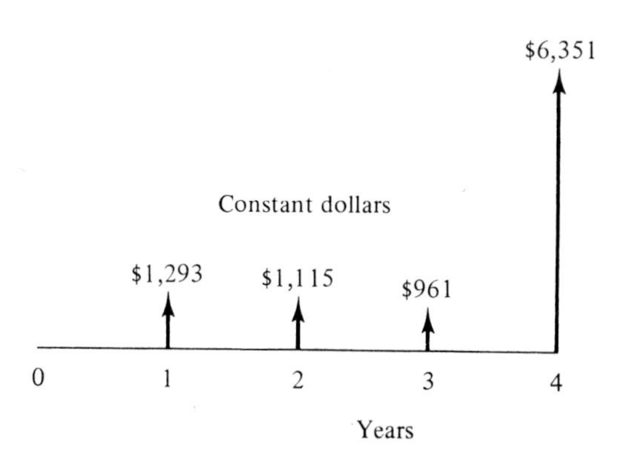

FIGURE 5.6. Remaining constant-dollar receipts for $10,000, 15% bond.

withdrawals that will allow her to live comfortably for 5 years beyond her retirement?

Using actual-dollar analysis, we first find the amounts in actual dollars that would be required at ages 66 through 70 to support her present life style. These calculations are presented in Table 5.4.

If the end-of-the-year convention is used, it is observed that at age 70 the woman requires $434,500 to purchase the same goods that she could buy at age 30 for $20,000. This difference represents a serious loss in purchasing power, and it becomes even more serious at higher rates of inflation.

The cash flow in Figure 5.7 reflects, *A*, the annual amount to be deposited (in actual dollars), and the amounts to be withdrawn after retirement. Because money has earning power, the annual amounts that must be saved so that they provide the withdrawals needed are computed.

To find the value *A* that must be saved each year, it is necessary to find the savings cash flow that is equivalent to the withdrawal cash flow. Since two equivalent cash flows can be equated at any point in time, the end of year 35 is selected for convenience. Since this is an actual-dollar analysis, the market interest rate of 12% is applied.

$$\begin{aligned} A\overset{F/A,12,35}{(431.664)} &= \$319{,}360\overset{P/F,12,1}{(0.8929)} + \$344{,}920\overset{P/F,12,2}{(0.7972)} \\ &\quad + \$372{,}500\overset{P/F,12,3}{(0.7118)} + \$402{,}300\overset{P/F,12,4}{(0.6355)} \\ &\quad + \$434{,}500\overset{P/F,12,5}{(0.5674)} \end{aligned}$$

$$A = \$3{,}075 \text{ per year.}$$

The $3,075 represents the actual dollars that would have to be deposited

TABLE 5.4.
FINDING ACTUAL DOLLARS REQUIRED TO MAINTAIN LIVING STANDARD

End of Year	Age	Dollars Required at Year *n* to Provide $20,000 per Year in Present Dollars with Inflation at 8% per Year
36	66	$\$20{,}000\overset{F/P,8,36}{(15.968)} = \$319{,}360$
37	67	$20{,}000\overset{F/P,8,37}{(17.246)} = 344{,}920$
38	68	$20{,}000\overset{F/P,8,38}{(18.625)} = 372{,}500$
39	69	$20{,}000\overset{F/P,8,39}{(20.115)} = 402{,}300$
40	70	$20{,}000\overset{F/P,8,40}{(21.725)} = 434{,}500$

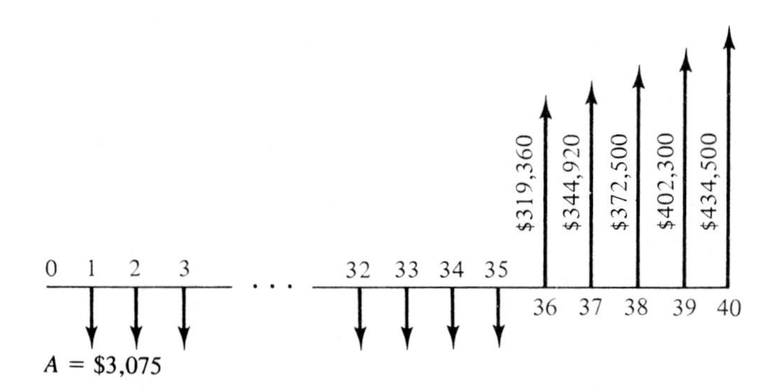

FIGURE 5.7. Savings and withdrawals in terms of actual dollars.

each year. Although the number of dollars deposited remains the same, the individual is giving up less purchasing power with each succeeding deposit. Because of inflation, a dollar in any one period will buy less than that same dollar in any earlier period.

To quantify the purchasing power of the actual dollars deposited, we could compute the constant-dollar amount (base year, the present) for each deposit. For example, the deposit made at the end of the 35th year would have a constant-dollar value of

$$F' = \$3{,}075\overset{P/F,8,35}{(0.0676)} = \$208.$$

Now suppose one desired to calculate the equal annual constant-dollar amounts that would be equivalent to the equal annual actual-dollar deposits. This calculation requires a constant-dollar analysis, but there are a variety of approaches that will yield the correct result. One approach is to convert each actual-dollar deposit to its constant-dollar equivalent on a year-by-year basis. Then convert these 35 constant-dollar payments to an equivalent equal annual series over 35 years, using the inflation-free rate. Unfortunately, since each of these constant-dollar payments is different in value, this approach is quite time consuming.

Another approach is to calculate the present equivalent of the actual-dollar series using the market rate of interest:

$$P = \$3{,}075\overset{P/A,12,35}{(\ 0.0189\ )} = \$25{,}140.$$

Since the $25,140 occurs at the base year, it can be considered either an actual-dollar or a constant-dollar amount based on $1/(1+f)^n = 1$ when $n = 0$ in Equation (5.4). Using a constant-dollar analysis the inflation-free rate is required. For this example it is

$$i' = \frac{1.12}{1.08} - 1 = 0.037 \text{ or } 3.7\%.$$

The equal annual constant-dollar series over 35 years that is equivalent to $25,140 at present is

$$A' = \$25{,}140\overset{A/P,3.7,35}{(\;0.0514\;)} = \$1{,}292 \text{ per year.}$$

An identical result can be achieved by realizing that *any* cash flow series in one domain can be converted to its equivalent cash flow in the other domain no matter what point in time is selected for the conversion. Finding a single-payment equivalent to the series in one domain, converting it to a single payment in the other domain, and then converting it to the desired series yields cash flows in both domains that are equivalent. An example of this type of calculation with conversion at $t = 35$ follows.

$$A' = \$3{,}075\overset{F/A,12,35}{(431.664)}\overset{P/F,8,35}{(0.0676)}\overset{A/F,3.7,35}{(0.01436)} = \$1{,}292.$$

Using a constant-dollar analysis shows that making the actual deposits of $3,075 over 35 years is equivalent to foregoing only $1,292 each year of present-day purchasing power for the next 35 years. This is a more realistic assessment of the level of sacrifice being made during the earning years so that the woman can enjoy her retirement years.

Although numerous public agencies and corporations utilize constant-dollar analysis as their primary approach, actual-dollar analysis is more common in engineering economy studies. The actual-dollar method is favored because it is more intuitive and therefore more easily understood. Also, it requires the use of an interest rate that is commonly available in the marketplace and therefore more easily determined. In contrast, the inflation-free interest rate is an abstract rate that usually must be computed, since it is not directly available from conventional financial sources. For the foregoing reasons, the presentation of results or discussion of an analysis is usually facilitated if the cash flows are stated in terms of actual dollars. In this text, unless specified otherwise, all cash flows are assumed to be in actual dollars and the interest rates are in terms of the market rate.

5.4.2 Geometric-Gradient-Series Applications

The geometric-gradient factor of Equation (3.12) is applicable when a dollar amount, F_1, (beginning at $t = 1$) is increasing at a known rate. This increase may be due to inflationary effects.

Consider a firm that has fuel costs of $20,000 this year. It expects these costs to increase at a rate of 9% per year: 7% because of inflation and 1.869% [(1.01869)(1.07) = (1.09)] because of an increase in the amount of fuel consumed. If the interest rate available for investment is 12%, what is the present single-payment equivalent of the next 8 years of fuel costs? This equivalent cost is found using the geometric gradient, where the first-year cost of $20,000 is increasing at the annual rate of 9%. With $g' = (1.12)/(1.09) - 1 = 0.0275$, the present equivalent cost is

$$P = \frac{\$20{,}000\overset{P/A,2.75,8}{(\;7.0943\;)}}{1.09} = \$130{,}170.$$

Suppose the firm wants a single amount at $t = 8$ equivalent to these 8 years of fuel costs. This amount in actual dollars is:

$$F = \$130{,}170\overset{F/P,12,8}{(\;2.476\;)} = \$322{,}300.$$

If the equivalent amount just determined is needed in terms of constant dollars (base $t = 0$), convert from actual dollars at $t = 8$ to constant dollars at $t = 8$, using the inflation rate of 7%:

$$F' = \$322{,}300\overset{P/F,7,8}{(0.5820)} = \$187{,}579.$$

The geometric gradient is also usable for constant-dollar cash flows. Assume that a geometrically growing *constant*-dollar series is increasing at the rate of 5% per year over 15 years. If $i = 14\%$ and $f = 6\%$ over the time span of the series and F_1 (in constant dollars) is \$3,000, what is the present equivalent of this series? Recall that Equation (3.13) is based on i being the interest rate at which money is moved through time, and g the rate of growth of the series. Since the geometric series in this example is in constant dollars, it is required that these constant dollars be moved through time at i', the inflation-free rate. Thus, i must be replaced by i' when finding g' from Equation (3.13).

To work the example, first find i' from Equation (5.7)

$$i' = \frac{(1 + i)}{(1 + f)} - 1 = \frac{(1.14)}{(1.06)} - 1 = 0.0755.$$

then from Equation (3.13) find

$$g' = \frac{(1.0755)}{(1.05)} - 1 = 0.0243$$

yielding

$$P = \frac{\$3{,}000\overset{P/A,2.43,15}{(\;12.4492\;)}}{1.05} = \$35{,}569.$$

5.4.3 Different Inflation Rates for Cash Flow Components

When an engineering economy study is considering inflation, different elements contributing to the overall cash flow are commonly found to inflate at different rates. Certain material costs may be experiencing price changes at one rate while other material costs are changing at a different rate. In addition, labor and overhead costs will more than likely be increasing at their own particular rates.

To assure that these different rates are properly accounted for in the analysis, determine the component cash flows in actual dollars and combine them to find the overall actual dollar cash flow.

Combining the *constant-dollar* cash flows for each component *will not* provide the proper result. Although there is a method for handling the analysis in terms of constant dollars, it is generally more difficult to make the proper calculation. Therefore, it is not described here.

A company is considering acquisition of a machine that produces a single product. This product requires Materials X and Y, and there is a component of direct labor charged for each item produced. The production process is expected to continue for 4 years, at which time the machine will be sold. Table 5.5 presents the constant-dollar cash flow for each component cash flow.

It is expected that the salvage value of the machine will inflate at 3% per year, while sales income is expected to experience inflation of 5% per year. Also, it is estimated that wages will be increasing at 8% per year, while the cost of Material X is expected to rise at 6% per year. Material Y is supplied on the basis of a long-term contract, and its cost will remain the same over the next 4 years. The actual-dollar cash flows for each of these components are presented in Table 5.6 along with the total actual-dollar cash flow associated with this project. For a market rate of 20%, the present equivalent amount for this project is

$$P = -\$100{,}000 + \$35{,}520\overset{P/F,20,1}{(0.8333)} + \$37{,}477\overset{P/F,20,2}{(0.6945)}$$

$$+ \$39{,}335\overset{P/F,20,3}{(0.5787)} + \$63{,}787\overset{P/F,20,4}{(0.4823)} = \$9{,}154.$$

TABLE 5.5.
COMPONENT CONSTANT-DOLLAR CASH FLOWS

End of Year	Annual Cash Flows (Constant dollars)				
	Capital Costs	Sales Income	Labor Costs	Material X Costs	Material Y Costs
0	−$100,000				
1		$40,000	−$3,000	−$1,000	−$2,000
2		40,000	− 3,000	− 1,000	− 2,000
3		40,000	− 3,000	− 1,000	− 2,000
4	20,000	40,000	− 3,000	− 1,000	− 2,000

TABLE 5.6.
COMPONENT ACTUAL-DOLLAR CASH FLOWS

	Annual Cash Flows (Actual dollars)					
End of Year	Capital Costs (3%)	Sales Income (5%)	Labor Costs (8%)	Material X Costs (6%)	Material Y Costs (0%)	Total Cash Flows
0	−\$100,000					−\$100,000
1		\$42,000	−\$3,240	−\$1,060	−\$2,000	35,520
2		44,100	− 3,499	− 1,124	− 2,000	37,477
3		46,305	− 3,779	− 1,191	− 2,000	39,335
4	22,510	48,620	− 4,081	− 1,262	− 2,000	63,787

The equivalent receipts exceed the equivalent disbursements on an actual-dollar basis. Therefore, the project appears to be economically sound if the estimates are realized.

5.4.4 Different Inflation Rates for Each Time Period

Analysis of the past performance of investments will usually require consideration of different inflation rates for each year. Assume that a past investment had receipts and disbursements as shown in Table 5.7 and that the rate of interest earned on the investment in terms of purchasing power at the end of 1975 is to be calculated.

The cash flow in constant dollars in Table 5.7 is developed by using each year's inflation rate in Table 5.1. Then the actual-dollar cash flow for each year is converted to constant dollars. For example, the conversion of the actual-dollar cash flow in 1978 to the constant-dollar amount in the same year is given by

$$F'_{1978} = \$411\overset{P/F,5.8,1}{(0.9452)}\overset{P/F,6.5,1}{(0.9390)}\overset{P/F,7.7,1}{(0.9295)} = \$339.$$

To find the rate of interest earned on an inflation-free basis, solve for i' from the expression

$$\$1{,}000 = \$388\overset{P/F,i',1}{(\quad)} + \$365\overset{P/F,i',2}{(\quad)} + \$339\overset{P/F,i',3}{(\quad)} + \$305\overset{P/F,i',4}{(\quad)} + \$268\overset{P/F,i',5}{(\quad)}.$$

Solving for i' using trial and error yields $i' = 21\%$. This rate can be compared to the market interest rate i that sets the equivalent receipts equal to the equivalent disbursements for the actual-dollar cash flow. The market rate is found as follows:

$$\$1{,}000 = \$411\overset{P/A,i,5}{(\quad)}$$

$$i = 30\%.$$

TABLE 5.7.
PAST INVESTMENT WITH MULTIPLE INFLATION RATES

End of Year	Cash Flow (actual dollars)	Inflation Rate for Year	Cash Flow (constant dollars)
1975	−$1,000		−$1,000
1976	411	5.8%	388
1977	411	6.5%	365
1978	411	7.7%	339
1979	411	11.3%	305
1980	411	13.5%	268

For the varying inflation rates shown in Table 5.7, this investment yields approximately two-thirds of the return in 1975 purchasing power that was realized from the actual dollars received. Inflationary effects usually cause an overstatement of earnings, and it is important that such overstatements be properly recognized.

When preparing engineering economy studies utilizing future estimates of the annual inflation rate, the most common approach is to estimate an average inflation rate over some specified time span. The aim is to gain insight with regard to the overall effect of inflation without having to contend with difficulties of estimating inflation rates for each individual year. Otherwise, instead of a single average inflation rate, $\bar{f}$, the inflation rate for each year, f_t, would have to be specified. Then the calculation to convert from constant dollars to actual dollars at $t = n$ would be expressed as

$$\text{actual dollars} = [(1 + f_1)(1 + f_2) \cdots (1 + f_{n-1})(1 + f_n)](\text{constant dollars}).$$

The effect of utilizing an average inflation rate rather than the individual yearly rates is revealed by the example in Table 5.7. The average annual inflation rate is found from the expression

$$(1 + \bar{f})^5 = (1.058)(1.065)(1.077)(1.113)(1.135)$$

$$\bar{f} = 8.92\%.$$

The constant-dollar cash flows based on the average inflation rate and the individual inflation rates are presented in Table 5.8. The percent differences indicate the problem of substituting an average inflation rate for the individual rate. In this particular example, by comparison with the use of individual inflation rates, the use of the average inflation rate understates (negative percentage) the constant-dollar value of cash flow. Thus, the possible inaccuracies of using an average inflation rate are evident. Unfortunately, in most analyses of future cash flows the difficulty of accurately estimating individual yearly inflation rates far

TABLE 5.8.
CONSTANT-DOLLAR CASH FLOWS FOR AVERAGE AND INDIVIDUAL INFLATION RATES

	Constant-Dollar Cash Flows		
Year	Individual Inflation Rates	Average Inflation Rate	Percent Difference
0	−$1,000	−$1,000	
1	388	377	−3.1%
2	365	347	−4.9%
3	339	318	−6.2%
4	305	292	−4.3%
5	268	268	0.0%

outweighs the value of any increased accuracy that results. For this reason, inflation rates specified for each year are usually limited to analyses based on cash flows that have occurred in the past.

5.4.5. Considering Deflation

When a currency will buy more goods or services than it has previously, the currency has experienced deflation. In this situation the value of the currency is increasing and prices are decreasing. A decrease in prices may be expressed as a negative inflation rate. Using the data in Table 5.2, observe that in 1949 the purchasing power of the dollar actually increased, so that the inflation rate for that year was

$$1 + \text{inflation rate (1949)} = \frac{1.387}{1.410} = 0.99$$

$$\text{inflation rate (1949)} = 0.99 - 1 = -0.01 \text{ or } -1\%$$

$$\text{deflation rate (1949)} = 1\%.$$

When converting actual dollars to constant dollars, Equation (5.4) developed for inflation also applies for deflation, but the rate of increase in prices is negative. If the rate of deflation over 2 years averages 4%, the constant-dollar equivalent 2 years hence of $100 in actual dollars is expressed as

$$\$100 \text{ (actual dollars)} \frac{1}{(1 - 0.04)^2} = \$108.51 \text{ (constant dollars)}.$$

For these circumstances the purchasing power of the $100 has increased in terms of what could have been purchased 2 years previously. Calculations similar to those made for inflation can be made for conditions of deflation by simply substituting a negative rate in the relationship.

5.4.6 Currency Exchange

Another means by which the purchasing power of a currency can increase or decrease is in the conversion from one country's currency to another's. Every day the rate of exchange of currencies fluctuates on the world currency exchange markets. These fluctuations reflect a myriad of economic conditions and in some instances can be quite drastic. Devaluations of more than 50% in a single day have been experienced in recent times. Where international firms are doing business in a number of countries, variations in the exchange rate can significantly affect the ultimate profitability of an investment in terms of the firm's base currency.

Table 5.9 gives a hypothetical listing of historical exchange rates between two currencies represented by a dollar and a kron. The exchange rate gives the amount of the other country's currency that one unit of the currency listed would purchase. (For example, in 1983, 1 dollar would purchase 5.0 krons, while 1 kron would purchase 0.20 dollars.)

Suppose an international firm has completed a project in the country where the kron is the official currency. The cash flow for the project is given in krons in Table 5.10. This table also presents the actual-dollar equivalent cash flow based on the exchange rates given in Table 5.9—that is, the value in dollars that would be received if the receipts were converted from krons to dollars at the time the payment occurred. The last column in Table 5.10 represents the original investment's equivalent receipts in terms of the purchasing power of the dollar at the end of 1980.

To demonstrate the conversion from actual krons to actual dollars and then to constant dollars, the year 1986 is used as an example. First, the actual krons are converted to actual dollars using the exchange rate provided in Table 5.9. This conversion for 1986 gives

TABLE 5.9.
HYPOTHETICAL EXCHANGE RATES

Date	Krons/Dollar	Dollars/Kron
1980	4.0	0.250
1981	4.5	0.222
1982	6.0	0.167
1983	5.0	0.200
1984	6.3	0.159
1985	6.1	0.164
1986	14.0	0.071
1987	13.9	0.072

TABLE 5.10.
CASH FLOW IN KRONS, ACTUAL DOLLARS, AND CONSTANT DOLLARS

Date	Kron (actual)	Dollars (actual)	Dollars (constant, base 1980)
1980	−$80,000	−$20,000	−$20,000
1981	20,000	4,440	4,023
1982	20,000	3,340	2,851
1983	20,000	4,000	3,308
1984	20,000	3,180	2,552
1985	20,000	3,280	2,512
1986	20,000	1,420	1,067
1987	20,000	1,440	1,036

$$\$1{,}420 = (20{,}000 \text{ krons})(0.071 \text{ dollars/kron}).$$

The calculation of the constant-dollar equivalent of $1,420 in 1986 where the base year is 1980 requires the data presented in Table 5.1. The result is found from the expression

$$\$1{,}067 = \$1{,}420\left(\frac{246.8}{328.3}\right).$$

By calculating the interest rate that sets the equivalent receipts equal to the equivalent disbursements, the interest rate per year being earned from the investment is found. In terms of krons, the investment earns an interest rate of i that satisfies the following expression:

$$80{,}000 \text{ krons} = (20{,}000 \text{ krons})(\overset{P/A,i,7}{\quad\quad})$$

$$i = 16.3\% \text{ (based on the kron cash flow).}$$

After converting krons to actual dollars, find the interest being earned by the following computation:

$$\begin{aligned}\$20{,}000 = {} & \$4{,}400(\overset{P/F,i,1}{\quad\quad}) + \$3{,}340(\overset{P/F,i,2}{\quad\quad}) + \$4{,}000(\overset{P/F,i,3}{\quad\quad}) \\ & + \$3{,}180(\overset{P/F,i,4}{\quad\quad}) + \$3{,}280(\overset{P/F,i,5}{\quad\quad}) + \$1{,}420(\overset{P/F,i,6}{\quad\quad}) \\ & + \$1{,}440(\overset{P/F,i,7}{\quad\quad})\end{aligned}$$

$$i = 1.6\% \text{ (based on the actual-dollar cash flow).}$$

Because there has been a loss in purchasing power of the kron compared to the dollar over the life of the investment, the conversion of krons to dollars has reduced the return realized. Computing the interest rate earned on the investment when the effects of inflation on the dollar are removed requires the use of the constant-dollar cash flow in Table 5.10. The rate of interest being earned in terms of 1980 purchasing power is found by solving for the interest rate that equates the equivalent disbursements with the equivalent receipts. Based on the constant-dollar cash flow, the expression yields

$$\$20{,}000 = \$4{,}023(\overset{P/F,i,1}{\quad}) + \$2{,}851(\overset{P/F,i,2}{\quad}) + \$3{,}308(\overset{P/F,i,3}{\quad})$$
$$+ \$2{,}552(\overset{P/F,i,4}{\quad}) + \$2{,}512(\overset{P/F,i,5}{\quad}) + \$1{,}067(\overset{P/F,i,6}{\quad})$$
$$+ \$1{,}036(\overset{P/F,i,7}{\quad})$$

$i = -4.2\%$ (based on the constant-dollar cash flow).

This is a situation where an investment in another country appears profitable in terms of its official currency (krons). However, if the investment's earnings are converted to another currency (dollars) at the prevailing exchange rate, the loss of purchasing power due to the comparative weakness of the kron becomes evident: the rate of yield is reduced from 16% to less than 2%. When the loss of purchasing power in the dollar due to inflation is considered, the investment actually earns a return that is negative. In other words, more purchasing power was invested in the project than the project ever returned.

This example illustrates that the flow of cash associated with an investment experiences a change in purchasing power due to two factors. First, when there is a conversion between different countries' currencies, the exchange rates determine the loss or gain in purchasing power. Second, owing to changes in prices in a given country's currency, there is also a change in purchasing power. *Since the objective is to make investments that ultimately increase purchasing power, those factors that affect real purchasing power must be accounted for in engineering economy studies.*

KEY POINTS

- Money has a time value that is significantly affected by both the earning power of money and the purchasing power of money.
- Historical change in prices for goods and services is represented by price indexes such as the Consumer Price Index or the Producer Price Index.
- Annual inflation rates or average annual inflation rates can be computed directly from the various price indexes.

- Definitions of i, i', and f are presented along with definitions of actual and constant dollars.
- Converting actual (constant) dollars to constant (actual) dollars at the same point in time requires the use of the inflation rate, f.
- Converting actual dollars from one point in time to another (working in the actual-dollar domain) requires the market rate of interest, i, while converting constant dollars to different points in time (working in the constant-dollar domain) requires the inflation-free interest rate i'.
- Actual-dollar analysis or constant-dollar analysis will result in the same conclusion about the preference for investment alternatives.

PROBLEMS

1. Given past historical trends of the inflation rate based on the CPI and your understanding of the present economic environment, estimate the annual inflation rate for the year after this current year.

2. Using Table 5.2, indicate those years for which there was a decrease in prices (deflation) given that the change in prices is based on:

a. Producer prices.

b. Consumer prices.

3. Calculate the annual rate of inflation from the Consumer Price Index for the following years.

a. 1987. Answer: 3.69%

b. 1990.

c. 1966.

d. 1974.

e. 1980.

4. Calculate the average annual rate of inflation based on the CPI from the end of:

a. 1975 through 1987. Answer: 6.43%

b. 1966 through 1980.

c. 1983 through 1991.

d. 1967 through 1990.

e. 1971 throgh 1989.

5. Utilizing the data in Table 5.2, which express the purchasing power of the dollar in various years, compute the inflation rate based on consumer prices for the following years.

a. 1983. Answer: 3.28%

b. 1990.

c. 1979.

d. 1968.

e. 1955.

6. For the following time periods find the average annual inflation rate from data in Table 5.2 that show the purchasing power of the dollar in terms of consumer prices.

a. End of 1973 through 1987. Answer: 6.94%

b. End of 1958 through 1979.

c. End of 1978 through 1985.

d. End of 1961 through 1991.

e. End of 1946 through 1986.

7. For the time spans listed, find the average annual inflation rate from the producer price data given in Table 5.2.

a. End of 1962 through 1987.

b. End of 1980 through 1990.

c. End of 1973 through 1980.

d. End of 1967 through 1984.

e. End of 1945 through 1982. Answer: 4.53%

8. Using the CPI figures in Table 5.1, calculate the constant-dollar value (in 1967 dollars) of the actual dollars received at the time indicated.

a. $800 at the end of 1990.

b. $4,000 at the end of 1977.

c. $4,000 at the end of 1985.

d. $20,000 at the end of 1981.

e. $6,000 at the end of 1980. Answer: $2,431

9. A person desires to receive an amount in actual dollars n years from the present that has the purchasing power at time n that $10,000 has at present. If the annual inflation rate is f, find this actual-dollar amount.

a. $n = 9, f = 7\%$. Answer: $18,385

b. $n = 50, f = 6\%$.

c. $n = 17, f = 11\%$.

d. $n = 20, f = 9\%$.

e. $n = 42, f = 3\%$.

10. Find the constant-dollar equivalent of a $3,000 payment n years from the present, if the annual inflation rate is, f. The constant-dollar base year is the present.

a. $n = 12, f = 6\%$.

b. $n = 10, f = 14\%$.

c. $n = 15, f = 10\%$.

d. $n = 50, f = 5\%$.

e. $n = 28, f = 3\%$. Answer: \$1,311

11. End-of-year payments of \$700 are to be received from an investment over the next 5 years. The annual inflation rate is or was 9%. Convert each of these actual-dollar payments to its constant-dollar equivalent if the constant-dollar base year is

a. $t = 0$ (the present).

b. $t = -3$ (3 years prior to the present).

c. $t = 2$ (2 years after the present).

12. An individual is scheduled to receive at \$40,000 distribution from a trust fund 8 years from the present. The inflation rate is expected to average 6% per year over that time. Find the constant-dollar equivalent to this payment if the constant-dollar base is

a. $t = 0$ (the present).

b. $t = 4$ (4 years after the present).

c. $t = -7$ (7 years prior to the present). Answer: \$16,691

13. A person owns a \$5,000, 12% bond. With payments made semiannually and 15 years until maturity, find the constant-dollar equivalent (base year, $t = 0$) for the following bond payments. The future inflation rate is expected to be 14% per year compounded semiannually.

a. 8th payment. Answer: \$174.60

b. 4th payment.

c. 20th payment.

d. 27th payment.

e. Last interest payment.

14. Solve Problem 13 for an inflation rate of 6% per year compounded semiannually.

15. The purchase of a home requires a \$100,000 loan, which is repaid in equal monthly payments over 30 years. If the inflation rate is $\frac{1}{2}\%$ per month and the loan rate is 12% per year compounded monthly, find the constant-dollar equivalent of the following payments. List the actual-dollar value and the constant-dollar value and assume that the constant-dollar base year is the present.

a. 213th payment. Answer: \$1,029; \$355.67

b. 240th payment.

c. 12th payment.

d. 96th payment.

e. Last payment.

16. Solve Problem 15 if the annual inflation rate is 1% per month and the loan rate is 9% per year compounded monthly.

17. The operating cost (primarily from consumption of electricity) of a refrigeration storage unit was $14,000 last year. Since the unit operates continuously, its power consumption is expected to remain the same in the future. If the cost of electrical power is predicted to increase at the rate of 8% annually, find the actual-dollar cash flow representing the operating costs of this unit over the next 5 years.

18. You presently have P dollars to purchase an asset that costs exactly P at the present and will increase at the inflation rate. However, you may invest P at an interest rate of i per year and postpone your purchase to some future date. Because the rate of inflation is f per year, you wish to calculate the advantage or disadvantage of postponing your decision. Find the rate per year at which postponing is beneficial (or not beneficial). That is, find x% in terms of i and f that indicates the annual rate at which the portion of the asset you can purchase increases (or decreases).

19. The rates of interest available in the marketplace for various years are shown below. Find the inflation-free rate for each of these years based on the inflation rates in Table 5.1.

a. 1981, i = 17%.

b. 1967, i = 8%.

c. 1975, i = 10%.

d. 1986, i = 7%. Answer: 5.00%

e. 1980, i = 20%.

20. If the inflation-free rate and the inflation rate for a period are given, find the rate that represents the market rate of interest for that period.

a. i' = 3%, f = 8%. Answer: 11.24%

b. i' = 6%, f = 4%.

c. i' = 10%, f = −3%.

d. i' = 5%, f = 9%.

e. i' = 4%, f = −8%.

21. For year 1, 2, 3, and 4 the inflation rate is predicted to be 5%, 10%, 13%, and 9%, respectively. If an investment is expected to earn annual yields on a constant-dollar basis of 5%, 4%, −2%, and 8%, what market rate of interest is the investment earning for each of these 4 years?

22. Over the next 3 years it is estimated that the annual inflation rate will be 8%. It is expected that the interest that can be earned from investment will be 10%, 9%, and 12% in the first, second, and third year, respectively. Find the inflation-free interest rates for each of these 3 years. Using the inflation-free rates for each year, compute the average annual inflation-free rate.

23. Two years ago the annual inflation rate was 12% and the annual interest rate available for investing was 20%. Last year these rates were 8% and

13%, respectively. Find the inflation-free rates for each of the last 2 years and then find the average inflation-free rate over that 2-year period. If the average market rate and the average inflation rate for the 2 years are computed first, what then is the average inflation-free rate?

24. A series of four end-of-year payments of $8,000 has been promised. If over the next 4 years the market interest rate is 12% per year and the annual inflation rate is 7%, find the present equivalent of this series using

a. Actual-dollar analysis.

b. Constant-dollar analysis.

25. A payment of $20,000 is to be received 10 years hence, followed by a $40,000 payment 17 years from the present. If over this time span the annual inflation rate is 5% while the expected annual market rate is 9%, calculate the present equivalent of these two payments using

a. Actual-dollar analysis. Answer: $17,692

b. Constant-dollar analysis.

26. The operating costs of a small electrical generating unit are expected to remain the same ($150,000 per year) if the effects of inflation are not considered. The best estimates indicate that the annual inflation-free rate of interest will be 4% and the annual inflation rate, 8%. If the generator is to be used 4 more years, what is the present equivalent of its operating costs? Solve using

a. Constant-dollar analysis.

b. Actual-dollar analysis.

27. An annuity provides for 10 consecutive end-of-year payments of $10,000, beginning one year from the present. The inflation rate for the next 10 years is estimated to be 12% compounded annually. If the inflation-free rate is 7% compounded annually, what is this annuity worth in terms of a single equivalent amount of present-day dollars? Calculate using

a. Actual-dollar analysis.

b. Constant-dollar analysis.

28. An individual has been offered two jobs with starting annual salaries of $30,000 and $32,000. It is expected that he will receive actual raises of 6% per year over the next 30 years and inflation will be 5% per year. If the market rate of interest is 15% per year, find the present equivalent that represents the difference between these two offers using

a. Actual-dollar analysis with the geometric-gradient factor. Answer: $20,296

b. Constant-dollar analysis with the geometric-gradient factor.

29. The annual maintenance costs of an electric pump this year are estimated to be $1,800. Since the level of maintenance is expected to remain the same in the future, these costs will be constant, assuming no inflation. If the pump's life is predicted to be 13 years, find the present equivalent of its maintenance costs when the annual inflation rate is 9% and the annual market rate of interest is 12%. Solve using

a. Geometric gradient. Answer: $17,846

b. Constant-dollar analysis.

30. A series of twenty constant-dollar payments beginning with $5,000 at the end of the first year are growing at the rate of 4% per year. If the inflation-free rate is 3% per year and the inflation rate is 6% per year, find the present equivalent of this series of payments using the geometric-gradient factor by

a. Constant-dollar analysis.

b. Actual-dollar analysis.

31. The purchase of a home requires a couple to borrow $110,000 at 12% per year compounded monthly. The loan is to be repaid in equal monthly payments over 30 years. The average monthly inflation rate is expected to be 0.4%.

a. What equal monthly payments in terms of constant dollars over the next 30 years is equivalent to the series of actual payments to be made over the life of the loan?

b. If this were a no-interest loan to be repaid in equal monthly payments over 30 years, what would be the monthly payments in actual dollars?

c. If the inflation rate exceeded the borrowing rate (say, $f = 15\%$ per year compounded monthly), find the answer to (a). Compare this result to that of (b).

32. a. A $10,000, 12% bond maturing in 10 year with interest paid quarterly is offered for sale for $8,021. If the market rate of interest is compounded quarterly, what is the effective rate per quarter earned on this bond? What is the nominal interest rate earned? What is the effective annual rate earned? Answer: 4%; 16%; 16.99%

b. If the inflation rate is 2% per quarter, what equal-payment series for 40 quarters in terms of constant dollars (base year, $t = 0$) is equivalent to the interest payments of $300 per quarter to be received from the bond? What is the constant-dollar value of the par value of the bond received at maturity? What is the present equivalent value of the bond when inflation is considered? Calculate the effective inflation-free interest rate being earned by this bond for the given rate of inflation. Compare these values to those in part (a). Answer: $216; $4,529; $8,021; 1.96%

33. A woman is planning to save a fixed percentage of her salary to buy a new car 5 years from now (at the end of the fifth year). The car presently costs $12,000, and its price is expected to increase at the rate of inflation, which is anticipated to be 6% per year. The first deposit will be made one year from now, and her present salary of $30,000 will be assumed to occur at the end of the present year. Subsequently, she expects her salary to increase at the rate of 8% per year. If she can earn 10% per year on her savings, what fixed percentage of her gross salary must be saved each year? Solve using actual-dollar analysis.

34. A firm has an option to purchase 4 years from now a parcel of land that presently is priced at $250,000. The option states that the price at that time will be adjusted for inflation, assuming inflation at the rate of 5% per year. The firm had net earnings last year of $1,000,000, which are assumed to have occurred at the present. The firm's earnings are expected to increase at the rate of 20% per year, and the rate at which they can be invested is 12% compounded annually. What fixed percentage of net earnings must be saved at the end of each of the next 4 years so that the land can be purchased 4 years from now?

35. Suppose a young couple with an 8-year-old son attempt to save for their son's college expenses in advance. Assuming that he enters college 10 years from the present, they estimate that an amount of $9,000 per year in terms of today's dollars will be required to support his college expenses for 4 years. It is also estimated that the future rate of inflation will be 8% per year, and they can invest their savings at 12% compounded annually. Determine the equal amount this couple must save each year until they send their son to college. College payments are made at the start of the school year. Answer: $4,198

36. A cable TV system is offering the following arrangement. You may pay for 10.5 months service in a lump sum at the present and make no other payments until the following year. The other option is to pay the regular amount at the end of each month for the next 12 months, with the payments beyond the first year being identical to those of the first option.

a. If the interest rate is 9% per year compounded monthly when there is no inflation, what inflation rate per month makes a subscriber indifferent to these two options?

b. If *f*, the inflation rate, is greater than the inflation rate computed in part (a), which option is favored?

37. Given below are two methods of financing the purchase of a $120,000 home. For each of these two financing arrangements find the equivalent cost at the present. Assume that the future inflation rate is $\frac{1}{2}$% per month and the future market rate of interest is 12% per year compounded monthly.

FINANCING A $120,000 HOME

	Interest-Free Mortgage	*Conventional Fixed-Rate Mortgage*
Interest rate	0%	13.5%
Term	5 years	30 years
Down payment	(33%) $40,000	(10%) $12,000
Amount financed	$80,000	$108,000
Monthly payment	$1,333	$1,237
Total cost of house	$120,000	$457,320

38. An individual is considering the purchase of life insurance which would provide $50,000 of benefits. Two policies providing the same coverage have been proposed which have different payment plans. Policy A requires end-of-year premiums of $280 for 25 years. (No premiums are paid after 25 years.) Beginning one year after the last payment, the policy will pay the policyholder five equal payments, each of which is 20% of the total amount paid in premiums. Thus with Policy A, the policyholder will receive in return all that was paid if he lives for 30 years.

Policy B requires $200 in end-of-year premiums for 30 years. With this policy a cash value will be accumulated so that the policyholder could withdraw $2,000 at the end of 30 years.

a. After 30 years no payments are made on either policy and the coverage will remain in effect. The policyholder believes the market interest rate and the inflation rate will average 9% and 4%, respectively, over the next 30 years. If the policyholder assumes that he will live more than 30 years, which policy should be selected? Answer: Select *B*

b. For Policy A, calculate, for the actual payments made to the policyholder the equivalent equal annual amount in *constant dollars* received at time $t = 26, 27, 28, 29$, and 30. The constant-dollar base year is the present ($t = 0$). Answer: $469

39. An individual is considering an investment in a retirement fund that earns 14% per year compounded semiannually. He has just celebrated his 40th birthday and he is planning to retire on his 65th. By making equal semiannual deposits of $2,000 up to and including his 65th birthday, what equal annual withdrawals in actual dollars could be made beginning on his 65th birthday, the last withdrawal occurring on his 75th birthday? If the annual inflation rate is 10% compounded semiannually, find the constant-dollar equal-annual series over the same 11 years that is equivalent to these withdrawals. (Constant-dollar base, 40th birthday.) Answer: $132,898; $7,727

40. An individual inherited a trust fund which will pay $10,000 at the end of 1989 and each following year including the end of year 2000. There will be 12 payments received. The interest rate over this period of time is expected to be 13% compounded annually while the inflation rate is 7% per year.

a. Find the *actual-dollar* single payment equivalent to this series of payments at the end of the year 2000.

b. Find the *constant-dollar* single payment equivalent to this series at the end of the year 2000. The base year is the *end* of 1990.

c. Find the *constant-dollar* equal annual series of payments from 1989 through 2000 equivalent to the actual dollar series of payments. Again the base year is the end of 1990.

41. Consider a project that has the following cost series for a 5-year period:

End of Year	*1*	*2*	*3*	*4*	*5*
Estimated future cost in constant dollars	$1,000	$1,000	$1,000	$1,000	$1,000

a. If the rate of inflation is 6% and the market rate is 10%, what is the present equivalent of this series? What is the equal-annual series of payments equivalent to this series in actual dollars?

b. If the rate of inflation is 15% and the market rate is 11%, what is the equivalent at the present of this series? What is the annual equivalent of this series in actual dollars?

42. A family borrowed $100,000 and purchased a home 20 years ago. The loan was for 30 years at 7% compounded annually. The savings and loan association where the money was obtained is now offering the following deal. A borrower can pay off a loan now by paying only 85% of the remaining balance. It is anticipated that the average market rate of interest will be 12% over the next 10 years. (Assume that payments on the loan are made annually and the 20th payment has just been made.) Determine whether the borrower should pay the loan off immediately. The inflation rate is 10% annually.

43. An individual has just purchased, for $8,750, a $10,000 bond with interest of 8% per year payable semiannually. The bond matures in 10 years, and the average inflation rate over the 10-year period is expected to be 4% compounded semiannually.

a. If interest can be earned at 10% compounded semiannually for the first 4 years and 6% compounded semiannually for the remaining years, what is the total amount that would be accumulated at the end of the tenth year from the investment of the bond receipts during the 10-year period? Answer: $21,124

b. What is the single amount at the end of year 10, in *constant* dollars, that is equivalent to the receipts received from the bond for the interest rates given in part (a)? (Assume the base year is the end of year 3.)

c. What equal semiannual payment series over the 10-year period, in *constant* dollars (base year 3), is equivalent to the receipts received from the bond? [The interest rates available to the investor are those indicated in part (a).]

44. Mr. Brown has just received his annual salary of $40,000. Under normal conditions he will retire 16 years from the present. His annual retirement income will be 60% of his last year's salary, with the first annual payment occurring at the moment of retirement. He expects to receive 20 equal annual retirement payments.

Because Mr. Brown served for three years in the military before being employed by this present firm, the current pension system allows him the option of purchasing 3 years of credit. That is, if he pays a single amount now he will be able to retire 3 years earlier. For this option he will receive 60% of his last year's salary beginning at the moment of retirement. In this case he expects to receive 23 equal annual payments.

He expects his salary to increase at the rate of 6% per year, and the expected average rate of inflation is 4%. If the market rate of interest is 10% over the entire time span, what is the maximum amount he would pay for the improved net advantage in *retirement income* if he retired early rather than at the normal time? (It is assumed that Mr. Brown would derive enough pleasure from his early retirement to offset the salary income that he would receive for the last 3 years if he did not retire early.)

45. A 20-MW power plant now under construction is expected to be in full commercial operation 2 years later. This plant is designed to be operated on distillate oil only. The fuel cost is a function of plant size, thermal conversion efficiency (heat rate), and plant utilization factor. However, since it is believed that the future price of oil will increase, the fuel cost in each year will be represented by the following expression:

$$F_n = (C)(H)(U)\left(\frac{8{,}760 \text{ hr/year}}{10^6}\right)P_n$$

where $P_n = P_{n-1}(1 + f)$, if P_n = price of fuel per million Btu in year n, and

F_n = annual fuel cost in the nth year ($/year);
C = plant size in kW (1 MW = 1,000 kW);
H = heat rate at operating conditions in Btu/kWh;
U = plant utilization factor;
f = average annual fuel inflation rate.

If the starting price for fuel during the first year of operation is $\$3.5/10^6$ Btu and this fuel cost is increased at the rate of 6% every year thereafter, what is the equivalent annual fuel cost for 10 years after the plant begins operation if the rate of interest is 12% per year? (Assume that $H = 9{,}300$ Btu/kWh, $U = 0.15$.)

46. An electric pump in a refinery operates continuously and its annual operating energy cost is \$1,500 per year if there is no inflation. Given below is the price index for electric energy over the next 8 years.

		Annual inflation rate
Present (1990)	100.00	
1991	107.00	7%
1992	111.28	4%
1993	122.41	10%
1994	124.86	2%
1995	131.10	5%
1996	141.59	8%
1997	145.84	3%
1998	163.34	12%

a. What is the average rate of inflation from the end of 1994 to 1998? Show your calculations.

b. If the base year is the end of 1990 (the present), what present amount is equivalent to the *first 4 years* (1991, 1992, 1993, and 1994) of electric energy costs? The annual market rate of interest is 10% for 1991, 8% for 1992, 9% for 1993, and 11% for 1994.

Part Three

ECONOMIC ANALYSIS OF ALTERNATIVES

In this part of the text, focus is on the process and some applications of the process for doing an economic analysis of alternatives. Part One provided the conceptual foundation and Part Two developed the needed mathematical tools. Bases for comparison are presented first as generally applicable criteria for choice. Next, a formal procedure is offered for forming and selecting among mutually exclusive alternatives. This is followed by methods for evaluating replacement alternatives and then for evaluating public activities. The last chapter of Part Three deals with investment alternatives that can only be compared properly by using break-even and optimization analysis.

6 Bases for Comparison of Alternatives

All decision criteria considered in this book incorporate some measure of equivalence, or basis for comparison, that expresses the real differences between investment alternatives.

> A **basis for comparison** is an index containing particular information about a series of receipts and disbursements representing an investment opportunity.

The reduction of alternatives to a common base is necessary so that apparent differences become real differences, with the time value of money considered. When expressed in terms of a common base, real differences become directly comparable and may be used for decision making. The most common bases for comparison are the present worth, annual equivalent, the capitalized equivalent, the future worth, the internal rate of return, and the payback period.

6.1 PRESENT WORTH

Currently, present-worth comparisons are the most widely utilized methods for considering the time value of money when making investment decisions. An understanding of the use and meaning of this method is basic to engineering economic analysis.

The **present worth** is a net equivalent amount at the present that represents the difference between the equivalent disbursements and the equivalent receipts of an investment's cash flow for a selected interest rate.

Letting F_t be the cash flow at time t, the present worth of an investment alternative at interest rate i with a life of n years can be expressed as

$$PW(i) = F_0(\overset{P/F,i,0}{\quad}) + F_1(\overset{P/F,i,1}{\quad}) + F_2(\overset{P/F,i,2}{\quad}) + \cdots + F_n(\overset{P/F,i,n}{\quad})$$

$$PW(i) = \sum_{t=0}^{n} F_t(\overset{P/F,i,t}{\quad}).$$

But since

$$(\overset{P/F,i,t}{\quad}) = (1 + i)^{-t},$$

then

$$\boxed{PW(i) = \sum_{t=0}^{n} F_t(1 + i)^{-t}.} \tag{6.1}$$

The range of interest rates for which the present worth has economic meaning extends from -1 to ∞. As i approaches -1, the factor $1/(1 + i)^t$ approaches ∞, causing the present-worth amount also to increase indefinitely. For $i = -1$ the present worth is not defined. Although in some instances negative interest rates are appropriate (see Section 5.4), most practical situations require interest rates that are positive. Therefore, in many plots of $PW(i)$, the range of i will be confined to $0 \leq i < \infty$.

The present worth has a number of features that make it suitable as a basis for comparison. First, it considers the time value of money according to the value of i selected for the calculation. Second, it concentrates the equivalent value of any cash flow in a single index at a particular point in time ($t = 0$). Third, a single unique value of the present worth is associated with each interest rate used, no matter what the investment's cash flow pattern may be.

By examining the $PW(i)$ function in Figure 6.1, based on the data in Table 6.1, considerable information useful for decision making can be ascertained about the investment opportunity. For the range of interest rates ($0 \leq i < 22\%$) it is observed that $PW(i)$ is positive, indicating that the equivalent receipts in the present exceed the equivalent disbursements. On the assumption that the cash flow estimates of Table 6.1 eventually prove to be correct, the significance of the $PW(i)$ function is that for a particular value of i, say 10%, the investment can be said to provide \$268 in net present equivalent "profit."

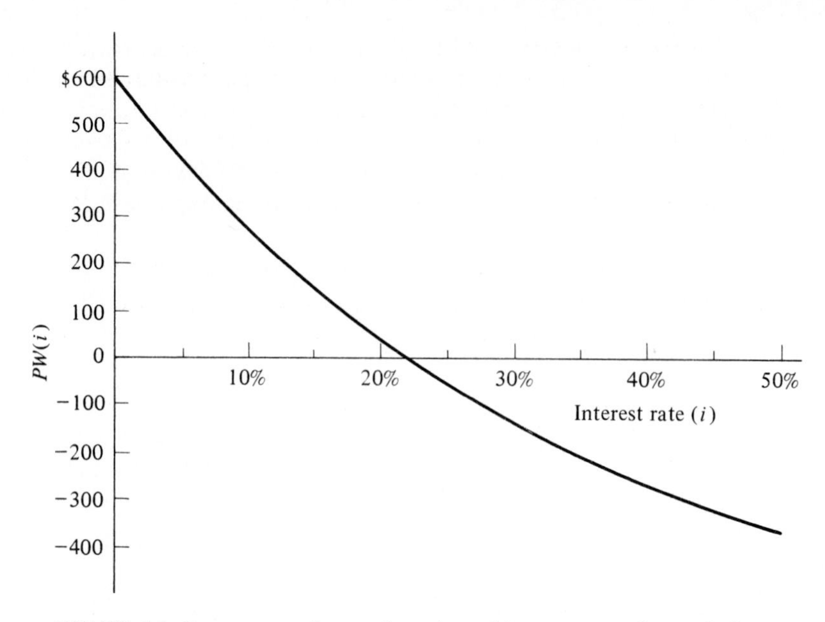

FIGURE 6.1. Present worth as a function of interest rate for cash flow in Table 6.1.

TABLE 6.1.
CALCULATION OF THE PRESENT WORTH FROM A CASH FLOW FOR A RANGE OF INTEREST RATES

End of Year	Cash Flow	i	$PW(i) = -\$1{,}000 + \$400\left\{\sum_{t=1}^{4}\left[\frac{1}{(1+i)^t}\right]\right\}$
0	−$1,000	0%	$600
1	400	10%	268
2	400	20%	35
3	400	22%	−3
4	400	30%	−133
		40%	−260
		50%	−358
		∞	−1,000

If $PW(i)$ had been extended for $i > 50\%$, the function would be asymptotic to $PW(i) = -\$1,000$, the value of F_0. This can be seen from Equation 6.1. As the interest rate is increased, each cash flow in the future is discounted to the present by a factor of the form $1/(1 + i)^t$. As i approaches infinity, these factors will approach zero for all points in time except $t = 0$. Thus, $F_0 = -\$1,000$ is the only cash flow that is not reduced to zero.

6.2 ANNUAL EQUIVALENT

The annual equivalent amount is a basis for comparison that has characteristics similar to the present worth. This similarity is evident when it is realized that any cash flow can be converted into a series of equal annual amounts by first calculating the present worth for the series and then multiplying the present worth by the factor $(\overset{A/P,i,n}{\quad})$.

The **annual equivalent** is the annual equivalent receipts less the annual equivalent disbursements of a cash flow.

Thus, the annual equivalent for interest rate i and n years can be mathematically defined as

$$\begin{aligned} AE(i) &= PW(i)(\overset{A/P,i,n}{\quad}) \\ &= \left[\sum_{t=0}^{n} F_t(1 + i)^{-t}\right]\left[\frac{i(1 + i)^n}{(1 + i)^n - 1}\right]. \end{aligned} \tag{6.2}$$

Two important features of this relationship need to be understood. First, if the values of i and n are fixed, the relationship reduces to $AE(i) = PW(i)$ times a constant. Therefore, when different cash flows are evaluated for a particular value of i and a particular value of n, the comparison of their annual equivalent amounts will yield the same relative results as those obtained from making the comparison on the basis of the present worth. That is, the ratio of the annual equivalents for two different cash flows will equal the ratio of the present worths for the respective cash flows.

Second, the values of $AE(i)$ and $PW(i)$ will be zero for the same value of i. Graphically this means that the intersections of the horizontal axis ($AE(i) = 0$) by the $AE(i)$ function will occur at the same value of i for which the $PW(i)$ function intersects the horizontal axis ($PW(i) = 0$). Thus, the present worth and the annual equivalent can be said to be *consistent* bases for comparison and they will yield the same selection of alternatives for fixed values of i and n.

6.2.1 Repeating Cash Flows

The $AE(i)$ basis of comparison will sometimes be preferred to $PW(i)$ because of computational advantages that may arise when it is assumed that a particular cash flow pattern will repeat itself. Figure 6.2 illustrates a repeating cash flow pattern that results from renewing an investment every 2 years. The $1,000 disbursement at the end of year 2 is actually the disbursement at the *beginning* of year 3 for another proposal that will return $400 and $900 at the end of year 3 and year 4, respectively. Therefore, it is only necessary to calculate the $AE(i)$ over the life of one cash flow cycle. When the cash flow cycle is repeated, the same $AE(i)$ will be repeated.

For the example in Figure 6.2, $AE(i)$ is calculated as follows:

$$AE(10) = \left[-\$1{,}000 + \$400(\overset{P/F,10,1}{0.9091}) + \$900(\overset{P/F,10,2}{0.8265})\right](\overset{A/P,10,2}{0.5762}) = \$61.93$$

Regardless of the number of times this cash flow cycle is repeated, the $AE(i)$ over 2 years, 4 years, 6 years, etc. would still be $61.93 per year.

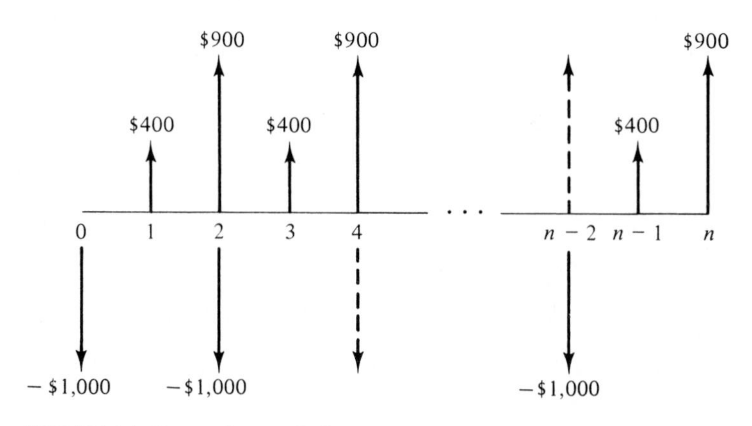

FIGURE 6.2. Repeating cash flow patterns.

6.3 FUTURE WORTH

The future worth can be found by converting the present worth of any investment to its equivalent at some future date. Because of its interchangeability with other measures of worth, its meaning must be understood.

The **future worth** represents the difference between the equivalent receipts and disbursements at some common point in the future.

This basis for comparison calculated at time n years from the present at a given interest rate, i, is

$$FW(i) = F_0(\overset{F/P,i,n}{\quad}) + F_1(\overset{F/P,i,n-1}{\quad}) + \cdots + F_{n-1}(\overset{F/P,i,1}{\quad}) + F_n(\overset{F/P,i,0}{\quad})$$

$$FW(i) = \sum_{t=0}^{n} F_t(\overset{F/P,i,n-t}{\quad}).$$

But since

$$(\overset{F/P,i,n-t}{\quad}) = (1 + i)^{n-t},$$

then

$$\boxed{FW(i) = \sum_{t=0}^{n} F_t(1 + i)^{n-t}.} \tag{6.3}$$

Another method of calculating the future worth is to first determine the present worth of the cash flow which is then converted to its future equivalent n years hence. Thus, the future worth can be expressed as

$$FW(i) = PW(i)(\overset{F/P,i,n}{\quad}).$$

Since $FW(i)$ is merely $PW(i)$ times a constant, the relative difference between alternatives on the basis of $PW(i)$ will be the same as the relative differences on the basis of $FW(i)$.

The future worth, annual equivalent, and present worth are consistent bases of comparison. As long as i and n are fixed and Alternatives A and B are being compared, the following relationships will hold:

$$\frac{PW(i)_A}{PW(i)_B} = \frac{AE(i)_A}{AE(i)_B} = \frac{FW(i)_A}{FW(i)_B}.$$

Because the $PW(i)$, $AE(i)$ and $FW(i)$ are all measures of equivalence, differing only by the points in time at which they are stated, it is evident that they provide consistent bases for comparison.

Therefore, any decision criterion that compares present equivalent amounts could just as well employ future equivalent amounts or annual equivalent amounts without affecting the outcome.

6.4 INTERNAL RATE OF RETURN

The internal rate of return measures a characteristic about investments that is quite different than the present worth types of measures. To understand this measure of worth, it is important to first understand how it is determined.

> The **internal rate of return (IRR)** is the interest rate that causes the equivalent receipts of a cash flow to equal the equivalent disbursements of that cash flow.

Another way of stating this concept is to define the IRR as the interest rate that reduces the present worth of a series of receipts and disbursements to zero. That is, the internal rate of return for an investment proposal is the interest rate i^* that satisfies the equation

$$0 = PW(i^*) = \sum_{t=0}^{n} F_t(1 + i^*)^{-t} \tag{6.4}$$

where the proposal has a life of n periods.

6.4.1 Computing the IRR

The computation of IRR generally requires a trial-and-error solution. For example, to calculate the IRR for the cash flow shown in Table 6.2 it is necessary to find the value of i^* that satisfies

$$\begin{aligned} 0 &= PW(i^*) \\ &= -\$1{,}000 - \$800(\overset{P/F,i^*,1}{\quad}) + \$500(\overset{P/A,i^*,4}{\quad})(\overset{P/F,i^*,1}{\quad}) + \$700(\overset{P/F,i^*,5}{\quad}). \end{aligned}$$

TABLE 6.2.
CASH FLOW PATTERN

End of Year t	Cash Flow F_t
0	−$1,000
1	−800
2	500
3	500
4	500
5	1,200

Instead of trying to solve for i^* directly from this equation, employ a *trial-and-error* solution. Try $i^* = 0\%$:

$$PW(0) = -\$1{,}000 - \$800(1) + \$500(4)(1) + \$700(1)$$

$$PW(0) = \$900.$$

With $PW(0) > 0$, examine the cash flow to see what rate should be tried next. Since all the positive cash flows are further in the future than the negative cash flows, an *increase* in the interest rate will reduce the present worth of the receipts more than the present worth of the outlays. Thus, the total $PW(i)$ will be *decreased* toward zero. Try $i = 12\%$:

$$PW(12) = -\$1{,}000 - \$800(\overset{P/F,12,1}{0.8929}) + \$500(\overset{P/A,12,4}{3.0374})(\overset{P/F,12,1}{0.8929}) + \$700(\overset{P/F,12,5}{0.5674})$$

$$PW(12) = \$39.$$

Since $PW(12)$ is still greater than zero, try a larger interest rate. With $i = 13\%$

$$PW/(13) = -1{,}000 - \$800(\overset{P/F,\,13,1}{0.8850}) + \$500(\overset{P/A,\,13,4}{2.9745})(\overset{P/F,\,13,1}{0.8850}) + \$700(\overset{P/F,\,15,5}{0.5428})$$

$$PW(13) = -\$12.$$

Thus, it is determined that the IRR will lie between 12% and 13%. By interpolation

$$i^* = 12\% + 1\%\left[\frac{39 - 0}{39 - (-12)}\right] = 12\% + 1\%\left[\frac{39}{51}\right] = 12.8\%.$$

The IRR is related to $PW(i)$ as shown in Figure 6.3. The value of i where $PW(i)$ intersects the horizontal axis is i^*. Note that Figure 6.3 was developed for the cash flow in Table 6.2. The solution it exhibits is the same as that found from the trial and error approach.

6.4.2 The Meaning of IRR

In economic terms, the IRR represents the percentage or rate earned on the *unrecovered* balance of an investment. The calculations displayed in Table 6.3 exhibit the fundamental meaning of IRR. Each cash flow can be viewed as an arrangement in which $1,000 is borrowed at a rate of 10% on the unpaid or unrecovered balance, with plans to reduce the unpaid balance to zero at the end

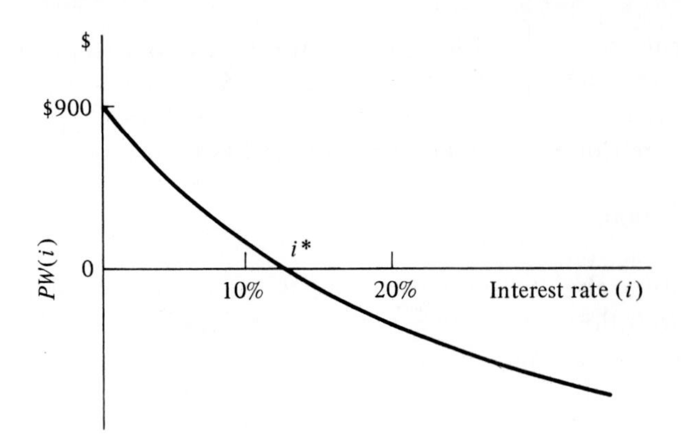

FIGURE 6.3. IRR and its realtionship to the present-worth amount.

TABLE 6.3.
THREE CASH FLOWS DEMONSTRATING THE FUNDAMENTAL MEANING OF IRR

End of Year	Cash Flow at End of Year t	Unrecovered Balance of Beginning of Year t	Interest Earned on the Unrecovered Balance During Year t	Unrecovered Balance at the End of Year t
		Alternative $A\,(i_A^* = 10\%)$		
t	$F_{A,t}$	U_{t-1}	$U_{t-1}(0.10)$	$U_t = U_{t-1}(1 + 0.10) + F_{A,t}$
0	−$1,000	–	–	−$1,000
1	400	−$1,000	−$100	−700
2	370	−700	−70	−400
3	240	−400	−40	−200
4	220	−200	−20	0
		Alternative $B(i_B^* = 10\%)$		
t	$F_{B,t}$	U_{t-1}	$U_{t-1}(0.10)$	$U_t = U_{t-1}(1 + 0.10) + F_{B,t}$
0	−$1,000	–	–	−$1,000
1	100	−$1,000	−$100	−1,000
2	100	−1,000	−100	−1,000
3	100	−1,000	−100	−1,000
4	1,100	−1,000	−100	0
		Alternative $C(i_C^* = 10\%)$		
t	$F_{C,t}$	U_{t-1}	$U_{t-1}(0.10)$	$U_t = U_{t-1}(1 + 0.10) + F_{C,t}$
0	−$1,000	–	–	−$1,000
1	0	−$1,000	−$100	−1,100
2	0	−1,100	−110	−1,210
3	0	−1,210	−121	−1,331
4	1,464	−1,331	−133	0

of four years. These cash flows could also represent the purchase of productive assets, where these assets will yield a return of 10% on the dollars unrecovered or "tied up" in the assets during their lifetime. Regardless of how these cash flows are viewed, the fact is that the IRR's of Alternatives A, B, and C are all 10%. Thus, $i_A^* = 10\%$, $i_B^* = 10\%$, and $i_C^* = 10\%$ will satisfy the following expressions:

$$0 = -\$1{,}000 + \$400(\overset{P/F,i_A^*,1}{\quad}) + \$370(\overset{P/F,i_A^*,2}{\quad}) + \$240(\overset{P/F,i_A^*,3}{\quad}) + \$220(\overset{P/F,i_A^*,4}{\quad})$$

$$0 = -\$1{,}000 + \$100(\overset{P/A,i_B^*,4}{\quad}) + \$1{,}000(\overset{P/F,i_B^*,4}{\quad})$$

$$0 = -\$1{,}000 + \$1{,}464(\overset{P/F,i_C^*,4}{\quad})$$

If U_t = the unrecovered balance at the end of period t, the unrecovered balance for any time period can be found from the recursive equation where

F_t = amount received at the end of period t;
i^* = interest rate earned on the unrecovered balance during period t (IRR);
U_0 = initial amount of loan or first cost of asset (F_0).

$$\boxed{U_t = U_{t-1}(1 + i^*) + F_t} \qquad (6.5)$$

The unrecovered balances related to each cash flow appear as negative values in Table 6.3, indicating that they are amounts owed by the borrower or amounts yet to be recovered by the lender.

All cash flows in Table 6.3 earn a return each period equal to their respective unrecovered balances multiplied by their respective IRR's. Alternative A earns 10% on \$1,000, \$700, \$400, and \$200; Alternative B earns 10% on \$1,000 for each of four periods; and Alternative C earns 10% on \$1,000, \$1,100, \$1,210, and \$1,331. The interest *amount* earned differs for each alternative. For A it is \$230, whereas it is \$400 for B, and \$464 for C. *In general it can be stated that the internal rate of return (IRR) is the interest rate earned on the unrecovered balance over an investment's life so that the unrecovered balance at the end of that time is zero.*

6.4.3 Polynomial Explanation of IRR

The expression

$$PW(i) = \sum_{t=0}^{n} F_t(1 + i)^{-t}$$

is an nth-degree polynomial. It may be studied by letting

$$x = \frac{1}{(1 + i^*)}$$

which gives

$$0 = F_0 + F_1x + F_2x^2 + F_3x^3 + \cdots + F_nx^n.$$

Although an nth-degree polynominal must have n roots, only certain ones will have practical meaning. The IRR, i^*, must lie in the interval $(-1 < i^* < \infty)$ to be economically relevant. For this interval of i^*, x must be a real, positive number $(0 < x < \infty)$, since, as

$$i^* \to \infty, \qquad x \to 0$$

and

$$i^* \to -1, \qquad x \to \infty.$$

Thus, only for real, positive roots will the polynomial yield solutions for the IRR that have an economic interpretation.

There are certain cash flows for which *no* IRR exists in the interval $(-1 < i < \infty)$. The most common example is when the cash flow consists of either all receipts or all disbursements, with the initial receipt or disbursement occurring at $t = 0$. In practice, investment proposals are frequently described by cost cash flows when the alternatives are assumed to provide the same service, benefit, or revenue. Since it is impossible to calculate directly a meaningful IRR for such a cash flow pattern, other methods must be utilized when this situation occurs.

Since the solution for the IRR of a cash flow with a life of n periods is the solution of an nth-degree polynomial, there exist various mathematical methods that systematically converge on the roots or values of i that satisfy such a polynomial. Computer solutions are becoming the dominant approach, wherein numerical solutions can be very accurately found.

6.4.4 Cash Flows with a Single IRR

Investment cash flows that have a present-worth function of the form presented in Figure 6.3 are assured to have a single IRR and

$$PW(i) > 0 \quad \text{for} \quad i < i^*$$

$$PW(i) = 0 \quad \text{for} \quad i = i^*$$

$$PW(i) < 0 \quad \text{for} \quad i > i^*$$

Because of the importance of this type of function it is necessary to have the means for predicting easily whether a particular cash flow will produce such a function.

To correctly apply the IRR decision rules presented in this book, the present-worth function must have the form of Figure 6.3.

Test 1, consisting of three conditions, guarantees that a cash flow will exhibit a function of the form shown in Figure 6.3. The three conditions, all of which must be satisfied, are

1. $F_0 < 0$
   (The first non-zero cash flow is a disbursement)
2. One change in sign in the sequence $F_0, F_1, F_2, \ldots, F_n$
   (The cash flow has an initial disbursement or a series of disbursements followed by a series of receipts.)
3. $PW(0) > 0$
   (The sum of all the receipts is greater than the sum of all the disbursements.)

Table 6.4 presents two cash flows (*A* and *B*) that satisfy the conditions of Test 1 and three cash flows (*C*, *D*, and *E*) that do not.

For Cash Flow *A* the sum of the receipts ($1,500) is greater than the sum of the disbursements ($1,000), and for *B* the sum of the receipts ($3,500) exceeds the sum of the disbursements ($2,500). Thus, these cash flows have a single rate of return with a present worth plot of the form shown in Figure 6.3. Most practical alternatives have estimated cash flow patterns similar to *A* and *B*, since most investments require an initial commitment of funds, followed by an income series resulting from the project's net yield.

TABLE 6.4.
FIVE CASH FLOW PATTERNS

	Cash Flow				
End of Year	*A*	*B*	*C*	*D*	*E*
0	−$1,000	−$1,000	0	−$2,000	−$1,000
1	500	− 500	−$3,000	0	4,700
2	400	− 500	1,000	10,000	− 7,200
3	300	− 500	1,900	0	3,600
4	200	1,500	− 800	0	0
5	100	2,000	2,720	−10,000	0

Although Cash Flow C does not meet the second condition of Test 1 (it has three changes in sign), it has a single IRR, $i^* = 20\%$, along with a $PW(i)$ function like Figure 6.3.

To determine if a cash flow having multiple sign changes has a function similar to Figure 6.3, a more complete test is provided. Test 2 requires that the following conditions be satisfied.

1. $F_0 < 0$.

2. Find an IRR, i^*, for the cash flow. For the known i^*, $U_t < 0$ for $t = 0, 1, 2, \ldots, n - 1$.

(The unrecovered balance evaluated at the known IRR must always be negative except at $t = n$, where $U_n = 0$.) Table 6.3 presented sample calculations of U_t.

For cash flow C, $t = 1$ is assumed to be $t = 0$ assuring that $F_0 < 0$. Next, calculate the unrecovered balance U_t, using Equation 6.5 at $i^* = 20\%$ for each period over the life, giving

$$U_0 = -\$3{,}000$$

$$U_1 = -\$3{,}000(\overset{F/P,20,1}{1.200}) + \$1{,}000 = -\$2{,}600$$

$$U_2 = -\$2{,}600(\overset{F/P,20,1}{1.200}) + \$1{,}900 = -\$1{,}220$$

$$U_3 = -\$1{,}220(\overset{F/P,20,1}{1.200}) + \$800 = -\$2{,}264$$

$$U_4 = -\$2{,}264(\overset{F/P,20,1}{1.200}) + \$2{,}717 = 0.$$

Since the values for $t = 0, 1, 2$ and 3 are negative, the second condition is satisfied.

Test 1 is much simpler to utilize than Test 2, since it requires only easily computed sums and observations of the changes in sign of the cash flow values. However, many cash flows that do not satisfy Test 1 may satisfy Test 2. Cash Flow C is an example of such a case. Thus, the more elaborate Test 2 has the ability to be more discriminating, thereby including a much wider variety of cash flows.

It may appear that cash flows having $F_0 > 0$ or $F_0 = 0$ will not satisfy the first condition. However, certain changes in a cash flow can *always* alter that cash flow so that $F_0 < 0$. Multiplying the polynomial

$$0 = \sum_{t=0}^{n} F_t(1 + i^*)^{-t}$$

by a nonzero constant has no effect on the value or values of i^* that satisfy this expression. Therefore, if $F_0 > 0$, multiply each F_t in the cash flow by -1

and

if $F_0 = 0$, find the first nonzero cash flow and calculate the present worth at its time of occurrence.

Using Cash Flow C in Table 6.4 and finding the value of i^* that sets the present worth equal to zero at $t = 0$

$$0 = -\$3{,}000(\overset{P/F,i^*,1}{\quad}) + \$1{,}000(\overset{P/F,i^*,2}{\quad}) + \$1{,}900(\overset{P/F,i^*,3}{\quad})$$
$$- \$800(\overset{P/F,i^*,4}{\quad}) + \$2{,}720(\overset{P/F,i^*,5}{\quad})$$

and at $t = 1$

$$0 = -\$3{,}000 + 1{,}000(\overset{P/F,i^*,1}{\quad}) + \$1{,}900(\overset{P/F,i^*,2}{\quad}) - \$800(\overset{P/F,i^*,3}{\quad})$$
$$+ \$2{,}720(\overset{P/F,i^*,4}{\quad})$$

gives identical results ($i^* = 20\%$). Since the second expression is the first expression multiplied by the constant $(\overset{P/F,i^*,1}{\quad})$, the IRR is unchanged. Therefore, using the second expression guarantees that the first nonzero cash flow is a disbursement without affecting the proposal's IRR.

6.4.5 Cash Flows with Multiple IRR's

Cash Flows D and E in Table 6.4 represent a more unusual class of cash flow patterns: disbursements, receipts, more disbursements, receipts, and so on. Such cash flows do not follow the pattern of the class of cash flows that include A and B, so there is no assurance that their present-worth functions will resemble that of Figure 6.3.

For decision-making purposes cash flows that have a unique IRR and behave similarly to the example shown in Figure 6.3 are much simpler to handle than cash flows with multiple IRR's. When multiple IRR's occur, questions arise such as, "Which IRR is the correct one?" and "Are the decision rules most frequently used for investment selection applicable here?"

The answer to both of these questions is that *when multiple IRR's are found there is no rational means for judging which of them is most appropriate for determining economic desirability. Since the most often applied methods are not designed to consider multiple IRR's, the most common practice is to avoid IRR as a basis for comparison when multiple IRR's occur.* Therefore, it is important that cash flows having multiple IRR's be identified early in the analysis.

A rule that can be helpful in identifying the possibility of multiple IRR's is Descartes' rule of signs for an nth-degree polynomial. This rule states that the number of real positive roots of an nth-degree polynomial with real coefficients is never greater than the number of changes of sign in the sequence of its coefficients

$$F_0, F_1, F_2, F_3, \ldots, F_{n-1}, F_n$$

and, if less, is always so by an even number. For example, the sequence of signs of the cash flows for *A* and *B* shown in Table 6.4 changes only once, while the sequence of signs for *C*, *D*, and *E* changes three, two and three times, respectively. The sequence of signs for *D* has one change from the initial negative value to positive at the end of year 2. (A zero cash flow can be considered signless for the purpose of applying the rule of signs.) The sequence of signs remains unchanged until the end of year 5, when it changes from positive back to negative.

The rule of signs indicates for cash flow *D* that the maximum possible number of positive, real roots is two. Figure 6.4 depicts the present-worth function for *D*, and we see that in fact this cash flow does have two distinct interest rates for which the present worth is zero. The two IRR's for *D* are 9.8% and 111.5%.

The decision maker, who develops a decision procedure based on the assumption that the alternatives being considered will have a present-worth function similar to Figure 6.3, must assure himself that the cash flows do indeed produce such a function. Application of Test 1 or Test 2 (any cash flow satisfying Test 1 will always satisfy Test 2) is the first step. If the cash flows do not pass either of these tests, then there is the possibility of multiple IRR's, and the next step is to plot $PW(i)$. Using the plot of the present-worth function, one can properly interpret the desirability of the proposed investment by observing the values of i for which $PW(i) > 0$. For Proposal *E* in Table 6.4 the present-worth plot shown in Figure 6.5 reveals that this proposal is desirable for values of i less than 20% and for values of i between 50% and 100%. To attempt to interpret multi-

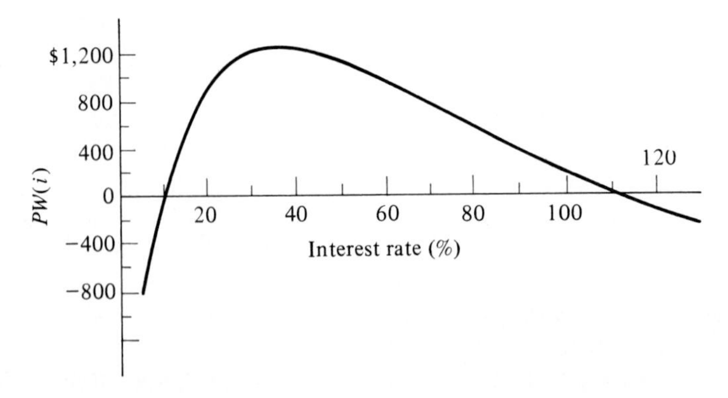

FIGURE 6.4. Present-worth function for cash flow *D* in Table 6.3.

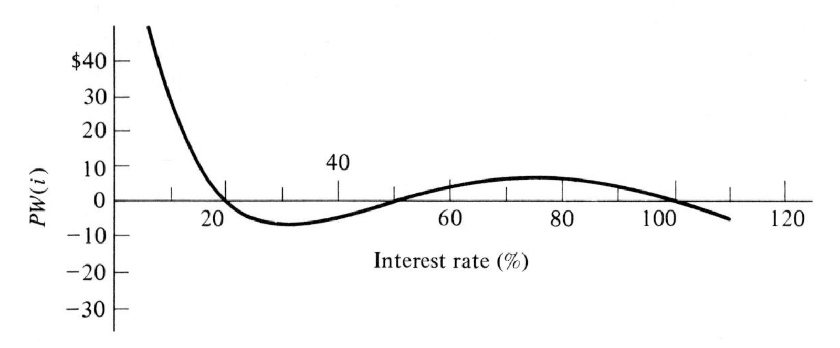

FIGURE 6.5. Present-worth function for cash flow E in Table 6.3.

ple IRR's using normal procedures will in most cases be meaningless without the supplementary use of the present-worth function.

It is a simple matter to construct a cash flow pattern that produces selected multiple IRR's. Suppose one desires to find a pattern that has IRR's of 20%, 50%, and 100%; one need only multiply the factors that yield these IRR's. The result is an equation that represents a future-worth calculation, equal to zero. An example of such a calculation is shown below.

$$FW(i^*) = 0 = [(1 + i) - 1.2][(1 + i) - 1.5][(1 + i) - 2.0]$$

$$0 = (1 + i)^3 - 4.7(1 + i)^2 + 7.2(1 + i) - 3.6.$$

Multiplying by −\$1,000 gives

$$0 = -\$1,000(1 + i)^3 + 4,700(1 + i)^2 - \$7,200(1 + i) + \$3,600.$$

The coefficients in this last equation are the individual cash receipts and disbursements of E described in Table 6.4. The $PW(i)$ plot in Figure 6.5 confirms that the IRR's of this cash flow are 20%, 50%, and 100%.

6.5 PAYBACK PERIOD

Expressions such as "This investment will pay for itself in less than three years" are common in business and industry and emphasize the tendency to evaluate assets in terms of a payback or payout period. This section presents methods for assessing the payback period of an investment.

6.5.1 Payback Without Interest

The payback period is probably the most popular method used by industry for assessing the economic desirability of an investment.

The **payback period without interest** is commonly defined as the length of time required to recover the first cost of an investment from the net cash flow produced by that investment for an interest rate of zero.

That is, if F_0 = first cost of the investment and if F_t = the net cash flow in period t, then the payback period is defined as the smallest value of n that satisfies the equation

$$\sum_{t=0}^{n} F_t \geq 0. \tag{6.6}$$

When comparing the payback period for investment proposals it is usually more desirable to have a short payback period than a longer one. A short payback period indicates that the investment provides revenues early in its life sufficient to cover the initial outlay. Thus, an investment with a short payback period can be viewed as having a higher degree of liquidity than one with a longer payback period. This quicker return of the capital invested also shortens the time span over which the investment is susceptible to possible economic loss.

For example, Table 6.5 presents the cash flows for three investment alternatives for each of which the payback period is 3 years. Analyses of these cash flows reveal that the payback period as a measure of investment desirability has serious shortcomings. Under normal circumstances these three proposals do not have equal economic desirability, although they have equal payback periods.

In general, the most serious deficiencies of the payback period are that it fails to consider

1. The time value of money.

2. The consequences of the investment following the payback period, including the magnitude and timing of the cash flows and the expected life of the investment.

Because of the limitations just mentioned the payback period tends to favor shorter-lived investments. Experience has generally indicated that this bias is unjustifiable and in many cases economically unsound.

Nevertheless, it must be said that the payback period does give some measure of the rate at which an investment will recover its initial outlay. For situations where there is a high degree of uncertainty concerning the future and a firm is interested in its cash position and borrowing commitments, the payback period can supply useful information about investments under consideration. As a result, this measure of investment desirability is frequently used to supplement the bases for comparison discussed earlier.

TABLE 6.5.
THREE ALTERNATIVES WITH A PAYBACK PERIOD OF THREE YEARS

End of Year	A	B	C
0	−$1,000	−$1,000	−$700
1	500	200	−300
2	300	300	500
3	200	500	500
4	200	1,000	0
5	200	2,000	0
6	200	4,000	0
Present Worth, $i = 0$	$PW(0)_A = \$600$	$PW(0)_B = \$7,000$	$PW(0)_C = \$0$
Payback Period	3 years	3 years	3 years

6.5.2 Payback with Interest

To include consideration of the time value of money when calculating the payback period, a method known as the *discounted payback period* may be used.

> **Payback with interest** determines the length of time required until the investment's equivalent receipts exceed the equivalent capital outlays.

Using F_t as defined previously, the discounted payback period is the smallest value, n', that satisfies the expression

$$\boxed{\sum_{t=0}^{n'} F_t(1 + i)^{-t} \geq 0.} \tag{6.7}$$

An example of this calculation is presented for Alternative A in Table 6.5, for an interest rate of 15%.

$$-\$1,000 + \$500(\overset{P/F,15,1}{0.8696}) + \$300(\overset{P/F,15,2}{0.7562}) + \$200(\overset{P/F,15,3}{0.6575})$$

$$+ \$200(\overset{P/F,15,4}{0.5718}) + \$200(\overset{P/F,15,5}{0.4972}) \geq 0.$$

$$\$7 \geq 0.$$

In this example $n'_A = 5$ is the shortest time period required for the equivalent cash receipts to exceed the capital investment.

Calculation of the discounted payback periods for Alternatives A, B, and C at an interest of 15% yields $n'_A = 5$ years; $n'_B = 4$ years; n'_C never recovers its investment costs. Table 6.5 shows that the payback period without interest is identical for all three alternatives, $n = 3$ years. However, when discounted payback

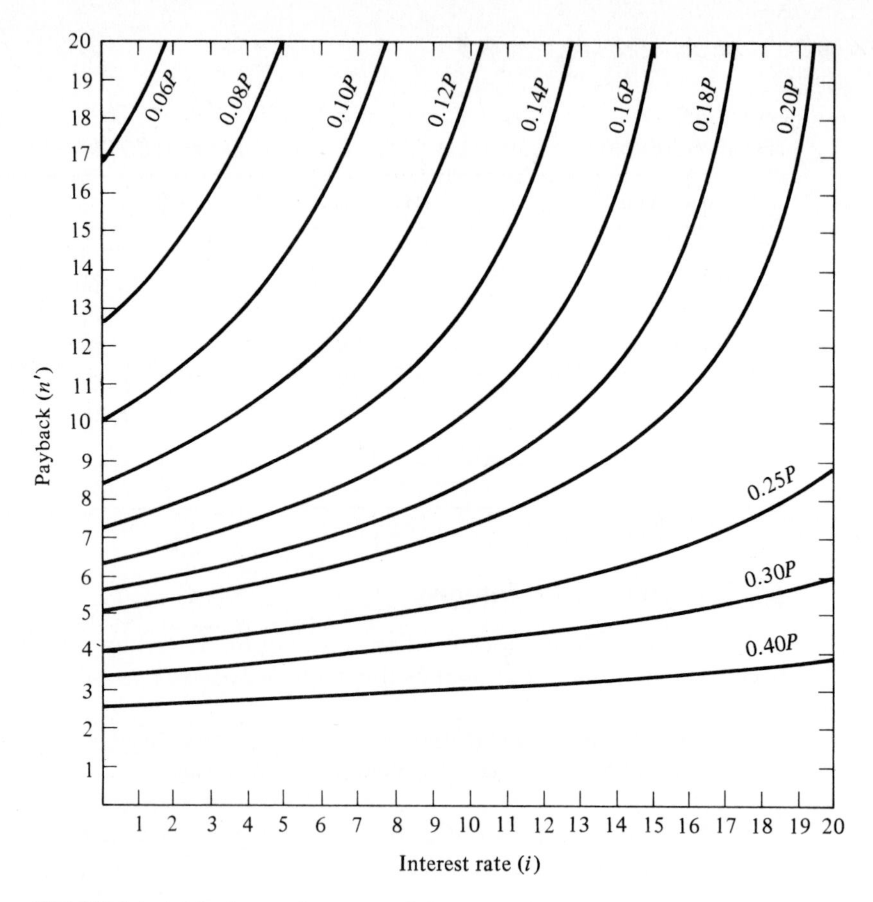

FIGURE 6.6. Payback as a function of the interest rate for selected annual recovery amounts.

is applied, the periods of payback for these three alternatives are longer and they vary considerably.

An interesting version of payback with interest is the simple case of a series of equal annual benefits from an investment of P dollars at $t = 0$. If it is assumed that A dollars are recovered each year, with A being a percentage of P, then the number of years required for payback can be found as a function of the interest rate i.

A set of curves, one for each of the selected percentages of P, can be developed. Let

$$A = \%P, \text{ where } 0 < \% < 100$$

and

$$A = P\left[\frac{i(1+i)^{n'}}{(1+i)^{n'}-1}\right].$$

Figure 6.6 exhibits n' as a function of i (for A as a percentage of P in the range of

6% to 40%). It illustrates the trade-off between the payback period and the interest rate.

Figure 6.6 is useful in determining the payback with interest for an investment with a life between 0 and 20 years. If any two of three variables (interest rate, life of the project, and desired payback as a percentage of the investment) are constant, the third variable can be determined by inspection. As an example, let $A = 0.12P$ and $i = 9\%$. This gives a payback of just over 16 years. By decreasing the interest rate to 8%, the payback period is reduced to 14.4 years. Similarly, an increase of the interest rate to 10% increases the payback period to 18.8 years.

6.6 CAPITALIZED EQUIVALENT AMOUNT

A special case of the present worth basis of comparison is the *capitalized equivalent, CE(i)*.

> The **capitalized equivalent** represents a basis of comparison that consists of finding a single amount at the present which, at a given rate of interest, will be equivalent to the net difference of receipts and disbursements if a given cash flow pattern is repeated in perpetuity.

This concept should not be confused with the accountant's concept of "capitalizing" an expenditure in the books of account. For the accountant an expenditure is capitalized if it is recorded as an asset or prepaid expense rather than being recorded as an expense at the time it is incurred.

To calculate the capitalized equivalent for an investment or a series of investments that are expected to produce cash flows from the present to infinity, the most common method is to first convert the actual cash flow into an equivalent cash flow, of equal annual amounts, A, that extends to infinity. Then the equal annual payments are discounted to the present:

$$CE(i) = PW(i) \text{ where the cash flow extends forever } (n = \infty)$$

$$CE(i) = A(\overset{P/A,i,\infty}{\quad}) = A\left[\frac{(1+i)^{\infty} - 1}{i(1+i)^{\infty}}\right]$$

$$= A\left[\frac{1 - \dfrac{1}{(1+i)^{\infty}}}{i}\right] = \frac{A}{i}.$$

Thus,

$$\boxed{CE(i) = \frac{A}{i}.} \qquad (6.8)$$

An intuitive understanding of this last relationship is obtained by considering what present-worth amount invested at i will enable an investor to periodically withdraw an amount A forever. If the investor withdraws more than amount A each period, he will be withdrawing a portion of the initial principal. If this initial principal is being partially consumed with each withdrawal, it will eventually be exhausted. However, when the amount being withdrawn each period equals the interest earned on the principal for that period, the principal remains intact. Thus, the series of withdrawals can be continued forever.

As an example of the use of capitalized equivalent suppose a philanthropic foundation is considering a gift to a city to build a park and to maintain it forever. Suppose that the annual interest rate that can be earned in perpetuity is 8% and the annual maintenance cost is expected to be \$16,000 per year for the first 15 years, increasing to \$25,000 per year after 15 years. What is the amount of the gift received at the present that will be required to assure continuing maintenance on the park?

$$CE(8) = \frac{\$16{,}000}{0.08} + \frac{\$9{,}000}{0.08}\overset{P/F,8,15}{(0.3153)}$$

$$= \$200{,}000 + \$35{,}471 = \$235{,}471.$$

With this calculation it is observed that \$200,000 received at the present will earn \$16,000 per year when invested at 8%, and this amount can be continued forever if the \$200,000 remains invested. The additional \$35, 471 when invested at 8% for 15 years will amount to

$$\$35{,}471\overset{F/P,8,15}{(3.172)} = \$112{,}515.$$

After that time, \$112,515 will earn \$9,000 per year at 8%, covering the additional maintenance costs forever.

6.7 CAPITAL RECOVERY WITH RETURN

An investment in an asset is expected to result in income sufficient not only to recover the amount of the original investment, but also to provide for a return on the funds tied up (unrecovered balance) in the asset at any time during its life.

> The **capital recovery with return, *CR(i)*,** *for any investment is the equal annual cash flow over its life equivalent to the capital costs of the investment represented by the initial outlay and the eventual salvage value.*

Two monetary transactions are associated with the procurement and eventual retirement of a capital asset: its first cost and salvage value. From these amounts, it is possible to derive a simple formula for the $AE(i)$ cost of the asset for use in engineering economy studies. Let

P = first cost of the asset;

F = estimated salvage value;

n = estimated service life in years;

$CR(i)$ = capital recovery with return.

The $AE(i)$ cost of the asset may be expressed as the $AE(i)$ first cost less the $AE(i)$ salvage value, or

$$CR(i) = P(\overset{A/P,i,n}{\quad}) - F(\overset{A/F,i,n}{\quad}). \tag{6.9}$$

But since

$$(\overset{A/F,i,n}{\quad}) = (\overset{A/P,i,n}{\quad}) - i$$

by substitution

$$CR(i) = P(\overset{A/P,i,n}{\quad}) - F\left[(\overset{A/P,i,n}{\quad}) - i\right]$$

and

$$\boxed{CR(i) = (P - F)(\overset{A/P,i,n}{\quad}) + Fi} \tag{6.10}$$

As an example of the use of this important formula, consider the following situation. An asset with a first cost of \$5,000 has an estimated service life of 5 years and an estimated salvage value of \$1,000. For an interest rate of 10% the capital recovery with return is

$$CR(10) = (\$5,000 - \$1,000)(\overset{A/P,10,5}{0.2638}) + \$1,000(0.10)$$

$$= \$1,155.20.$$

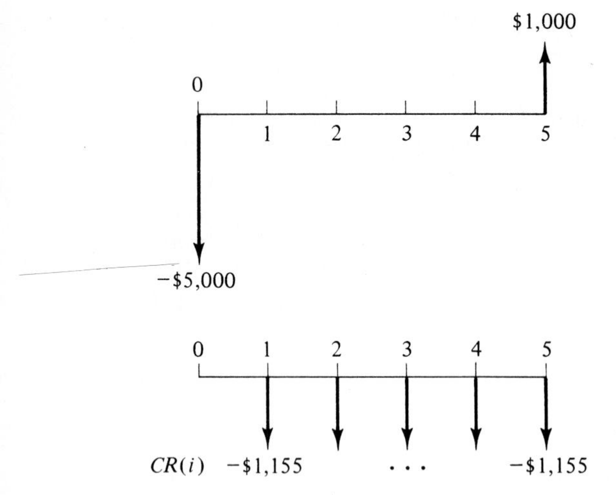

FIGURE 6.7. Capital recovery with return as an equivalent to an asset's loss in value.

The equivalence between the asset's actual loss in value and the annual cost representing that loss (capital recovery with return) is illustrated in Figure 6.7.

6.8 PROJECT BALANCE

The conventional bases for comparison discussed in Sections 6.1 through 6.7 have been widely adopted by those involved in the economic analysis of alternatives. Other methods exist, however, for measuring the economic worth of alternatives. These methods provide new insights and additional information not available from the conventional bases previously described. One such method is based on the concept of the *project balance.*

6.8.1 The Concept of Project Balance

Whereas each of the conventional bases for comparison of alternatives consolidates certain economic facts about a cash flow into a single index, the project balance describes the equivalent loss or profit of a cash flow as a function of time.

> The **project balance** is a time profile that measures the net equivalent amount of dollars tied up or committed to the project at each point in time over the life of the cash flow.

If the cash flow is unexpectedly terminated at the end of time t, the project balance $PB(i)_t$ identifies the equivalent loss or profit associated with the cash flow at that time.

To demonstrate the idea of project balance, the cash flow presented in Figure 6.8 will be used. For this example, let the interest rate be 20%.

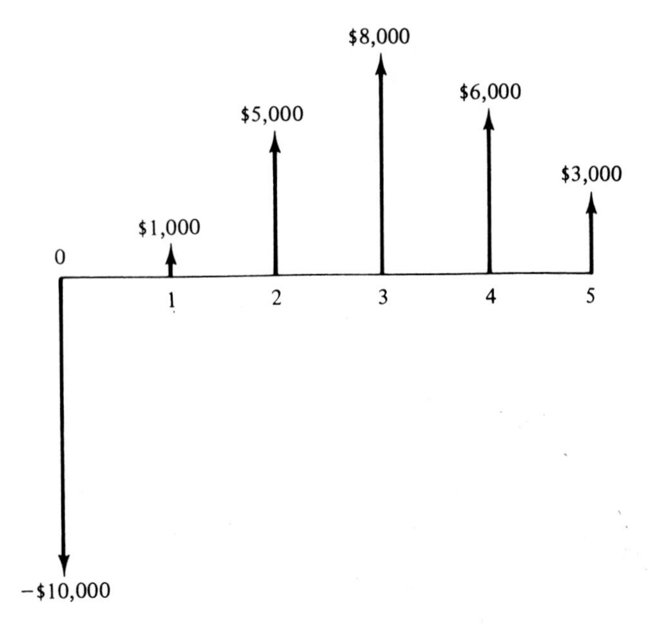

FIGURE 6.8. Cash flow for project balance example.

If at any time before the receipt of \$1,000 at time $t = 1$ the investment is terminated, only the initial investment will be lost (it is assumed that the cost of having money tied up in the investment is not incurred unless it is committed for the entire period). The project balance at the present is

$$PB(20)_0 = -\$10{,}000.$$

The investment at time $t = 1$ has an accumulated commitment of \$12,000, which consists of the initial outlay and the associated cost of having \$10,000 tied up in this investment for one period. However, this accumulated commitment is immediately reduced by the cash receipt of \$1,000 at $t = 1$. The project balance at $t = 1$ is

$$PB(20)_1 = -\$10{,}000(1.20) + \$1{,}000 = -\$11{,}000.$$

If the investment is terminated at $t = 1$ or before $t = 2$, it will experience a loss of \$11,000. The project balance at $t = 1$ becomes the amount committed to that investment at the beginning of the next period, so that at $t = 2$ the amount tied up in the investment is

$$PB(20)_2 = -\$11{,}000(1.20) + \$5{,}000 = -\$8{,}200.$$

This amount represents the cost of having \$11,000 committed at the beginning of the second period and the receipt of \$5,000 at the end of that period.

At the end of the third period ($t = 3$) the project balance is

$$PB(20)_3 = -\$8{,}200(1.20) + \$8{,}000 = -\$1{,}840.$$

For the last two periods of the investment's life the project balance is

$$PB(20)_4 = -\$1{,}840(1.20) + \$6{,}000 = \$3{,}792$$

and

$$PB(20)_5 = \$3{,}792(1.20) + \$3{,}000 = \$7{,}550.$$

If the investment alternative survives to its expected terminal date ($t = 5$), the future worth of the investment is the terminal project balance, $PB(20)_5$. This fact is confirmed by the expression for the future worth of the cash flow:

$$\begin{aligned} FW(20) &= -\$10{,}000(\overset{F/P,20,5}{2.488}) + \$1{,}000(\overset{F/P,20,4}{2.074}) + \$5{,}000(\overset{F/P,20,3}{1.728}) \\ &\quad + \$8{,}000(\overset{F/P,20,2}{1.440}) + \$6{,}000(\overset{F/P,20,1}{1.200}) + \$3{,}000 \\ &= \$7{,}550. \end{aligned}$$

Defining the project balance mathematically based on the previous example yields the following recursive relationship:

$$\boxed{PB(i)_t = (1 + i)PB(i)_{t-1} + F_t \qquad \text{for } t = 1, 2, \ldots, n} \tag{6.11}$$

where $PB(i)_0 = F_0$. Hence F_t is the cash receipts (+) or disbursements (−) at time t, and the duration of the cash flow is n. Note that Equation (6.11) is the same as Equation (6.5) except that the interest rate used in Equation (6.5) is the IRR, i^*, while the interest rate used in Equation (6.11) is selected by the analyst.

Another expression for the project balance can be developed by making substitutions in Equation (6.11). At $t = 0$

$$PB(i)_0 = F_0$$

so that at $t = 1$

$$PB(i)_1 = (1 + i)PB(i)_0 + F_1 = F_0(1 + i) + F_1$$

giving at $t = 2$

$$PB(i)_2 = (1 + i)PB(i)_1 + F_2 = F_0(1 + i)^2 + F_1(1 + i) + F_2$$

so that at any time $t = T$

$$PB(i)_T = (1 + i)PB(i)_{T-1} + F_T = F_0(1 + i)^T + F_1(1 + i)^{T-1} + \cdots + F_T.$$

Therefore, the project balance at any time T can be defined as

$$PB(i)_T = \sum_{t=0}^{T} F_t(1 + i)^{T-t} \qquad \text{for } T = 0, 1, 2, \ldots, n. \tag{6.12}$$

If the project balance associated with an investment is plotted, a visual description is produced that provides insight regarding four important characteristics of the investment. An example based on the cash flow described in Figure 6.8 is presented in Figure 6.9. This time-profile of the equivalent profit or loss that would be incurred if the investment were terminated prematurely is referred to as the *project balance diagram*.

6.8.2 Four Elements of Project Balance

The four important characteristics depicted by the project balance diagram are

1. The net future worth of the investment, $PB(20)_5$.
2. The time when the equivalent committed dollars switch from negative to nonnegative, $t = 4$.

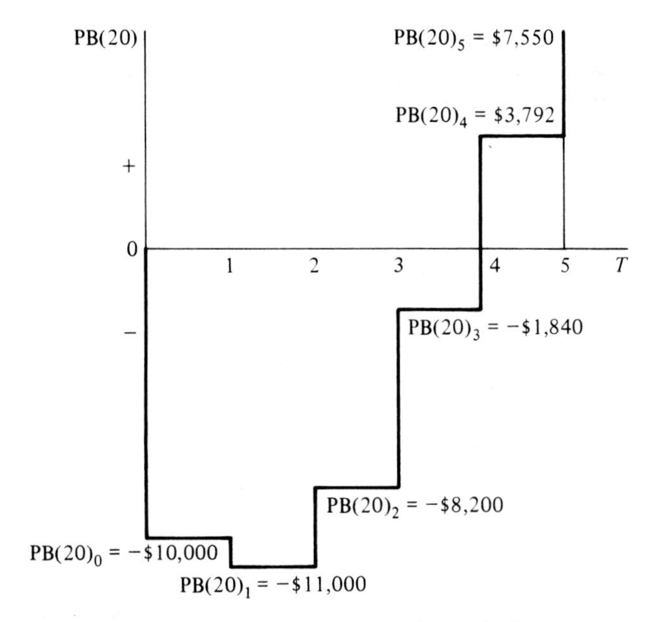

FIGURE 6.9. Project balance diagram for cash flow in Figure 6.8.

3. The net equivalent committed dollars exposed to risk of loss—the area where $PB(i)_t$ is negative.
4. The net equivalent dollars earned—the area where $PB(i)_t$ is positive.

In general, if the life of the investment is n and the point in time at which the project balance becomes positive is n', the conventional form of the project balance is as shown in Figure 6.10.

Although there are numerous cash flow patterns for which it would differ, for most practical investment options the project balance diagram will follow the general pattern shown here. Also, this general profile can be utilized as a visual aid in understanding the four important elements displayed by project balance diagram, as discussed below.

1. *The net future worth, $PB(i)_n$.* As shown earlier, $PB(i)_n$ measures the equivalent future worth of the investments receipts less the equivalent future worth of disbursements evaluated at interest rate i. This element of information can be used just as present worth or annual equivalent is used in conventional analysis, since $PW(i) = PB(i)_n/(1 + i)^n$.

2. *The payback period, n'.* A valuable piece of information associated with any investment concerns the time at which the investment is judged to have no risk of loss. This condition occurs when the $PB(i)_t$ becomes nonnegative at $t = n'$. Generally, if the project is terminated after n', no effective economic loss will be incurred. Of course, it is possible to have more than one n' for a single project, but the project balance diagram easily accommodates this situation.

3. *The exposure to risk of loss.* The project balance describes, period by period, how much committed capital is exposed to risk of loss if the project is terminated before n'. In Figure 6.10, since receipts are received prior to n', the amount of

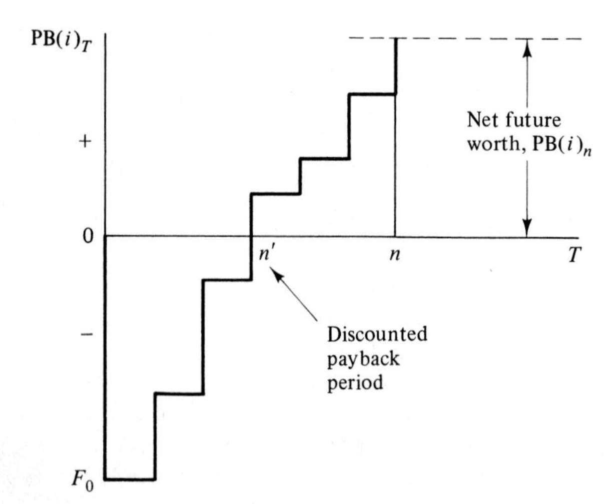

FIGURE 6.10. A general project balance diagram.

possible loss increases for the first period and then decreases with the increase of time. Based on the cash flow in Figure 6.8, we see that for the first period the exposure to risk of loss actually increases, even though there was a receipt of \$1,000 at $t = 1$. This situation correctly shows that the cost of having the funds committed to the investment during the first period exceeds the revenues received during that period. With the project balance diagram, the magnitude of the capital exposed to risk of loss and the rate at which this exposure to possible loss is reduced become immediately evident.

4. *The profit potential.* Once the investment becomes profitable (i.e., the equivalent receipts exceed the equivalent disbursements), the project balance diagram indicates the magnitude of profits expected and the rate at which they will be accumulated. If the investment in Figure 6.8 is held until the end of the fourth year, Figure 6.9 indicates that if the investment can be continued one more year, an additional \$3,758 = (\$7,550 − \$3,792) in equivalent profit will be realized. This information can be quite useful when making decisions about the retirement of assets.

To demonstrate the additional information that can be provided by project balance, examine Figure 6.11, which presents four different cash flows and their associated project balance diagrams. These cash flows all have the same future worth (\$1,200) when they are evaluated at $i = 25\%$. The project balance diagrams in Figure 6.11 indicate that, although these cash flows have identical future worths, their exposure to risk of loss and the point at which they first earn a profit are usually not the same.

Cash Flow 1, which receives its cash receipts more quickly, has the shortest discounted payback period ($n' = 2$ years). Cash Flows 2 and 3 have discounted payback periods of 3 years, while Cash Flow 4 has the longest payback period at $n' = 4$ years. If any cash flow were terminated before its respective discounted payback period, it would suffer a loss.

Equally important, the exposure to the risk of loss can be easily observed by examining the project balance diagram. Cash Flow 4 clearly has the greatest risk of loss, as the amount committed continues to grow until the large receipt is received at $t = 4$. On the other hand, cash flows 2, 3, and 4 have a decreasing risk of loss the longer the investment is continued. Of the four flows, cash flow 1 has the least risk of loss.

Also, the accumulation of profit occurs earlier for cash flow 1. This rate of profit accumulation diminishes as one moves from project balance diagrams 2 to 3 to 4. As mentioned earlier, this information may be extremely useful when attempting to decide when a profitable endeavor should be terminated.

It has been shown that the project balance concept can provide information concerning three distinct characteristics of an investment: (1) discounted payback period, (2) exposure to risk of loss, and (3) rate of profit accumulation. Since these characteristics are not normally provided by conventional $PW(i)$ $AE(i)$, or $FW(i)$ analyses, the project balance can provide new insights into the

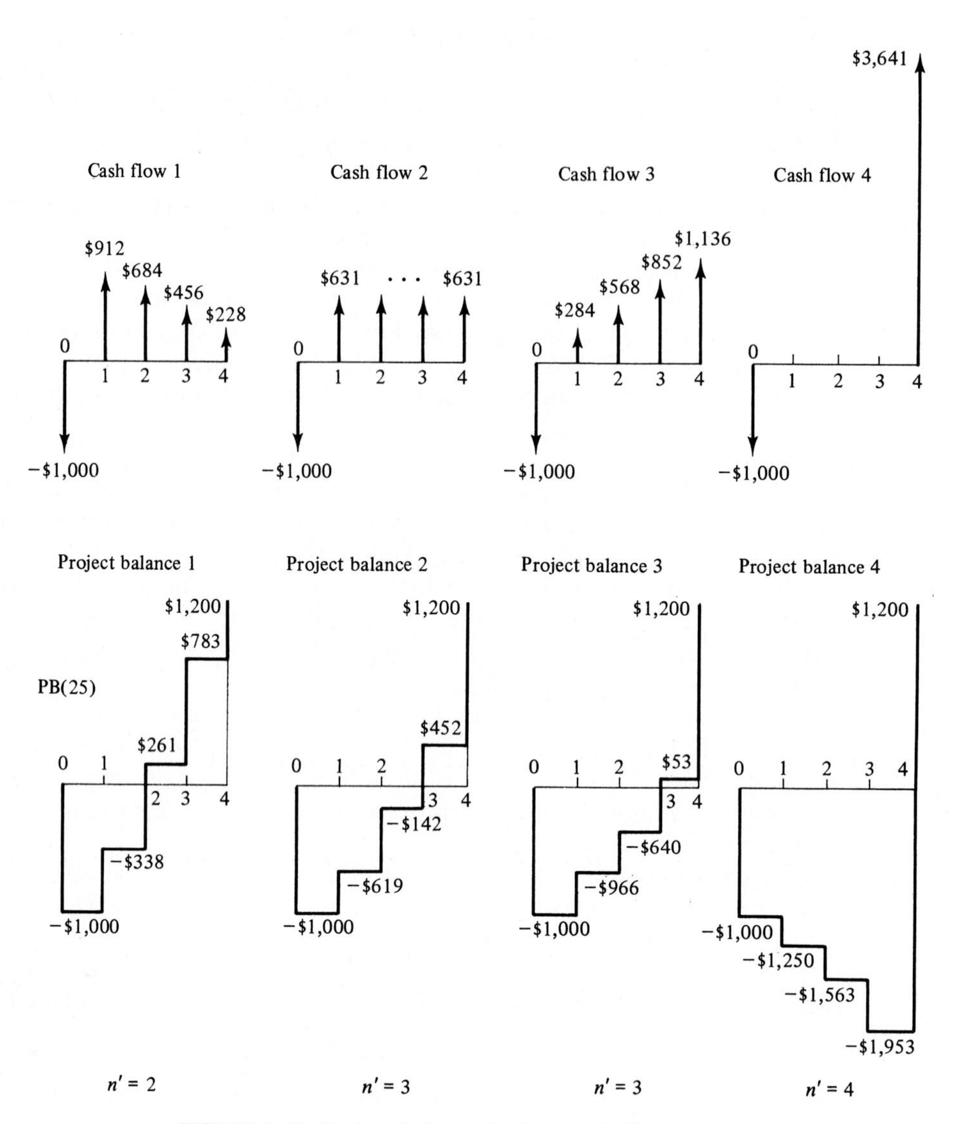

FIGURE 6.11. Project balance for four cash flow patterns.

behavior of investment alternatives. With the use of computer spreadsheets and graphics, these project balance profiles can be easily computed and displayed.

6.8.3 Project Balance and Inflation

When it is important to include the effects of inflation in the project balance calculation, it is necessary to convert the time-dependent values $PB(i)_t$ to values that reflect constant purchasing power as explained in Section 5.3.

Let $PB'(i')_T$ = project balance at constant purchasing power at $t = T$;

f = annual inflation rate;

i = market interest rate;

i' = inflation-free-rate.

Then

$$PB'(i')_T = \sum_{t=0}^{T} F_t(1 + f)^{-t}(1 + i')^{T-t} \qquad \text{for } T = 0, 1, 2, \ldots, n. \qquad (6.13)$$

For the above calculation, the inflation-free rate is found from $i' = [(1 + i)/(1 + f)] - 1$. With these adjustments the project balance provides the exposure to risk of loss in terms of constant purchasing power ($t = 0$ is considered to be the base year for the constant-dollar calculation). Also, the net future worth, $PB'(i')_{n'}$, will be in terms of constant purchasing power.

KEY POINTS

- There are a number of indexes or bases for comparison that summarize different types of information about the economic desirability of investment cash flows. These bases for comparison include present worth, annual equivalent, future worth, the internal rate of return, the payback period, and the project balance.
- Calculation of the $AE(i)$ when cash flows are repeated requires only the calculation of the $AE(i)$ for the first cycle of the repeated cash flows.
- The unrecovered balance, U_t, measures the balance remaining for a cash flow evaluated at the internal rate of return, i^*. The project balance is based on the same formula as unrecovered balance, except that the interest rate utilized is selected external to the cash flows being evaluated.

- If a cash flow has multiple IRR's, the conventional decision rules for selecting a preferred alternative may be invalid. Therefore, IRR should not be applied when an investment cash flow has multiple IRR's.
- The capitalized equivalent, $CE(i)$, measures the present worth of a series of payments extending forever.
- Capital recovery with return, $CR(i)$, measures the annual equivalent cost of the capital associated with the procurement of an asset and its retirement.
- Payback with and without interest do not reflect the consequences of cash flows occurring after the payback period.
- Project balance provides important insights regarding an investment's desirability not normally recognized by the other conventional measures of worth discussed in this chapter.

PROBLEMS

1. An asset with an expected life of 4 years is to be purchased for $5,000. It is estimated that it will produce annual revenues of $2,000 while incurring annual expenses of $400. At retirement at the end of the fourth year it is estimated that the asset will be sold for $900. At that time it is expected that $700 will have to be paid to remove the asset from the plant. The interest rate is 8% compounded annually

   a. Find the present worth of the receipts and the present worth of the disbursements and subtract the equivalent disbursements from the equivalent receipts.

   b. Find the net cash flow for this investment and calculate the net present worth from this cash flow. Compare your result with that of part (a).

2. An investor is considering a business opportunity with the anticipated receipts and disbursements shown below. Calculate the net cash flow representing this investment, and then find the present worth of the net cash flow for an interest rate of 12%. Answer: −$10,103

End of Year	Disbursements	Receipts
0	250,000	$ 0
1	45,000	20,000
2	60,000	60,000
3	15,000	135,000
4	0	180,000
5	0	110,000

3. Graph the $PW(i)$ as a function of i for the following cash flows.

a.

End of year	*0*	*1*	*2*	*3*	*4*
Cash flow	−8,000	$2,000	$2,000	$2,000	$2,000

b.

End of year	*0*	*1*	*2*	*3*	*4*
Cash flow	$1,000	$1,500	$2,000	−$2,500	−$3,000

c.

End of year	*0*	*1*	*2*	*3*	*4*
Cash flow	−$2,000	$7,800	−$10,000	$4,290	0

d.

End of year	*0*	*1*	*2*	*3*	*4*
Cash flow	$500	−$500	$500	−$500	$1,000

Answers:

(a) $PW(i) = -\$8{,}000 + \$2{,}000(\overset{P/A,i,4}{\quad})$; (c) $PW(i) = -\$2{,}000 + \$7{,}800(\overset{P/F,i,1}{\quad}) - \$10{,}000(\overset{P/F,i,2}{\quad}) + \$4{,}290(\overset{P/F,i,3}{\quad})$

4. For the following cash flow, calculate and graph the present worth as a function of the interest rate.

End of year	*0*	*1*	*2*	*3*	*4*	*5*	*6*	*7*
Cash flow	−$50,000	$5,000	$8,000	$11,000	$14,000	$17,000	$20,000	$23,000

a. Assume the interest rate is compounded annually.

b. Assume the interest rate is compounded continuously.

c. Repeat for the annual equivalent amount.

d. Repeat for the future-worth amount.

5. Calculate and graph the present worth as a function of the interest rate for the following cash flow.

End of year	*0*	*1*	*2*	*3*	*4*	*5*
Cash flow	−$10,000	$2,500	$3,500	$4,500	$5,500	$6,500

a. Assume the interest rate is compounded annually.

b. Assume the interest rate is compounded continuously.

c. Repeat for the annual equivalent.

d. Repeat for the future worth.

6. Given the cash flow shown below, find the $PW(i)$ as a function of i and graph the results.

End of year	*0*	*1*	*2*	*3*
Cash flow	−$10,000	$6,000	$5,000	$3,000

a. Assume the interest rate is compounded annually.
b. Assume the interest rate is compounded continuously.
c. Repeat for the $AE(i)$.
d. Repeat for the $FW(i)$.

7. Investments A and B have the net cash flows given below.

End of year	*0*	*1*	*2*	*3*	*4*
A	−$250	$ 75	$ 75	$175	$150
B	−$250	$150	$150	$ 75	$ 75

a. Compare the present worth of A with the present worth of B for an interest rate of 5%. Which has the higher value? Answer: A
b. If the interest rate is 15%, which has the higher value? Answer: B
c. On the same axis, graph the present worth of each investment as a function of the interest rate.

8. Two investment proposals, C and D, have the same first cost of $5,000. The only receipt received from C is a single amount of $21,500, 8 years from the time the investment is undertaken. Proposal D has 8 equal annual receipts of $3,000 beginning one year after the initial investment.

a. Compare $PW(10)_C$ with the $PW(10)_D$. Which is larger?
b. Compare $PW(20)_C$ with the $PW(20)_D$. Which is larger?
c. On the same axis graph the present worth for each of these proposals as a function of interest rate. If the present worths are evaluated at the same interest rate, is there any value of i for which $PW(i)_c > PW(i)_D$?

9. An asset is expected to generate $10,000 each year for the next 8 years. Find $PW(8)$ and $PW(20)$ for this cash flow. From $PW(8)$ calculate $FW(8)$, and from $PW(20)$ determine $FW(20)$. What can be said about $AE(8)/AE(20)$, $PW(8)/PW(20)$, and $FW(8)/FW(20)$?

10. An automobile owner is concerned about the increasing cost of gasoline. He feels that the cost of gasoline will be increasing at the rate of 8% per year over this year's price of 35¢ per liter. Experience with his car indicates that he averages 12 kilometers per liter of gasoline. Since he expects to drive an average of 20,000 kilometers each year, what is the present worth of the cost of fuel for this individual for the next 4 years? Answer: $PW(12) = \$1,974$ What is the annual equivalent cost of fuel over this period of time? The interest rate is 12%.

11. Two investment opportunities are expected to produce the following receipts and disbursements.

	End of year			
Alternative	0	1	2	3
A1	−$1,000	$1,100	$1,210	$1,130
A2	−$1,200	$1,100	$1,210	$1,330

Determine the ratios PW_{A1}/PW_{A2}, AE_{A1}/AE_{A2}, FW_{A1}/FW_{A2} for an interest rate of (a) 0%, (b) 10%. Answers: 1.0; 1.028

12. Consider the cash flows given below and assume that $i = 10\%$.

	End of Year				
Cash Flow	0	1	2	3	4
A	−$100	$50	$50	$50	$50
B	−$100	40	40	60	60

Calculate the present-worth amounts, annual equivalents, and future-worth amounts of these two cash flows. Next calculate PW_A/PW_B, AE_A/AE_B, and FW_A/FW_B. Then compare these ratios. What important observation can be made from this comparison?

13. Find the internal rate of return for the following cash flows.

	Year				
	0	1	2	3	4
(a)	−$10,000	$4,000	$4,000	$4,000	$4,000
(b)	− 800	100	200	300	400
(c)	− 1,000	− 1,000	1,000	1,000	1,000
(d)	− 100	25	25	25	20
(e)	− 2,000	800	400	700	500

14. Find the IRR for the following cash flows.

	Year				
	0	1	2	3	4
(a)	−$5,000	$2,500	$2,500	$2,500	$2,500
(b)	− 400	100	100	100	100
(c)	− 400	200	400	600	800
(d)	− 1,200	250	500	500	500
(e)	− 3,500	1,600	800	1,400	1,200
(f)	1,000	− 900	500	500	500

Answers (a) 34.9%, (c) 82.6%; (f) there is no IRR.

15. A VCR can be purchased for $500. If a credit plan calls for a 20% down payment of the initial cost plus 12 end-of-month payments of $40, what are the monthly and the effective annual rates of return charged by the plan? Answer: 3% per month, 42.6% per year.

16. A silver mine can be purchased for $1,500,000. On the basis of estimated production, an annual net income of $389,000 is foreseen for the next 15 years. After 15 years, the mine will probably be worthless. What annual IRR is in prospect?

17. A VLSI testing station was designed and fabricated for $80,000 with the estimate that it would result in an inspection-cost saving of $10,500 per year for 15 years. With a zero salvage value at the end of 15 years, what was the expected IRR? Actually, the VLSI station became obsolete after 6 years and was sold for $20,000. What was the actual IRR? Answers: 9.97%; 0.9%

18. An individual presently owns an oil lease. The payments from this lease beginning at the end of this year are $10,000, decreasing at the rate of 5% each year. There are 15 annual payments to be received over the life of the lease. A investor offers to purchase the lease for a single payment of $174,642 payable 10 years from the present. Find the IRR on this investment. Answer: 13%

19. An investor deposits $10,000 per year for 4 years, with the first deposit made 1 year from the present. One year after the last deposit the investor makes the 1st withdrawal of $10,000. One year later the second withdrawal is 5% smaller than the first payment withdrawn. The third withdrawal one year later is 5% less than the second withdrawal. There are a total of 15 annual withdrawals, each being 5% less than the previous one.

a. Find the effective annual IRR being earned on this investment to the nearest percent.

b. If the dollars invested and withdrawn in part (a) are in actual dollars and the inflation rate for the 19-year time span of the investment is 9% per year, what is the inflation-free IRR earned on this investment?

20. If a cash flow is composed of a series of disbursements in its early life followed by a series of receipts that total to an amount greater than the disbursements, what can be said about the relationship between the IRR and the $PW(i)$?

21. Graph the present worth of the following cash flows as a function of the interest rate. What is the IRR for each cash flow? Answer: (a) 10%, 33%

	Year		
	0	1	2
(a)	−$35,000	$85,000	−$51,150
(b)	− 4,000	8,920	− 4,968
(c)	1,200	− 3,600	6,000
(d)	− 4,000	8,400	4,000

22. Graph the $PW(i)$ as a function of i for each cash flow shown below. The range of interest-rate values for which these graphs should be drawn extends from 0% to 50%. From the graphs determine the internal rates of return for each cash flow.

	End of Year				
	0	1	2	3	4
(a)	−$ 3,000	$ 7,500	$ 4,620	$ 0	$ 0
(b)	− 10,000	50,000	− 93,500	77,500	− 24,024
(c)	− 2,000	0	0	20,000	− 20,000
(d)	− 5,000	19,500	− 25,150	10,725	0
(e)	− 500	1,800	− 2,160	864	0
(f)	− 6,000	6,000	− 1,800	6,000	0
(g)	− 4,000	1,000	− 2,000	3,600	3,000

23. Apply Test 1 and Test 2 to the cash flows in Problem 22 and determine whether the present-worth function for the cash flows is of the form shown in Figure 6.3.

24. For the cash flows described in Problem 22 indicate the ranges of i, the interest rate, for which the cash flow is economically desirable and economically undesirable.

25. Find the cash flow pattern that has the multiple internal rates of return given below.

a. 20%, 40%, 50%, and 60%.

b. 10%, and 50%.

c. 15%, 40%, and 100%.

d. 20%, 20%, and 20%.

e. 5%, 25%, and 125%.

26. Develop a cash flow having a present-worth function that intersects the horizontal axis only once, although Descartes' rule of signs indicates that the cash flow will have a maximum number of positive, real roots that exceeds one.

27. Compare payback without interest to payback with interest for the following cash flows, assuming $i = 15\%$.

Cash Flow (End of Year)

	0	1	2	3	4	5
(a)	−$1,000	$ 300	$ 300	$ 300	$ 300	—
(b)	−$5,000	$1,000	$2,000	$3,000	$4,000	—
(c)	−$3,000	$3,000	$2,000	$1,000	—	—
(d)	−$1,000	$ 150	$ 150	$ 150	$ 150	$ 400
(e)	−$2,000	−$1,000	−$ 500	$ 600	$1,200	$2,400
(f)	$1,000	$1,000	−$1,300	−$ 700	$ 500	$1,000

28. An investment of $100,000 is to yield an annual net benefit of $20,000. Use Figure 6.6 to find the payback period in years if $i = 10\%$; if $i = 15\%$. Rework analytically using the applicable formula and compare the result with that found graphically.

29. Compare payback without interest to payback with interest for the following cash flows if $i = 10\%$; if $i = 30\%$.

Cash Flow (End of Year)

	0	1	2	3	4	5
(a)	−$1,000	$ 200	$ 200	$ 200	$ 200	$ 200
(b)	−$8,000	−$2,000	$11,000	$9,000	$7,000	—
(c)	−$5,000	$1,400	$ 1,400	$1,400	$1,400	$ 1,400
(d)	−$3,000	$1,000	$ 2,000	$2,000		—
(e)	$2,000	$3,000	−$ 4,000	−$2,000	−$1,900	$ 1,000
(f)	−$6,000	$ 0	$ 0	$ 0	$ 0	$20,000

Answer for (c): Without interest = 4; with interest at 10% = 5; with interest at 30%, > 5.

30. Using the algebraic form of the $(\overset{P/A,i,n}{\quad})$ factor, find the value of that factor when $n = \infty$. This result allows for the calculation of the $CE(i)$, which assumes an equal annual series that extends forever.

31. Calculate the present worth for the following cost cash flows that extend forever. The interest that can be earned into perpetuity is 7%.

	End of Year					
	1	2	3	4	. . .	∞
(a)	$22,000	$22,000	$22,000	$22,000	. . .	$22,000
(b)	11,000	11,000	12,000	12,000	. . .	12,000
(c)	5,000	5,000	5,000	2,000	. . .	2,000
(d)	15,000	20,000	20,000	20,000	. . .	20,000
(e)	0	0	3,000	3,000	. . .	3,000

32. A tunnel to transport water through a mountain range requires periodic maintenance. If the maintenance costs are as shown below for each 6-year maintenance cycle, what is the capitalized equivalent of these expenses? The interest rate is 7%

End of Year	1	2	3	4	5	6
Maintenance costs	$35,000	$35,000	$35,000	$45,000	$45,000	$45,000

33. A woman is considering giving an endowment to a university in order to provide payments of $5,000, $4,000, $3,000, and $2,000, respectively, at the end of the first, second, third, and fourth quarters during a year. If the interest rate is 12% compounded quarterly, what is the capitalized equivalent that must be deposited now so that the quarterly payments can be repeated forever? Answer: $117,900

34. The operating and maintenance costs of a public park are estimated to be $9,000 at the beginning of each year, based on the purchasing power of a dollar at the present. The MARR is 16% compounded annually. If inflation is expected to be 6% per year and the park is to be in existence in perpetuity, how much would a philanthropist have to deposit at the present to assure that all these costs can be paid?

a. Solve using constant-dollar analysis.

b. Solve using the geometric-gradient factor.

35. By depositing $2,000 at the present, an individual wishes to generate a series of annual amounts that begin with $100 one year from the present and grow at the rate of 10% per year. If this series continues forever, what interest rate is being earned on the deposit? Answer: 15%

36. A firm requires power shovels for its open-pit mining operation. This mining equipment, with a first cost of $500,000, has an estimated salvage value of $70,000 at the end of 10 years' service. If the firm uses a rate of interest of 16% for the project evaluation, how much must be earned on an equivalent annual basis so that the firm recovers its invested capital plus earns a return on the capital committed to the equipment during its lifetime?

37. A mechanical robot that is part of a welding operation on an automobile assembly line costs $750,000. Because of the specialized functions it performs its expected useful life is only 6 years, while its estimated salvage value when retired is anticipated to be approximately $40,000. What is the capital recovery with return for this investment if the interest rate is 20%? What annual labor savings must be realized over the robot's life to justify this expenditure? Answer: $221,500

38. A college freshman wants to buy a portable computer for her engineering class and hopes to resell it at the time of graduation. She can purchase the unit at $1,750 with an estimated salvage value of $250 at the end of 4 years. For an interest rate of 14%, find the equivalent annual cost of this investment using capital recovery with return. Answer: $549.80

39. In problem 38, assume that the salvage value of the computer is zero. Find the $AE(i)$ cost using Figure 6.6. Compare the result with that found by formula.

40. A company can invest in one of two mutually exclusive alternatives. The life of each alternative is estimated to be 5 years, with the following initial investments and salvage values. Assume that the rate of interest is 17%.

	Alternative	
	A	*B*
Investment	−$10,000	−$12,000
Salvage value	1,500	3,500

a. What is the $CR(i)$ for each alternative?

b. Determine the salvage value at the end of the project life for Alternative *B* that will result in the same capital recovery with return for both alternatives.

41. For the cash flows in Problem 27, plot the project balance for each period over the life of the cash flow. Compute the future worth of these cash flows using interest factors and compare to $PB(15)_5$.

42. Compute the project balances at each point during the life of the cash flows in Problem 29. Determine the discounted payback period and the future worth of the investment from the project balance values.

43. Give an interpretation of the following project balance profiles. Identify the payback with interest, the future worth, and the exposure to risk of loss.

Investment	Project Balance (End of Year)					
	$PB(i)_0$	$PB(i)_1$	$PB(i)_2$	$PB(i)_3$	$PB(i)_4$	$PB(i)_5$
(a)	−$5,000	−$ 3,600	−$1,500	$1,200	$3,000	$3,500
(b)	−$3,000	$ 200	$ 200	$ 200	$ 200	$ 200
(c)	−$7,000	−$11,000	−$6,000	−$2,000	−$1,000	−$ 500
(d)	$6,000	$ 7,500	$5,300	$2,300	$2,000	1,500
(e)	−$4,000	$ 1,000	$1,500	−$2,000	−$1,100	$1,800
(f)	−$2,000	−$ 3,000	−$4,000	−$5,000	−$6,000	$3,500

44. The project balance profiles for proposed investment projects are presented below. Determine the initial cash flow, the discounted payback period, and the future worth for each project. Explain your assessment of the exposure to risk of loss and the rate of profit accumulation as represented by the project balance profile.

Project	Project Balance (End of Year)					
	$PB(i)_0$	$PB(i)_1$	$PB(i)_2$	$PB(i)_3$	$PB(i)_4$	$PB(i)_5$
(a)	−$ 9,000	−$5,000	−$2,500	−$ 1,500	−$ 800	$ 500
(b)	−$15,000	$1,000	$1,500	$ 2,000	$2,000	$1,500
(c)	−$ 5,000	−$5,500	−$8,300	−$10,000	−$1,000	−$1,500
(d)	−$ 6,000	−$6,000	$ 500	$ 1,000	$1,000	$9,000
(e)	−$12,000	$8,000	$8,200	$ 8,400	$8,700	$9,000
(f)	−$ 4,000	$1,000	−$4,000	$ 500	−$4,000	−$1,000

7

Decision Making Among Alternatives

Previous chapters provided fundamentals for calculating economic equivalence and set forth various bases for the comparison of alternatives. This chapter focuses on a systematic procedure for decision making among alternatives in accordance with established criteria.

> A **decision criterion** is a rule or procedure that prescribes how to select investment alternatives so that certain objectives can be achieved.

The degree to which these objectives are realized depends on the efficacy of the decision criterion. It is important that the strengths and weaknesses of the most commonly used decision criteria be fully understood. Accordingly, this chapter examines decision criteria with respect to the fundamentals of economic equivalence.

7.1 TYPES OF INVESTMENT PROPOSALS

As discussed in Section 1.4.3 engineering proposals represent the means for altering the physical environment to achieve particular objectives. In this section the more general term *investment proposal* is adopted to include not only engineering proposal but any economic arrangement that can be described by a cash flow.

An **investment proposal** is a single undertaking or project being considered as an investment possibility.

Whether proposals are independent of each other or whether they are dependent in some way determines the selection process by which one proposal will be judged superior to another proposal. The following paragraphs define independent and dependent proposals and provide guidelines for identifying such proposals.

7.1.1 Independent Proposals

Even though few proposals in a firm are truly independent, for practical purposes it is common to assume that certain proposals are independent. Usually, if proposals are functionally different and there are no obvious dependencies between them, it is reasonable to consider the proposals to be independent.

A proposal is said to be **independent** when the acceptance of a proposal from a set has no effect on the acceptance of any of the other proposals in the set.

For example, proposals concerning the purchase of a numerically controlled milling machine, a security system, office furniture, and fork lift trucks would, under most circumstances, be considered independent.

7.1.2 Dependent Proposals

For many decision situations, a group of proposals will be related to one another in such a way that the acceptance of one of them will influence the acceptance of the others. Such interdependencies among proposals occur for a variety of reasons.

Proposals are said to be **mutually exclusive** if the proposals contained in the set of proposals being considered are related so that the acceptance of one proposal from the set precludes the acceptance of any of the others.

Mutually exclusive proposals usually occur when a decision maker is attempting to fulfill a need and there are a variety of proposals, each of which will satisfy that need.

For example, a road-building contractor may require additional earthmoving capability. There may be a number of types of equipment, each of which could perform the function desired. Although these proposals may have different first costs and different operating characteristics, they are still considered to be mutually exclusive for decision-making purposes, since the selection of one precludes the selection of the others.

Another type of relationship between proposals arises from the fact that once some initial project is undertaken, a number of other auxiliary investments become feasible as a result of the initial investment. Such auxiliary proposals are called *contingent* proposals.

> A **contingent proposal** is one whose acceptance is dependent on the acceptance of some prerequisite proposal, whose acceptance in turn is independent of acceptance of the contingent proposal.

Thus, the purchase of computer software is contingent on the purchase of computer hardware. The construction of the third floor of a building is contingent on the construction of the first and second floors. A contingent relationship is a one-way dependency between proposals.

When there are limitations on the amount of money available for investment and the initial cost of all the proposals exceeds the money available, financial interdependencies are introduced between proposals. These interdependencies are usually complex, and they will occur whether the proposals are independent, mutually exclusive, or contingent. Thus, whenever a budget constraint is imposed on some decision problem, interdependencies that may not be obvious are being introduced.

7.1.3 Investment Alternative

It is important to distinguish an investment proposal from an *investment alternative.*

> An **investment alternative** is a decision option that represents a course of action.

According to this definition and the definition of an investment proposal, an investment alternative can consist of a group or set of proposals. It can also represent the option of rejecting all proposals under consideration, an action commonly referred to as "doing nothing."

7.2 FORMING MUTUALLY EXCLUSIVE ALTERNATIVES

Engineering proposals can be independent, mutually exclusive, or contingent, and additional interdependencies among them can exist if there is a limited amount of money to invest. To devise special rules to include each of these different relationships in a decision criterion would make the procedure complicated and difficult to apply.

In order to provide a simple method of handling various types of proposals and to provide insight into formulations of the decision problem, a general approach is presented. This approach requires that all proposals be arranged so that the selection decision involves only the consideration of cash flows for *mutually exclusive alternatives.*

TABLE 7.1.
FORMING INVESTMENT ALTERNATIVES FROM PROPOSALS

	Proposals		
Alternatives	$P1$	$P2$	Action
$A0$	0	0	Do Nothing
$A1$	1	0	Accept $P1$
$A2$	0	1	Accept $P2$
$A3$	1	1	Accept $P1$ and $P2$

A general procedure for forming mutually exclusive alternatives from a given set of proposals is based on an enumeration of all possible combinations of the proposals. For example, if two proposals ($P1$ and $P2$) are being considered, four mutually exclusive investment alternatives exist, as shown in Table 7.1. Note that a binary variable, $X_j = 0$ or 1, is used to indicate proposal rejection or acceptance.

Generalization of the procedure used in Table 7.1 for k proposals, $k = 1, 2, 3, \ldots$ leads to a number of alternatives A, given by

$$A = 2^k.$$

A 0–1 matrix exhibiting all possible alternatives (rows) is shown in Table 7.2. Let $1, 2, 3, \ldots, k-1, k$ designate proposals (columns) from left to right. Moving down $P1$, the first column ($k = 1$), place a single zero followed by a single one alternating until all alternatives have been assigned a zero or one. Next, go to $P2$, the second column ($k = 2$). Move down the column and place 2^{k-1} zeros followed by 2^{k-1} ones, alternating until each alternative has an entry. For the third column, $P3$, $k = 3$ so that $2^{k-1} = 4$. Thus, moving down that column, place 4 zeros followed by 4 ones, repeating until all alternatives have an assigned value. This process is repeated for each proposal (column), and the results are presented in Table 7.2. Since all the entries in row 1 will be zero, $A0$ represents the Do Nothing alternative.

The approach just presented makes possible the consideration of a variety of proposal relationships in a single form: the mutually exclusive alternative. Therefore, any decision criteria that are selected to make decisions about mutually exclusive alternatives can also accommodate proposals that are independent, mutually exclusive, or contingent if the proposals are arranged into mutually exclusive alternatives. In addition, the imposition of a budget constraint can also be easily incorporated into the decision process. Thus, a conceptually simple and consistent procedure is available for a wide range of investment situations.

As an example application, suppose that four engineering proposals are being considered with cash flows over a 10-year planning horizon as shown in

TABLE 7.2.
GENERAL 0–1 MATRIX OF INVESTMENT ALTERNATIVES

Investment Alternatives	Proposals					
	*P*1	*P*2	*P*3	. . .	*P*(*k* − 1)	*P*(*k*)
*A*0	0	0	0	. . .	0	0
*A*1	1	0	0		0	0
*A*2	0	1	0		0	0
*A*3	1	1	0		0	0
*A*4	0	0	1		0	0
*A*5	1	0	1		0	0
.	.					.
.	.					.
.	.					.
$A(2^k - 2)$	0	1	1		1	1
$A(2^k - 1)$	1	1	1	. . .	1	1

Table 7.3. Proposals *P*1 and *P*2 are mutually exclusive, proposal *P*3 is contingent on proposal *P*1, and proposal *P*4 is contingent on proposal *P*2. The budget limit is $100,000.

Given four proposals, $A = 2^4$ or 16 possible mutually exclusive investment alternatives to consider. These alternatives are enumerated by following the procedure in Table 7.2, which gives Table 7.4.

Next, the feasibility of each mutually exclusive alternative must be tested. The tests to be applied derive from dependencies among the proposals. First, composite cash flows are calculated for each alternative from the cash flows in Table 7.3 and the involved proposals, as exhibited in Table 7.4. These are shown in Table 7.5.

Finally, infeasible alternatives are identified so they can be eliminated from further consideration. Table 7.6 summarizes the results from a search for infeasible alternatives. Remaining are investment alternatives *A*0, *A*1, *A*2, and *A*10, all of which are feasible and mutually exclusive. Selecting the best from among these is the purpose of subsequent sections.

TABLE 7.3.
CASH FLOWS IN THOUSANDS FOR FOUR PROPOSALS

Cash Flows	Proposals			
	*P*1	*P*2	*P*3	*P*4
Initial investment	$30	$22	$82	$70
Net annual benefit	8	6	18	14
Salvage value	3	2	7	4

TABLE 7.4.
MATRIX OF INVESTMENT ALTERNATIVES FOR FOUR PROPOSALS

Investment Alternative	Proposals			
	$P1$	$P2$	$P3$	$P4$
$A0$	0	0	0	0
$A1$	1	0	0	0
$A2$	0	1	0	0
$A3$	1	1	0	0
$A4$	0	0	1	0
$A5$	1	0	1	0
$A6$	0	1	1	0
$A7$	1	1	1	0
$A8$	0	0	0	1
$A9$	1	0	0	1
$A10$	0	1	0	1
$A11$	1	1	0	1
$A12$	0	0	1	1
$A13$	1	0	1	1
$A14$	0	1	1	1
$A15$	1	1	1	1

TABLE 7.5.
COMPOSITE CASH FLOWS IN THOUSANDS

Investment Alternative	Initial Investment	Net Annual Benefit	Salvage Value
$A0$	$ 0	$ 0	$ 0
$A1$	30	8	3
$A2$	22	6	2
$A3$	52	14	5
$A4$	82	18	7
$A5$	112	26	10
$A6$	104	24	9
$A7$	134	32	12
$A8$	70	14	4
$A9$	100	22	7
$A10$	92	20	6
$A11$	122	28	9
$A12$	152	32	11
$A13$	182	40	14
$A14$	174	38	13
$A15$	204	46	16

TABLE 7.6.
IDENTIFYING INFEASIBLE ALTERNATIVES

Investment Alternative	Alternative Feasible?	Reasons for Infeasibility
*A*0	Yes	
*A*1	Yes	
*A*2	Yes	
*A*3	No	*P*1 and *P*2 mutually exclusive
*A*4	No	*P*3 conditional on *P*1
*A*5	No	Budget constraint
*A*6	No	*P*3 conditional on *P*1 and budget constraint
*A*7	No	*P*1 and *P*2 mutually exclusive and budget constraint
*A*8	No	*P*4 contingent on *P*2
*A*9	No	*P*4 contingent on *P*2
*A*10	Yes	
*A*11	No	*P*1 and *P*2 mutually exclusive and budget constraint
*A*12	No	*P*3 conditional on *P*1, *P*4 conditional on *P*2, and budget constraint
*A*13	No	*P*4 conditional on *P*2 and budget constraint
*A*14	No	*P*3 conditional on *P*1 and budget constraint
*A*15	No	*P*1 and *P*2 mutually exclusive and budget constraint

For problems involving only a few proposals the general technique just presented for arranging various types of proposals into mutually exclusive alternatives can be computationally practical. For larger numbers of proposals, however, the number of mutually exclusive alternatives becomes quite large and this approach becomes computationally cumbersome. For instance, the maximum number of mutually exclusive alternatives, N, that can be obtained where

S = the maximum number of sets of proposals that are independent;

M_j = the maximum number of proposals within each set j where each proposal in that set is mutually exclusive;

is

$$N = \prod_{j=1}^{S} (M_j + 1) = (M_1 + 1)(M_2 + 1)(M_3 + 1) \ldots (M_s + 1)$$

Suppose a decision maker had the following proposals to consider:

$A1 \quad A2 \quad A3 \quad A4 \quad A5 \quad A6$
$B1 \quad B2 \quad B3$
$C1 \quad C2 \quad C3 \quad C4$
$D1 \quad D2$

where the proposals in each row are mutually exclusive and the set of proposals in each row is independent from any other set or row of proposals. That is, proposals *A*1 and *A*2 are mutually exclusive but proposals *A*1 and *B*1 are independent. The maximum number of mutually exclusive alternatives that can be obtained from the group of proposals in this example is found to be 420:

$$S = 4 \text{ (the number of rows)}$$

$$M_1 = 6, \quad M_2 = 3, \quad M_3 = 4, \quad M_4 = 2$$

so that

$$N = (M_1 + 1)(M_2 + 1)(M_3 + 1)(M_4 + 1)$$

$$= (7)(4)(5)(3) = 420.$$

Thus, for this relatively small problem the number of mutually exclusive alternatives is sizable.

Using the procedure in Table 7.6 for the 15 proposals just considered would require the enumeration of $2^{15} = 32{,}768$ alternatives. However, due to the mutually exclusive relationships there are only 420 feasible alternatives as calculated for *N*. Thus, substantial saving in time and effort can be realized by avoiding complete enumeration.

To solve these types of problems manually requires ingenuity on the part of the analyst. For small problems (a couple of hundred alternatives or less) the analyst should list only the feasible alternatives based on the mutually exclusive relationships. In some cases, an examination of the measure of worth for each individual proposal will allow for immediate elimination of certain proposals from consideration. For example, if the $PW(i)_j < 0$ and the objective is to maximize total present worth, then proposal j can be eliminated before the listing of feasible alternatives.

When there are limited funds, an additional strategy is to first solve the problem as if there were no limit on the funds available to invest. Using this assumption reduces the problem to finding the best proposal in each set of mutually exclusive proposals. For the above problem four separate decisions can be made regarding the proposals designated *A*, *B*, *C*, and *D*. The remainder of this chapter is devoted to the decisions rules necessary to find the best alternative in a set of mutually exclusive alternatives.

7.3 ELEMENTS OF DECISION CRITERIA

This section presents three elements of decision criteria that require special attention in the process of comparing mutually exclusive alternatives. These are

(1) the differences between alternatives, (2) the minimum attractive rate of return. and (3) the Do Nothing alternative.

7.3.1 Differences Between Alternatives

In Section 1.5, it is pointed out that examining the differences between alternatives is sufficient to judge the superiority of one alternative compared to another. This concept is basic to the decision criteria to be presented in this chapter and is central to all economic comparisons. No concept presented in this book is more fundamental.

When comparing mutually exclusive alternatives, it is the difference between them that is relevant for determining the economic desirability of one compared to the other.

The reason the difference between alternatives is so fundamental is demonstrated in Table 7.7. To compare Alternatives *A*1 and *A*2, it is sufficient to examine the cash flow that represents the difference between them. The cash flow representing Alternative *A*2 can be viewed as being the sum of two separate and distinct cash flows, as shown in Figure 7.1. One of these is identical to the cash flow of Alternative *A*1. The other represents the difference between *A*1 and *A*2. To decide which of the two alternatives is economically superior, it is sufficient to use the following simple decision rule.

If cash flow ($A2 - A1$) is economically *desirable*,
Alternative *A*2 is preferred to Alternative *A*1.
If cash flow ($A2 - A1$) is economically *undesirable*,
Alternative *A*1 is preferred to Alternative *A*2.

If the cash flow representing the differences between the alternatives is economically desirable, then Alternative *A*2, consisting of a cash flow that is the sum of a cash flow like Alternative *A*1 and a desirable cash flow, is clearly economically superior to *A*1. On the other hand, if the difference between alternatives is con-

TABLE 7.7.
DIFFERENCES BETWEEN MUTUALLY EXCLUSIVE ALTERNATIVES

	Alternatives		
End of Year	*A*1	*A*2	Cash Flow Difference ($A2 - A1$)
0	−$1,000	−$1,500	−$500
1	800	700	−100
2	800	1,300	500
3	800	1,300	500

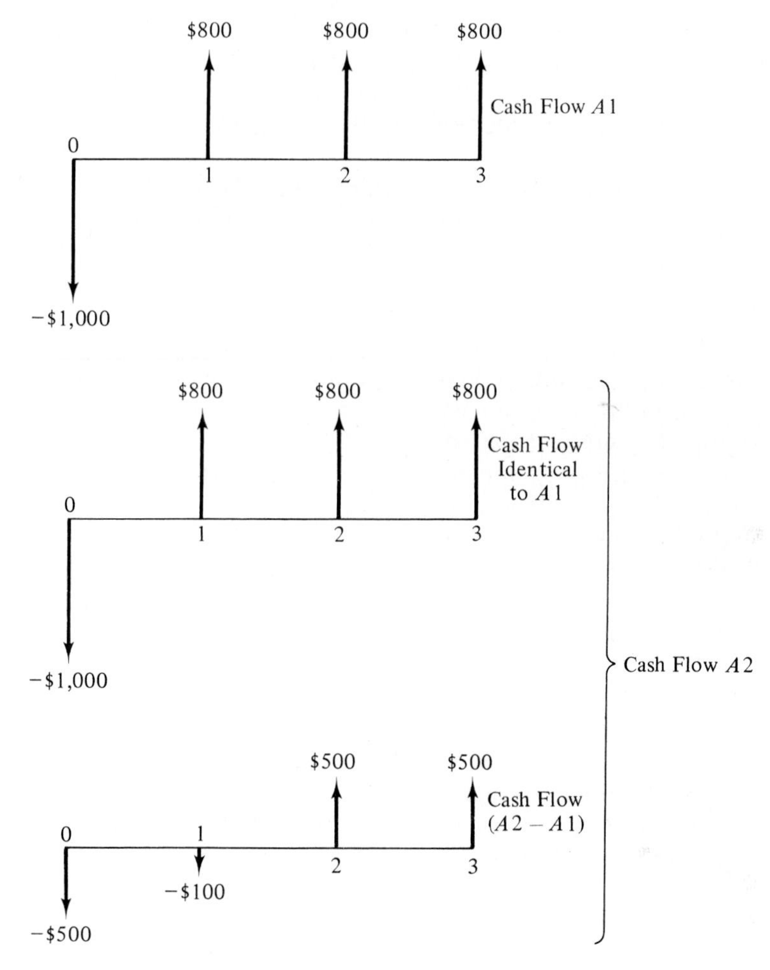

FIGURE 7.1. Cash flow difference between two alternatives.

sidered to be undesirable, Alternative $A2$ is inferior to $A1$. This argument is fundamental to most decision processes.

For the example cited in Table 7.7, the decision to undertake Alternative $A2$ rather than $A1$ requires an additional or incremental investment of \$500 now and \$100 one year hence. The extra receipts expected from the extra investment are \$500 at the end of years 2 and 3. Do the extra receipts justify the extra investment? This is the question that must be answered to determine which of the two alternatives is economically more desirable.

7.3.2 Minimum Attractive Rate of Return (MARR)

The decision criteria discussed in subsequent sections have as their objective the maximization of equivalent profit, given that all investment alternatives must yield a return that exceeds some *minimum attractive rate of return*. This cut-off rate is usually the result of a policy decision made by the management of the firm. Thus, the firm identifies a target rate of interest that represents a "floor" rate, or lowest rate of return that will be considered acceptable.

The **minimum attractive rate of return (MARR)** is a cut-off rate representing a yield on investments that is considered minimally acceptable.

The minimum attractive rate of return can be viewed as a rate at which the firm can always invest, since it has a large number of opportunities that yield such a return. Thus, whenever any money is committed to an investment proposal, an opportunity to invest that money at the MARR has been foregone. For this reason the minimum attractive rate of return is sometimes considered to be an "opportunity" cost, as shown in Figure 7.2.

Over the years there has been much discussion about how to select the MARR. Unfortunately, there is yet to be offered a completely satisfactory method for precisely determining this rate. Because the rate that is selected represents the firm's profit objectives, it is usually based on the judgment of the firm's senior management. This judgment is in turn based on the management's view of the firm's future opportunities along with the firm's financial situation.

If the MARR selected is too high, many investments that have good returns may be rejected. On the other hand, a rate that is too low may allow the acceptance of proposals that are marginally productive or result in economic loss. Thus, when choosing a MARR there is a trade-off between being too selective or not being selective enough.

One method for selecting a MARR is to examine the proposals available for investment and to identify the maximum rate that can be earned if the funds are

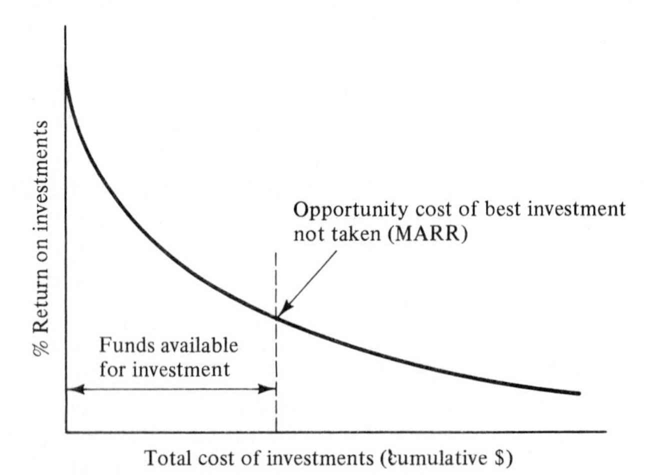

FIGURE 7.2. The MARR as an opportunity cost.

not invested in them. For example, an individual should avoid selecting a MARR that is less than the interest rate banks are paying on savings accounts. This is because the individual *always* has the opportunity to invest at the bank rate regardless of his other investment opportunities. For this reason the Do Nothing alternative (which represents the return earned if all proposals under consideration are rejected) assumes that all available funds are invested at the MARR.

Another consideration in choosing a MARR relates to the rationing of the scarce resource, investment capital. For example, a large firm may want to assure that funds allocated to various divisions within the firm are used effectively. If there is considerable variance in the quality of proposals produced by one division compared to another, the appropriate MARR will prevent investing in unproductive proposals. This allows for the redistribution of the uninvested funds to the divisions that do have high-return proposals.

This concept of capital rationing can also be applied to investment decisions to be made over some time span. The fact that there are business cycles produces fluctuations in the quality of investment proposals available at various points in time. The proper selection of the MARR can prevent investing in marginally productive proposals during the "down" years. These unspent funds can then be made available for financing the higher-quality proposals that are available in the "up" years.

The MARR should not be confused with the cost of capital: a composite rate that represents the cost of providing money from external sources through the sale of stock, the sale of bonds, and by direct borrowing. Normally the minimum attractive rate of return is substantially higher than the cost of capital. Where a firm's cost of capital may be 12%, its minimum attractive rate of return may be 20%. This difference occurs because few firms are willing to invest in projects that are expected to earn slightly more than the cost of capital, owing to the risk elements in most projects and because of uncertainty about the future.

7.3.3 The Do Nothing Alternative

In many engineering economy studies it is assumed that if the funds available are *not* invested in the projects being considered, they will be invested in the Do Nothing alternative. The Do Nothing alternative does not mean that the funds would be "hidden under a mattress," thereby yielding no return.

> The **Do Nothing alternative** means that the investor will "do nothing" about the projects being considered and that the funds made available by not investing will be placed in investments that yield an IRR equal to the MARR.

For the Do Nothing alternative the result of these assumptions is summarized as

$$i_{A0}^{*} = \text{MARR}. \tag{7.1}$$

Because the rate of return is defined as the interest rate that causes the present-worth, annual equivalent, or future-worth amounts to equal zero, for the Do Nothing alternative it follows that

$$PW(\text{MARR})_{A0} = 0$$

$$AE(\text{MARR})_{A0} = 0$$

$$FW(\text{MARR})_{A0} = 0.$$

These expressions indicate that when the Do Nothing alternative is evaluated at the MARR, its equivalent profit is always zero. This fact simplifies the comparison of alternatives, because the cash flow pattern of the Do Nothing alternative does not have to be known and it can be assumed for computational purposes that the cash flows associated with the Do Nothing alternative are zero.

Many an error has been committed in the comparison of investment alternatives by the omission of the Do Nothing alternative. When the alternatives being considered are described by both their receipts and disbursements, it is important that the option to invest at the MARR also be considered. Too frequently this option is ignored or forgotten, and investment decisions that are less profitable than "doing nothing" are selected. This misuse of investment funds can be easily prevented by recognizing that there is usually the option of not investing in the alternatives being considered.

When the receipts from a number of alternatives being compared are assumed to be identical, it is common to describe the cash flows of the alternatives by displaying only their costs. In this case the Do Nothing alternative will *not* be considered, because it is based on the assumption that *both* its receipts and disbursements are invested at the MARR, yielding a net equivalent profit of zero.

7.4 PRESENT WORTH ON INCREMENTAL INVESTMENT

It is shown in Section 7.3.1 that it is the differences between mutually exclusive alternatives that are relevant for decision-making purposes. The criterion of present worth on incremental investment demonstrates this rule, since it requires that the incremental differences between alternative cash flows actually be calculated.

In comparing one alternative to another, first determine the cash flow representing the difference between the two cash flows. Then the decision whether to select a particular alternative rests on the determination of the economic desirability of the additional increment of investment required by one alternative over the other. The incremental investment is considered to be desirable if it yields a return that exceeds the MARR. If

$$
\boxed{\begin{array}{l} PW(i)_{A2-A1} > 0\text{: accept } A2 \\ PW(i)_{A2-A1} \leq 0\text{: reject } A2 \text{ and accept } A1. \end{array}} \qquad (7.2)
$$

To apply this decision criterion to a set of mutually exclusive alternatives such as shown in Table 7.8, proceed as follows:

1. List the alternatives in ascending order based on their initial capital outlays. This is already done for Table 7.8 data.

2. Select as the initial "current best" alternative the one that requires the smallest first cost. In most cases the initial "current best" alternative will be Do Nothing, as in this example. The exclusion of the Do Nothing alternative can lead to the investment of a scarce resource, money, in unproductive activities—that is, activities that yield a return that is less than the MARR.

3. Compare the initial "current best" alternative and the first "challenging" alternative. The challenger is always the next highest alternative in order of first-year cost that has not been previously involved in a comparison. The comparison is accomplished by examining the differences between the two cash flows. If the present worth of the incremental cash flow evaluated at the MARR is greater than zero, the challenger becomes the new "current best" alternative. If the present worth is less than or equal to zero, the "current best " alternative remains unchanged, and the challenger in the comparison is eliminated from consideration. The new challenger is the next alternative in order of first-year cost that has not been a challenger previously. Then the next comparison is made between the alternative that is the "current best" and the alternative that is currently the challenger.

4. Repeat the comparisons of the challengers to the "current best" alternative as described in step 3. These comparisons are continued until every alternative other than the initial "current best" alternative has been a challenger. The alternative that maximizes present worth and provides a rate of return that exceeds the MARR is the last "current best" alternative.

TABLE 7.8.
CASH FLOWS REPRESENTING FOUR MUTUALLY EXCLUSIVE ALTERNATIVES

	Alternatives			
End of Year	$A0$	$A1$	$A2$	$A3$
0	\$0	−\$5,000	−\$8,000	−\$10,000
1–10	0	1,400	1,900	2,500

Steps 3 and 4 lead to the following calculations for the alternatives being considered in Table 7.8. Assume that the MARR is equal to 15%.

The first comparison to be made in this example is between Alternative $A1$ (the first challenger) and the Do Nothing alternative (the initial "current best" alternative). The subscript notation in $PW(15)_{A1-A0}$ indicates the present worth is for the cash flow representing the difference between Alternative $A1$ and Do Nothing.

$$PW(15)_{A1-A0} = -\$5{,}000 + \$1{,}400(\overset{P/A,\,15,10}{5.0188}) = \$2{,}026.$$

Note that when comparing an alternative to the Do Nothing alternative, the cash flow representing the incremental investment is the same as the cash flow on the total investment.

Because the present worth of the differences between the cash flows is greater than zero ($2,026), Alternative $A1$ becomes the new "current best" alternative as dictated by step 3. The second challenger becomes $A2$. Alternative $A2$ is then compared to $A1$ on an incremental basis as follows:

$$PW(15)_{A2-A1} = -\$3{,}000 + \$500(\overset{P/A,\,15,10}{5.0188}) = -\$490.$$

Since this value is negative, Alternative $A2$ is dropped from further consideration and $A1$ remains the "current best" alternative. The third challenger is Alternative $A3$. Comparing the "current best" with the next challenger yields.

$$PW(15)_{A3-A1} = -\$5{,}000 + \$1{,}100(\overset{P/A,\,15,10}{5.0188}) = \$521.$$

The present worth on the additional investment required by Alternative $A3$ over $A1$ is positive, and therefore that increment is economically desirable. Thus, $A3$ becomes the "current best" alternative, and the list of alternatives has been exhausted so that there is no new challenger possible. *According to step 4, when all challengers have been considered, the "current best" alternative is the one that maximizes present worth and provides a return greater than the MARR.* Therefore, Alternative $A3$ is the best selection from the set of alternatives shown in Table 7.8.

The present worth on incremental investment criterion is also appropriate for comparing alternatives described by cash flows that exclude earnings. Table 7.9 presents the cash flows for a set of alternatives where it is assumed that the services they provide are identical. Since these cash flows do not reflect the income produced by these alternatives, the Do Nothing alternative is not meaningful i.e., no one would ever accept a cash flow that resulted in only disbursements if the opportunity to not invest was available. Thus, it is assumed to be mandatory to select one of the alternatives listed in Table 7.9.

TABLE 7.9.
NET CASH FLOWS FOR FOUR ALTERNATIVES PROVIDING THE SAME SERVICE

	Alternatives			
End of Year	*B*1	*B*2	*B*3	*B*4
0	−$10,000	−$12,000	−$12,000	−$15,000
1	−2,500	−1,500	−1,200	−400
2	−2,500	−1,500	−1,200	−400
3	1,000*	1,500	1,500	3,000

* *Positive values arise as the result of the salvage value received when the asset is sold at the end of its life.*

Applying the present worth on incremental investment criterion to the alternatives shown in Table 7.9 produces the following calculations for a MARR = 10%.

$$PW(10)_{B2-B1} = -\$2{,}000 + \$1{,}000(\overset{P/A,10,2}{1.7355}) + \$500(\overset{P/F,10,3}{0.7513}) = \$111.$$

Since this value is positive, *B*2 becomes the "current best" alternative and *B*1 is eliminated from further consideration. The next comparison pits Alternative *B*3 against *B*2 as

$$PW(10)_{B3-B2} = \$0 + \$300(\overset{P/A,10,2}{1.7355}) + \$0 = \$520.$$

Thus, Alternative *B*3 is accepted and *B*2 is dropped from consideration. Now compare Alternative *B*4 to *B*3, giving

$$PW(10)_{B4-B3} = -\$3{,}000 + \$800(\overset{P/A,10,2}{1.7355}) + \$1{,}500(\overset{P/F,10,3}{0.7513}) = -\$484.$$

The present worth of the increment (*B*4 − *B*3) is negative, and therefore Alternative *B*4 is unacceptable. It is eliminated from further consideration. Since there are no more alternatives to become challengers, the decision process is completed and *B*3, the last "current best" alternative, is the overall best.

Substitution of the annual equivalent or the future worth for the present worth as the basis for comparison for incremental decision making will lead to consistent solutions. The relationships shown below represent the decision rules for applying incremental analysis using annual equivalent and future worth:

$$\boxed{\begin{array}{l} AE(i)_{A2-A1} > 0\text{: accept } A2 \\ AE(i)_{A2-A1} \leq 0\text{: accept } A1. \end{array}} \qquad (7.3)$$

$$\boxed{\begin{array}{l} FW(i)_{A2-A1} > 0\text{: accept } A2 \\ FW(i)_{A2-A1} \leq 0\text{: accept } A1. \end{array}} \tag{7.4}$$

7.5 IRR ON INCREMENTAL INVESTMENT

The criterion of IRR on incremental investment is based on the same type of incremental analysis applied to the previous criterion, present worth on incremental investment. After the alternatives are arranged in increasing order of first-year cost, the incremental cash flows are determined by the incremental procedure described for present worth on incremental investment. If an incremental cash flow cannot pass Test 1 or Test 2 (an indication of possible multiple IRR's as discussed in Section 6.4.4), the IRR method *should not* be applied. Instead, it would be appropriate to substitute present worth on incremental investment.

Once it is assured that the incremental cash flow being considered has a present-worth function like that shown in Figure 6.3, then the only difference in the decision rules between these two criteria is the decision rule in step 3 that determines whether an increment of investment is economically desirable. For the IRR on incremental investment criterion the decision rule is

$$\boxed{\begin{array}{l} i^*_{A2-A1} > \text{MARR: accept } A2 \\ i^*_{A2-A1} \leq \text{MARR: reject } A2 \text{ and accept } A1. \end{array}} \tag{7.5}$$

To apply IRR on an incremental basis it is *necessary first to rank the alternatives in order by increasing first-year cost* and then to select the initial "current best" alternative. Using the set of alternatives in Table 7.8, steps 3 and 4 of the incremental analysis procedure require the following calculations. Find the value i^* so that the equation representing the present worth of the incremental cash flow is set equal to zero. Again the MARR is assumed to be 15%. For the increment $A1 - A0$.

$$0 = -\$5{,}000 + \$1{,}400(\overset{P/A,i,10}{\quad})$$

$$i^*_{A1-A0} = 25.0\%.$$

Because $i^*_{A1-A0} > 15\%$, $A1$ becomes the initial "current best" alternative and Do Nothing is dropped from further consideration. Next, compare Alternative $A2$ to $A1$. For the increment $A2 - A1$,

$$0 = -\$3{,}000 + \$500(\overset{P/A,i,10}{\quad})$$

$$i^*_{A2-A1} = 10.5\%.$$

Because $i^*_{A2-A1} < 15\%$, Alternative $A1$ remains the "current best" and $A2$ is rejected. Then compare $A3$ to $A1$, the "current best" alternative. For increment $A3 - A1$,

$$0 = -\$5{,}000 + \$1{,}100(\overset{P/A,i,10}{\quad})$$

$$i^*_{A3-A1} = 17.6\%.$$

Alternative $A3$ becomes the "current best" and $A1$ is removed from consideration. Since $i^*_{A3-A1} > 15\%$ and all the alternatives have been compared, $A3$, the last "current best" alternative, is the optimum solution. This is the same solution given by the present worth on incremental investment criterion.

In general, the criteria discussed to this point will yield identical solutions for most types of investment decision problems. *However, if the incremental cash flow being analyzed has multiple IRR's, then the decision rules just applied may not give results consistent with the other decision criteria.* The present worth plots in Figures 7.3 and 7.4 show dashed curves representing the value of the present worth of the incremental investment between two alternatives. (The curve representing the $PW(i)_{B-A}$ is simply the difference between the two curves representing

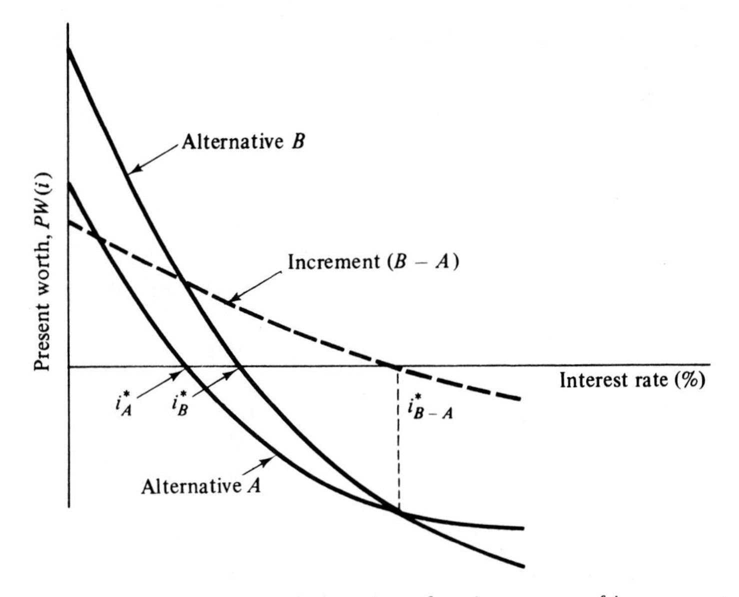

FIGURE 7.3. Present-worth function of an increment of investment.

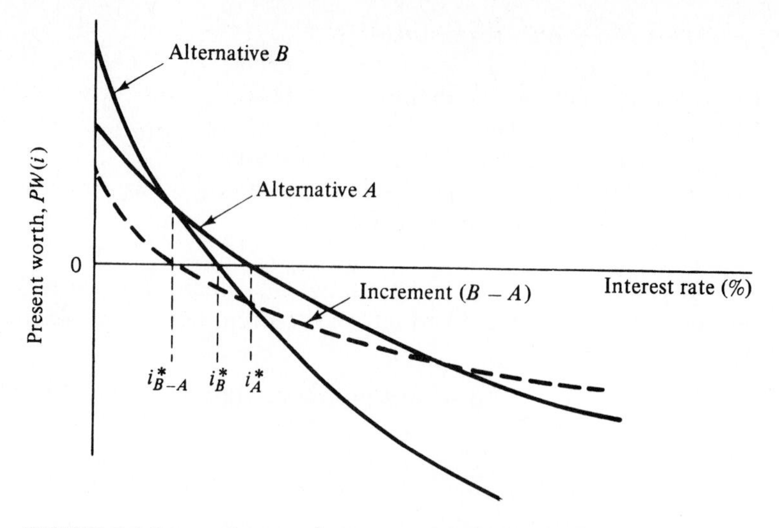

FIGURE 7.4. Internal rates of return on total investment and incremental investment.

$PW(i)_B$ and $PW(i)_A$.) In both figures, the $PW(i)_{B-A}$ curve intersects the horizontal axis only once, assuring that there is only one incremental IRR. These two figures reflect the type of investment alternatives most frequently encountered. Thus, for those alternatives with incremental cash flows satisfying Test 1 or 2 discussed in Section 6.4.4, the criteria that have been discussed will select the alternative that maximizes total present worth and provides a return greater than the MARR.

The application of IRR on incremental investment for cost only cash flows requires no changes in the criterion's decision rules. For example, the set of alternatives described in Table 7.9 are analyzed as follows. After the alternatives are placed in order of increasing first-year cost, each increment of investment is compared to the "current best" alternative. In this case the initial "current best" alternative is $B1$ and the MARR is 10%.

$$0 = PW(i)_{B2-B1} = -\$2{,}000 + \$1{,}000(\overset{P/A,i,2}{\quad}) + \$500(\overset{P/F,i,3}{\quad})$$

$$i^*_{B2-B1} = 13.5\% \qquad (B_2 \text{ becomes "current best" alternative})$$

$$0 = PW(i)_{B3-B2} = \$0 + \$300(\overset{P/A,i,2}{\quad}) + \$0$$

$$i^*_{B3-B2} = \infty \qquad (B_3 \text{ becomes "current best" alternative})$$

$$0 = PW(i)_{B4-B3} = -\$3{,}000 + \$800(\overset{P/A,i,2}{\quad}) + 1{,}500(\overset{P/F,i,3}{\quad})$$

$$i^*_{B4-B3} = 1.5\% \qquad (B_3 \text{ remains "current best" alternative}).$$

Alternative $B3$ is the best selection, as has been shown in Section 7.4.

One method for calculating the incremental rate of return for two alternatives is based on first finding the incremental cash flow between the alternatives and then computing the rate of return associated with this cash flow. We can understand this approach by first examining the general formulation of incremental IRR for two alternatives, $A1$ and $A2$.

$$0 = \sum_{t=0}^{n} (F_{A2,t} - F_{A1,t})(1 + i^*_{A2-A1})^{-t}. \tag{7.6}$$

This formulation corresponds to the method used throughout this section. Here the incremental cash flows are determined and then the incremental IRR is found from the incremental cash flow.

The other method for calculating incremental IRR arises from the fact that Equation 7.6 can be rewritten as

$$\sum_{t=0}^{n} F_{A2,t}(1 + i^*_{A2-A1})^{-t} = \sum_{t=0}^{n} F_{A1,t}(1 + i^*_{A2-A1})^{-t}.$$

By finding the interest rate that sets the two cash flows equal to each other, we also find i^*_{A2-A1}. As an example, using Alternatives $B1$ and $B2$ in Table 7.9 we can determine i^*_{B2-B1} from the expression

$$\begin{aligned} PW(i^*_{B2-B1})_{B1} &= PW(i^*_{B2-B1})_{B2} \\ -\$10{,}000 - \$2{,}500&\left(\overset{P/A,i,2}{\quad}\right) + \$1{,}000\left(\overset{P/F,i,3}{\quad}\right) \\ &= -\$12{,}000 - \$1{,}500\left(\overset{P/A,i,2}{\quad}\right) + \$1{,}500\left(\overset{P/F,i,3}{\quad}\right). \end{aligned}$$

Solving for i from this expression yields $i^*_{B2-B1} = 13.5\%$, which agrees with the previous result. Note that the same result could be obtained from

$$AE(i^*_{B2-B1})_{B1} = AE(i^*_{B2-B1})_{B2} \tag{7.7}$$

where the $AE(i)$ for each alternative is calculated over its particular life.

7.6 COMPARISONS BASED ON TOTAL INVESTMENT

Present worth, annual equivalent, and future worth on total investment approaches are presented in this section. Although computationally different, comparisons based on total investment will select the mutually exclusive alternative that is most desirable. Consistency between the incremental and total investment approaches is illustrated in the paragraphs that follow.

7.6.1 Present Worth on Total Investment

The *present worth on total investment* criterion is one of the most frequently used methods for selecting an investment alternative from a set of mutually exclusive alternatives. Since the stated objective is to choose the alternative with the maximum present worth, the rules for this criterion are rather simple. If

$$\boxed{\begin{array}{l} PW(i)_{A2} > PW(i)_{A1}\text{: accept } A2 \\ PW(i)_{A2} \leq PW(i)_{A1}\text{: accept } A1 \end{array}} \tag{7.8}$$

To demonstrate its computational simplicity, this criterion is applied to the mutually exclusive alternatives described in Table 7.8. Using a MARR = 15%, calculation of the $PW(i)$ amounts gives

$$PW(15)_{A0} = \$\ 0$$

$$PW(15)_{A1} = -\$\ 5{,}000 + \$1{,}400(\overset{P/A,15,10}{5.0188}) = \$2{,}026$$

$$PW(15)_{A2} = -\$\ 8{,}000 + \$1{,}900(\overset{P/A,15,10}{5.0188}) = \$1{,}536$$

$$PW(15)_{A3} = -\$10{,}000 + \$2{,}500(\overset{P/A,15,10}{5.0188}) = \$2{,}547.$$

The maximum value of the $PW(i)$ amounts for these four alternatives is $PW(15)_{A3} = \$2{,}547$. Although for this example the alternative selected happened to have the largest first cost, it is possible for alternatives with the smaller first costs to have present worths greater than those with the larger first costs. For example, if Alternative $A3$ is excluded from consideration, Alternative $A1$ has a larger present worth than $A2$, even though it requires less initial outlay.

For MARR = 10% the $PW(i)$ calculations for the alternatives presented in Table 7.9 are as follows:

$$PW(10)_{B1} = -\$10{,}000 - \$2{,}500(\overset{P/A,10,2}{1.7355}) + \$1{,}000(\overset{P/F,10,3}{0.7513}) = -\$13{,}587$$

$$PW(10)_{B2} = -\$12{,}000 - \$1{,}500(\overset{P/A,10,2}{1.7355}) + \$1{,}500(\overset{P/F,10,3}{0.7513}) = -\$13{,}476$$

$$PW(10)_{B3} = -\$12{,}000 - \$1{,}200(\overset{P/A,10,2}{1.7355}) + \$1{,}500(\overset{P/F,10,3}{0.7513}) = -\$12{,}956$$

$$PW(10)_{B4} = -\$15{,}000 - \$\ \ 400(\overset{P/A,10,2}{1.7355}) + \$3{,}000(\overset{P/F,10,3}{0.7513}) = -\$13{,}440$$

The negative values represent the $PW(i)$ costs associated with the four alternatives. Alternative $B3$ minimizes these costs.

At first glance it appears that the criterion, present worth on total investment, violates the basic decision rule that requires considering the differences between alternatives. (See Section 7.3.1.) The fact that such differences are reflected in the comparison of the present worths on total investment becomes clear in the following example. Suppose there are two mutually exclusive alternatives, $A1$ and $A2$, as shown in Table 7.10.

Alternative $A2$ can be visualized as consisting of a cash flow identical to $A1$ plus the cash flow representing the difference between $A2$ and $A1$. The cash flow portion of $A2$ that is identical to $A1$ will have the same present worth as $A1$. Therefore, the only difference between the present worths for the total investment $A1$ and the total investment $A2$ is represented by the present worth for the incremental cash flow $(A2 - A1)$($1,191 − $495 = $696).

It has been shown by examples that both the incremental and total investment methods lead to the same solution. In fact, it is easy to prove generally that both methods will yield the same selection of alternatives. All that is required is to show that

$$PW(i)_{A2} - PW(i)_{A1} = PW(i)_{A2-A1}. \tag{7.9}$$

By definition, for alternative j

$$PW(i)_j = \sum_{t=0}^{n} F_{j,t}(1 + i)^{-t}$$

so that

$$\begin{aligned}
PW(i)_{A2} - PW(i)_{A1} &= \sum_{t=0}^{n} F_{A2,t}(1 + i)^{-t} - \sum_{t=0}^{n} F_{A1,t}(1 + i)^{-t} \\
&= F_{A2,0} - F_{A1,0} + F_{A2,1}(1 + i)^{-1} - F_{A1,1}(1 + i)^{-1} + \cdots \\
&\quad + F_{A2,n}(1 + i)^{-n} - F_{A1,n}(1 + i)^{-n} \\
&= F_{A2-A1,0} + F_{A2-A1,1}(1 + i)^{-1} + \cdots + F_{A2-A1,n}(1 + i)^{-n} \\
&= \sum_{t=0}^{n} F_{A2-A1,t}(1 + i)^{-t} \\
&= PW(i)_{A2-A1}.
\end{aligned}$$

TABLE 7.10.
DIFFERENCES BETWEEN TWO ALTERNATIVES

End of Year	$A1$	$A2$	$A2 - A1$
0	−$1,000	−$1,500	−$500
1–5	500	900	400
$PW(20)$	$495	$1,191	$696

Thus, the relationship between the decision rules of the two present worth criteria should be evident. If the objective is to maximize present worth and $PW(i)_{A2} > PW(i)_{A1}$, the present worth on total investment criterion selects Alternative $A2$. If $PW(i)_{A2} > PW(i)_{A1}$, then $PW(i)_{A2-A1}$ must be positive, and the decision rule for the present worth on incremental investment criterion is to accept $A2$ rather than A_1 if $PW(i)_{A2-A1} > 0$.

The validity of this basic relationship for different types of cash flows can be observed by using cash flows from Tables 7.8 and 7.9.

$$PW(15)_{A2} - PW(15)_{A1} = \$1{,}536 - \$2{,}026 = -\$490 = PW(15)_{A2-A1}$$

$$PW(10)_{B2} - PW(10)_{B1} = -\$13{,}476 - (-\$13{,}587) = \$111 = PW(10)_{B2-B1}.$$

In the second example, costs must be signified by negative signs, so that desirable increments of investment will be positive.

7.6.2 Annual Equivalent and Future Worth on Total Investment

It was shown in Chapter 6 that the $PW(i)$, $AE(i)$, and the $FW(i)$ are consistent bases for comparing alternatives. Therefore, if

$$\boxed{\begin{array}{l} AE(i)_{A2} > AE(i)_{A1}\text{: accept } A2 \\ AE(i)_{A2} \leq AE(i)_{A1}\text{: accept } A1 \end{array}} \qquad (7.10)$$

and if

$$\boxed{\begin{array}{l} FW(i)_{A2} > FW(i)_{A1}\text{: accept } A2 \\ FW(i)_{A2} \leq FW(i)_{A1}\text{: accept } A1. \end{array}} \qquad (7.11)$$

By applying the $AE(i)$ on total investment criterion or the $FW(i)$ on total investment criterion to the alternatives in Table 7.8. the same conclusion, select Alternative $A3$, is reached.

$$AE(15)_{A0} = \$\ 0$$

$$AE(15)_{A1} = -\$\ 5{,}000(\overset{A/P,15,10}{0.1993}) + \$1{,}400 = \$404$$

$$AE(15)_{A2} = -\$\ 8{,}000(\overset{A/P,15,10}{0.1993}) + \$1{,}900 = \$306$$

$$AE(15)_{A3} = -\$10{,}000(\overset{A/P,15,10}{0.1993}) + \$2{,}500 = \$507$$

or

$$FW(15)_{A0} = \$\ 0$$

$$FW(15)_{A1} = -\$\ 5{,}000(\overset{F/P,15,10}{4.046}) + \$1{,}400(\overset{F/A,15,10}{20.304}) = \$\ 8{,}196$$

$$FW(15)_{A2} = -\$\ 8{,}000(\overset{F/P,15,10}{4.046}) + \$1{,}900(\overset{F/A,15,10}{20.304}) = \$\ 6{,}210$$

$$FW(15)_{A3} = -\$10{,}000(\overset{F/P,15,10}{4.046}) + \$2{,}500(\overset{F/A,15,10}{20.304}) = \$10{,}300$$

An examination of the calculations of the $FW(i)$ indicates that the receipts from the investment are actually invested at the minimum attractive rate of return from the time they are received to the end of the life of the alternative. Thus it is said that $FW(i)$ calculations *explicitly* consider the investment or "reinvestment" of the future receipts generated by investment alternatives. Because the three decision criteria just discussed are consistent and lead to the same selection of alternatives, it follows that use of the $PW(i)$ and the $AE(i)$ *implicitly* assumes the investment or "reinvestment" of alternatives' receipts at the MARR.

7.6.3 Rank on Rate of Return and its Deficiencies

The decision rules of this criterion are accurately described by its name: rank on rate of return (RORR). The IRR is calculated for each proposal, and then the proposals are ranked in descending order of IRR. The decision rule is to move down the ranked proposals, accepting each until there are no more proposals with an IRR greater than the MARR. Although this criterion is well known, it has two important deficiencies.

1. *Ranking on the rate of return will guarantee to select the set of proposals that maximize the total present worth only if all the proposals are independent and there is no limitation on the money available for investments.* The fact that rank on rate of return is a reliable criterion only for the conditions just described limits its usefulness in practice.

If mutually exclusive relationships are introduced or a budget constraint limits the money available for investment, the ranking scheme falters. As soon as interdependencies exist between the proposals, it becomes necessary to compare one alternative to another.

2. *When comparing mutually exclusive alternatives, selecting the alternative with the highest rate of return on its total cash flow* may not *lead to the alternative that will maximize the total present worth at the MARR.*

The IRRs for the *total* cash flows of the alternatives in Table 7.8 are

$$i^*_{A0} = 15\%, \quad i^*_{A1} = 25\%, \quad i^*_{A2} = 19.9\%, \quad i^*_{A3} = 21.9\%.$$

If Alternative $A1$ is selected because it has the maximum rate of return, the total present worth will *not* be maximized for a MARR = 15%. It has already been shown in Section 7.4 that Alternative $A3$ will maximize the present worth for that MARR.

7.6.4 Considering Inflation

When comparing alternatives that are expected to be pursued during periods of inflation, the methods described in Chapter 5 should be applied. Suppose the cash flows for the four alternatives presented in Table 7.9 have been estimated in terms of constant dollars. If the inflation rate is 9%, the cash flows in Table 7.11 represent the investments' cash flows after they have been transformed to actual dollars at the 9% rate. For example, the actual cost at the end of year 2 for Alternative $B1$ will be

$$F = -\$2{,}500(\overset{F/P,9,2}{1.188}) = -\$2{,}970.$$

To compare these four alternatives, compute the present worth of each, using the MARR, which for this problem is 10%.

TABLE 7.11.
CASH FLOWS ADJUSTED FOR INFLATION

	Alternatives			
End of Year	$B1$	$B2$	$B3$	$B4$
0	−$10,000	−$12,000	−$12,000	−$15,000
1	−2,725	−1,635	−1,308	−436
2	−2,970	−1,782	−1,426	−475
3	1,295	1,943	1,943	3,885

$$PW(10)_{B1} = -\$10{,}000 - \$2{,}725(\overset{P/F,10,1}{0.9091}) - \$2{,}970(\overset{P/F,10,2}{0.8265}) + \$1{,}295(\overset{P/F,10,3}{0.7513})$$

$$= -\$13{,}959.$$

$$PW(10)_{B2} = -\$12{,}000 - \$1{,}635(\overset{P/F,10,1}{0.9091}) - \$1{,}782(\overset{P/F,10,2}{0.8265}) + \$1{,}943(\overset{P/F,10,3}{0.7513})$$

$$= -\$13{,}499.$$

$$PW(10)_{B3} = -\$12{,}000 - \$1{,}308(\overset{P/F,10,1}{0.9091}) - \$1{,}426(\overset{P/F,10,2}{0.8265}) + \$1{,}943(\overset{P/F,10,3}{0.7513})$$

$$= -\$12{,}908.$$

$$PW(10)_{B4} = -\$15{,}000 - \$436(\overset{P/F,10,1}{0.9091}) - \$475(\overset{P/F,10,2}{0.8265}) + \$3{,}885(\overset{P/F,10,3}{0.7513})$$

$$= -\$12{,}870.$$

For an inflation rate of 9%, we see that $B4$ is the least-cost alternative. In contrast, for *no inflation* the least-cost alternative was $B3$.

The procedure just described requires considerable computational effort, but it produces intermediate values that can be easily related to the investor's actual experience. When using computer spreadsheets this method is the one most analysts adopt.

The constant-dollar method in Chapter 5 provides the same results with less calculation. Here i', the inflation-free rate, is found from Equation (5.7). Then the estimates in constant dollars are converted directly to the $PW(i)$. An example of this calculation follows.

$$1 + i' = \frac{1 + i}{1 + f} = \frac{1.10}{1.09} = 1.009174$$

$$i' = 0.917\%.$$

For Alternative $B1$ the calculation of present worth for $i' = 0.917\%$ gives

$$PW(0.917)_{B1} = -\$10{,}000 - \$2{,}500(\overset{P/A,0.917,2}{1.97281}) + \$1{,}000(\overset{P/F,0.917,3}{0.97297})$$

$$= -\$13{,}959.$$

This amount is identical to $PW(10)_{A1}$ after the cash flows were inflated at 9%.

Applying the constant-dollar approach to the alternatives in Table 7.8 yields the following results if the inflation rate is 10% and the MARR is 15%. (It is assumed that the cash flows in Table 7.8 are stated in terms of constant dollars.)

$$i' = \frac{1+i}{1+f} - 1 = \frac{1.15}{1.10} - 1 = 4.55\%$$

$$PW(4.55)_{A1} = -\$\ 5{,}000 + \$1{,}400(\overset{P/A,4.55,10}{7.8933}) = \$6{,}051$$

$$PW(4.55)_{A2} = -\$\ 8{,}000 + \$1{,}900(\overset{P/A,4.55,10}{7.8933}) = \$6{,}997$$

$$PW(4.55)_{A3} = -\$10{,}000 + \$2{,}500(\overset{P/A,4.55,10}{7.8933}) = \$9{,}733.$$

With inflation considered, Alternative $A3$ is preferred over the other two. (This same conclusion was reached when no inflation was assumed.) However, Alternative $A2$ is preferred to $A1$ for a 10% rate of inflation, as shown in Section 7.6.1, assuming no inflation causes this preference to be reversed. Therefore, it is essential that the effects of inflation be included in the analysis.

7.7 COMPARING ALTERNATIVES WITH UNEQUAL LIVES

All examples up to this point have demonstrated the application of the various decision criteria based on alternatives that have equal service lives. Often it is required that alternatives having unequal service lives be compared. In such situations it is necessary to make certain assumptions about the service interval so that the techniques of decision making just discussed are applicable.

When comparing alternatives with unequal lives, the principle that all alternatives under consideration must be compared over the same time span is basic to sound decision making.

The time spans over which alternatives are considered must be equal so that the effect of undertaking one alternative can be considered identical to the effect of undertaking any of the others. Clearly, the direct comparison of an alternative with a 5-year life and one with an 11-year life fails to consider the possible investments that could be undertaken during the 6 years following the shorter-lived alternative.

The time period over which the alternatives are to be compared is usually referred to as the *study period* or planning horizon. This study period, denoted by n^*, may be set by company policy or it may be determined by the time span over which reasonably accurate cash flow estimates can be made. Also, the length of the lives of the alternatives being studied can be a basis for determining the study period. For example, the study period might be the life of the shortest-lived alternative or perhaps that of the longest-lived alternative. If it is not prescribed by existing policy, the final selection of the study period is usually left to the analyst.

Because alternatives being compared must be judged over the same study period, various assumptions are made to place alternatives with unequal lives within the same study periods. This section presents three different methods of calculation when alternatives have unequal lives, and it discusses the assumptions associated with each method. Although each method is valid for its assumptions, different results can occur, since the assumptions for each method are basically different.

It is necessary to distinguish between service alternatives and revenue alternatives, since Methods 1 and 2 are applicable to both types, while Method 3 is appropriate only for revenue alternatives.

> **Service alternatives** are described by their capital costs and their other cost cash flows. It is assumed that each service alternative provides over its lifetime an identical service on a per-year basis and thus the revenue cash flows are not usually shown.

Examples of service alternatives are shown in Table 7.12.

TABLE 7.12.
TWO SERVICE ALTERNATIVES WITH UNEQUAL LIVES

End of Year	$A1$	$A2$
0	−$15,000	−$20,000
1	− 6,000	− 2,000
2	− 6,000	− 2,000
3	− 6,000	− 2,000
4	− 6,000	—
5	− 6,000 + 3,000	—

Revenue alternatives are described by all the revenue and cost cash flows associated with that alternative.

Examples of revenue alternatives are shown in Table 7.14.

7.7.1 Method 1 (Estimation of Required Cash Flows)

- Appropriate for service alternatives and revenue alternatives.
- $n^* > 0$. (The study period may be selected to be *any* time span.)

Assumption: Actual costs and revenues will equal the values estimated.

1. **Alternative life > study period ($n > n^*$).** For this case the salvage value for the alternative extending beyond the study period, n^*, must be directly estimated. The result of this direct estimation is shown in Figure 7.5. (The cash flow represents the alternative as if it were terminated or disposed of at the end of the study period.) It is assumed that $n^* = 3$, and

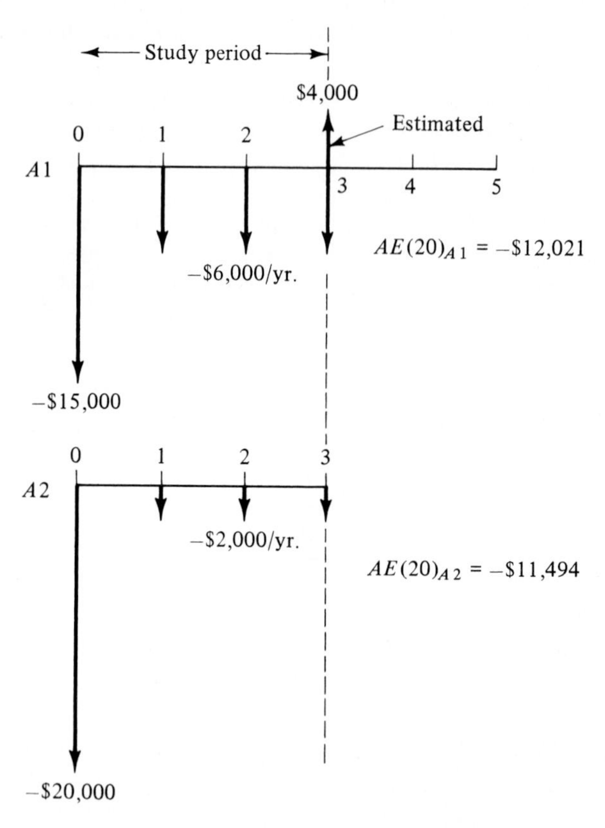

FIGURE 7.5. Direct estimate of salvage value if Alternative $A1$ is disposed of at $t = 3$ (Method 1).

it is estimated that if Alternative $A1$ in Table 7.12 is sold after 3 years, the salvage value would be $4,000.

The comparison of Alternatives $A1$ and $A2$ based on $AE(i)$ on total investment discussed in Section 7.6.2 yields

$$AE(20)_{A1} = -\$15{,}000(\overset{A/P,20,3}{0.4747}) - \$6{,}000 + \$4{,}000(\overset{A/F,20,3}{0.2747})$$

$$= -\$12{,}021$$

$$AE(20)_{A2} = -\$20{,}000(\overset{A/P,20,3}{0.4747}) - \$2{,}000$$

$$= -\$11{,}494.$$

The advantage of Alternative $A2$ over $A1$ is $527 per year for 3 years when this approach is used.

Other measures of worth such as $PW(i)$ and $FW(i)$ may be applied either on total investment or incrementally. Since the modified cash flows in Figure 7.5 have equal lives, any of the criteria discussed in Sections 7.5 and 7.6 are appropriate.

2. **Alternative life < study period ($n < n^*$).** For this case the cash flows necessary to provide the additional service from the end of the alternative's life to the end of the study period, n^*, must be directly estimated. It is assumed that to provide the same service resulting from Alternative $A1$ in Table 7.12 at $t = 4$ and $t = 5$ will require costs of $3,000 per year for the last two years of Alternative $A2$'s life. The adjusted cash flows are presented in Figure 7.6. The comparison of these two alternatives based on $AE(i)$ on total investment gives

$$AE(20)_{A1} = -\$15{,}000(\overset{A/P,20,5}{0.3344}) - \$6{,}000 + \$3{,}000(\overset{A/F,20,5}{0.1344})$$

$$= -\$10{,}613$$

$$AE(20)_{A2} = -\$20{,}000(\overset{A/P,20,5}{0.3344}) - \$2{,}000 - \$1{,}000(\overset{F/A,20,2}{2.200})(\overset{A/F,20,5}{0.1344})$$

$$= -\$8{,}984.$$

The advantage of Alternative $A2$ over $A1$ is $1,629 per year for 5 years.

7.7.2 Method 2 (Calculate the $AE(i)$ of the Capital Costs and the $AE(i)$ of All Other Costs and Revenues Over Each Alternative's Life)

- Appropriate for service alternatives and revenue alternatives
- $n^* > 0$. (The study period may be selected to be *any* time span.)

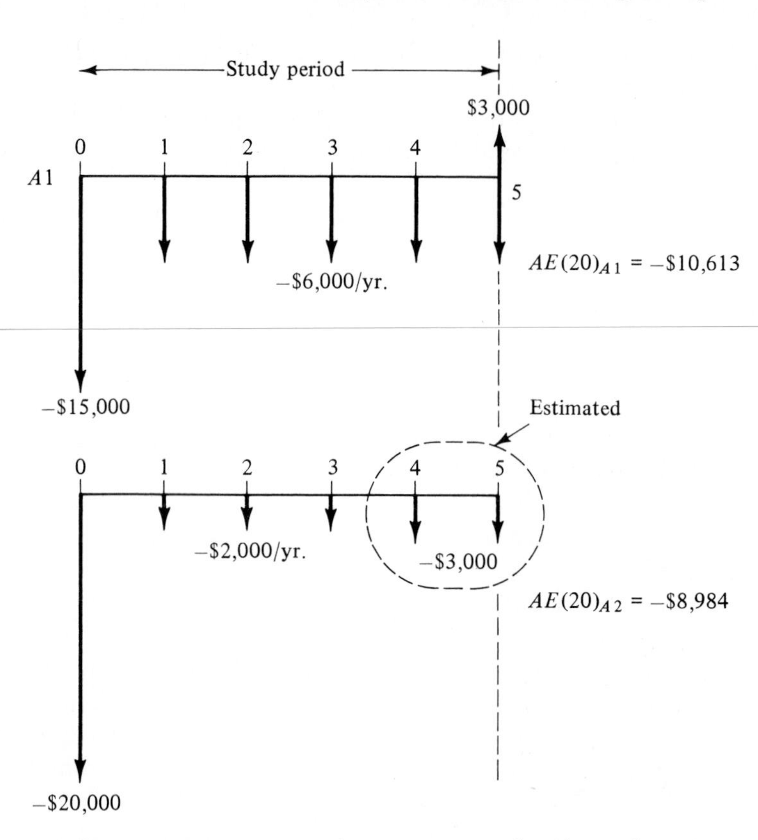

FIGURE 7.6. Direct estimate of operating costs for Alternative $A2$ at $t = 4$ and $t = 5$ (Method 1).

Assumptions:

a. For a study period longer than an alternative's life the alternative will be followed by repeating identical alternatives until the study period is equal or exceeded (*Repeatability*).

b. If the alternative or its repeated replacements exceed the study period, there is an implied salvage value associated with the capital costs of the alternative at the end of the study period. Only if the study period equals the alternative's life or equals some integer multiple of that life will there be no implied salvage value.

Because of assumption (a) any alternative will always equal or exceed the study period. In Section 6.2.1 it was demonstrated that when sequences of cash flows are repeated, it is only necessary to calculate the $AE(i)$ for the first sequence to find the $AE(i)$ over all the sequences. Therefore the solution procedure for Method 2 is to calculate the $AE(i)$ of all the cash flows over each alternative's life. Again using the alternatives described in Table 7.12, these calculations are

$$AE(20)_{A1} = -\$15{,}000(\overset{A/P,20,5}{0.3344}) - \$6{,}000 + \$3{,}000(\overset{A/F,20,5}{0.1344})$$

$$= -\$10{,}613$$

$$AE(20)_{A2} = -\$20{,}000(\overset{A/P,20,3}{0.4747}) - \$2{,}000$$

$$= -\$11{,}494.$$

Since the $AE(i)$ for each of the two alternatives can be assumed to be repeated, the study period can be selected to be any length of time. The following interpretations for some selected study periods are the result of the assumptions of Method 2. The three study periods of $n^* = 3$, $n^* = 7$, and $n^* = 15$ and the repeated alternatives are shown in Table 7.13.

n* = 3 years. (This study period is at the end of the life of Alternative $A2$.)

TABLE 7.13.
TWO SERVICE ALTERNATIVES WITH IDENTICAL REPLACEMENTS FOR THREE DIFFERENT STUDY PERIODS

End of Year	Alternative $A1$			Alternative $A2$	
0	−$15,000			−20,000	
1	−6,000			−2,000	
2	−6,000			−2,000	
3	−6,000	($n^* = 3$)		−2,000	−20,000
4	−6,000			−2,000	
5	−6,000	−15,000	+3,000	−2,000	
6	−6,000			−2,000	−20,000
7	−6,000	($n^* = 7$)		−2,000	
8	−6,000			−2,000	
9	−6,000			−2,000	−20,000
10	−6,000	−15,000	+3,000	−2,000	
11	−6,000			−2,000	
12	−6,000			−2,000	−20,000
13	−6,000			−2,000	
14	−6,000			−2,000	
15	−6,000	+3,000		−2,000	
		($n^* = 15$)			

In this case it would be stated that the cost advantage of Alternative $A1$ over $A2$ is \$881 per year for 3 years.

There is an implied salvage value for Alternative $A1$ of \$9,128. [See Equation (7.12) and the calculations of implied salvage value following these examples.] The implied salvage value is not used to decide among economic alternatives. It is merely a result of the $AE(i)$ calculations, and it can be utilized to judge the appropriateness of Method 2 to model the actual situation. (For example, suppose it is well known that Alternative A1 will never have a salvage value greater than \$3,000 at any time during its life. But the analysis based on Method 2 assumptions indicates a salvage value for \$9,128, a clear contradiction. Thus the assumptions of Method 2 do not fit the facts and should not be used. Method 1 could be applied in this instance.)

n^* = 7 years. (This study period requires Alternative $A1$ to be repeated once and Alternative $A3$ to be repeated twice.)

The cost advantage of Alternative $A1$ over Alternative $A2$ is again \$881 per year, but in this case for 7 years.

Since both alternatives exceed the study period, the implied salvage values for alternatives $A1$ and $A2$ are \$11,453 and \$14,505, respectively. [As an exercise, calculate these values using Equation (7.12).]

n^* = 15. (This study period is at the common multiple of lives for Alternative $A1$ and Alternative $A2$.)

The cost advantage of Alternative $A1$ over Alternative $A2$ continues to be \$881 per year, but for this case the advantage occurs over 15 years.

Since both alternatives terminate at year 15, the implied salvage values are not defined. In this situation the actual salvage values of \$3,000 and \$0 are used for Alternatives A1 and A2, respectively. As just shown for the case when $n^* = 15$, there is no implied salvage value when the study period equals the alternative's life. This follows from the term's definition:

> The **implied salvage value** represents the unused capital costs of an investment alternative at some point in time prior to the end of its expected life.

To calculate the implied salvage value, as depicted in Figure 7.7, find F_{n^*} so that the $AE(i)$ over the full service life of the alternative will equal the $AE(i)$ over the study period. This expression for Alternative $A1$ when $n^* = 3$ is

$$\overbrace{-\$15{,}000(\underset{A/P,20,5}{0.3344}) + \$3{,}000(\underset{A/F,20,5}{0.1344})}^{AE(i)\text{ over full service life}} = \overbrace{-\$15{,}000(\underset{A/P,20,3}{0.4747}) + F_{n^*}(\underset{A/F,20,3}{0.2797})}^{AE(i)\text{ over study period}}$$

$$F_{n^*} = \$9{,}128$$

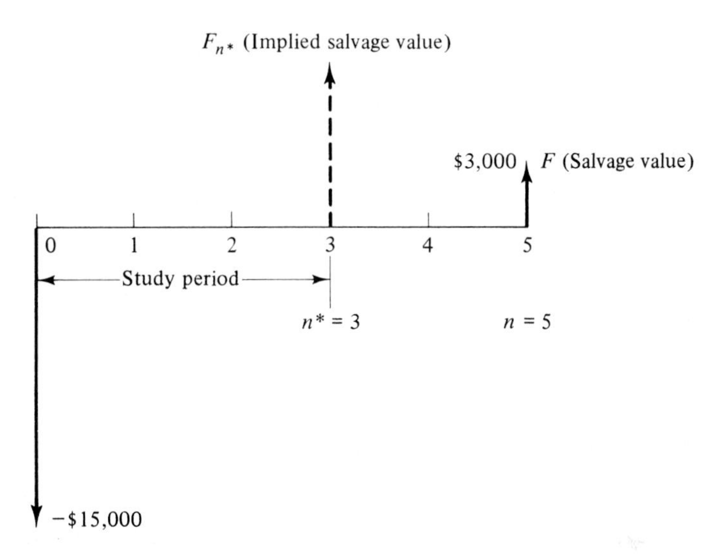

FIGURE 7.7. The implied salvage value.

If the following expression uses this implied salvage value to calculate the $AE(i)$ for Alternative $A1$, the annual equivalent for 3 years is

$$AE(20)_{A1} = -\$15{,}000(\overset{A/P,20,3}{0.4747}) - \$6{,}000 + \$9{,}128(\overset{A/F,20,3}{0.2747})$$

$$= -\$10{,}613 \text{ per year.}$$

This result agrees with the earlier calculation of the $AE(20)_{A1}$ for its service life of 5 years.

A direct method for finding the implied salvage value for a service life n and a study period n^* is given by the expression

$$F_{n^*} = CR(i)(\overset{P/A,i,n-n^*}{\quad}) + F(\overset{P/F,i,n-n^*}{\quad}). \tag{7.12}$$

Here $CR(i)$ is the annual equivalent cost of capital recovery with return and F is the estimated salvage value at the end of the service life. Using this expression for the example just considered gives

$$F_{n^*} = [\$15{,}000(\overset{A/P,20,5}{0.3344}) - \$3{,}000(\overset{A/F,20,5}{0.1344})](\overset{P/A,20,2}{1.5278}) + \$3{,}000(\overset{P/F,20,2}{0.6945})$$

$$= \$9{,}128.$$

From this expression we see that the implied salvage value of an asset consists of two components: (1) the single-payment equivalent of the loss of capital that will be incurred after the study period, and (2) the equivalent value at the end of the study period of the salvage value to be received at $t = n$. The sum of these two components represents the total equivalent value remaining in the investment at the end of its study period.

7.7.3 Method 3

[Calculate the $PW(i)$ or $FW(i)$ of each alternative over its particular life]

- Appropriate revenue alternatives only
- $n^* \geq$ life of longest-lived alternative. (Alternatives must have lives shorter than or equal to the study period.)

Assumption: All cash flows will be reinvested at an interest rate, *i*, until the end of the study period.

An example of the analysis dictated by this assumption is presented for the two alternatives in Table 7.14. Comparing these alternatives over a 6-year study period at an interest rate of 15% yields

$$FW(15)_{B1} = -\$40{,}000(\overset{F/P,15,6}{2.313}) + \$14{,}000(\overset{F/A,15,6}{8.754}) = \$30{,}036$$

$$FW(15)_{B2} = \left[-\$70{,}000(\overset{F/P,15,4}{1.749}) + \$25{,}000(\overset{F/A,15,4}{4.993}) + \$15{,}000\right](\overset{F/P,15,2}{1.323})$$

$$= \$23{,}078.$$

The future-worth calculation is presented so that the assumption regarding the investment of the net receipts of Alternative $B2$ at the MARR of 15% until the end of year 6 is explicitly shown.

TABLE 7.14.
TWO REVENUE ALTERNATIVES WITH A 6-YEAR STUDY PERIOD

End of Year	*B*1	*B*2
0	−$40,000	−$70,000
1	14,000	25,000
2	14,000	25,000
3	14,000	25,000
4	14,000	25,000 + 15,000
5	14,000	
6	14,000	

The present-worth calculation yields

$$W(15)_{B1} = -\$40{,}000 + \$14{,}000(\overset{P/A,15,6}{3.7845})$$

$$= \$12{,}983$$

$$PW(15)_{B2} = -\$70{,}000 + \$25{,}000(\overset{P/A,15,4}{2.8550}) + \$15{,}000(\overset{P/F,15,4}{0.5718})$$

$$= \$9{,}952$$

If it is assumed that $B1$ and $B2$ are repeated every 6 years by identical alternatives, then the $FW(i)$ amounts can be used to find the $AE(i)$ amounts from

$$AE(15)_{B1} = \$30{,}036(\overset{A/F,\,15,6}{0.1142}) = \$3{,}430$$

$$AE(15)_{B2} = \$23{,}078(\overset{A/F,\,15,6}{0.1142}) = \$2{,}636.$$

The annual equivalent calculations shown for Method 2 are not the same as those for method 3. The approaches and results are different because the assumptions are different. Note that, as expected, the ratios of these calculations are the same.

$$\frac{PW(15)_{B1}}{PW(15)_{B2}} = \frac{AE(15)_{B1}}{AE(15)_{B2}} = \frac{FW(15)_{B1}}{FW(15)_{B2}} = 1.30$$

Thus, for the assumption that the cash flows will be reinvested to the end of the study period these three measures of investment worth are consistent.

The reader must be cautioned. Because of the different assumptions used for Method 2 and Method 3 their calculation should be applied only in situations where their assumptions are valid. Remember that Method 3 *cannot* be correctly applied in the following two circumstances.

1. Service alternatives are being evaluated.

For example, the following calculations are incorrect for comparing service alternatives $A1$ and $A2$ presented in Table 7.12 when $n^* = 5$.

$$PW(20)_{A1} = -\$15{,}000 - \$6{,}000(\overset{P/A,20,5}{2.9906}) = -\$31{,}738$$

$$PW(20)_{A2} = -\$20{,}000 - \$2{,}000(\overset{P/A,20,3}{2.1065}) = -\$24{,}213.$$

Such a calculation and comparison implies that for years 4 and 5 Alternative $A2$ will provide a service or income equal to that of $A1$ at no cost. For this reason, the use of the $PW(i)$ comparison just described will indicate that $A2$ is less expensive than $A1$. However, Method 2 has already demonstrated that

$$AE(20)_{A1} = -\$10{,}613$$

$$AE(20)_{A2} = -\$11{,}494.$$

These $AE(i)$ amounts are based on $n^* = 3$ years or the common-multiple-of-lives assumption. Therefore, to correctly compare these alternatives on the basis of $PW(i)$ for $n^* = 5$ years, calculate

$$PW(20)_{A1} = -\$10{,}613(\overset{P/A,20,5}{2.9906}) = -\$31{,}738$$

$$PW(20)_{A2} = -\$11{,}474(\overset{P/A,20,5}{2.9906}) = -\$34{,}374.$$

This result will lead to the selection of Alternative $A1$ as the best option, which agrees with the earlier results obtained using Method 2.

2. Alternatives having lives longer than the study period are being evaluated. If for alternatives $B1$ and $B2$ shown in Table 7.14 the study period is selected to be 4 years ($n^* = 4$), Method 3 is not applicable. The following calculations are *not* correct.

$$PW(15)_{B1} = -\$40{,}000 + \$14{,}000(\overset{P/A,15,4}{2.8550})$$

$$= -\$30$$

$$PW(15)_{B2} = -\$70{,}000 + \$25{,}000(\overset{P/A,15,1}{2.8550}) + \$15{,}000(\overset{P/F,15,4}{0.5718})$$

$$= \$9{,}952$$

Why? Because the above calculation includes all the initial cost in the $PW(i)$ calculation so that Alternative $B1$ is being required to allocate *all* \$40,000 of its capital costs to the first 4 years and zero capital cost to the last 2 years of its 6-year life. This allocation violates the principle that an alternative's capital costs should be related to its actual service life. Thus any alternative with a study period shorter than its actual life is forced to recover all of its initial investment over a period shorter than its expected life. This approach is patently unfair to longer-lived alternatives.

KEY POINTS

- All investment options can be rearranged into mutually exclusive alternatives.

- The fundamental rule, on which the comparison of mutually exclusive alternatives is based, requires that the difference between alternatives be evaluated.
- The minimum attractive rate of return (MARR) is a cut-off rate that represents the yield on investments the firm considers minimally acceptable. This rate should reflect the opportunity to invest if the investments under consideration are not undertaken.
- The Do Nothing alternative, $A0$, represents the option to invest at an interest rate equal to the MARR. Thus the $PW(\text{MARR})_{A0} = 0$.
- $PW(i)$, $AE(i)$, and $FW(i)$ can be correctly applied on total investment or incremental investment. The internal rate of return, i^*, will provide the same results as the investment criteria just mentioned only if it is applied on incremental investment.
- When comparing alternatives with unequal lives the comparison must be made over equal periods of time. Three different methods are presented for placing such alternatives over equal periods. Each method is based on a different set of assumptions, and there is no assurance that the three methods will provide the same conclusions.

PROBLEMS

1. A company is considering Proposals 1, 2, and 3, one of which must be adopted. Proposal 1 is contingent on Proposal 2, and Proposals 2 and 3 are mutually exclusive.

a. List all mutually exclusive investment alternatives.

b. Designate alternatives as feasible or infeasible, giving reasons for infeasibility. Answer: $A2$, $A3$, and $A4$ are feasible.

2. A firm has four independent proposals under consideration. For the following conditions develop the matrix of investment alternatives for Proposals 1, 2, 3, and 4.

a. No dependency relationships.

b. Proposal 1 and 2 are mutually exclusive.

c. Proposal 4 is contingent on Proposal 3.

d. Proposal 2 is contingent on Proposal 3.

e. Proposals 2, 3, and 4 are mutually exclusive.

3. A company is considering Engineering Proposals 1, 2, and 3, with cash flows estimated to be

	*P*1	*P*2	*P*3
Investment	$800,000	$600,000	$400,000
Project life	8 years	8 years	8 years
Annual revenue	$450,000	$400,000	$300,000
Annual cost	$200,000	$180,000	$150,000
Salvage value	$100,000	$ 80,000	$ 60,000

Proposals 1 and 2 are mutually exclusive and Proposal 3 is contingent upon Proposal 2. The budget limit is $1,000,000.

a. Develop the matrix of investment alternatives, indicate which ones are not feasible, and give reasons for the infeasibility. Answer: *A*3, *A*4, *A*5, and *A*7 are not feasible.

b. Develop the cash flow profiles for feasible investment alternatives.

4. A firm is considering Engineering Projects 1, 2, and 3, with cash flows estimated to be as follows:

	Project 1	*Project 2*	*Project 3*
Investment	$4,000	$7,000	$8,000
Project life	5 years	5 years	5 years
Annual revenue	$4,500	$6,000	$5,500
Annual cost	$1,500	$1,800	$2,000
Salvage value	$1,000	$1,500	$1,000

Project 1 and 2 are mutually exclusive and Project 3 is contingent upon Project 2. The budget limit is $10,000.

a. Develop the matrix of investment alternatives.

b. Indicate which alternatives are not feasible and give reasons for the infeasibility.

c. Develop the cash flow profiles for feasible alternatives.

5. A company is considering three proposals, 1, 2, and 3. The firm must implement one or more of the proposals to resolve an urgent situation. Proposals 1 and 3 are mutually exclusive and Proposal 2 is contingent on 1. Cash flows for the proposals over a 10-year planning horizon are given below. The firm has a budget limit of $850,000. The MARR is 20%.

	*P*1	*P*2	*P*3
Initial investment	$300,000	$450,000	$600,000
Salvage value	$ 50,000	$ 50,000	$100,000
Annual receipts	$350,000	$350,000	$450,000
Annual disbursements	$200,000	$100,000	$200,000

a. List all possible investment alternatives and identify feasible alternatives. Give appropriate reasons for discarding alternatives that are not feasible.

b. Develop the cash flow profiles for feasible alternatives.

c. Use the present worth on incremental investment method to find the best alternative. Answer: *A*3

6. A needed service can be purchased for $102 per unit. The same service can be provided by equipment that costs $100,000 and that will have a salvage value of $25,000 at the end of 10 years. Annual operating expense will be $5,500 per year plus $31 per unit.

a. If these estimates are correct, what will be the incremental IRR on the investment if 400 units are provided per year? Answer: 20%

b. What will be the incremental IRR on the investment if 250 units are provided per year?

c. If the firm requiring this service has a MARR of 12%, what would be the alternative to select for the service levels in parts (a) and (b)?

7. An industrial firm can purchase a special machine for $22,000. A down payment of $2,500 is required, and the balance can be paid in 5 equal year-end installments at 14% interest on the unpaid balance. As an alternative the machine can be purchased for $19,000 in cash. If the firm's MARR is 20%, determine which alternative should be accepted. Use the present worth on incremental investment approach.

8. Given below are the cash flows for two mutually exclusive investment alternatives.

Year	*A*1	*A*2
0	−$10,000	−$10,000
1	4,000	5,000
2	2,000	2,000
3	2,000	2,000
4	2,000	2,000
5	7,353	5,779

a. Find the internal rate of return on the incremental cash flow ($A2 - A1$). Using the decision rule for incremental rates of return, indicate which alternative will maximize equivalent wealth for an MARR of 16%.

b. Calculate the present worth on total investment for each alternative for MARR = 16% and select the best alternative.

c. Calculate the present worth on incremental investment for each alternative for MARR = 16% and select the best alternative.

d. Are the results in parts (a), (b), and (c) the same? If not, check the incremental cash flow using Test 2 from Section 6.4.4. What are your conclusions?

9. Two mutually exclusive investment plans have the cash flows described as follows:

	Year			
Plan	0	1	2	3
A	−$10,000	$5,000	$5,000	$5,000
B	− 12,000	6,100	6,100	6,100

For a MARR of 25%

a. Determine present worth on incremental investment.

b. Determine IRR on incremental investment.

c. Which alternative would you recommend? Explain.

10. It is estimated that the annual heat loss cost in a small power plant is $5,200. Two competing proposals have been formulated that will reduce the loss. Proposal *A* will reduce heat-loss cost by 60% and will cost $3,000. Proposal *B* will reduce heat-loss cost by 55% and will cost $2,500. If the interest rate is 8%, and if the plant will benefit from the reduction in heat loss for 10 years, which proposal should be accepted?

a. Use present worth on incremental investment. Answer: $1,245

b. Use annual equivalent on incremental investment. Answer: $185

c. Use IRR on incremental investment. Answer: 51.2%

11. Three mutually exclusive alternatives requiring different investments are being considered. The life of all three alternatives is estimated to be 20 years with no salvage value. The MARR that is considered acceptable is 15%. The cash flows representing these three alternatives are shown below.

	*A*1	*A*2	*A*3
Investment	−$60,000	−$30,000	−$100,000
Net income per year	10,692	6,162	19,238
Return on total investment	17.1%	20.0%	18.6%

Find the investment that should be selected using (a) IRR on incremental investment, (b) present worth on incremental investment, and (c) present worth on total investment. Answer: *A*3

12. A firm has identified three viable but mutually exclusive investment alternatives. The life of all three alternatives is estimated to be 5 years with negligible salvage value. The MARR is 7%.

	A1	A2	A3
Investment	\$10,000	\$14,000	\$17,000
Annual net income	2,638	3,884	4,600
Return on total investment	10%	12%	11%

Find the alternative that should be selected by using the following:

a. IRR on incremental investment.

b. Present worth on incremental investment.

13. The following mutually exclusive alternatives were the only feasible ones obtained from a set of proposals. All cash flows are in thousands and the MARR is 15%.

	End of Year		
Alternative	0	1–6	6
*A*0	0	0	0
*A*1	−600	220	80
*A*2	−500	150	60

a. *PW* on total investment.

b. *AE* on total investment.

c. *FW* on total investment.

14. Select the most attractive alternative among those given below. All cash flow amounts are in thousands and the MARR is 20%.

	End of Year		
Alternative	0	1–8	8
*A*0	0	0	0
*A*1	−600	220	80
*A*2	−1000	370	140
*A*3	−500	150	100

a. *PW* on total investment. Answer: *A*2

b. *AE* on total investment. Answer: *A*2

c. *FW* on total investment. Answer: *A*2

15. Which is the most attractive alternative among the three itemized below? All amounts are in thousands and the MARR is 25%.

	End of Year		
Alternative	0	1–10	10
*A*0	−600	250	100
*A*1	−300	150	50
*A*2	−750	400	100

a. *PW* on total investment.

b. *AE* on total investment.

c. *FW* on total investment.

16. Of the three investment alternatives given, only one can be selected.

	Year				
Alternative	0	1	2	3	4
*D*1	−$1,510	$ 900	$ 900	$ 900	$ 900
*D*2	− 3,360	1,550	1,550	1,550	1,550
*D*3	− 9,800	3,500	3,500	3,500	3,500

Find the range of values of the MARR for which each alternative is most economical. Plot $PW(i)$ vs i and then use incremental IRR.

17. The T Toy Corporation would like to decide to which new product, out of four current candidates, they should commit their resources for R&D, production, and subsequent distribution. The following table shows the estimated cash flows for each one of the proposed products. If the MARR for this company is 10%, which product should be selected?

	Year				
Product	0	1	2	3	4
1	−$ 80,000	−$ 80,000	−$ 70,000	$160,000	$160,000
2	− 150,000	50,000	50,000	55,000	55,000
3	− 200,000	− 200,000	180,000	180,000	180,000
4	− 380,000	20,000	170,000	170,000	170,000

a. Use internal rate of return on incremental investment. Answer: 3

b. Use present worth on total investment. Answer: 3

18. An electronic components manufacturer is considering the introduction of several new products. However, because of the large capital outlays required, it has been decided that only one of these new products will eventually be marketed. Since the technology is expected to change rapidly, it is believed that after 6 years the demand for these products will be minimal.

The following cash flows describe the five alternatives being investigated. (Negative cash flows are capital outlays.)

	Year (cash flows in millions of dollars)						
Product	0	1	2	3	4	5	6
A	−$ 8.0	$2.4	$2.4	$2.4	$ 2.4	$ 2.4	$ 2.4
B	− 10.0	4.0	4.0	4.0	4.0	4.0	4.0
C	− 14.0	6.7	6.7	6.7	6.7	0.0	0.0
D	− 9.0	− 8.0	8.6	8.6	8.6	8.6	8.6
E	− 10.0	− 3.0	− 3.0	− 2.0	15.6	15.6	15.6

If the MARR for this firm is 30%, which product should be introduced?

a. Solve using present worth on incremental investment.

b. Solve using present worth on total investment.

c. For an inflation rate of 10% which new product would be the best investment? Assume the cash flows describing the outlays and revenues are in time zero constant dollars.

19. A firm is considering the purchase of a new machine to increase the productivity of an existing production process. Of all the machines considered, management has narrowed them to those represented by the following cash flows:

Machine	*Initial Investment*	*Annual Operating Cost*
1	$100,000	$22,500
2	110,000	20,540
3	125,000	17,082
4	130,000	15,425
5	150,000	11,274

If each of these machines provides the same service for 8 years and the minimum attractive rate of return is 12%, which machine should be selected? Solve by using the rate of return on incremental investment. Compare this result to the result obtained by applying the $AE(i)$ on total investment.

20. A wholesale distributor is considering the construction of a new warehouse to serve a geographic region that he has been unable to serve until now. There are six cities where the warehouse could be built. After extensive study the expected income and costs associated with locating the warehouse in each city has been determined.

City	*Initial Cost*	*Net Annual Income*
A	$1,000,000	$407,180
B	1,120,000	444,794
C	1,260,000	482,377
D	1,420,000	518,419
E	1,620,000	547,771
F	1,900,000	562,476

The life of the warehouse is estimated to be 15 years. If the MARR is 12%, where should the wholesaler locate his warehouse?

a. Solve using an incremental approach.

b. Solve using a total-investment approach.

c. What city would be selected if the alternative that maximized rate of return on total investment were used? Does this conform to the results in part (a) or part (b)?

d. If the income estimates are in constant dollars and the inflation rate is 9% per year, which location should be selected? Answer: Select City *E*.

21. A shipping firm is considering the purchase of a materials handling system for unloading ships at the dock. The firm has reduced its choice to five different systems, all of which are expected to provide the same unloading speed. The initial costs and the operating costs estimated for each system are described below.

System	*Initial Cost*	*Annual Operating Expenses*
A	$650,000	$ 91,810
B	780,000	52,569
C	600,000	105,000
D	750,000	68,417
E	720,000	74,945

The life of each system is estimated to be 5 years, and the firm's MARR is 15%. If the firm must select one of the materials handling systems, which one is the most desirable?

a. Solve using the total-investment approach.

b. Solve using an incremental approach.

c. Assuming the cost estimates are in constant dollars and the annual inflation rate is expected to be 9%, which system is preferred?

22. The state highway department is considering six routes for a new highway. Listed below are the estimated construction costs, maintenance costs, and the user cost associated with each route.

Route	Construction Costs per Kilometer	Annual Maintenance Costs per Kilometer	Annual Users Cost per Kilometer
*A*1	$1,500,000	$9,789	$450,000
*A*2	1,687,500	9,225	437,262
*A*3	1,875,000	8,682	427,104
*A*4	2,100,000	7,977	400,326
*A*5	2,250,000	6,639	370,524
*A*6	2,437,500	6,399	356,094

The life of the highway is expected to be 25 years with no salvage value. If the initial interest rate is 8%, which highway route is most desirable?

a. Solve using incremental analysis.

b. Solve using total investment analysis. Answer: *A*5

c. If the estimated costs are in constant dollars and the annual inflation rate is 8%, which route is favored? Answer: *A*6

23. An engineering student who will soon receive his B.S. degree is contemplating continuing his formal education by working toward an M.S. degree. The student estimates that his average earnings for the next 6 years with a B.S. degree will be $40,000 per year. If he can get an M.S. degree in one year, his earnings should average $44,000 per year for the subsequent 5 years. His earnings while working on the M.S. degree will be negligible and his additional expenses to be paid out over this year will be $10,000. The student estimates that his average per-year earnings in the three decades following the initial 6-year period will be $42,000, $45,000, and $49,000 if he does not stay for an M.S. degree. If he receives an M.S. degree his earnings per year in the three decades can be stated as $42,000 + x, $45,000 + x, and $49,000 + x. For an interest rate of 13% find the value of x for which the extra investment in formal education will pay for itself. Answer: $8,833

24. The heat loss through the exterior walls of a building costs $21,500 per year. Insulation that will reduce the heat-loss cost by 93% can be installed for $13,500, and insulation that will reduce the heat-loss cost by 89% can be installed for $9,000. If the building is to be used for 8 years and if the interest rate is 12%, which alternative would you recommend?

25. The STORE Company is considering the acquisition of a materials handling system to increase the productivity of their existing warehouses. Besides the Do Nothing alternative, which is denoted by 0, three new alternatives are under consideration: *E*1, *E*2, and *E*3. Their required investments are $360,000, $380,000, and $405,000, respectively, and they all have the same expected life. The table below gives the incremental IRR's for all possible incremental comparisons. If the MARR is 12%, what system should be selected? Show your comparisons. Answer: *E*2

Y \ X	0	*E*1	*E*2	*E*3
	Incremental IRR, Y − X			
*E*1	20%	—		
*E*2	21%	25%	—	
*E*3	19%	16%	10%	—

26. A manufacturing plant and its equipment are insured for $700,000. The present annual insurance premium is $0.96 per $100 of coverage. A sprinkler system with an estimated life of 20 years and no salvage value at the end of that time can be installed for $18,000. Annual operation and maintenance cost is estimated at $360. Taxes are 0.8% of the initial cost of the plant and equipment. If the sprinkler is installed and maintained, the premium rate will be reduced to $0.38 per $100 of coverage.

a. What will be the incremental IRR if the sprinkler system is installed? Answer: 18.8%

b. If the MARR is 15%, which alternative should be selected?

27. Every year the stationery department of a large concern uses 1,200,000 sheets of paper with three holes drilled for binding and 250,000 sheets that have the corners rounded. At present the drilling and corner cutting are done by a commercial printing establishment at a cost of $0.35 and $0.30 per thousand sheets, respectively. Two alternatives are being considered. Alternative *A* consists of the purchase of a paper drill for $2,000, and Alternative *B* consists of the purchase of a combination paper drill and corner cutter for $2,800. Obviously the two alternatives do not provide equal service. The following data apply to the two machines.

	Drill	*Combined Drill and Cutter*
Life	15 years	15 years
Salvage value	$150	$200
Annual maintenance	35	46
Annual space charge	30	30
Annual labor to drill	330	240
Annual labor to cut corners	—	100
Interest rate	25%	25%

a. Alter one or the other of the alternatives given above so that they may be compared on an equitable basis. Calculate the equivalent annual cost of each of the revised alternatives.

b. What other alternative or alternatives should be considered? Determine its annual cost.

28. A 100-horsepower motor is required to power a large-capacity blower. Two motors have been proposed with the following engineering and cost data.

	Motor A	*Motor B*
First cost	\$9,000	\$8,000
Anticipated life	12 years	12 years
Salvage value	0	0
Efficiency $\frac{1}{2}$ load	85%	83%
Efficiency $\frac{3}{4}$ load	94%	87%
Efficiency full load	90%	88%
Hours use per year at $\frac{1}{2}$ load	800	800
Hours use per year at $\frac{3}{4}$ load	1,000	1,000
Hours use per year at full load	600	600

Power cost per kilowatt-hour is \$0.10. Annual maintenance, taxes, and insurance will amount to 1.6% of the original cost. Interest is 10%.

a. What is the $AE(i)$ for each motor? Answer: \$15,895; \$16,409.

b. What will be the IRR on the additional amount invested in Motor *A*? Answer: 66%

29. In a hydroelectric development under consideration, the question to be decided is the height of the dam to be built. The function of the dam is to create a head of water. Because of the width of the proposed dam site at different elevations, heights of the dam under consideration are 173, 194, and 211 feet; costs for these heights are estimated at \$22,320,000, \$27,840,000, and \$36,240,000, respectively. The capacity of the power plant is based on the minimum flow of the stream of 1,760 cubic feet per second. This flow will develop $[(h \times 1{,}760 \times 62.4) \div 550]0.75$ horsepower, where h equals the height of the dam in feet. A horsepower-year is valued at \$372. The cost of the power plant, including building and equipment, is estimated at \$2,160,000 for the building and \$408 per horsepower of capacity for the equipment. To be conservative, the useful life of the dam, buildings, and power equipment is estimated at 40 years with no salvage value. Annual maintenance, insurance, and taxes on the dam and buildings are estimated at 2.8% of first cost while these costs are 4.7% of first cost for equipment. Operation costs are estimated at \$456,000 per year for each of the alternatives. Determine the rate of return for each height and the rate of return on the added investment for each added height. To which height should the dam be built if 10% is required on all investments?

30. A sales manager has received 11 proposals for future expenditures from the 4 sales districts in his region. The proposals listed below are expected to span 10 years. To determine the acceptability of investment proposals the sales manager uses a MARR of 12%. The proposals from each district are designated by a different letter. The acceptance of a proposal from one district does not affect the acceptance of proposals from the other districts unless money is limited. The proposals related to a particular district are mutually exclusive, so it is impossible to select more than one proposal from a particular district. What proposal or proposals should the sales manager select if the money available for investment is (a) unlimited, (b) $700,000, (c) $450,000, and (d) $350,000? Answer (a): *Q*4, *R*2, *S*2, and *T*2

Proposals	*Initial Costs*	*Net Annual Revenues*
*Q*1	$100,000	$ 19,925
*Q*2	120,000	24,695
*Q*3	130,000	26,688
*Q*4	140,000	29,488
*R*1	150,000	35,778
*R*2	180,000	41,755
*S*1	200,000	32,550
*S*2	240,000	57,245
*S*3	300,000	48,825
*T*1	400,000	95,408
*T*2	500,000	123,415

31. The production manager of a plant has received the sets of proposals listed below from the supervisors of three independent production activities. The proposals related to a particular production activity are identified by the same letter and they are mutually exclusive. If the proposals are expected to have a life of 8 years with no salvage value and the MARR is 15%, what proposals should be selected if the amount of money available for investment is (a) unlimited, (b) $40,000, and (c) $20,000?

Proposal	*Initial Investment*	*Net Annual Income*
*A*1	$10,000	$3,004
*A*2	20,000	6,530
*A*3	30,000	7,970
*B*1	5,000	1,006
*B*2	10,000	5,312
*B*3	15,000	6,209
*B*4	20,000	7,077
*C*1	15,000	4,506
*C*2	30,000	7,829

32. A company is considering a group of research proposals that are related to either Product *A*, Product *B*, or Product *C*. It has been decided that *one proposal will be selected from each set related to a particular product.* The research proposals concerned with Product *A* are identified by the letter *A*, those concerned with Product *B* by the letter *B*, and so on. The company expects the research to extend over a 5-year period. In the past the company has considered a return on investment of at least 19% to be satisfactory. Since it is believed that all projects will be equally beneficial to the company, only the costs related to each project are shown below. If the money available is (a) unlimited, (b) $90,000, and (c) $75,000, what proposals should be selected?

Proposal	*Initial Cost*	*Annual Expenses*
A1	$40,000	$ 8,100
A2	62,000	2,134
B1	20,000	8,200
B2	25,000	6,528
B3	30,000	5,115
C1	15,000	16,200
C2	20,000	15,013
C3	35,000	7,840
C4	50,000	5,530

33. Consider the four investment projects listed below. Projects having the same letter are mutually exclusive, while those having different letters are independent. The MARR is 15%.

Project	*Investment Required*
*A*1	$ 9,000
*A*2	12,000
*B*1	8,000
*B*2	10,000

The following table lists the incremental rates of returns for all possible incremental comparisons. The Do Nothing alternative is not feasible. Using incremental analysis (show your comparisons), find the alternative that maximizes the equivalent profit if there is only $20,000 in funds available for investment. Remember to order the alternatives in order of first-year cost. Answer: Select *A*1 and *B*2.

Incremental IRR for $Y - X$

		Alternatives X							
		*A*1	*A*2	*B*1	*B*2	(*A*1, *B*1)	(*A*1, *B*2)	(*A*2, *B*1)	(*A*2, *B*2)
Alternatives Y	*A*1	—							
	*A*2	16%	—						
	*B*1	40%	20%	—					
	*B*2	30%	8%	35%	—				
	(*A*1, *B*1)	13%	10%	17%	9%	—			
	(*A*1, *B*2)	18%	18%	24%	17%	35%	—		
	(*A*2, *B*1)	14%	13%	16%	14%	16%	4%	—	
	(*A*2, *B*2)	17%	17%	22%	16%	22%	22%	35%	—

34. It is estimated that a manufacturing concern's needs for storage space can be met by providing 240,000 square feet of space at a cost of $15 per square foot now and an additional 60,000 square feet of space at a cost of $55,000 plus $15 per square foot of space 6 years hence. A second plan is to provide 300,000 square feet of space now at a cost of $14 per square foot. Either installation will have zero salvage value when retired some time after 6 years. If taxes, maintenance, and insurance cost $0.15 per square foot and the interest rate is 12%, which plan should be adopted when using a 6 year study period? Answer: Plan 1

35. A small telephone company is considering two mutually exclusive proposals for the installation of automatic switching equipment. Alternative *A*1 includes a future expansion, will cost $360,000 now, will have a 12-year service life, and will entail annual operating cost of $95,000. At the end of year 5 the system will be expanded at a cost of $300,000. This additional equipment will have a service life of 10 years and will increase operating costs by $60,000 per year. Alternative *A*2 is to install a system with full capacity. This system will require an initial investment of $580,000, will have a 12-year service life, and will entail annual operating costs of $110,000 the first 5 years and $170,000 thereafter. If the service will be required for at least the next 10 years, determine which alternative should be implemented, using the following methods. The MARR is 15%.

- **a.** Method 1 (Assume that the salvage values of all the proposals are zero throughout their lives). Use a 10-year study period.
- **b.** Method 2 (Assume that the salvage values for the proposals are zero at the end of the life of each piece of equipment). Use a 10-year study period and indicate the implied salvage values.

36. A firm has two alternatives for improvement of its current production system. The data are as follows:

	Machine A	*Machine B*
Initial installment cost	$15,000	$25,000
Annual operating cost	$8,000	$6,500
Service life	5 years	8 years
Salvage value	$0	$0

Determine the i^* on the extra investment in Machine B and select the best alternative for an interest rate of 15%. Assume repeatability and use a 40-year study period. [Use Method 2 and see Equation (7.7).] Answer: 20%; Select B

37. A firm is considering the purchase of one of two new machines. The data on each are given below.

	Machine A	*Machine B*
Initial cost	$3,400	$6,500
Service life	3 years	6 years
Salvage value	$100	$500
Net operating cost after taxes	$2,000/year	$1,800/year

If the firm's MARR is 12%, which alternative should be selected when using the following methods?

a. Annual equivalent cost approach (Method 2).

b. Present-worth comparison (Method 2).

c. Incremental IRR comparison (Method 2).

38. A refinery can provide for water storage with a tank on a tower or a tank of equal capacity placed on a hill some distance from the refinery. The cost of installing the tank and tower is estimated at $102,000. The cost of installing the tank on the hill, including the extra length of service lines, is estimated at $83,000. The life of the two installations is estimated at 40 years, with negligible salvage value for either. The hill installation will require an additional investment of $9,500 in pumping equipment, whose life is estimated at 20 years with a salvage value of $500 at the end of that time. Annual cost of labor, electricity, repairs, and insurance incident to the pumping equipment is estimated at $1,000. The interest rate is 15%.

a. Compare the $PW(i)$ cost of the two plans.

b. Compare the two plans on the basis of $AE(i)$ cost.

c. Compare the two plans on the basis of their $CE(i)$ costs.

39. An investor is considering the purchase of one of two bonds. The first is a 12% bond maturing in 8 years and paying interest semiannually. It is being offered for $800. The second bond matures in 10 years and pays $70 every six months. Both bonds have a par value of $1,000. For the conditions given below, what offering price for the second bond will make the investor indifferent to which bond is purchased? The investor desires to earn 16% per year compounded semiannually.

a. The study period is 8 years. (Use Method 2.)

b. The study period is 10 years. (Use Method 3.)

40. Two alternatives are being considered regarding construction of a new high-voltage transmission line. Alternative I would build the transmission towers and the line at a capacity of 230 kVA, which is expected to be adequate for 15 years. After 15 years the 230-kVA lines would be removed and 560-kVA lines placed on the existing towers. Alternative II would build the transmission towers and the 560-kVA lines immediately. Given below are the pertinent data on the costs of these facilities.

	Present Cost	*Expected Service Life*	*Expected Salvage Value*
Transmission towers	$15,000,000	55 years	0 after 30 years
230-kVA lines	$ 8,000,000	25 years	10% of first cost regardless of age at retirement
560-kVA lines	$12,000,000	35 years	10% of first cost regardless of age at retirement

The cost of 560-kVA lines will inflate at the rate of 10% per year. The MARR is 15%. Make the analysis to determine which alternative is least expensive for a 35-year study period, using Method 1.

41. The ATLAS Corporation is currently studying the four investment proposals given below. Proposals with the same letter (e.g., *C*1, *C*2) are mutually exclusive, while those with different letters are independent. All proposals have zero future salvage values, and the MARR is 20%. Which proposal or proposals should be accepted?

Proposal	*Initial Investment*	*Annual Net Revenues*	*Expected Life (years)*
*C*1	$200,000	$ 76,000	6
*C*2	260,000	80,000	9
*C*3	280,000	100,000	6
*X*1	100,000	50,000	6
*X*2	170,000	60,000	9

a. Solving using the shortest-lived asset as the study period, with no budget limitation.

b. Solve using the longest-lived asset as the study period, with no budget limitation. Answer: *C*2, *X*2, when using Method 3.

c. Solve parts (a) and (b) when the budget is restricted to $375,000.

d. Solve parts (a) and (b), assuming the above values are in constant dollars and the inflation rate is 8% per year.

42. The PLASCO Corporation is considering six investment proposals.

Proposal	*Required Initial Investment*	*Net Annual Revenue*	*Expected Life (years)*
*A*1	$1,200,000	$ 400,000	40
*B*1	1,500,000	450,000	35
*C*1	2,400,000	820,000	45
*C*2	2,600,000	840,000	38
*D*1	3,800,000	1,200,000	30
*D*2	5,000,000	1,500,000	35

The net salvage value at the end of the life of each proposal is expected to be zero, and the pretax MARR used by Plasco is 20%. Which proposal or proposals should be accepted if there is no limitation on the amount of money available? Which if the budget is limited to $8,000,000? List any assumptions that you make. Proposals with the same letters are mutually exclusive (e.g., *C*1, *C*2). Answers: *A*1, *B*1, *C*1, *D*2; *A*1, *C*1, *D*1

8 Evaluating Replacement Alternatives

Mass production has been found to be an economical means for satisfying human wants. However, mass production requires the employment of large quantities of capital assets, many of which because of wearout, inadequacy, obsolescence, or other reasons must be replaced from time to time. The failure to continuously upgrade these assets can result in serious loss of operating efficiency. A sound program of replacement analysis can be vital to the financial success of an enterprise.

The methods of analysis presented in Chapter 7 are appropriate for comparing the cash flows of replacement alternatives. However, certain concepts and techniques in replacement analysis, such as sunk cost, economic life, and unused value, require special attention. This chapter presents these concepts and provides examples of their application in evaluating replacement alternatives.

8.1 THE GENERAL NATURE OF REPLACEMENT ANALYSIS

To facilitate the discussion of the principles involved in replacement analysis, it is necessary to introduce two important terms commonly used by practitioners involved in replacement analyses.

The **defender** is an existing asset being considered for replacement.

The **challenger** is the asset proposed to be the replacement.

Because the economic characteristics of the defender and the challenger are usually so dissimilar, special attention is required when these two options are compared. One obvious point is that the duration and the magnitude of cash flows for old existing assets and new assets are quite different. New assets characteristically have high capital costs and low operating costs. The reverse is usually true for assets being considered for retirement. Thus, capital costs for an asset to be replaced may be expected to be low and decreasing while operating costs are usually high and increasing.

8.1.1 Basic Reasons for Replacement

There are two basic reasons for considering the replacement of a physical asset:

1. Physical deterioration
2. Obsolescence

Physical deterioration refers only to changes in the physical condition of the asset itself. *Obsolescence* is used here to describe the effects of changes in the environment external to an asset. Physical deterioration and obsolescence may occur independently or jointly.

Physical deterioration may lead to a decline in the value of service rendered, increased operating cost, increased maintenance cost, or a combination. For example, physical deterioration may reduce the capacity of a bulldozer to move earth and consequently reduce the value of the service it can render. Fuel consumption may rise, thus increasing its operating cost, or the physical deterioration may necessitate increased expenditure for repairs.

Obsolescence occurs as a result of the continuous improvement of the tools of production. Often, improvement occurs so rapidly that it is cost effective to replace a physical asset in good operating condition with an improved unit. In some cases, the activity for which a piece of equipment has been used declines to the point that it becomes advantageous to replace it with a smaller unit. In either case, replacement is due to obsolescence and necessitates disposing of the present asset in order to allow for the employment of the more efficient unit. Therefore, obsolescence is characterized by changes external to the asset and is used as a distinct reason in itself for replacement where warranted.

8.1.2 Replacement Should Be Based on Economic Factors

When the success of an economic venture depends upon profit, replacement should be based upon the economy of future operation. Although this goal is sound, motives other than economy often enter into analysis concerned with the replacement of assets.

Part of the reluctance to replace physically satisfactory but economically inferior units of equipment has roots in the fact that a decision to replace has much greater import than a decision to continue with the old. A decision to replace is a commitment for the life of the replacing equipment. But a decision to continue with the old is usually only a deferment of a decision to replace that may be reviewed at any time when the situation seems clearer. Also, a decision to continue with old equipment that results in a loss will usually result in less censure than a decision to replace it with new equipment that results in an equal loss.

The economy of scrapping a functionally efficient unit of productive equipment lies in the conservation of effort, energy, material, and time resulting from its replacement. The unused remaining utility of an old unit is sacrificed in favor of savings in prospect with a replacement. Consider, by way of illustration, a shingle roof. Even a roof that has many leaks will have some utility as a protection against the weather and may have many sound shingles in it. The remaining utility could be made use of by continual repair. But the excess of labor and materials required to make a series of small repairs over the labor and materials required for a complete replacement may exceed the utility remaining in the roof. If so, labor and materials can be conserved by a decision to replace the roof.

8.2 DESCRIBING REPLACEMENT ALTERNATIVES

The method of treating cash flows relative to an existing asset should be the same as that used in treating cash flows relative to a possible replacement. In *both* cases only the present and the future costs of the asset should be considered and *past costs should be disregarded.* Given the definition of past costs as sunk cost in Section 2.4.5 there are two important principles that guide the identification of the *appropriate* cash flows for replacement studies.

The value of the defender that should be used in a study of replacement is not what it cost when originally purchased but what it is worth at the present time. Thus, the defender's sunk cost, which has no effect on future action, is irrelevant to any decision about the defender.

The description of the defender's initial cash flow is based on the fundamental concept of the *opportunity foregone.* We have seen in previous chapters that when money is invested in a productive asset, the cost of the opportunity foregone represented by the interest costs is associated with each alternative.

When an explicit decision is made to retain rather than dispose of an asset already owned, an opportunity to receive compensation has been foregone. This economic opportunity foregone must also be associated with the alternative that caused it to be lost.

To assure that these two principles are satisfied in replacement problems, this chapter utilizes a concept referred to as the *outsider's viewpoint.*

8.2.1 Proper Treatment of Sunk Costs in a Replacement Analysis

An example will be used to illustrate the correct method of evaluating replacements where sunk costs are involved. Suppose that Machine *A* was purchased 4 years ago for $2,200. It was estimated to have a life of 10 years and a salvage value of $200 at the end of its life. Its operating expense has been found to be $700 per year, and it appears that the machine will serve satisfactorily for the balance of its estimated life. Presently a salesman is offering Machine *B* for $2,400. Its life is estimated at 10 years and its salvage value at the end of its life at $300. Operating costs are estimated at $400 per year.

The operation for which these machines are used will be carried on for many years in the future. Equipment investments are expected to justify a 15% MARR in accordance with the policy of the company concerned. The salesman offers to take the old machine in on trade for $600, reflecting its current market value.

In order to make a proper comparison of alternatives the analysis may be undertaken from the standpoint of a person who has a need for the service that Machine *A* or Machine *B* will provide but owns neither (an outsider). The outsider finds that Machine *A* can be purchased for $600 and Machine *B* for $2,400. This analysis of which to buy will not be biased by the past, since the outsider was not part of the original transaction for Machine *A* and, therefore, will not be forced to admit a loss. With this *outsider viewpoint,* the cash flow representing the retain option (keep Machine *A*) begins at the present and considers only present and future cash flows. The past cost, the $2,200 paid 4 years ago for Machine *A*, is not relevant to the outsider who does not own Machine *A*. Just as this sunk cost is ignored by an outsider, it should also be disregarded by the owner when making economic decisions about the future. Therefore, the outsider viewpoint assures that sunk costs are not included in a replacement analysis, as shown in Figure 8.1.

Because these service alternatives have different lives, Method 2 discussed in Section 7.8 is applicable. A study period of 6 years is assumed.

The $AE(i)$ cost to continue with Machine *A* for 6 years is calculated as shown in the following equation:

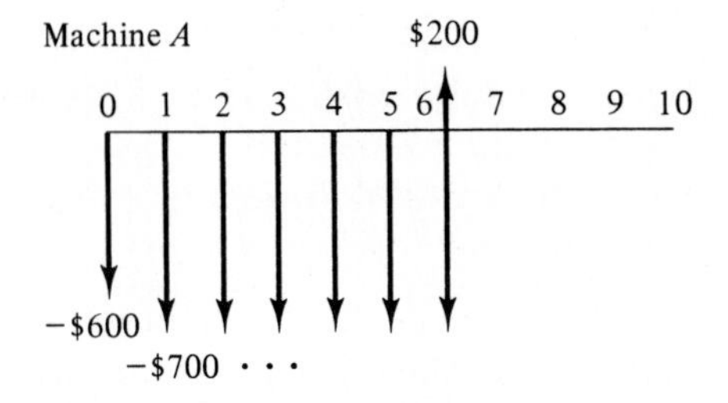

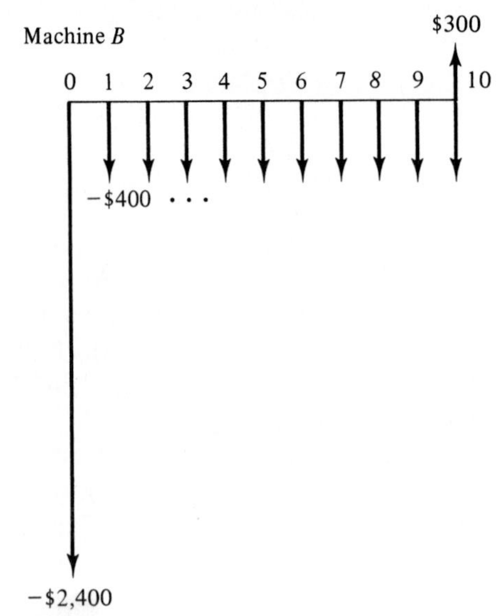

FIGURE 8.1. Outsider viewpoint for a replacement problem.

$CR(15)_A = (\$600 - \$200)(\overset{A/P,15,6}{0.2642}) + \$200(0.15)$	\$136
Annual operating cost	700
	\$836

The $AE(i)$ cost to dispose of Machine A, purchase Machine B, and use it for 10 years is calculated as follows.

$CR(15)_B = (\$2{,}400 - \$300)(\overset{A/P,15,10}{0.1993}) + \$300(0.15)$	\$464
Annual operating cost	400
	\$864

If the alternative to continue with Machine A is adopted, the annual savings prospect for the next 6 years is \$864 − \$836 = \$28. For the next 4 years thereafter the amount of savings will depend upon the characteristics of the machine that might have been purchased 6 years from the present to replace Machine A. If it is assumed that Machine A will be replaced after 6 years by a machine identical to Machine *B*, the $AE(i)$ costs of the two alternatives will be the same after the first 6 years. In that event, the 6-year study period is reasonable.

8.2.2 Comparison Based on Outsider Viewpoint

The outsider's viewpoint may be used for other types of comparisons. For example, suppose a small manufacturing firm located in the southeastern United States is considering the replacement of its natural gas heating unit. During a year 2.488×10^9 Btu are required to keep this well-insulated facility at design temperatures. The following alternatives are being considered.

Alternative $A1$: Remove existing furnace and purchase steam.
Alternative $A2$: Continue to use existing heating unit.
Alternative $A3$: Remove existing unit and purchase new heating system.

The data required for the analysis at a MARR = 20% are as follows:

	Purchase Steam	*Existing Heating Unit*	*New Heating System*
Original cost	$0	$10,000 (10 years ago)	$40,000
Present salvage	—	−$3,000 (cost of removal)	—
Future salvage*	$0	−$3,000 (cost of removal)	$0
Efficiency	—	0.72	0.85
Annual maintenance cost*	$1,500	$2,500	$800
Present fuel cost per 100,000 Btu	—	$1.05	$1.05
Annual cost of steam*	$39,000		
Expected life	10 years	5 years	12 years

** Estimated in constant dollars with base year $t = 0$.*

It is estimated that the price of natural gas will increase at an annual rate of 10%. The future cost of removal for the existing heating system along with the maintenance costs for all three alternatives are also expected to increase at the rate of 10% per year. The contract for the purchase of steam specifies that the cost will be increased at the rate of 10% per year. The contract assures that steam will be provided for at least 10 years if desired, but that the purchaser can discontinue service at no penalty after 4 years.

The negative salvage values represent the cost of removal of the existing heating unit. There is no initial outlay for the steam option, as the supplier will at no charge convert the present system to steam after the existing heating unit has been removed.

The cash flows based on the outsider's viewpoint for these three alternatives are presented in Figure 8.2 in terms of constant dollars. Remember that for

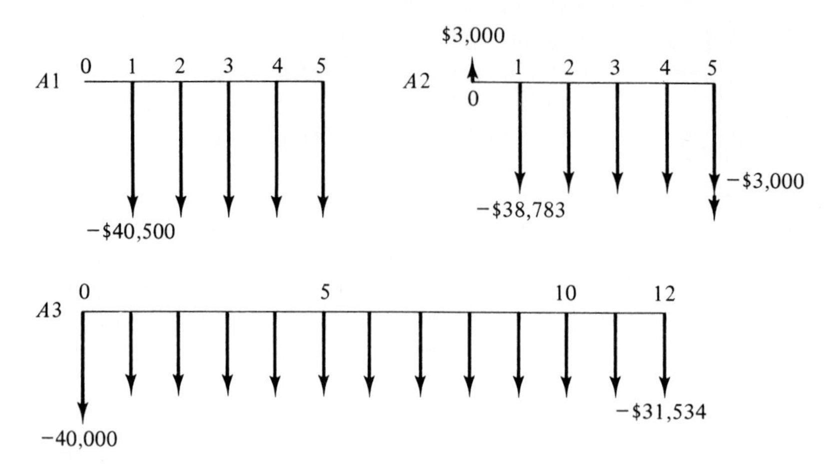

FIGURE 8.2. Alternatives at *A*1, *A*2, and *A*3 in constant dollars.

each alternative the outsider must possess the same assets that the firm would have. If the firm obtains or retains an asset then the outsider must also possess this asset. If the firm disposes of an asset, then the outsider must not possess this asset.

In Figure 8.2 Alternative *A*1 indicates no effect on the cash flow at $t = 0$. From the outsider's view, the purchase of steam requires no initial investment, and the disposal of the furnace has no effect on the initial outlay. Because the outsider does not own the furnace initially, the act of disposal as indicated or Alternative *A*1 requires no action by the outsider.

For Alternative *A*2 the negative salvage value indicates that the removal costs exceed the present market value of the unit. The decision by the firm to *retain* this unit provides the firm the opportunity to avoid the immediate $3,000 expenditure (a saving) associated with the cost of removal. This cost of removal is postponed until the end of this unit's estimated life, as seen in Figure 8.2.

Alternative *A*3 in Figure 8.2 represents the firm obtaining the new heating system and disposal of the existing unit. This alternative is represented by the outsider acquiring that system at its market value with no effect associated with the disposal of the existing furnance.

The annual fuel costs are calculated for the two alternatives using natural gas as follows:

Existing heating unit:

$$\text{annual fuel costs} = (2.488 \times 10^9 \text{ Btu})(\$1.05/100{,}000 \text{ Btu})/0.72$$

$$= \$36{,}283 \text{ per year.}$$

New heating system:

$$\text{annual fuel costs} = (2.488 \times 10^9 \text{ Btu})(\$1.05/100{,}000 \text{ Btu})/0.85$$

$$= \$30{,}734 \text{ per year.}$$

The $AE(i)$ cost comparison of these three alternatives is based on a 5-year study period. Because the inflationary effects are all at the same rate, the comparison is based on a constant-dollar analysis. $i' = (1.20)/(1.10) - 1 = 0.0909$ or 9.09%. *If the inflation rates had been different for any elements of this problem, then the analysis would have to be accomplished in terms of actual dollars.* (See Section 5.4.3).

Alternative A1 (Purchase steam):

$CR(9.09)_{A1}$	=	0
Annual cost of steam	=	−$39,000
Annual maintenance costs	=	−$ 1,500
Total $AE(9.09)$ cost (constant dollars)		−$40,500

Alternative A2 (retain existing heating unit):

$CR(9.09)_{A2} = \$3{,}000(\overset{A/P,9.09,5}{0.2577}) - \$3{,}000(\overset{A/F,9.09,5}{0.1668})$	=	$ 273
Annual fuel cost	=	−$36,283
Annual maintenance costs	=	−$ 2,500
Total $AE(9.09)$ cost (constant dollars)		−$38,510

Alternative A3 (purchase heating unit):

$CR(9.09)_{A3} = -\$40{,}000(\overset{A/P,9.09,12}{0.1403})$	=	−$ 5,612
Annual fuel cost	=	−$30,734
Annual maintenance costs	=	−$ 800
Total $AE(9.09)$ costs (constant dollars)		−$37,146

In this example we find that the purchase of the new heating unit (Alternative $A3$) provides the least-cost option. This conclusion is based on a study period of 5 years. By use of the outsider viewpoint we have properly associated the opportunity costs with each alternative and we have ignored sunk costs.

8.3 REPLACEMENT ANALYSIS FOR UNEQUAL LIVES

Because most replacement decisions are concerned with the replacement of old assets by new ones, the economic alternatives being examined are seldom of equal duration. In Section 7.8 the basic principle for comparing alternatives with unequal lives requires that such alternatives be evaluated over equal periods of time.

As an illustration of the use of different study periods with Method 2, consider the following example. A certain operation is now being carried on with Machine *E*, whose present salvage value is estimated to be $2,000. The future life of Machine *E* is estimated at 5 years, at the end of which its salvage value is estimated to be zero. Operating costs with Machine *E* are estimated at $1,200 per year. It is expected that Machine *E* will be replaced after 5 years by Machine *F*, whose initial cost, life, final salvage value, and annual operating costs are estimated to be, respectively, $10,000, 15 years, zero, and $600.

The desirability of replacing Machine *E* with Machine *G* is being considered. Machine *G*'s estimated initial cost, life, final salvage value, and annual operating costs are estimated to be, respectively, $8,000, 15 years, zero, and $900. The interest rate is taken to be 10%. Detailed investment and cost data for Machines *E*, *F*, and *G* are given in Table 8.1.

8.3.1 Analysis Based on a 15-Year Study Period, Recognizing Implied Salvage Value

Because of the difficulty of making further estimates into the future, a study period of 15 years coinciding with the life of Machine *G* is selected. This will necessitate calculations that consider an alternative that has a life longer than the study period. Using Method 2 reflects an implied salvage value as discussed in Section 7.8.

Under Plan I, the study period embraces 5 years of service with Machine *E* and 10 years of service with Machine *F*, whose useful life extends 5 years beyond the study period. Thus, an equitable allocation of the costs associated with Machine *F* must be made for the period of its life coming within and after the study period. In assuming that annual costs associated with this unit of equipment are constant during its life, the present worth cost of service during the study period may be calculated as follows.

The annual equivalent cost for Machine *F* during its life is equal to

$$AE(10)_F = \$10{,}000(\overset{A/P,10,15}{0.1315}) + \$600 = \$1{,}915.$$

The present worth cost of 15 years of service in the study period is equal to

$$\begin{aligned} PW(10)_{\mathrm{I}} &= \$2{,}000 + \$1{,}200(\overset{P/A,10,5}{3.791}) + \$1{,}915(\overset{P/A,10,10}{6.1446})(\overset{P/F,10,5}{0.6209}) \\ &= \$13{,}856. \end{aligned}$$

TABLE 8.1.
ANALYSIS BASED ON A SELECTED STUDY PERIOD

	End of Year	Plan I Machine Investment	Plan I Operating Costs	Plan II Machine Investment	Plan II Operating Costs
Fifteen-year study period	0	Machine *E*, \$2,000		Machine *G*, \$8,000	
	1		\$1,200		\$900
	2		1,200		900
	3		1,200		900
	4		1,200		900
	5	Machine *F*, \$10,000	1,200		900
	6		600		900
	7		600		900
	8		600		900
	9		600		900
	10		600		900
	11		600		900
	12		600		900
	13		600		900
	14		600		900
	15		600		900
	16		600		
	17		600		
	18		600		
	19		600		
	20		600		

By distributing the first cost of Machine *F* over the entire 15 years of its estimated service life, the calculations reflect that 5 years (years 16 through 20) of Machine *F*'s value is unused for the 15-year study period assumed. On this basis the implied salvage value remaining in Machine *F* at the end of the study period may be calculated as a matter of interest from Equation (7.12).

$$F_n{}^* = \$10{,}000(\overset{A/P,10,15}{0.1315})(\overset{P/A,10,5}{3.791}) = \$4{,}985.$$

Under Plan II, the life of Machine *G* coincides with the study period. The $PW(i)$ cost of 15 years service in the study period is equal to

$$PW(10)_{11} = \$8{,}000 + \$900(\overset{P/A,10,15}{7.606}) = \$14{,}845.$$

On the basis of $PW(i)$ costs of \$13,856 and \$14,845 for a study period of 15 years, Plan I should be chosen.

8.3.2 Analysis on the Basis of a 5-Year Study Period

Lack of information often makes it necessary to use rather short study periods. For example, the characteristics of the successor to Machine *E* in Table 8.1 might be vague. In that case a study period of 5 years might be selected to coincide with the estimated retirement date of Machine *E*.

The $AE(i)$ cost of continuing with Machine *E* during the next 5 years is

$$AE(10)_E = \$2{,}000(\overset{A/P,10,5}{0.2638}) + \$1{,}200 = \$1{,}728.$$

The $AE(i)$ cost of Machine *G* based on a life of 15 years is

$$AE(10)_G = \$8{,}000(\overset{A/P,10,15}{0.1315}) + \$900 = \$1{,}952.$$

The \$224-per-year cost advantage of Machine *E* over Machine *G* can be interpreted in two ways. First, the retention of Machine *E* will produce such savings for a 5-year period if it is assumed that if Machine *G* is purchased it would be sold for its implied salvage value at the end of 5 years.

Second, the above calculations could be interpreted as the cost of repeating Alternative *E* and Alternative *G* over a common multiple of lives (15 years in this case). Such an assumption may not be reasonable, since Alternative *E* is an older, partially used asset. Thus, it is improbable that Alternative *E* will be replaced by an asset with identical performance characteristics. Usually for replacement comparisons the repeatability assumption of Method 2 is not justified.

In general, the longer the study period, the more significant the results. But the longer the study period, the more likely it is that estimates are in error. Thus, the selection of a study period must be based on estimate and judgment.

8.4 THE ECONOMIC LIFE OF AN ASSET

The preceding section has discussed the types of analyses that may be applied when the service life is known. In many instances, however, the length of time a particular asset will be retained is only conjecture. Since replacement analyses are usually sensitive to the lives assumed, it is prudent to consider the economic consequences of holding an asset for various lengths of time. Of particular interest is that length of time within the service life of any asset referred to as the economic life of the asset.

> The **economic life of an asset** is the time interval that minimizes the asset's total equivalent annual costs or maximizes its equivalent annual net income.

The economic life is also referred to as the minimum-cost life or the optimum replacement interval. The concept of economic life is relevant in both replacement and new investment studies.

8.4.1 Finding the Economic Life of an Asset

Because in replacement analysis the defender and challengers should be compared on their economic characteristics, we must consider carefully how to calculate the economic life of an asset. The analysis simply involves the calculation of the total $AE(i)$ cost at the end of each year in the asset's life. Selection of the total $AE(i)$ cost that is a minimum would specify a minimum-cost life for the asset. The application of this approach is demonstrated by the following example.

The economic future of an asset whose first cost is $3,000, with decreasing salvage values and with operating costs beginning at $1,000 and increasing by $700 each year for an interest rate of 12%, is shown in Table 8.2. To find this asset's economic life, it is necessary to identify the relevant cash flows associated with retaining the asset 1, 2, 3, or 4 years. These cash flows are depicted in Figure 8.3, and they are the basis for the annual equivalent calculations shown in Table 8.2. These $AE(i)$ costs are

$$n = 1: \text{AE}(12) = (\$3{,}000 - \$1{,}500)\overset{A/P,12,1}{(1.1200)} + \$1{,}500(0.12) + \$1{,}000 + \$700\overset{A/G,12,1}{(0.0000)} = \$2{,}860$$

$$n = 2: \text{AE}(12) = (\$3{,}000 - \$1{,}000)\overset{A/P,12,2}{(0.5917)} + \$1{,}000(0.12) + \$1{,}000 + \$700\overset{A/G,12,2}{(0.4717)} = \$2{,}633$$

$$n = 3: \text{AE}(12) = (\$3{,}000 - \$500)\overset{A/P,12,3}{(0.4164)} + \$500(0.12) + \$1{,}000 + \$700\overset{A/G,12,3}{(0.9246)} = \$2{,}748$$

$$n = 4: \text{AE}(12) = (\$3{,}000 - \$0)\overset{A/P,12,4}{(0.3292)} + \$0(0.12) + \$1{,}000 + \$700\overset{A/G,12,4}{(1.3589)} = \$2{,}938$$

TABLE 8.2.
TABULAR CALCULATION OF ECONOMIC LIFE

End of Year	Salvage Value When Asset Retired at Year n	Operating Costs During Year n	$AE(i)$ Cost of Capital Asset When Retired at Year n	$AE(i)$ Cost of Operating for n Years	Total $AE(i)$ Cost When Asset Retired at Year n
1	$1,500	$1,000	$1,860	$1,000	$2,860
2	1,000	1,700	1,303	1,330	2,633*
3	500	2,400	1,101	1,647	2,748
4	0	3,100	987	1,951	2,938

* *Economic life.*

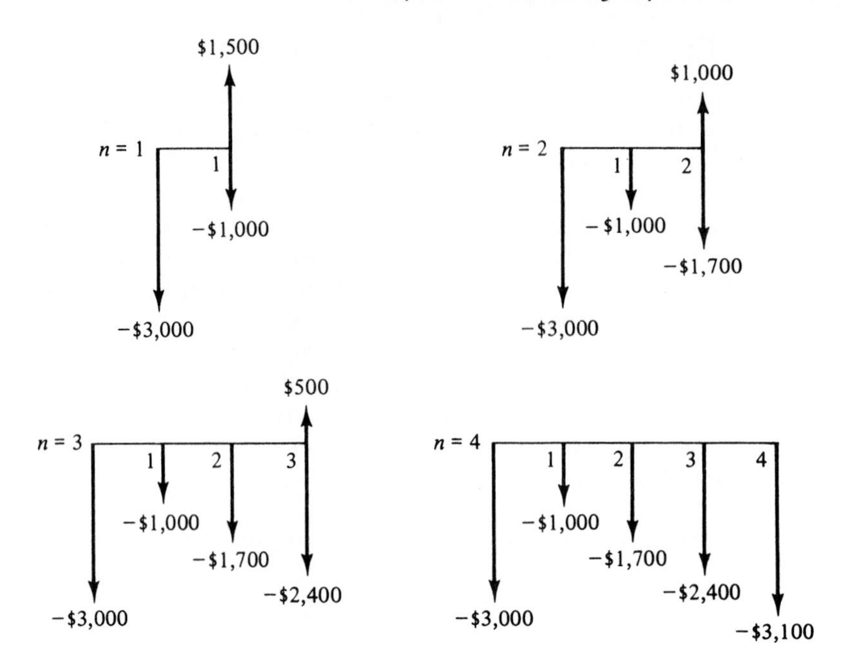

FIGURE 8.3. Cash flows of asset retained 1, 2, 3, or 4 years.

and the economic life for this asset is 2 years. If the asset were sold after 2 years, it would have minimum $AE(i)$ cost of $2,633 per year, and that is the life that is most favorable for comparison purposes.

Table 8.2 illustrates the tabular method for determining the economic life of an asset. Study of the annual equivalent operating costs, the capital recovery costs, and the resulting total annual equivalent cost curve in Figure 8.4 indicates the relationships that lead to an economic life occurring somewhere between the shortest possible life and the asset's service life.

In addition to the general procedure just discussed, there are two special situations for which the economic life can be discovered without lengthy calculations.

Case I. One of these situations occurs whenever an asset's annual operating costs remain *constant* over its life while its future salvage values remain *unchanged.* For this case, the longer an asset is retained in service, the lower is its total $AE(i)$ cost. Therefore, the economic life for such an asset corresponds to its service life.

In many instances an existing asset will have a future salvage value of zero with no expectations of any change due to continued use. If such an asset has constant operating costs over its life, the appropriate strategy is to retain it as long as possible.

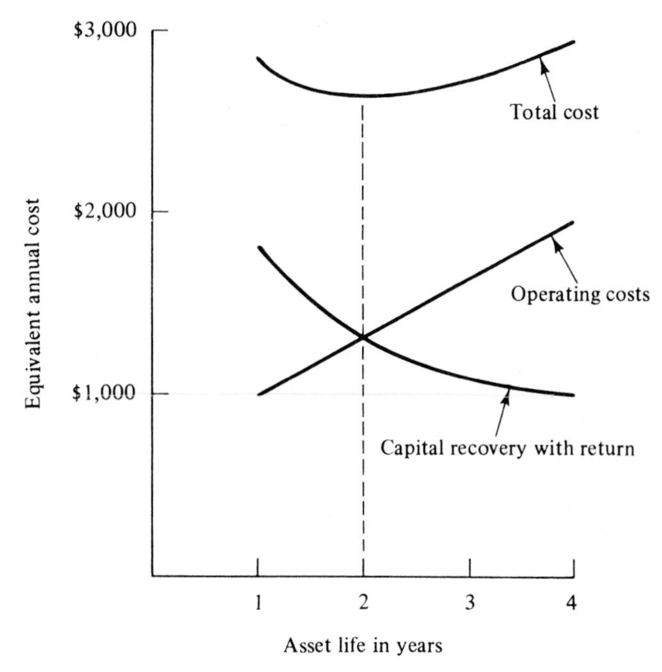

FIGURE 8.4. Economic life of an asset.

To present an example of this case, suppose that an existing asset has a present salvage value of $1,000 but its salvage value is expected to be $400 if it is retired at any time in the future. At 12% interest the total $AE(i)$ costs for each of its possible lives are presented in Table 8.3.

Case II. Another special situation occurs if the initial outlay and future salvage values *always equal* each other and the annual operating and maintenance

TABLE 8.3.
ECONOMIC-LIFE CALCULATION FOR CASE I EXAMPLE

End of Year n	Salvage Value When Asset Retired at Year n	Operating Costs During Year n	$AE(i)$ Costs of Capital Recovery When Asset Retired at n Year	$AE(i)$ Cost of Operating for n Years	$AE(i)$ Cost When Asset Retired at Year n
0	$1,000	—	—	—	—
1	400	$1,500	$720	$1,500	$2,220
2	400	1,500	403	1,500	1,903
3	400	1,500	298	1,500	1,798
4	400	1,500	246	1,500	1,714*

* *Economic life.*

costs are *always increasing*. For this set of circumstances, the economic life is the shortest possible life, namely, 1 year (or one period, depending upon the frequency of the replacement studies). This fact is evident from the relationship describing the total $AE(i)$ costs of the asset for any year.

$$\text{Total } AE(i) \text{ costs} = CR(i) + AE(i) \text{ operating costs.}$$

The capital-recovery portion of total costs will be constant for any asset for which $P = F$ no matter how long the asset is in service, since

$$CR(i) = (P - F)(\overset{A/P,i,n}{\quad}) + Fi.$$

The $AE(i)$ operating costs for an asset will be ever increasing as long as each year's operating expense is greater than the proceeding year's expense. Thus, for these two conditions the total $AE(i)$ costs will be minimized for the shortest time the asset might be reasonably retained.

An example of this case is presented in Table 8.4 for an asset having a present salvage value of $300 with all future salvage values equal to that figure. With the annual operating expenses increasing from $1,000 to $1,300 in increments of $100 and for an interest rate of 12% the economic life is 1 year. The calculations are described in Table 8.4. This pattern of cash flows occurs most frequently when an asset is worth nothing now and will be worth nothing if it is retired in the future.

TABLE 8.4.
ECONOMIC-LIFE CALCULATIONS FOR CASE II EXAMPLE

End of Year n	Salvage Value When Asset Retired at Year n	Operating Costs During Year n	Annual Equivalent Costs of Capital Recovery When Asset Retired at Year n	Annual Equivalent Cost of Operating for n Years	Total Annual Equivalent Cost When Asset Retired at Year n
0	$300	—	—	—	—
1	300	$1,000	$36	$1,000	$1,036*
2	300	1,100	36	1,048	1,084
3	300	1,200	36	1,094	1,130
4	300	1,300	36	1,138	1,174

* *Economic life.*

8.5 REPLACEMENT DECISIONS AND ASSUMPTIONS

The economic impact of a replacement decision is dependent on the study period selected for the analysis. If the study period chosen is the life of the shortest-lived asset (usually the defender), the method of analysis discussed in Section 7.8 is applicable. However, there are many other assumptions that can be applied in replacement studies.

In this section explicit assumptions are made regarding the type and frequency of replacement that follow the retirement of the defender and the current challengers. Four different assumptions are to be considered, and solution methods are presented. These assumptions range from simplistic to very general, and they demonstrate that the more definite the knowledge of future events, the more precise the analysis.

Suppose that a firm is considering three different assets that provide identical service. One is the defender *D*1, an asset presently owned. There are two other assets being considered as the possible replacement for *D*1. These challengers, are identified as *C*1 and *C*2. The three investment alternatives are mutually exclusive, as the level of service requires only one of these assets to be used at any one point in time. The cash flows associated with *D*1, *C*1, and *C*2 are presented in Table 8.5. It is observed that the service life for *D*1 is 5 more years. For *C*1 and *C*2 the service lives are 6 years and 7 years, respectively. However, each of these assets can be disposed of at any time prior to the end of its service life.

8.5.1 Assume Replacement of Each Alternative by Identical Alternatives

This assumption regarding future replacements requires that each current alternative will be succeeded by a sequence of assets identical to that alternative. To establish a study period it can be assumed that the cash flows are repeated until a common multiple of lives is reached. (See Section 7.8, Method 2.) If it is assumed that the actual salvage value will equal the implied salvage value, any time span may be selected for the study period. For the example represented by *D*1, *C*1, and *C*2 in Table 8.5, the possible alternatives with their sequences of replacements are represented in Figure 8.5.

The result is $5 + 6 + 7 = 18$ mutually exclusive alternatives. These alternatives must be compared over equal periods of time, the objective being to minimize the $AE(i)$ over the study period selected. In this case, the study period might be selected at the common multiple of lives, 210 years, or at any other point in time. If a study period is selected where some replacements extend beyond it, Method 2 presented in Section 7.8 must be employed and the assumptions of implied salvage value are in effect. Notice that because the cash flows are assumed to repeat, the $AE(i)$ has only to be calculated over the initial service

TABLE 8.5.
DATA ASSOCIATED WITH DEFENDER AND TWO CHALLENGERS

	Alternatives								
	*D*1			*C*1			*C*2		
End of Year *n*	Salvage Value When Asset Retired at Year *n*	Operating Costs During Year *n*	Total Annual Equivalent Cost When Asset Retired at Year *n*	Salvage Value When Asset Retired at Year *n*	Operating Costs During Year *n*	Total Annual Equivalent Cost When Asset Retired at Year *n*	Salvage Value When Asset Retired at Year *n*	Operating Costs During Year *n*	Total Annual Equivalent Cost When Asset Retired at Year *n*
0	$1,200	—	—	$ 1,800**	$ —	—	$2,500**		
1	900	$ 300	$ 780*	0	$ 107	$2,177	0	$150	$3,025
2	600	900	1,038	0	314	1,310	0	150	1,688
3	800	500	852	0	521	1,083	0	150	1,245
4	500	550	875	0**	728	1,012	0	150	1,025
5	300	800	905	0	935	1,000*	0	150	896
6	—	—	—	0	1,142	1,017	0	150	811
7	—	—	—	—	—	—	0	150	751*

* *Economic life*

** *Initial investment*

$D1_t$ = Defender with life t years $\quad t = 1, 2, 3, 4, 5$
$C1_t$ = Challenger 1 with life t years $\quad t = 1, 2, 3, 4, 5, 6$
$C2_t$ = Challenger 2 with life t years $\quad t = 1, 2, 3, 4, 5, 6, 7$

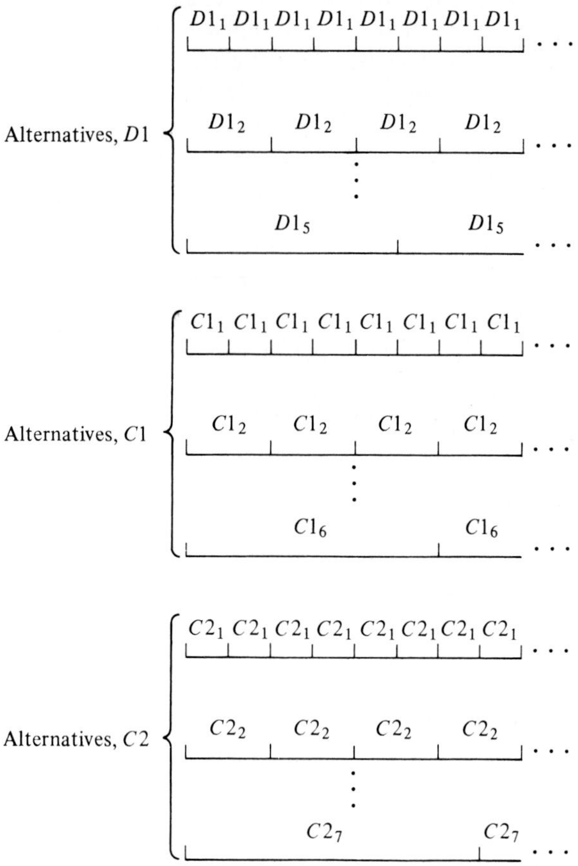

FIGURE 8.5. Alternatives assuming replacement by identical alternatives.

life of each of the 18 current alternatives, $D1_1$, $D1_2$, . . . , $C1_1$, . . . , $C2_1$, . . . , $C2_7$. (See Section 6.2.1.)

By finding the economic lives for the defender, $D1$, and the two challengers, $C1$ and $C2$ from Table 8.5 and comparing their $AE(i)$ amounts based on those lives, the proper conclusion is reached. This comparison is as follows.

$$AE(15)_{D1_1} = \$780 \text{ per year}$$

$$AE(15)_{C1_5} = \$1{,}000 \text{ per year}$$

$$AE(15)_{C2_7} = \$751 \text{ per year.}$$

Thus, for the set of assumptions used in this section the least expensive alternative is $C2_7$ for any study period selected between 1 year and infinity.

8.5.2 Assume No Replacement of Current Alternatives

This situation usually occurs when a defender and challenger are being compared and there is no knowledge of their possible replacements. For this assumption, the life of the shortest-lived asset becomes the study period. The alternatives for the example in Table 8.5 are shown in Figure 8.6, and the study period is seen to be 1 year.

Section 7.8 describes the analysis of alternatives for this type of comparison where the lives of the assets are longer than the study period. Method 2 requires that the $AE(i)$ be calculated over the service life of each asset, implying a salvage value at the end of the study period. Then the $AE(i)$ amounts based on the economic life of each alternative can be compared directly.

$$AE(15)_{D1_1} = \$780 \text{ per year}$$

$$AE(15)_{C1_5} = \$1{,}000 \text{ per year}$$

$$AE(15)_{C2_7} = \$751 \text{ per year.}$$

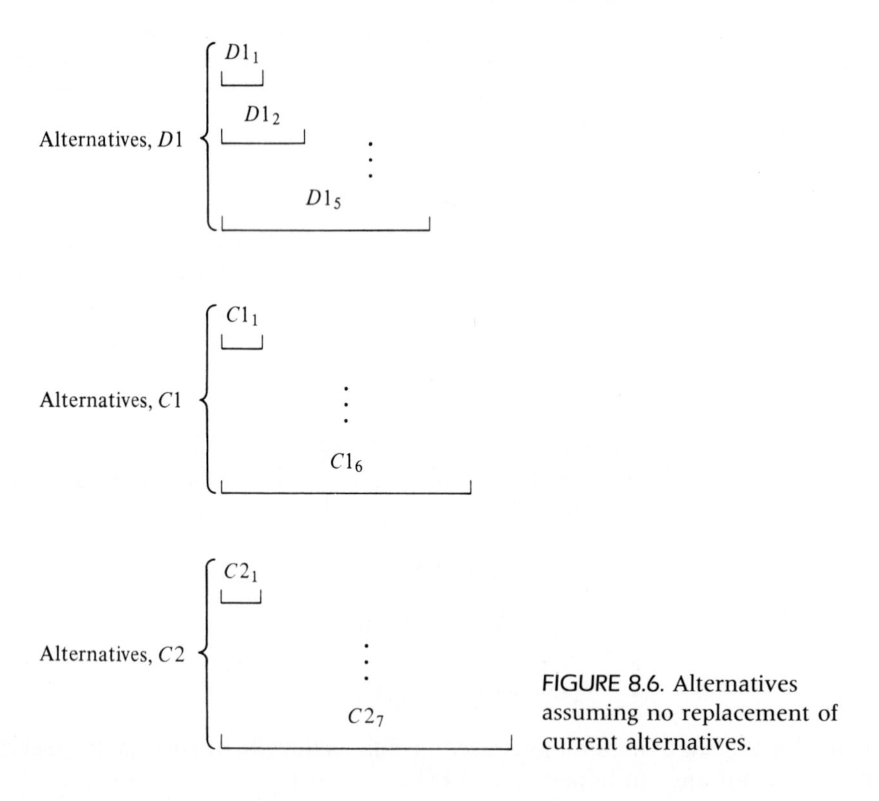

FIGURE 8.6. Alternatives assuming no replacement of current alternatives.

It was shown for the repeated-lives assumption that comparing the $AE(i)$ values associated with each asset's economic life produces the correct selection. When it is assumed that there is *no* replacement of current alternatives, the identical decision rule based on economic life again makes the proper selection, Alternative $C2_7$.

8.5.3 Assume Replacement of Each Alternative by the Best Challenger

This assumption differs from those mentioned earlier in this section in that it presumes that all *future* challengers will be identical to the best *current* challenger. *The best challenger is the current challenger among all the available current challengers that minimize the equivalent costs.* The problem of identifying the least costly alternative out of the 13 alternatives associated with $C1$ and $C2$ is straightforward. Whichever alternative is selected, it will be replaced in the future by itself. Since the $AE(i)$ represents a repeated cash flow, the challenger that minimizes the $AE(i)$ is the best challenger. Using the figures in Table 8.5, the best challenger is $C2_7$. Now the comparison reduces from the 18 alternatives shown in the previous discussion to the 6 alternatives represented in Figure 8.7.

Condition 1: Since the $AE(i)$ amount is identical for each alternative beyond the life of the longest-lived current alternative, the study period can be selected at 7 years ($C2_7$ is the longest-lived alternative in Figure 8.7). For this problem, which is essentially a comparison of $C2_7$ and $D1$, observe the plot of $AE(i)$ for each option in Figure 8.8.

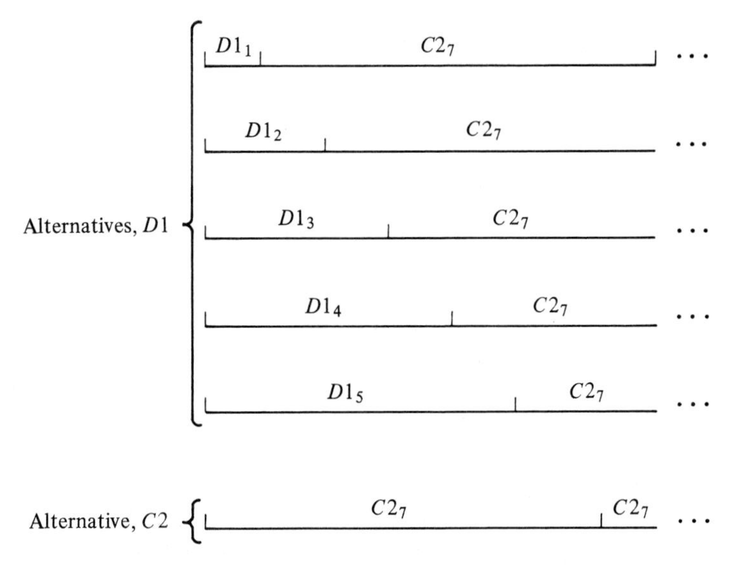

FIGURE 8.7. Alternatives when replacement by best challenger is assumed.

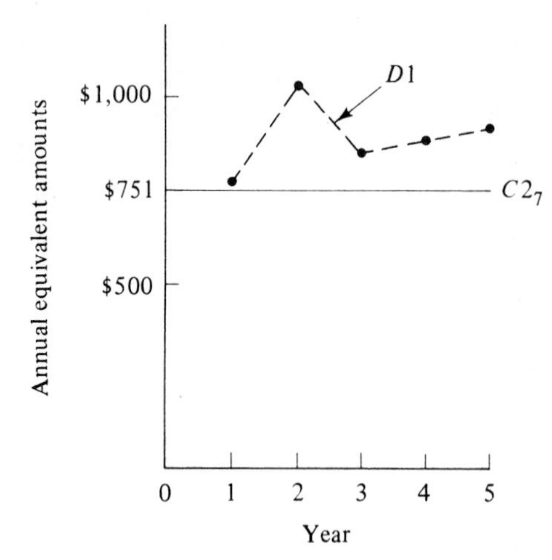

FIGURE 8.8. $AE(i)$ for $D1$ and $C2_7$.

For this special *condition*, when *all* the $AE(i)$ costs of the defender are *greater* than the $AE(i)$ costs of the best challenger, there are no circumstances that would favor the defender. Even the use of the minimum $AE(i)$ cost at the economic life of one year for $D1$ still favors the challenger. Thus, comparing the alternatives based on the economic life of $D1$ and $C2$ yields the proper choice. Following this procedure and using the $AE(i)$ values from Table 8.5 gives

$$AE(15)_{D1_1} = \$780 \text{ per year}$$

$$AE(15)_{C2_7} = \$751 \text{ per year.}$$

Again, $C2_7$ is the preferred alternative.

Condition 2: Now suppose that the only challenger is $C1$, and it is being compared to $D1$ ($C2$ is not available). To find the best $C1$ alternative, go to Table 8.5 and find the time period that minimizes the $AE(i)$ amount. The time period is 5 years for an $AE(i)$ amount of $1,000 per year. Assuming that the best challenger alternative, $C1_5$, will succeed the current alternatives being considered, the present worths for a 5-year study period are calculated. These present worths are presented in Table 8.6, and two examples of their calculation follow. For alternative $D1_4$

$$PW(15)_{D1_4} = [\$1{,}200 - \$500\overset{P/F,15,4}{(0.5718)} + \$300\overset{P/F,15,1}{(0.8696)} + \$900\overset{P/F,15,2}{(0.7562)}$$

$$+ \$500\overset{P/F,15,3}{(0.6575)} + \$550\overset{P/F,15,4}{(0.5718)} + \$1{,}000\overset{P/F,15,5}{(0.4972)}]$$

$$= \$2{,}996 \tag{8.1}$$

TABLE 8.6.
PRESENT WORTHS FOR CURRENT ALTERNATIVES REPLACED BY $C1_5$ FOR A 5-YEAR STUDY PERIOD

	Alternative	
End of Year	*D*1	*C*1
0	—	—
1	$3,161	$4,376
2	3,414	3,857
3	3,013	3,542
4	2,996	3,386
5	3,033	3,352*

* *Best challenger*

or, more conveniently, using values from Table 8.5

$$PW(15)_{D1_4} = [\$875\overset{P/A,15,4}{(2.8550)} + \$1{,}000\overset{P/F,15,5}{(0.4972)}] = \$2{,}996$$

For alternative $C1_3$

$$PW(15)_{C1_3} = [\$1{,}800 + \$107\overset{P/F,15,1}{(0.4972)} + \$314\overset{P/F,15,2}{(0.7562)} + \$521\overset{P/F,15,3}{(0.6572)}$$
$$+ \$1{,}000\overset{P/A,15,2}{(1.6257)}\overset{P/F,15,3}{(0.6575)}]$$

$$= \$3{,}542 \tag{8.2}$$

or, more conveniently, using values from Table 8.5

$$PW(15)_{C1_3} = [\$1{,}083\overset{P/A,15,3}{(2.2832)} + \$1{,}000\overset{P/A,15,2}{(1.6257)}\overset{P/F,15,3}{(0.6575)}]$$

$$= \$3{,}542.$$

The correct decision is to select Alternative *D*1 and keep it 4 years, when it is assumed it will be replaced by $C1_5$, the best challenger. This conclusion is based on a 5-year study period and the *PW*(15) data in Table 8.6. The same result occurs for any study period greater than 5 years. If it happens that the maximum life of the defender exceeds the life of the best current challenger, the longer of the two lives should be the basis for the study period.

Will the same choice be made if the decision is based on the comparison of the $AE(i)$ amounts at the economic life of $D1$ and $C1$? Although this approach has produced the correct conclusions for previous assumptions and conditions, in this case the answer is no! Selecting the preferred alternative by comparing the $AE(i)$ amounts at the economic life of $D1$ and $C1$ yields

$$AE(15)_{D1_1} = \$780$$

$$AE(15)_{C1_5} = \$1{,}000.$$

Based on these values, it appears that the superior option is to retain $D1$ one more year and replace it by $C1_5$. This conclusion is *incorrect,* as seen from the present worths in Table 8.6. A closer examination of Tables 8.5 and 8.6 for $D1$ indicates that for time periods of 1, 3, 4 and 5 years, the lowest $AE(i)$ amount *does not* produce the lowest present worth. This troublesome condition can occur if more than one of the $AE(i)$ amounts of the defender are less than the $AE(i)$ for the best challenger. Figure 8.9 illustrates an example of this condition.

Examination of the information in Figure 8.9 indicates that if the defender is kept exactly 2 years, the best challenger alternative, $C1_5$, is preferred. If the defender is retained either 1, 3, 4, or 5 years, the defender is the most economic option. Unfortunately, it is not possible to determine from the information in Figure 8.9 the most desirable length of time to retain the defender. *Thus, for Condition 2, where the best challenger is assumed to be used for any future replacement and more than one AE(i) amount of the defender is less than that of the challenger, do not*

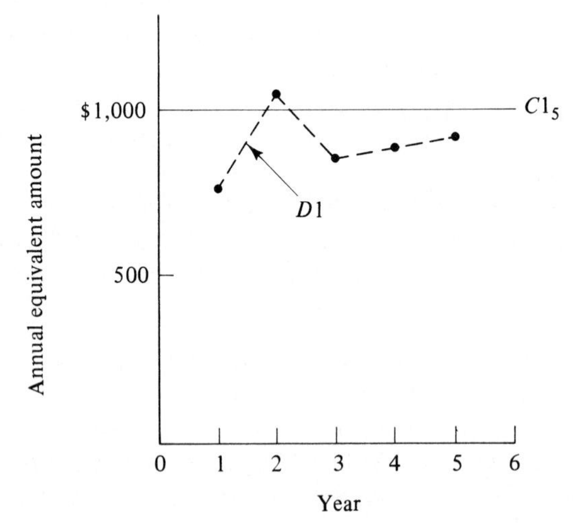

FIGURE 8.9. $AE(i)$ for $D1$ and $C1_5$.

directly compare the annual equivalent amounts defined by the alternative's economic lives.

For Condition 2 the correct approach is to compare the $PW(i)$ amounts for each alternative as calculated in Table 8.6. These $PW(i)$ amounts are based on the implied-salvage-value assumptions of Method 2 discussed in Section 7.8. (That is, for any alternative extending beyond the study period it is assumed that the actual salvage is the implied salvage at the end of the study period.)

8.5.4 Assume Replacement of Each Alternative by Dissimilar Challenger

The assumptions of the previous three sections are rather restrictive, considering all the forces that are at work in the economy. With technological innovations, new types of equipment are made available, and this future equipment may be substantially different from the equipment presently available. In addition, inflation and other financial effects will distort the cash flows of future acquisitions. Therefore, it is unlikely to expect that all future challengers will be similar.

In this section, the inclusion of dissimilar future challengers allows for the consideration of many additional effects. This more realistic view of replacement is intended to provide more pragmatic solutions to this rather complicated problem. Unfortunately, these complications preclude the use of simple assumptions and easy solution methodologies discussed in the previous sections. Because of the variety of possible sequences of replacements, a technique must be used that can consider all the possible combinations of future replacements. The following discussion provides a framework for the formulation of this type of replacement problem.

Suppose the defender $D1$, described in Table 8.5, is presently 2 years old and has 5 more years in its service life. It can be replaced at the end of any year. Each year there will be one challenger available, and each challenger which is defined by $C1$ in Table 8.5 has a maximum service life of 6 years. Each future challenger may be replaced at any age. All the possible replace (R) and keep (K) options for the defender and the challengers over a 10-year period are illustrated by the network shown in Figure 8.10.

Each of the possible paths, moving from left to right, from node 1 to any of the nodes at year 10 represents one possible sequence of replacements over a 10-year period. The heavy line shown on Figure 8.10 indicates the following sequence of actions.

$D1_2$ $C2_5$ $C7_3$

$C2_t$ = Challenger available at year 2 and retained t years
$C7_t$ = Challenger available at year 7 and retained t years

The defender is kept two years, and it is replaced by the new challenger available at year 2. This challenger is kept 5 years, and it is replaced at the end of

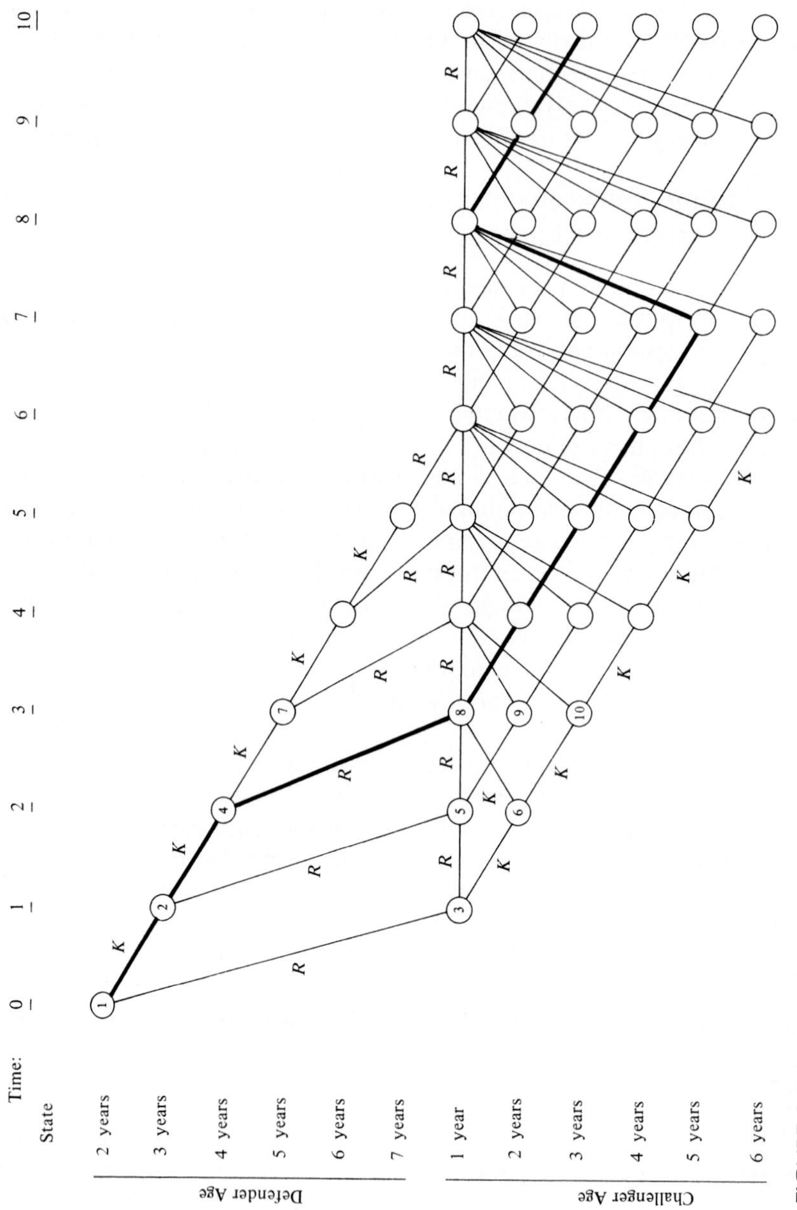

FIGURE 8.10. Network describing replacement alternatives.

year 7. Its replacement is the new challenger available at year 7. At the end of the 10th year, this asset is still in service at age 3.

The number of paths in a network such as Figure 8.10 increases exponentially as the number of challengers and the length of their service life increases. Therefore, it is generally not possible to examine all the possible combinations explicitly in order to identify the least costly path. Fortunately, there are efficient computational algorithms available to solve problems of this type.

If the network is small so that complete enumeration of all paths is practical, the total $PW(i)$ of the costs associated with each path is calculated. To facilitate this total-cost calculation or the calculations required by any of the algorithms that can be applied, it is necessary to know the incremental change in total $PW(i)$ that occurs when moving from one node to any of its adjacent nodes.

To demonstrate the economic information that must be on the paths connecting each of the nodes, it is assumed that $D1$ has the cash flow defined in Table 8.5. All current and future challengers have cash flows identical to $C1$ in Table 8.5.

The $PW(i)$ amounts that should be placed between adjacent nodes are calculated from the cash flows in Figure 8.11. Notice that the present worths are all calculated at $t = 0$, even though a particular cash flow represents the cost of going from t to $t + 1$ when $t > 0$.

Node
1 to 2

$$PW(15) = \$1{,}200 + (\$300 - \$900)(\overset{P/F,15,1}{0.8696}) = \$678$$

Node
2 to 4

$$PW(15) = [\$900 + (\$900 - \$600)(\overset{P/F,15,1}{0.8696})](\overset{P/F,15,1}{0.8696}) = \$1{,}010$$

Node
4 to 7

$$PW(15) = [\$600 + (\$500 - \$800)(\overset{P/F,15,1}{0.8696})](\overset{P/F,15,2}{0.7572}) = \$256$$

Node
1 to 3

$$PW(15) = \$1{,}800 + \$107(\overset{P/F,15,1}{0.8696}) = \$1{,}893$$

Node
3 to 6

$$PW(15) = \$314(\overset{P/F,15,2}{0.7562}) = \$237.$$

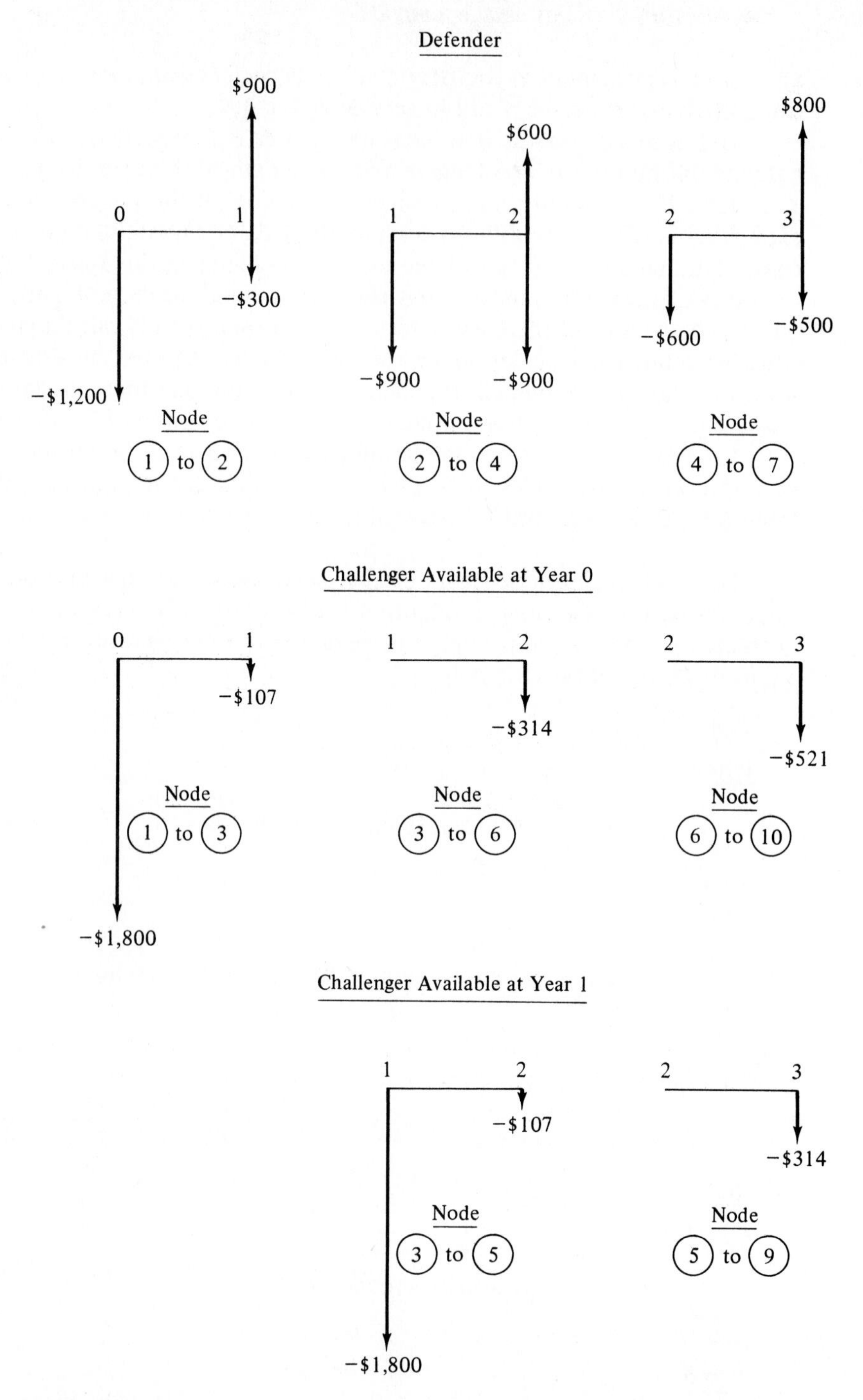

FIGURE 8.11. Cash flows representing incremental costs between adjacent nodes in Figure 8.10.

After all the incremental $PW(i)$ costs are calculated, they are placed on the paths connecting the appropriate nodes. Then, using a shortest-path algorithm, dynamic programming, or complete enumeration the best of all the possible sequences of replacements is identified.[1]

Usually the most urgent decision concerns whether to keep the defender or switch to the current challenger. (This is the decision that must be made at present.) Using the approach described here, it is possible to modify assumptions about the future and to test the sensitivity of the current decision concerning the defender and current challenger to these changes. Although the solution techniques are beyond the scope of this book, it is important to realize that it is possible to deal analytically with these very complex replacement problems.

8.6 RETIREMENT OR ABANDONMENT DECISIONS

In many instances, an asset may still be productive, but the firm must decide whether to continue the activity or to eliminate it. Such decisions are sometimes referred to as abandonment decisions. Strict retirement decisions concern only the defender; so for discussion purposes the defender $D1$ described in Table 8.5 is used.

The alternatives for the retirement decision are

Immediate disposal.
Keep $D1$, 1 year.
Keep $D1$, 2 years.
Keep $D1$, 3 years.
Keep $D1$, 4 years.
Keep $D1$, 5 years.

To select the preferred alternative, find the one that maximizes the $PW(i)$ of net income. (Note that for strict retirement decisions the receipts and disbursements must be known.) To use the example in Table 8.5, assume that $D1$ has an annual revenue of $1,000 per year. The present worths for each of these alternatives is calculated since the study period is 5 years and Method 3 is assumed.

$$PW(15) = \$1,200 \quad \text{(Immediate disposal produces a revenue of \$1,200.)}$$

$$PW(15)_{D1_1} = (-\$300 + \$900 + \$1,000)\overset{P/F,15,1}{(0.8696)} = \$1,391$$

[1] *These techniques are presented in Hillier and Lieberman,* Introduction to Operations Research, *4th ed., Holden-Day, 1986, Chapters 10 and 11.*

$$PW(15)_{D1_2} = -\$300\overset{P/F,15,1}{(0.8691)} + (-\$900 + \$600)\overset{P/F,15,2}{(0.7562)} + \$1{,}000\overset{P/A,15,2}{(1.6257)}$$

$$= \$1{,}138$$

$$PW(15)_{D1_3} = \$1{,}538$$

$$PW(15)_{D1_4} = \$1{,}556$$

$$PW(15)_{D1_5} = \$1{,}519$$

Examination of these results indicates that keeping $D1$ exactly 4 years is best. If $D1$ is kept 1, 3, 4 or 5 years it will provide more equivalent profit than if it is retired immediately. Keeping $D1$ for 2 years is an inferior option to receiving the $1,200 revenue generated by the present salvage value of the asset.

Frequently, an asset that is being considered for retirement does not have a well-identified present salvage value. If the asset is old or technologically inferior, there may be no market for it, and thus an accurate assessment of its value may be difficult. In this case, it is appropriate to impute a salvage value for the asset by considering its present economic value to the firm. Assessing the equivalent revenues generated minus the equivalent cost incurred will provide some sense of this value.

Another approach is to determine the current market value that would have to be expended if this asset were not available. For example, electric power transformers are frequently removed from one location, sent to storage, and later installed at another location. Although it appears that claiming a transformer from storage has no cost, consider the chain of events if there were no transformers in storage. Each transformer required would have to be purchased at the current market price; so it can be argued that value of a transformer in storage is its market price.

8.7 EXAMPLES OF REPLACEMENT PROBLEMS

The primary reasons leading to replacement were discussed earlier in Section 8.1. These reasons included physical deterioration and obsolescence. Obsolescence is a factor for replacement when an asset is outmoded because of changes external to the asset. Such changes might include the availability of assets with improved efficiency and inadequacy due to the need for greater capacity. Three examples are presented in the following discussion to illustrate the analysis of replacement problems in such situations.

8.7.1 Replacement Because of Improved Efficiency

As an illustration of the analysis involving replacement because of obsolescence, consider the following example. A manufacturer produces a hose coupling consisting of two parts. Each part is machined on a turret lathe purchased 13 years ago for $8,300 including installation. A new turret lathe is proposed as a replacement for the old. Its installed cost will be $25,000.

The production times per 100 sets of parts with the new and old machine are as follows.

Part	*Present Machine*	*New Machine*
Connector	2.92 hours	2.39 hours
Swivel	1.84 hours	1.45 hours
Total	4.76 hours	3.84 hours

The company's sales of the hose couplings average 40,000 units per year and are expected to continue at approximately this level. Machine operators are paid $17 per hour. The old and the proposed machines require equal floor space. The proposed machine will use power at a greater rate than the present one, but since it will be used fewer hours, the difference in cost is not considered worth figuring. This is also considered true of general overhead items. Interest is to be taken at 12%. The salesman for the new machine has found a small shop that will purchase the old machine for $1,200. The prospective buyer estimates the life of the new machine at 10 years and its salvage value at 10% of its installed cost of $25,000. The old turret lathe is estimated to be physically adequate for 2 more years and to have a salvage value of $250 at the end of each of the next two years.

Assume that any future replacements for the defender and challenger are not to be considered, and the study period is selected to be 2 years. This study period is chosen since the economic life for the present turret lathe is 2 years. The $AE(i)$ cost if the defender, D, is kept will be

$CR(12)_D = (\$1{,}200 - \$250)\overset{A/P,12,2}{(0.5917)} + \$250(0.12)$	$ 592
Direct labor, $(4.76 \div 100)(40{,}000)(\$17)$	32,368
	$32,960

The $AE(i)$ cost of operation if the new turret lathe, C, is purchased will be

$CR(12)_C = (\$25{,}000 - \$2{,}500)\overset{A/P,12,10}{(0.1770)} + \$2{,}500(0.12)$	$ 4,283
Direct labor, $(3.84 \div 100)(40{,}000)(\$17)$	26,112
	$30,395

The $AE(i)$ in favor of the new machine is $2,565 for 2 years. It should be noted that the new machine will be used $(3.84 \div 100) \times 40{,}000 = 1{,}536$ hours per year. No cognizance is taken of the fact that it is available for use many more hours per year; the unused capacity is of no value until used. Since, however, the additional capacity is potentially of value and may prove a safeguard against inadequacy, it should be considered an irreducible in favor of the new machine.

8.7.2 Replacement Because of Inadequacy

A physical asset that is inadequate in its capacity to perform the required services is a logical candidate for replacement. When there is inadequacy, a usable piece of equipment, often in excellent condition, is on hand. Often the desired increased capacity can be met only by purchasing a new unit of equipment of the desired capacity. In many cases, however, such as with pumps, motors, generators, and fans, the increased capacity desired can be met by purchasing a unit to supplement the present machine, should this alternative prove more desirable than purchasing a new unit of the desired capacity.

The method of comparing alternatives when inadequacy is the principal factor is illustrated by the following example. One year after a 10 hp motor has been purchased to drive a coal belt conveyor, it is decided to double the length of the belt. The new belt requires 20 hp. The needed power can be supplied either by adding a second 10 hp motor or by replacing the present motor with a 20 hp motor.

The present motor cost $700 installed and has a full-load efficiency of 86%. An identical motor can now be purchased and installed for $800. A 20 hp motor having an efficiency of 90% can be purchased and installed for $1,400. The present 10 hp motor will be accepted as $550 on the purchase price of the 20 hp motor. Electric current costs $0.08 per kWh, and the conveyor system is expected to be in operation 2,000 hours per year.

Maintenance and operating costs other than electric power for each 10 hp motor are estimated at $100 per year and for the 20 hp motor at $150 per year. Taxes and insurance are taken as 1% of the purchase price. Interest will be at the rate of 14%. The service of the new 10 hp motor in the present application is taken as 10 years. The 20 hp motor is expected to have a 12-year life. The present motor will be considered to have a remaining life of 10 years. All the motors have a salvage value of 20% of their original cost at any future time.

It is noted that the conveyor will continue to be needed for many years to come. Therefore, for analysis purposes it is to be assumed that the power drive for the alternatives presently under consideration will be replaced by an identical 20 hp motor when any future replacement occurs. This situation represents the *replacement of each alternative by the best challenger* as discussed earlier.

First, calculate the $AE(i)$ costs of the current alternatives to determine whether the conditions of Figure 8.8 or Figure 8.9 hold. With constant operating costs and all future salvage values equal for each motor, Case I conditions exist. (See Section 8.4.) Thus the economic life of each 10 hp motor is conveniently found to be 10 years, their service lives. For the same reasons the economic life for the 20 hp motor is 12 years.

Alternative *A* will involve the purchase of the 20 hp motor for $1,400 and the disposal of the present motor for $550. Any future replacement of the 20 hp motor will be by a 20-hp motor having identical costs. Using the outsider's viewpoint, the $AE(i)$ cost for this alternative's economic life is computed as

$CR(14)_A = (\$1{,}400 - \$280)(\overset{A/P,14,12}{0.1767}) + \$280(0.14)$	$ 237
Power cost, $\dfrac{20 \text{ hp}}{0.90 \text{ eff}} \times \dfrac{0.746 \text{ kW}}{\text{hp}} \times \dfrac{\$0.08}{\text{kWh}} \times 2{,}000$ h.	2,652
Maintenance and operating cost	150
Taxes and insurance, $1,400 × 0.01	14
Total $AE(14)_A$	$3,053

Alternative *B* will involve the purchase of an additional 10 hp motor for $800 to be combined with the present motor. Future replacement of these motors will be by a 20 hp motor having costs identical to the present challenger. The $AE(i)$ cost for this alternative is found as follows.

Present 10 hp motor:

$CR(14) = (\$550 - \$110)(\overset{A/P,14,10}{0.1917}) + \$110(0.14)$	$ 100
Power cost, $\dfrac{10}{0.86} \times 0.746 \times \$0.08 \times 2{,}000$	1,388
Maintenance and operating cost	100
Taxes and insurance, $700 × 0.01	7

New 10 hp motor:

$CR(14) = (\$800 - \$160)(\overset{A/P,14,10}{0.1917}) + \$160(0.14)$	145
Power cost, $\dfrac{10}{0.86} \times 0.746 \times \$0.08 \times 2{,}000$	1,388
Maintenance and operating cost	100
Taxes and insurance, $800 × 0.01	8
Total $AE(14)_B$	$3,236

Because the total $AE(i)$ cost of \$3,236 for Alternative B for its economic life exceeds the total $AE(i)$ cost of \$3,053 for Alternative A's economic life, we have the conditions shown in Figure 8.8. Therefore, there is *no* time period for which Alternative B could ever be less costly than Alternative A. The economic choice for this example indicates that Alternative A is superior for a study period of 10 years, or for that matter any range of years. (The implied-salvage-value of Method 2 is being used in this case.) On the basis of this analysis, the advantage of replacing the 10 hp motor rather than supplementing it is equivalent to \$3,236 less \$3,053 or \$183 per year over 10 years.

If the $AE(i)$ calculations had yielded an outcome where more than one $AE(i)$ amount for the defender is less costly than the $AE(i)$ for the challenger (e.g. see Figure 8.9) additional calculations would have to be made. These $PW(i)$ calculations would have been necessary to determine the best length of time to retain Alternative B.

8.7.3 Replacement Because of Growing Demand

In many actual situations the existing asset becomes inadequate as time passes. Usually this occurs when there is growing demand that ultimately exceeds the capacity of the existing asset. The following example demonstrates this situation and the analysis required.

An electric utility is experiencing a growth in demand for electric power in a geographic area serviced by its existing transmission system. To accommodate this growth, it is anticipated that if the existing transmission line (Line A) is to continue to be used, a new line (Line B) must be built 7 years from the present. The addition of Line B to Line A will double the capacity of the transmission system to the service area.

Another alternative being considered is to build immediately a single new line (Line C) with a capacity double that of the existing line. Since a new route will be used for Line C, the existing Line A will be removed and the right-of-way will be sold. The following data, given in *constant* dollars, applies to the transmission lines indicated. The base year for the constant dollars is the present.

Line A was completed 12 years ago at a cost of \$2,000,000. Its remaining life is estimated to be 20 years with a cost of removal of \$500,000 and revenue from right-of-way sales of \$1,400,000 at that time. If Line A is removed *now*, the cost of removal is \$500,000 while the sale of right-of-way will provide income of \$400,000. Annual maintenance costs will be \$20,000 at the end of the first year, increasing by \$10,000 each following year.

Line B will be constructed 7 years from the present at a cost of \$6,000,000. It is estimated that this line will last 35 years, having a net salvage of \$2,000,000 at that time. Annual maintenance costs will be \$10,000 per year.

Line C will be built now at a cost of \$4,000,000, which includes construction and right-of-way acquisition. This line is expected to have a 50-year life with a net salvage value at that time of \$7,000,000. Its maintenance costs will be \$9,000 per year.

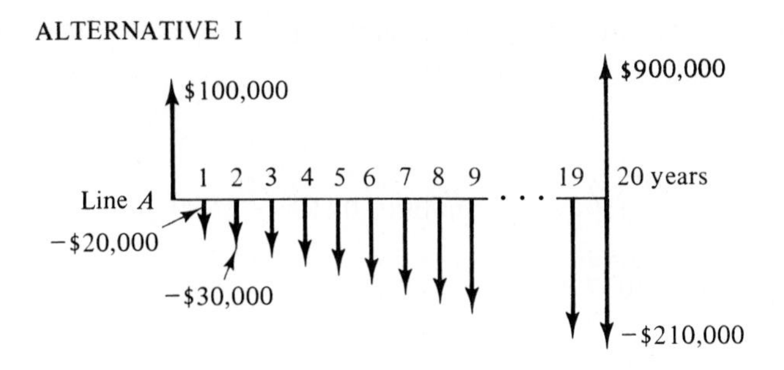

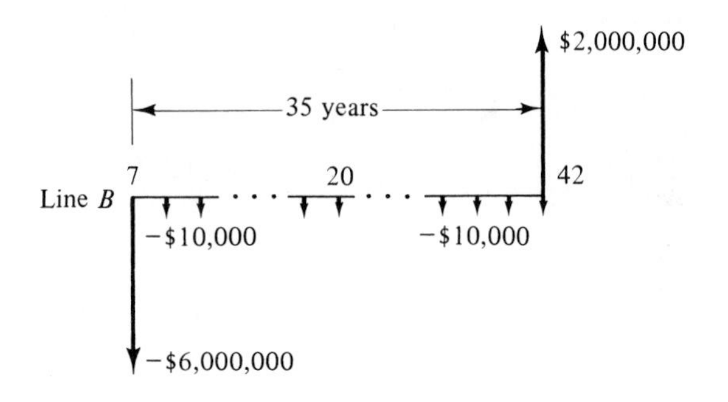

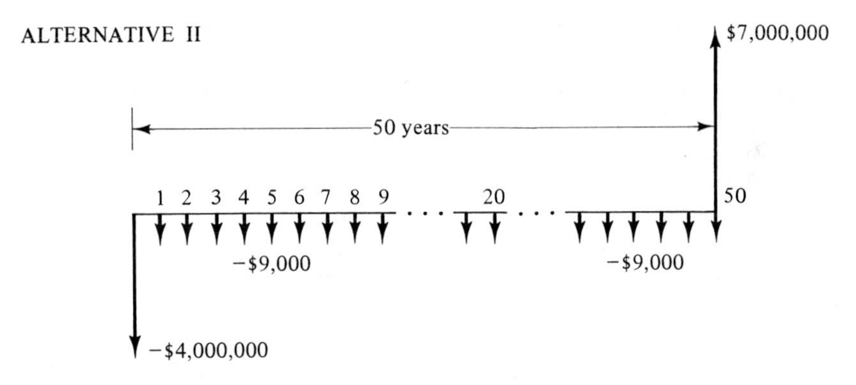

FIGURE 8.12. Cash flows for Alternatives I and II.

The cash flows for Alternative I (keeping Line *A* and building Line *B*) and Alternative II (disposing of Line *A* and building Line *C*) are presented in Figure 8.12. In this example a study period of 20 years is selected, and it is assumed that there is no information regarding the future replacements for Line *A*, *B*, and *C*. In addition it is known that each line, if selected, will be kept in place for its full functional life. The utility uses an inflation-free MARR of 8%, and it is assumed that all costs will increase at the same rate of inflation. The calculations are as follows:

Alternative I

$$AE(8)_A = \$100{,}000(\overset{A/P,8,20}{0.1019}) + \$900{,}000(\overset{A/F,8,20}{0.0219}) - [\$20{,}000 + \$10{,}000(\overset{A/G,8,20}{7.037})]$$

$$= \$ - 60{,}470 \text{ per year}$$

$$AE(8)_B = [-\$6{,}000{,}000(\overset{A/P,8,35}{0.0858}) + \$2{,}000{,}000(\overset{A/F,8,35}{0.0058}) - \$10{,}000](\overset{F/A,8,13}{21.495})(\overset{A/F,8,20}{0.0219}) = -\$241{,}584 \text{ per year}$$

$$AE(8)_{\text{I}} = AE(8)_A + AE(8)_B = -\$60{,}470 - \$241{,}584 = -\$302{,}054 \text{ per year.}$$

Alternative II

$$AE(8)_C = -\$4{,}000{,}000(\overset{A/P,8,50}{0.0818}) + \$7{,}000{,}000(\overset{A/F,8,50}{0.0018}) - \$9{,}000$$

$$= -\$323{,}600 \text{ per year}$$

$$AE(8)_{\text{II}} = -\$323{,}600 \text{ per year.}$$

Note that the implied-salvage-value assumption of Method 2 discussed in Section 7.8 is used for Lines B and C, as their lives extend beyond the study period. For Line B only that portion of its $AE(i)$ cost occurring within the study period is charged to the cost of Line B in the analysis. In this example, Alternative I minimizes the $AE(i)$ cost.

KEY POINTS

- The need for replacement may be due to physical deterioriation or obsolescence.
- Once the proper cash flows are identified for a replacement problem, the methods presented in Chapter 7 for selecting the most economic alternative are appropriate.
- Sunk costs are irrelevant for decision purposes when describing the cash flows associated with a defender.
- The opportunity cost associated with the decision to retain the defender is its current market value. This cost must be allocated over the defender's remaining lifetime.

- Use of the outsider's viewpoint assumes that sunk costs are ignored, and it associates the proper opportunity cost with the defender and its life.
- The economic life is the life of an asset that is economically most favorable to that asset. (That is, it is the length of time that minimizes the equivalent cost or maximizes the equivalent return for an asset.)
- When considering current investments, different assumptions can be made regarding their future replacements. These assumptions and the correct selection procedures are summarized as follows:
    1. *Replacement of each alternative by identical alternatives.* Solution method: Find the economic life of each asset and compare $AE(i)$ based on each asset's economic life.
    2. *No replacement of current alternative.* Solution method: Find the economic life of each asset and compare $AE(i)$ based on each asset's economic life.
    3. *Replacement of each alternative by the best challenger.* Solution method: If more than one life for the defender is more favorable than the best challenger on the basis of $AE(i)$ comparison, $PW(i)$ comparisons must be applied. [See Equations (8.1) and (8.2).] Otherwise, find the economic life of each asset and compare $AE(i)$ based on each asset's economic life.
    4. *Assume replacement of each alternative by dissimilar challengers.* Solution method: Develop the network of all possible replacement sequences, place the appropriate $PW(i)$ values on each arc of the network, and solve using a dynamic programming algorithm.
- Retirement or abandonment decisions require both the receipt and disbursement cash flows, and these cash flows must be compared on a $PW(i)$ basis using the assumptions of Method 3 described in Section 7.8.

PROBLEMS

1. A machine was purchased 4 years ago for $12,000. Its present market value is $5,000, and its operating expenses are expected to continue at $1,000 a year. A second-hand machine costing $2,000 is available, but its operating expenses are expected to be $1,600 per year. It is anticipated that the machines will be in service for 6 more years with $1,000 salvage value for the present machine and zero for the second-hand machine.
    a. Using the "outsider viewpoint," the IRR approach, and a MARR of 15%, find the best course of action for a 6-year study period.
    b. If the cost estimates are in constant dollars and the annual inflation rate is expected to be 8%, which machine should be selected?

2. A machine was purchased 5 years ago for $13,000. Its present value is $6,500, and operating expenses are expected to be $1,250 this year, increasing at a rate of 10% per year thereafter. A second-hand machine costing $2,800 is available, but its operating expenses are estimated to be $2,400 per year. It is anticipated that the machines will be in service for 6 more years with a salvage value of $1,600 for the present machine and zero for the second-hand machine.

a. Find the best course of action if the interest rate is 12%.

b. Since the values given above are in *actual* dollars, and the expected rate of inflation is 8%, and the inflation-free MARR is 10%, determine the best action.

3. A manufacturer is considering the purchase of an automatic lathe to replace one of two turret lathes. The turret lathes were purchased 8 years ago at a cost of $4,400. The automatic lathe can be purchased for $18,400, and the turret lathe can be sold for $1,500. Other pertinent data are as follows:

	Turret Lathes	*Automatic Lathe*
Annual output, Part *A*	45,000 each	90,000
Annual use other than on Part *A*	400 hours each	0
Production, units of Part *A* per hour	38 each	88
Labor (one man per machine)	$14.20 per hour	$12.40 per hour
Estimated annual maintenance	$1,350 each	$2,100
Power cost per hour of operation	$0.25	$0.40
Taxes and insurance, 2% of	Current value	Original cost
Space charges per year	$350 each	$570
Future salvage value	$400 each	$1,200

The turret lathes are in good mechancial condition and may be expected to serve an additional 8 years before maintenance rises appreciably. Their salvage value will probably never drop below $400 each. If one of the turret lathes is replaced by the automatic lathe, it is assumed that the automatic lathe will be used to produce the entire 90,000 units of Part *A* and that the remaining turret lathe will be used 800 hours per year on work other than Part *A*. If the interest rate is 16%, how many years will be required for the automatic lathe to pay for itself? Assume future replacement of each alternative by identical alternatives and use the outsider viewpoint.

4. Two external disk storage units for a computer cost $175,000 when the computer was purchased 3 years ago. New technology has made available an improved set of storage units that can decrease the processing time of the computer system by 18%. The manufacturer of the new storage units offers to allow 30% of the old units' first cost as a trade-in value. The new storage units cost $380,000. It is anticipated that the present computer system will be completely replaced in 4 years by a new-generation computer. The salvage values of the old and new storage units at that time are estimated to be $30,000 and $45,000, respectively. The computer will operate 8 hours a day for 21 days per month. If computer time saved is valued at $250 per hour and the interest rate is 15% compounded monthly, should the existing storage units be replaced? Maintenance costs are considered to be the same for both old and new units. If it is anticipated that the operating time is to be 10 hours per day, would this change the decision? Base your analysis on equivalent monthly costs.

5. Three years ago a delivery van was purchased for $14,000. Because the operating expenses are expected to be $0.35 per mile this year, inflating at the rate of 9% per year thereafter, the owner is considering the purchase of a new van. The new van can be purchased for $18,000, and the dealer will give $4,000 trade-in on the old one. The new van is expected to have operating expenses of $0.33 per mile this year, and these expenses are expected to inflate at 5% per year. In 5 years it is expected that the salvage value for the older van will be zero while the newer van will be worth $5,000. MARR is 18%.

   a. If the delivery activity logs 30,000 miles each year, determine the least-cost alternative for a 5-year study period.

   b. If the costs in the problem are in terms of constant dollars (the increasing operating costs are not the result of inflation), which option should be selected when the inflation rate is expected to be 10% per year? Answer: Old Van, $PW(18) = \$49{,}941$; New Van, $PW(18) =$ $56,029

6. A bottler of beverages purchased a bottling machine 2 years ago for $320,000. At that time it was estimated to have a service life of 8 years with no salvage value. Annual operating costs of the machine are expected to continue at $66,000. A new bottling machine is being considered that would cost $400,000 but would match the output of the old machine for an annual operating cost of $40,000. The new machine's service life is 6 years with no salvage value. An allowance of $95,000 would be made for the old machine on the purchase of the new machine. The interest rate is 15%.

   a. List the receipts and disbursements for the next 6 years if the old machine is retained; if the new machine is purchased. Compare the present worth of receipts and disbursements. Answer: Old favored

   b. Take the "outsider viewpoint" and calculate the $AE(i)$ cost for each of the two alternatives.

   c. Should the new machine be purchased? Answer: No

7. A municipality 6 years ago purchased a pump for its sewage treatment plant at $6,000. This pump had annual operating costs of $3,200 and these are expected to continue. This pump is expected to continue to operate satisfactorily for 6 additional years, at which time it may be expected to have negligible salvage value. The municipality has an opportunity to purchase a new pump for $8,500. The new pump is estimated to have a life of 6 years, negligible salvage value at the end of its life, and an annual operating cost of $1,400. If the new pump is purchased, the old pump will be sold for $1,600. The interest rate is 12%. With no knowledge of future replacements and a 6-year study period, what option should be selected? Answer: $AE(12)_D = \$3{,}589$; $AE(12)_C = \$3{,}467$

8. Two years ago a centrifugal pump driven by a direct-connected induction motor was purchased to meet a need for a flow of 2,000 gallons of water per minute for an industrial process. The unit cost $3,300 and had an estimated salvage value of $600 at the end of 6 years of use. It consumed electric power at the rate of 47 kW and was used 12 hours per day for 300 days a year. The process for which the water is required has been changed, and in the future a flow of only 800 gallons per minute is needed 3 hours per day for 300 days a year. At the decreased flow both pump and motor are relatively inefficient and the power consumption is 36 kW.

A new pump of a capacity conforming to future needs will cost $2,350. This new unit will have an estimated life of 8 years with an estimated salvage value of $400 at the end of that time. Its power consumption will be 19 kW. The present unit can be sold for $1,000. The original estimates of useful life and salvage value are still believed to be reliable. Insurance, taxes, and maintenance are estimated at 6% of the original cost for both units. The cost of power is $0.052 per kilowatt-hour and the MARR is 10%.

a. Should the old unit be retained or should the new unit be purchased if the study period is 4 years? Answer: Old Unit, $AE(10) = \$2{,}069$; New Unit, $AE(10) = \$1{,}436$.

b. At what number of hours per day for 300 days a year with a requirement of 800 gallons of water per minute will the future $AE(i)$ cost of the two units be equal?

9. A chemical processing plant secures its water supply from a well, which is equipped with a 6-inch, single-stage centrifugal pump that is currently in good condition. The pump was purchased 3 years ago for $3,000 and had an expected life of 10 years. Because of design improvements, the demand for a pump of this type is such that its present value is only $1,200. It is anticipated that 7 years from now the pump will have a trade-in value of $400. An improved pump of the same type can now be purchased for $5,800 and will have an estimated life of 10 years with a trade-in value of $300 at the end of that time.

The pumping demand is 225 cubic feet per minute against an average head of 200 feet. The old pump has an efficiency of 75% when furnishing the demand above. The new pump has an efficiency of 81% when furnishing the same demand. Power costs $0.084 per horsepower-hour, and either pump must operate 2,400 hours per year. Based on a 7-year study period and no knowledge of future replacements, do the improvements made in design justify the purchase of a new pump if interest of 15% is required? Horsepower = $h \times F \times 62.4 \div 550$, where h is the head in feet and F is the flow in cubic feet per second. Answer: Old pump, AE(15) = $23,127; new pump, AE(15) = $22,322

10. A company is studying two options concerning an existing special-purpose machine whose present net salvage value is estimated to be $10,000. The first option considers retaining the machine for its remaining service life of 4 years and immediately supplementing it with a small new machine whose first cost, life, salvage value, and operating costs are $7,000, 6 years, $3,000, and $1,500 per year, respectively. It is estimated that the old machine will have annual operating costs of $2,000, and a salvage value of $4,000 at the end of its life. The second option consists of having no machines, in which case the desired service will be subcontracted at a cost of $7,000 per year. Using a study period of 4 years and a MARR of 12%, which option should be selected? Answer: Sell existing machine.

11. Five years ago a conveyor system was installed in a manufacturing plant at a cost of $27,000. It was estimated at the time of purchase that the system, which is still in good condition, would have a useful life of 20 years. Annual operating costs are $1,350. The number of parts to be transported has doubled and will continue at the higher rate for the rest of the life of the system. An identical system can be installed for $22,000, or a system with a 20-year life and double the capacity can be installed for $31,000. Annual operating cost is expected to be $2,500. The present system can be sold for $6,500. Either of the three systems will have a salvage value at retirement of 10% of original cost. For (a) and (b) compare the two alternatives for obtaining the required services on the basis of $AE(i)$ over a 15-year study period, recognizing any implied salvage value in the systems at the end of that time.

a. The interest rate is 15%.

b. Assume that the values in the problem are in constant dollars with the inflation rate expected to be 6% and the market rate of interest 15.54%.

12. Six years ago an ore-crushing unit was installed at a mine at a cost of $86,000. Annual operating costs for this unit are $3,540. This unit is estimated to have a remaining life of 6 years. The amount of ore to be handled is to be doubled and is expected to continue at this higher rate for at least 20 years. A unit that will handle the same amount of ore and have the same annual operating costs as the one now in service can be installed for $100,000. A unit with double the capacity of the one now in use can be installed for $155,000; its life is estimated at 10 years and its annual operating costs at $6,120. The present realizable value of the unit now in use is $32,000. All units under consideration will have an estimated salvage value at retirement age of 12% of the original cost. The MARR is 15%.

Compare the two possibilities of providing the required service on the basis of $AE(i)$ cost over a study period of 6 years, recognizing implied salvage value remaining in the unit at the end of that time.

13. An apartment complex installed a furnace 8 years ago at a cost of $20,000 to provide the necessary heating. Because of new additions, the existing unit will supply only 50% of the forecasted heating requirements. Currently, this old furnace has a salvage value of $9,000, and it is estimated that it can be used for an additional 8-year period, with annual operating cost of $7,000, and a zero salvage value at the end of that time.

If the old furnace is retained, a new unit with the same capacity will be purchased at a cost of $18,000. It is estimated that its service life, final salvage value, and operating costs are, respectively, 10 years, $3,000, and $6,000 per year.

Another possibility is to buy a new furnace having the capacity of the two small units. This new unit has an initial cost of $32,000, estimated life of 12 years with $4,000 salvage value at the end of that time, and annual operating costs of $11,000.

a. Define all the alternatives that can be evaluated with the information given in the statement of the problem.

b. If the MARR is 15%, and using a study period of 8 years, which alternative should be selected? Answer: $AE(15)_{A1}$ = $18,446; $AE(15)_{A2}$ = $18,879; $AE(15)_{A3}$ = $16,766

14. A company is planning to replace a stand-by generator that is inadequate for current requirements. The owned generator will be sold immediately at a net salvage of $3,000. Three different generators are under consideration to replace the existing one; these alternatives are summarized below:

	Alternatives		
	HD	*WE*	*GE*
Initial cost	$20,000	$17,000	$19,000
Life (years)	5	7	6
Salvage value	$4,000	$2,000	$2,900
Operating cost per hour ($/h)	$500	$700	$590

It is estimated that the generator when used for stand-by purposes will be utilized an average of 15 hours per year, and the cost in lost production if no stand-by equipment is used is $2,000 per hour. The MARR is 12%. Which generator, if any, should be selected? Answer: Select *HD*.

15. A construction company is in the process of making a decision concerning a 4-year-old bulldozer it presently owns. Because of excessive use and lack of adequate maintenance in the past, it is estimated that if the bulldozer is not sold it will be able to provide only 40% of the earthmoving requirements over its remaining life of 8 years, with a $3,000 salvage value at the end of that time. Its annual operating costs (including maintenance) are $14,000. If the old bulldozer is retained, it will have to be complemented with a smaller bulldozer whose initial cost, life, salvage value, and annual operating cost are $10,000, 10 years, $5,000, and $2,500, respectively.

It is possible to perform a major overhaul to the existing bulldozer at a cost of $16,000. With this overhaul the existing bulldozer will be able to handle the future earthmoving requirements. Operating costs will be $12,000 per year, and the remaining service life will be increased to 10 years with a salvage value of $7,000 at the end of that time.

Another alternative being considered is the purchase of a new bulldozer at a cost of $45,000, in which case the existing unit will be sold for $15,000. The new unit will have an estimated life of 10 years, $6,000 final salvage value, and operating cost of $10,500 per year. If the MARR is 15%, which alternative is more economical?

16. A university is considering upgrading its computing capabilities. The computer that is now owned was purchased 2 years ago at a cost of $500,000. It was anticipated at the time of the purchase that annual operating costs would be $80,000, and that after 6 years of use the computer would be inadequate and therefore would be sold for $90,000. The existing unit has now a salvage value of $180,000 and, if retained for 4 more years, will have a salvage value of $40,000 at the end of that time. Its operating costs will increase at a rate of 3% per year from the current level of $80,000, and it will have to be supplemented immediately with a medium-size computer having initial cost, life, salvage value, and operating costs of $100,000, 5 years, $30,000, and $19,000, respectively.

A new, larger computer with the desired capacity can be bought for $420,000. It is estimated that it will have a service life of 5 years, final salvage value of $120,000, and operating costs of $50,000 per year.

As an alternative to the above options, there is a possiblity of leasing a computer with sufficient capacity for a 4-year period. This alternative will require an initial payment of $10,000 and will have total lease costs of $140,000 payable at the beginning of each year.

If the MARR is 12%, indicate the preferred alternative for a 4-year study period.

17. A hydroelectric plant utilizing a continuous flow of 12 cubic feet of water per second with an absolute head of 860 feet was built 4 years ago. The 18-inch pipeline in the system cost $102,000 for pipe, installation, and right-of-way, and has a loss of head due to friction of 85 feet. Additional water rights have been acquired that will result in a total of 24 cubic feet per second of water flow. The following plans are under consideration for utilizing the total flow. Plan *A*: Use the present pipeline. This will entail no additional expense but will result in a total loss of head due to friction of 300 feet owing to the increased velocity of the water. Plan *B:* Add a second 18-inch pipeline at a cost of $78,000. The loss of head for this line will be 85 feet. Plan *C:* Install a 26-inch pipeline at a cost of $101,000 and remove the existing line, for which $8,800 can be realized. The loss of head due to friction for the 26-inch pipeline will be 67 feet.

The energy of the water delivered to the turbine is valued at $84 per horsepower year, where horsepower $= h \times F \times 62.4 \div 550$. In this equation, h is the head in feet and F is the flow in cubic feet per second. Insurance and taxes amount to 3% of first cost. Operating and maintenance costs are essentially equal for all three plans. The interest rate is 12%. If all lines, including the one now in use, will be retired in 35 years with no salvage value, what is the comparable $AE(i)$ costs of the three alternatives?

18. An engineer is considering the replacement of a 10-year-old special-purpose pump. The pump is used to provide 14,000 gallons per minute of a liquid having specific gravity of 1.35 at 12 feet total dynamic head.

The existing pump now has a salvage value of $5,000 and operates at a 70% efficiency. If this unit is retained for 2 more years, it will be sold for $2,000 and replaced by a new pump at a cost, including installation, of $20,000. It is estimated that the new unit, which will be installed at the end of 2 years, will operate with an efficiency of 85% and will have a service life of 10 years with $5,000 salvage value at the end of that time.

Two other alternatives are under consideration. These alternatives require that the old pump be replaced immediately by either of the pumps listed below.

	Pump B	*Pump C*
Initial cost	$15,000	$24,000
Installation cost	$1,500	$1,500
Service life	15 years	14 years
Salvage value	$4,500	$4,000
Efficiency	80%	88%

Maintenance costs are assumed to be the same for all alternatives, and power is available at $0.08 per kilowatt-hour. Operation will be 24 hours per day for 300 days per year.

For a MARR of 15%, which alternative should be selected? [*Note:* Energy consumption in kilowatt-hours = head in feet × gal/min × specific gravity × number of hours × 0.746/(3,960 × efficiency).]

19. A special milling machine is being installed at a first cost of $10,000. Maintenance cost is estimated to be $5,500 for the first year and will increase by 6% each year. If interest is 12% and the salvage value is $2,000 at any time, for what service life will $AE(i)$ cost be a minimum? Answer: 7 years

20. The maintenance cost of a certain machine is zero the first year and increases by $200 per year for each year thereafter. The machine cost $2,000 and has no salvage value at any time. Its annual operating cost is $1,000 per year. If the interest rate is 10%, what life will result in minimum $AE(i)$ cost? Solve by trial and error, showing yearly costs in tabular form.

21. A special-purpose machine is to be purchased at a cost of $20,000. The following table shows the expected annual operating costs, maintenance costs, and salvage values for each year of service. If the rate of interest is 10%, what is the economic life for this machine?

Year of Service	*Operation Cost for Year*	*Maintenance Cost for Year*	*Salvage Value at End of Year*
1	$ 2,000	$ 200	$10,000
2	3,000	300	9,000
3	4,000	400	8,000
4	5,000	500	7,000
5	6,000	600	6,000
6	7,000	700	5,000
7	8,000	800	4,000
8	9,000	900	3,000
9	10,000	1,000	2,000
10	11,000	1,100	1,000

22. A large telephone cable is being considered as a candidate for retirement. If sold for scrap, based on present copper prices, the value to be received would exceed the cost of removal by $200,000. However, in the future it is anticipated that because of estimated changes in future copper prices and the cost of removal, the net revenue at retirement will remain fixed over time at $130,000. If maintenance costs are expected to remain constant over the remaining life of this cable, what is its economic life?

23. A 10-year-old truck had operating costs last year of $3,200, and it is expected that these costs will increase at the rate of 4% per year if there is no inflation. Presently the truck can be sold for $300 for scrap. It is anticipated that if there is no inflation this $300 would be received if the truck were sold at any time in the future. Find the economic life of the truck if the inflation rate is 9% per year and the market interest rate is 12%.

24. The trade-in value of an existing machine is $42,000, and it has no salvage value at any future time. Similar machines had shown a remaining service life of 5 years. The operating cost for this year is $9,000, and it is predicted to increase each year at the annual inflation rate of 8%. For the annual inflation rate of 8%, the fourth-year operating cost is $\$9{,}000 \times (1 + 0.08)^3$. If the market rate MARR is 13% annually, at what service life will the total $AE(i)$ costs in actual dollars be a minimum?

25. A producer of electronic components is considering the replacement of a machine required in its circuit-board assembly operation. The existing machine used in this operation originally cost $10,000 when purchased 7 years ago. The company is considering the purchase of a new, more efficient, machine currently costing $14,000. The existing machine can be sold immediately to a subsidiary for $6,000. If the existing machine is not replaced, its remaining service life is estimated to be 4 years. The new machine has a service life of 15 years. It is anticipated that the future salvage values for either machine will be zero. The operating and maintenance costs are expected to be as follows:

	Year								
Machine	1	2	3	4	5	6	. . .	14	15
Existing	$2,500	$3,000	$3,500	$4,000	—	—	. . .	—	—
New	200	800	1,400	2,000	2,600	3,200	. . .	8,000	8,600

If either machine can be disposed of at any time and there is no knowledge of future replacements, what is the company's best course of action? Using a MARR of 18%, make an economic comparison based on the economic life of each alternative.

26. A manufacturer of cans and packaging for the food industry is considering the replacement of some of its current production equipment. A new plan is to install equipment in an existing plant facility to produce a new two-piece, thin-steel container that consumes less energy and metal than the conventional three-piece soldered cans. The equipment presently in operation was installed 5 years ago at a cost of $100,000 and can presently be sold for $35,000. Because of the rapid obsolescence of production equipment as customers switch to lighter, more economical containers, the future salvage value of the present equipment is expected to decline by $4,000 a year. If the present equipment is retained one more year, its operating costs are expected to be $65,000, with increases of $3,000 a year thereafter. The new equipment will cost $130,000 installed. Its economic life is predicted to be 8 years with a salvage value of $10,000. Annual operating disbursements will be $49,000. If the firm's MARR is 15%, make a recommendation as to the desirability of installing the new equipment. Assume there is no knowledge of possible future replacements and indicate the study period. Answer: Old, $AE(15)$ for 1 year = $74,250; New, $AE(15)$ for 8 years = $77,248

27. A hospital is considering the replacement of one of its artificial kidney machines. The machine being considered for replacement cost $38,500 four years ago. If the present machine is kept one more year, its operating and maintenance costs are expected to be $27,000. Operating and maintenance expenses in the second and third years are expected to be $28,000 and

$32,000, respectively. The new machine, if purchased now, will cost $46,000 and will have an annual operating expense of $21,000. Its economic life is anticipated to be 6 years, and its salvage value at that time is estimated to be $11,000. The company selling the new machine will allow $10,000 on the old machine for trade-in. If the hospital delays its purchase for 1, 2, or 3 years, the trade-in value on the existing machine is expected to decrease to $8,000, $6,000, and $3,000, respectively. If the interest rate is considered to be 15%, what decision would be most economical? Use an $AE(i)$ comparison, assuming there is no knowledge of future replacements.

28. The replacement of a machine is being considered by the ABC Co. The new improved machine will cost $30,000 installed and will have an estimated economic life of 12 years and $2,000 salvage. It is estimated that annual operating and maintenance costs will average $1,000 per year. Four years ago the present machine had an original cost of $20,000. When purchased, its service life was estimated to be 10 years with a salvage of $5,000. An offer of $7,000 has just been received for the present machine. Its estimated costs for the next 3 years are shown below:

Year	*Salvage Value at End of Year*	*Operating Maintenance Costs During Year*
1	$4,000	$3,000
2	3,000	3,400
3	2,000	6,000

Using interest at 15%, make an $AE(i)$ comparison to determine whether or not it is economical to make the replacement. Assume there is no knowledge regarding future replacements for the defender or the current challenger. Indicate the study period used. Answer: Keep existing machine; study period, 2 years

29. Two years ago a soft-drink distributor purchased a materials handling system for his warehouse. Originally the system cost $80,000. It was anticipated that in 7 years from the time of this purchase the warehouse would be inadequate and therefore would be sold. The materials handling equipment that is now owned is expected to have no market value in the future. Annual operating expenses for the existing system are projected to increase through time. These annual expenses in order are $2,000, $10,000, $18,000, $26,000, and $34,000 for the remaining 5 years of its service life. A firm selling materials handling equipment is presently offering a new system costing $70,000 and is offering $40,000 for trade-in of the old equipment. This new system is expected to cost $8,000 per year to operate over its service life of 5 years. What is the economic life for the new system if its prospective salvage value is zero at any future time? For a

MARR = 8%, prepare a table showing the *AE*(*i*) costs for both alternatives for each of the 5 years these systems might be used. What are your conclusions, assuming there is no knowledge of future replacements?

30. A shipping company that is engaged mainly in the transportation of coal from nearby mines to steel mills is concerned about the replacement of one of its old steamships with a new diesel-powered ship. The old steamship was purchased 15 years ago at a cost of $250,000. If the steamship is sold now, the fair market value is estimated to be $15,000. Once a steamship reaches this age, no major changes in its future salvage value are expected. Currently, the annual operating and maintenance costs for the steamship amount to approximately $200,000 next year, and these costs are expected to increase at a rate of $15,000 a year thereafter. As an alternative to retaining the steamship, a new diesel-powered ship can be purchased at the quoted price of $470,000. In addition to this initial investment, the company would need to invest another $30,000 in a basic spare-parts inventory. The annual operating and maintenance costs and its salvage values are estimated in thousands of dollars as follows.

	End of Year														
	1	2	3	4	5	6	7	8	9	10	11	12	13	14	15
O&M	100	100	100	130	140	150	160	170	180	190	200	210	220	230	240
Engine overhaul						50						50			
Salvage value	430	370	320	280	250	220	190	160	130	100	70	50	40	30	20

The useful life of the diesel ship is estimated to be 15 years. The expenditure for a general engine overhaul is treated as an expense under the firm's current accounting practice. Assume the spare-parts inventory funds are working capital that is fully recoverable. If the firm's MARR is 12%, what decision should be made?

31. A textile firm is considering the replacement of its 3-year-old knitting machine, which has a current market value of $8,000. Because of a rapid change in fashion styles, the need for the existing machine is expected to last only 5 more years. The estimated operating costs and its salvage values for the old machine are given as follows.

	End of Year				
	1	2	3	4	5
Operating cost	$1,000	$1,500	$2,000	$2,500	$3,000
Salvage value	6,000	4,000	3,000	2,000	0

As an alternative, a new improved machine is available on the market at a price of $10,000 and has an estimated useful life of 6 years. The pertinent cost information can be summarized as follows.

	End of Year					
	1	2	3	4	5	6
Operating cost	$ 700	$ 900	$1,100	$1,300	$1,500	$3,000
Salvage value	8,000	6,000	4,000	2,000	0	0

It is assumed that there is no knowledge about future replacements. If the rate of interest is 15%, determine which alternative should be selected and how long the selected machine should be kept in service. State the study period used. Answer: Purchase new machine; study period, 4 years

32. A small manufacturing company leases a building for machining of metal parts used in their final product. The annual rental of $100,000 is paid in advance on January 1. The present lease runs until December 31, 1999, unless terminated by mutual agreement of both parties. The owner wishes to terminate the lease on December 31, 1995, and offers the company $40,000 if it will comply with the request. If the company does not agree, the lease will remain in effect at the same rate until the 1999 termination date. The company owns a suitable building lot and has a firm contract for construction of a building for a total cost of $1,700,000 to be completed and paid for at the end of 1995.

If the company elects to stay in the leased building, it will spend $80,000, $65,000, and $70,000 in 1996, 1997, and 1998, respectively, on the facility with no salvage value resulting. It is estimated that operating expenses will be $3,500 less per year in the new building for a comparable level of output. Taxes, insurance, and maintenance will cost 3.5% of the first cost of the building per year. The life of the building is estimated to be 25 years and the interest rate is 6%. The decision is to be made during 1995 on the basis of the present worth of the two plans as of December 31, 1995.

a. What is the present worth of the two plans as of December 31, 1995, if the study period is 25 years and it is assumed that all future replacements are by the current challenger? (Recognize implied salvage value. See Table 8.1.)

b. The company considers the privilege of waiting 4 years to build to have a present worth of $90,000. On the basis of this fact, and the results of part (a), which plan should be adopted?

33. An existing asset was purchased 3 years ago for $3,000. It is expected that it may continue in operation for another 4 years, and there is a buyer who is currently willing to pay $1,800 for it.

Two current challengers, $C1$ and $C2$, would cost $2,500, and $3,600, respectively, and their cost estimates and future market values are

	Operating Expenses		Market Value	
	Challenger		Challenger	
End of Year	C_1	C_2	C_1	C_2
1	$ 250	$ 255	$1,130	$0
2	400	255	620	0
3	900	400	200	0
4	1,500	800	0	0
5	1,600	1,200	0	0
6	—	1,600	—	0

Operating costs and the estimated salvage values for the existing asset are shown below.

	End of Year			
	1	2	3	4
Operating costs	$1,425	$ 250	$380	$600
Salvage values	450	$1,050	710	200

If the firm's MARR is 16% and future replacement of current alternatives is assumed to be the best current challenger

a. What is the minimum total $AE(i)$ cost for the best challenger?

b. What is the most economic of the current options available for the assumptions stated?

34. A company is studying the replacement of an existing data transmission terminal with two more advanced terminals. Estimated investment costs, salvage values, and operating and maintenance costs are

	Present Terminal		T1		T2	
End of Year	O & M Costs	Salvage Value	O & M Costs	Salvage Value	O & M Costs	Salvage Value
0	—	$400	—	$4,200*	—	$5,600*
1	$2,390	0	$ 900	1,200	$1,700	2,100
2	3,000	0	1,665	1,200	1,700	1,500
3	3,400	0	2,400	1,200	1,700	500
4	—	—	3,100	1,200	1,700	0
5	—	—	—	—	1,700	0

* *Original cost*

Assuming that $i = 18\%$ and all future challengers are identical to the best current challenger

a. Determine the $AE(i)$ cost of each challenger at its economic life.

b. Which of the current options shall be selected?

35. Machine D has a present market value of $5,700. This amount will decrease to $1,710, $1,100, and $285 in 1, 2, and 3 more years, respectively. After 3 years its salvage value is zero. Its estimated future operating and maintenance costs are

	End of Year				
	1	2	3	4	5
Operating and Maintenance costs	$1,200	$1,400	$4,300	$3,200	$3,800

There are three possible challengers to replace Machine D, and they have the following anticipated operating costs and salvage values.

	Operating Costs			Salvage Value		
	Challenger			Challenger		
End of Year	No.1	No.2	No.3	No.1	No.2	No.3
0	—	—	—	$8,100*	$8,600*	$9,500*
1	$ 900	$ 860	$ 855	7,700	7,500	8,200
2	900	860	900	7,500	6,400	6,900
3	900	920	945	7,300	5,300	5,600
4	980	970	990	6,900	4,200	4,300
5	1,180	1,030	1,035	6,500	3,100	3,000
6	1,700	1,380	1,080	5,700	2,000	1,700
7	—	—	1,125	—	—	400

* *Initial investment cost.*

Based on an annual cost comparison, using a 12% interest rate, and assuming that all future challengers will be identical to the best current challenger

a. Determine whether Machine D should be replaced.

b. Suppose that the only currently available challengers are No. 2 and No. 3. Should the existing machine be replaced?

36. The present delivery truck has a market value of $4,100. Its annual operating costs will be $1,025 for the remaining 2 years of its useful life. A new truck of advanced design will have annual operating expenses of $800 for the next 2 years, and then these costs will increase by $400 each year thereafter. The new truck is priced at $7,985 and is expected to have a useful life of 4 years. The future market value of both trucks is expected to decrease annually by 25% of the previous year's value. The required MARR is 15%.

Assuming that each year a new model truck will be available and that it will have the same costs and service life as the new truck presently available.

a. Show in a network format similar to Figure 8.7 the sequence of possible decisions over a planning period of 6 years, with all trucks to be retired at that time.

b. Indicate on the network the following sequence of actions.

1. $D1_2 - C2_3 - C5_1$
2. $D1_1 - C1_3 - C4_1 - C5_1$
3. $D1_2 - C2_4$
4. $D1_0 - C0_4 - C4_2$

c. Show the incremental present-worth amounts associated with the lines connecting the nodes representing the defender options and the challenger options for the challenger acquired at $t = 0$.

37. Expected cash flows for an existing testing machine are estimated as shown below.

	End of Year							
	0	1	2	3	4	5	6	7
Operating cost	—	$1,600	$1,700	$2,000	$2,400	$2,900	$3,500	$3,800
Salvage value	$2,500	2,000	1,500	1,000	500	0	−500	−1,000

Owning the testing machine saves a rental expense of $3,000 per year. Assuming an interest rate of 18%, find the best retirement option.

38. A company is considering the retirement of an existing machine. This machine is expected to produce receipts of $8,000 in year 1, $9,000 in year 2, and $10,000 in year 3. Subsequent receipts will be $11,500 per year. The machine requires increasing amounts of operating and maintenance cost each year, and its salvage value decreases over time as shown below.

End of Year	*Operating and Maintenance Cost*	*Salvage Value*
0	—	$5,500
1	$ 4,000	4,000
2	5,000	3,000
3	6,000	2,000
4	7,000	1,100
5	9,000	700
6	11,000	0

If the interest rate is 20%, determine the retirement age of the machine.

39. Your engineering staff has come up with the following revised estimates for an old machine tool.

	End of Year				
	1	2	3	4	5
Gross income	$6,350	$6,350	$6,350	$6,350	$6,350
Manufacturing cost	2,550	3,550	4,550	5,550	6,550
Salvage value	3,300	2,300	2,000	0	0

The current market value of the old machine is $3,800. Assuming an interest rate of 16%, determine for how long the machine tool should be kept in service.

9

Evaluating Public Activities

The standards by which private enterprise evaluates its activities are markedly different from those that apply to public activities. In general, private activities are evaluated in terms of profit; public activities, primarily in terms of the general welfare.

The government of the United States and its subdivisions engage in innumerable activities—all predicated upon promotion of the general welfare. So numerous are the services available to individual citizens, associations, and private enterprises that books are required to catalog them. This chapter presents concepts and methods of analysis applicable to the evaluation of such activities.

9.1 GENERAL WELFARE AIM OF GOVERNMENT

A national government is a superorganization to which all agencies of the government and all organizations in a nation, including lesser political subdivisions such as states, counties, cities, townships, and school districts as well as private organizations and individuals, are subordinate. Each citizen, if he chooses, may have a voice in the election of the policymaking group, the Congress of the United States.

It is a basic tenet that the purpose of government is to serve its citizens. The chief aim of the United States as stated in its Constitution is the *national defense* and the *general welfare* of its citizens. For convenience in discussion, these aims may be considered to be embraced by the single term—general welfare. This simply stated aim is, however, very complex. To discharge it perfectly requires that the desires of each citizen be fulfilled to the greatest extent and in equal degree with those of every other citizen.

Since the general welfare is the aim of the United States, it follows that the aims of the subordinate political subdivisions such as states and cities must conform to the same general objective, regardless of what other specific aims they may have.

9.1.1 The General Welfare Aim as Seen by the Citizenry

Since each citizen may have a voice, if he will exercise it, in a government, the objectives of the government stem from the people. For this reason the objectives taken by the government must be presumed to express the objectives necessary for attainment of the general welfare of the citizenry. This must be so, for there is no superior authority to decide the issue.

Thus, when the United States declares war, it must be presumed that this act is taken in the interest of the general welfare. Similarly, when a state votes highway bonds, it also must be presumed to do so in the interest of the general welfare of its citizens. The same reasoning applies to all activities undertaken by any political subdivision; for, if an opposite view is taken, it is necessary to assume that people collectively act contrary to their wishes.

Broadly speaking, the final measure of the desirability of an activity of any governmental unit is the judgment of the people in that unit. Also, it must be clear that governmental activities are evaluated by a summation of judgments of individual citizens, whose basis for judgment has been the general welfare as each sees it. The objectives of most governmental activities appear to be primarily social in nature, although economic considerations are often a factor. Public activities are proposed, implemented, and judged by the same group—the people of the governmental unit concerned.

The situation of the private enterprise is quite different. Those in control of private enterprise propose and implement services to be offered to the public, which judges whether the services are worth their cost. To survive, a private business organization must, at least, balance its income and costs; thus profit is of necessity a primary objective. For the same reason a private enterprise is rarely able to consider social objectives except to the extent that they improve its competitive position.

9.1.2 The General Welfare Aim as Seen by the Individual

Public activities are evaluated by a summation of judgments of individual citizens, each of whose basis for judgment has been the general welfare as he sees it. Each citizen is the product of his unique heredity and environment; his home, cultural patterns, education, and aspirations differ from those of his neighbor. Because of this and the additional fact that human viewpoints are seldom logically determined, it is rare for large groups of citizens to see eye-to-eye on the desirability of proposed public activities.

The father of a family of several active children may be expected to see more point to expenditures for school and recreational facilities than to those for a street-widening program to enhance the value of downtown property. It is not difficult for a person to extol the value of aviation to his community if a proposed airport will increase the value of his property or if he expects to receive the contract to build it. Many public activities have no doubt been strongly supported by a few persons primarily because they would profit handsomely thereby.

But it is incorrect to conclude that activities are supported only by those who see in them opportunity for economic gain. For example, schools and recreational facilities for youth are often strongly supported by people who have no children. Many public activities are directed to the conservation of national resources for the benefit of future generations.

It is clear that the benefits of public activities are very complex. Some that are of great general benefit may spell ruin for some persons, and vice versa. Lack of knowledge of the long-run effect of proposed activities is probably the most serious obstacle to the selection of those activities that can contribute most to the general welfare.

9.2 THE NATURE OF PUBLIC ACTIVITIES

Governmental activities may be classified under the general headings of protection, enlightenment and cultural development, and economic benefits. Included under *protection* are such provisions as the military establishments, police forces, the system of jurisprudence, flood control, welfare and health services. Under *enlightenment and cultural development* are such services as the public school system, the Library of Congress and other publicly supported libraries, publicly supported research, the postal service, and recreation facilities. *Economic benefits* include harbors and canals, power development, flood control, research and information service, and regulatory bodies.

The list above, although incomplete, shows that there is much overlapping in classification. For example, the educational system is considered by many to contribute to the protection, the enlightenment, and the economic benefit of people. Consideration of the purposes of governmental activities as suggested by the classification above is necessary in considering the pertinency of economic analysis to public activities.

9.2.1 Impediments to Efficiency in Public Activities

There are two major impediments to efficiency in public activities. *First, the person who pays taxes and receives services has little or no knowledge about the value of the transactions occurring between him and his government.* Therefore, he has no practical way of evaluating what he receives in return for his tax payment. His tax payments go into a common pool and lose their identity. The taxpayer, with few exceptions, receives nothing in exchange at the time or place at which he pays his taxes on which to base a comparison of the worth of what he pays in and what benefits he will recevie as the result of his payment.

In addition, the recipients of the products of tax-supported activities cannot readily evaluate the products in reference to what they cost. Where no direct payment is exchanged for products, a person may be expected to accept them on the basis of their value to the recipient only. Thus, the products of govermental activities will tend to be accepted even though their value to the recipient is less than the cost to produce them.

The second deterrent to efficiency in governmental activities is the lack of competitive forces required to instill sufficient concern about the efficient use of resources. This situation is the natural consequence of certain characteristics of government and the working environment within government. Probably the most important of these characteristics is that government avoids the pressures of a market mechanism that would induce greater efficiencies in its activities. Thus, government will not "go out of business" if resources are inefficiently applied. In fact, the federal government has the unique ability to spend more than it receives.

Since governmental units are exclusive franchises, the taxpayer has no choice as to which unit he must pay. Thus, he does not have an opportunity to evaluate the effectiveness of tax units on the basis of comparative performance nor an opportunity to patronize what he believes to be the most efficient unit.

In addition, the costs resulting from poor decisions are not recovered from the pocket of those responsible. Few direct economic pressures are felt, and seldom are promotions and salary increases related to efficiency. In many instances inefficient alternatives are considered because the particular government agency is overly concerned with its own continuance.

9.2.2 Multiple-Purpose Projects

Many projects undertaken by government have more than one purpose. A good example arises in connection with the public lands, such as a forest reserve. For example, suppose that a new road is being planned for a certain section of a national forest. Since public land is managed under the multiple-use concept, several benefits will result if the road is put into service. Among these are scenic driving opportunities, camping opportunities, improved fire protection, and ease of timber removal.

Justifying a public works project is normally easier if the project is to serve several purposes. This is especially true if the project is very costly and must rely upon support from several groups. The forest road will probably appeal to persons who like to drive and camp; the U.S. Forest Service, which is responsible for fire control; and the timber industry, which will be granted contracts to harvest timber resources from time to time.

A number of problems arise in connection with multiple-use projects. Foremost among these is the problem of evaluating the aggregate benefit to be derived. What is the benefit of a scenic drive or a camping trip? How can the benefit of improved accessibility for fire protection be measured? What is the benefit of easier timber removal? Each of these questions must be answered in quantitative monetary terms if an economic analysis of the project is to be performed. They must also be answered so that each group that benefits can share in the project's cost.

A second problem arising from multiple-purpose projects is the possibility for conflict of interest between the purposes. These conflicts frequently become political issues. A primary motive of every public servant is to get elected or reelected. By demonstrating that direct benefits have been obtained for the parties concerned, a candidate obtains votes. Because the desire is to show that the direct benefits are not very costly, there is a tendency to allocate project costs to those benefit categories that are deemed essential by all. For example, a major portion of the cost of the road may be allocated to the U.S. Forest Service under the category of improved fire protection. Then it is easy to show that the project is desirable to the general public owing to its low cost in connection with scenic driving and camping opportunities.

9.3 FINANCING PUBLIC ACTIVITIES

Two basic philosophies in the United States greatly influence the collection of funds and their expenditure by governmental subdivisions. These are collection of taxes on the premise of *ability to pay* and the expenditure of funds on the basis of *equalizing opportunity* of citizens. Application of the ability-to-pay viewpoint is clearly demonstrated in our income and property tax schedules. The equalization-of-opportunity philosophy is apparent in federal assistance to lesser subdivisions to help them provided improved educational and health programs, highway systems, old-age assistance, and the like.

Because of these two basic tenets of taxation, there often is litle relationship between the benefits an individual receives and the amount he pays for public activities. This is in large measure true of such major activities as government itself, military and police protection, welfare, the highway system, and most educational activities.

9.3.1 Methods of Financing

Funds to finance public activities are obtained through (1) the assessment of various taxes, (2) borrowing, and (3) charges for services. Federal receipts are derived chiefly from corporation, individual, excise, and estate taxes and from duties on imports. State income includes corporate, individual, gasoline, sales, and property taxes and vehicle licensing fees. Cities rely on income, sales, property taxes, and license fees.

Selling bonds is a common method of raising funds for a wide variety of governmental activities. Borrowing at the state and local level is usually confined to financing capital improvement projects or self-supporting activities. These "municipal" bonds are usually exempt from federal income taxes and, therefore, generally pay lower interest than federal and corporate bonds. This tax advantage encourages those in high-income tax brackets to invest and provides the "municipality" a low-cost source of funds.

There are numerous types of bonds, but the two most common are (1) general obligation bonds and (2) revenue bonds. The *general obligation bond* is secured by the issuer's credit and taxing power. Thus, the bondholder's risk is lessened because he has the taxing power of the government pledged to meeting the interest payments of the bond. These bonds generally offer the greatest security and the lowest interest rates. General obligation bonds usually require a vote of the citizens within the taxing authority before they may be issued, and the support required ranges from two-thirds majority to a simple majority. School bonds are normally general obligation bonds.

Revenue bonds are backed by the anticipated revenues to be generated by the project being financed. This type of bond is limited to revenue-producing projects such as toll roads and bridges, housing authorities, and water and sewer systems. Because of the increased risk (the project may fail to produce sufficient revenues) revenue bonds normally have higher interest rates than general obligation bonds.

Most activities financed by the federal government receive their money from the general fund. This fund is supported from various taxes and borrowings (Treasury bonds, notes, and bills). Thus, at the federal level it is more difficult to identify the source of the funds that are available for investment.

Considerable income on some governmental levels is derived from fees collected for services. Examples on the national level are incomes from the postal services and from sale of electricity from hydroelectric projects. On the city level, incomes are derived from supplying water and sewer services and from levies on property owners for sidewalks and pavements adjacent to their property.

9.3.2 Relating Benefits to the Cost of Financing

Many user taxes are structured so that there is a relationship between the benefits derived from the project and the project cost. The most obvious of these user-related taxes are those that provide the revenue for state highway projects. Highway-user taxation is designed to recover from the highway users those costs that can be appropriately identified with them. One concept considers that operating expenses provide an accurate assessment of services received. That is, the more one drives, the more use one makes of the highway system. The gasoline tax, which is based on this concept, certainly provides revenues in relationship to the amount of use. Of course, vehicles with lower fuel consumption pay less in comparison to the less efficient vehicles.

A second approach requires that the differential costs of providing for different classes of vehicles be considered. That is, if heavier vehicles are to use the roadway, it may have to be built thicker (at additional cost) and the rate of wear and tear will be increased. For example, suppose a state has 1,000 miles of paved highways that have been designed for heavy vehicles. The characteristics of the highway surface necessary to carry the various types of vehicles in the state are shown below.

Class of Vehicle	*Surface Thickness (Inches)*	*Cost per Mile*	*Incremental Cost*
Passenger cars	5.5	$2,200,000	$2,200,000,000
Light trucks	6.0	2,400,000	200,000,000
Medium trucks	6.5	2,580,000	180,000,000
Heavy trucks	7.0	2,740,000	160,000,000

With the vehicle registration in the state shown below

Passenger cars	2,000,000
Light trucks	200,000
Medium trucks	50,000
Heavy trucks	20,000
Total	2,270,000

the following scheme will allocate the incremental cost of construction to the class of vehicles responsible for those costs.

Allocation of Increment per Vehicle	*Passenger Cars*	*Light Trucks*	*Medium Trucks*	*Heavy Trucks*
$2,200,000,000/2,270,000	$969	$ 969	$ 969	$ 969
200,000,000/270,000		741	741	741
180,000,000/70,000			2,571	2,571
160,000,000/20,000				8,000
Total	$969	$1,710	$4,281	$12,281

If it is desired to collect taxes on the basis of the cost of service, a suitable tax plan must be devised. This may be accomplished by assessing a fuel tax and a vehicle license tax of proper amounts.

9.4 PUBLIC ACTIVITIES AND ENGINEERING ECONOMY

Engineering is a major factor in nearly all public activities because of the high levels of technological input required. It is evident that the complexity of systems and projects being undertaken by the federal and local governments is increasing instead of decreasing.

The contribution of engineering to the nation's space program, national defense, pollution control, urban renewal, and highway construction is well established. Engineering input to these activities involves engineers as employees of governmental agencies and as consultants to these agencies. Thus, public activities are a concern of all engineers as citizens and of many engineers as outlets for their talents.

The engineering process described in Chapter 1 involved the phases of determination of objectives, identification of strategic factors, determination of means, evaluation of engineering proposals, and assistance in decision making. Each of these phases is applicable in public activities. The main modification required is the substitution of benefit or general welfare for profit.

For example, suppose that a municipality has under consideration two projects, one a swimming pool and the other a library. It has resources for one or the other, but not for both. The selection cannot be made on the basis of profit, since no profit is in prospect; it must be made on the basis of which will contribute most to the general welfare as expressed by the citizens of the community, perhaps by a vote.

As a second example, consider the development of a weapons system to aid in national defense. Often several technically feasible systems are under consideration. In this case it is desirable to evaluate the effectiveness of each with the thought that weapons system effectiveness and the general welfare are related. By considering the cost of each system in relation to its effectiveness, a basis for choice is established.

It should be noted that evaluations of public activities in terms of the general welfare encompass both the benefits to be received from and the cost of the proposed activity. No matter how subjective an evaluation of the contribution of an activity to the general welfare may be, its cost may often be determined quite objectively. It may be fairly simple to determine the immediate and subsequent costs for the swimming pool, the library, or the weapons systems. A knowledge of the costs in prospect for benefits to be gained may be expected to result in sounder selection of public activities.

9.5 BENEFIT-COST ANALYSIS

Because of the spectacular growth in the size of government and the absence of competitive pressures for the more efficient use of government resources, there is an increased need to understand the economic desirability of using these resources. The general decision problem is to use the available resources in such a manner that the general welfare of the citizenry is maximized. To help accomplish this goal many agencies in federal, state, and local governments have relied on methods that in some manner quantitatively measure the desirability of particular programs and projects. Of these methods, the most widely utilized is a method referred to as *benefit-cost analysis.*

When applying benefit-cost analysis, the measure of a project's contribution toward the general welfare is normally stated in terms of the benefits "to whomsoever they may accrue" and the cost to be incurred. In order for a project to be considered desirable the benefits must exceed the costs or the ratio of benefits to costs must be larger than one. Otherwise, the government unit would be derelict in its responsibility by applying public resources in a manner that would produce a net decrease in the general welfare of its citizenry.

9.5.1 Point of View Used to Evaluate Public Projects

As in all economy studies, it is crucial that any alternative being considered be analyzed from the *proper point of view.* Otherwise, the description of the alternative will fail to represent all the significant effects associated with that alternative. Thus, the general rule is to assume a point of view that includes all the important consequences of the project being considered. This point of view can be geographical or it may be restricted to classes of people, organizations, or other identifiable groups.

Usually, the easiest method is to identify who is to receive the benefits and who is to pay for them. The point of view that encompasses these two groups is the one that should be selected. Listed below are some examples of particular projects and the point of view that seems most reasonable.

Point of View	*Project*
National	Interstate highway system, major water-resource projects, mass-transit systems
Regional	Regionally funded air-quality control projects
State	State-funded educational programs, state highway programs

County	County-funded medical services
Municipality	City-funded water-supply system, parks, fire protection
Governmental agency	Agency purchase of communications and computing equipment

For practical reasons there is usually a tendency to reduce the scope of the problem under consideration. To analyze on a national basis the effects of building a new library financed from city funds represents an extreme attempt to consider the most far-reaching effects of this project. On the other hand, to analyze an urban mass-transit system that is primarily funded by the federal government on the basis of the direct benefits and costs to the municipality is to erroneously understate the true costs of the system. Unfortunately, many state and local governments have the view that funds supplied by outside sources are "free" funds. The result is that actions are taken that provide benefits to some at the expense of others with no *net* improvement in the general welfare.

When alternatives are evaluated in the private sector, the costs and benefits of the alternatives are based on the viewpoint of the firm or organization making the analysis. When applied by a governmental unit, such a point of view can lead to a misleading description of alternatives. That is, if a state highway department analyzes highway improvements from its point of view instead of from the state's point of view, many effects will fail to be included in the description of alternatives.

9.5.2 Benefits and Costs Considered with or Without the Project

Another important consideration is to develop a basis of reference for identifyng the impact of the project on the nation or any other subunit involved. Thus, for any project it is important to observe what the state of the nation or subunit would be *with* and *without* the project. This base of reference provides the framework for identifying all the important benefits and costs associated with the project.

It should be recognized that this approach is not the same as examining the state of things before and after the project is installed. For example, the improvements in navigation of an inland waterway may increase the growth of barge traffic. However, if some of this growth would have been expected without the improvements, it is unfair to credit the total change in traffic to the project.

It is the change that is attributable to the project itself that is of primary importance when describing the benefits and costs.

Because benefit-cost analyses are intended to assist in the allocation of resources, promotion of the general welfare must reflect the numerous objectives of society. While the economic betterment of the people is one important objective, others include the desire for clean air and water, pleasant surroundings, and personal security.

Some of the benefits and disservices ("disbenefits") associated with these multiple objectives can be stated in economic terms while others cannot. It is important that those benefits that have a market value be represented in monetary terms. It is equally important that those benefits for which there is no market value be included in the analysis. However, it is improper to force the statement of noneconomic objectives in terms of monetary value. For example, it would be misleading to value a grove of hardwood trees in a park on the basis of the number of board-feet of lumber contained in these trees and the market price for that lumber.

9.5.3 Selecting an Interest Rate

Not to consider interest in the evaluation of public activities is equivalent to considering a future benefit equal to a present similar benefit. This appears to be contrary to human nature. Therefore, when considering future economic benefits and costs it is appropriate that an interest rate properly reflecting the time value of money be utilized. *This rate should reflect at least the government's cost of borrowed money.*

Since activities financed through taxation require payments of funds from citizens, the funds expended for public activities should result in benefits comparable with those which the same funds would bring if expended in private ventures. It is almost universal for individuals to demand interest or its equivalent as an inducement to invest their private funds. *To maintain public and private expenditures on a comparative basis, it seems logical that the interest rate selected should represent the opportunity foregone when taxes are paid. That is, the interest rate should reflect the rate that could have been earned if the funds had not been removed from the private sector.*

Some public activities are financed in whole or in part through the sale of services or products. Examples of such activities are electric power developments, irrigation projects, housing projects, and toll bridges. Many such services could be carried on by private companies and are in general in competition with private enterprise. Again, since private enterprise must of necessity consider interest, it seems logical to consider the opportunity forgone in the benefit-cost analysis of public activities that compete in any way with private enterprise.

The interest rate to use in an economy study of a public activity is a matter of judgment. The rate used should not be less than that paid for funds borrowed for the activity. In many cases, particularly where the activity is comparable or competitive with private activities, the rate used should be comparable with that used in private evaluations.

9.5.4 The Benefit-Cost Ratio

A popular method for deciding upon the economic justification of a public project is to compute the benefit-cost ratio. This ratio may be expressed as

$$BC(i) = \frac{\text{Benefits to the public}}{\text{Cost to the government}}$$

where the benefits and the costs are $PW(i)$ or $AE(i)$ amounts computed using the cost of money. Thus, the BC ratio reflects the users' equivalent dollar benefits and the sponsors' equivalent dollar cost. If the ratio is 1, the equivalent benefits and the equivalent cost are equal. This represents the minimum justification for an expenditure by a public agency.

Considerable care must be exercised in accounting for the benefits and the costs in connection with benefit-cost analysis. Benefits are defined to mean all the advantages, less any disadvantages, to the users. Many proposals that embrace valuable benefits also result in inescapable disadvantages. It is the net benefits to the users that are sought. Similarly, costs are defined to mean all costs, less any savings, that will be incurred by the sponsor. Such savings are not benefits to the users but are reductions in cost to the government. It is important to realize that adding a number to the numerator does not have the same effect as subtracting the same number from the denominator of the BC ratio. Thus, incorrect accounting for the benefits and costs can lead to a ratio that may be misinterpreted. Therefore, the benefit-cost ratio is normally defined as

$$\boxed{BC(i) = \frac{\text{Equivalent benefits}}{\text{Equivalent costs}}} \tag{9.1}$$

where

Benefits = All the advantages, less disadvantages to the user

Costs = All the disbursements, less any savings to the sponsor.

The sign convention used in benefit-cost analysis requires that

Benefits = (+) advantages, receipts, savings

(−) disadvantages, disbursements, losses

Costs = (+) disbursements, losses

(−) savings, receipts.

Following these conventions consistently is necessary to assure that the benefit-cost ratio exceeds 1 when the equivalent benefits are greater than the equivalent costs.

To better understand the implications of this definition, the equivalent costs are divided into two components: (1) the equivalent capital initially invested by the sponsor, (2) the equivalent annual operating and maintenance costs less any annual revenue produced by the project. This redefinition gives

I = equivalent capital invested by the sponsor;

C = net equivalent annual costs to the sponsor;

B = net equivalent benefits to the user.

The benefit-cost ratio can then be expressed as

$$\boxed{BC(i) = \frac{B}{I + C}} \tag{9.2}$$

For any project to remain under consideration, its benefit-cost ratio must exceed 1. Therefore, the first test of a project is to determine whether it is minimally acceptable by observing whether or not the equivalent benefits exceed the equivalent costs. We see below that using such a criterion will eliminate all those projects whose net equivalent amount is less than zero.

If

$$BC(i) > 1$$

then

$$\frac{B}{I + C} > 1$$

giving

$$B - (I + C) > 0.$$

An alternative method of expressing the benefit-cost ratio will appear in some benefit-cost analyses. Although the most widely accepted definition is the one in equation 9.2, it is important to understand the relationship between these two ratios. The only difference is that the second ratio reflects the net benefits less the annual costs of operation of the project divided by the investment cost. This is expressed

$$BC'(i) = \frac{B - C}{I}. \tag{9.3}$$

The advantage of defining the benefit-cost ratio in this manner is that it provides an index indicating the net gain expected per dollar invested.

Again, it is required that for a project to remain under consideration the alternative benefit-cost ratio must be larger than 1.

If

$$BC'(i) > 1$$

then

$$\frac{B - C}{I} > 1$$

giving

$$B - (I + C) > 0.$$

Therefore, Equations (9.2) and (9.3) will lead to the same conclusion on the initial acceptability of the project (as long as $I + C$ and I are greater than zero).

As an example of the application of benefit-cost analysis and the *BC* ratio consider the following situation, which is of interest to a state highway department. The accidents involving motor vehicles on a certain highway have been studied for a number of years. The calculable costs of such accidents embrace lost wages, medical expenses, and property damage. On the average there are 35 nonfatal accidents and 240 property damage accidents for each fatal accident. The average equivalent present cost of these three classes of accidents is calculated as

Fatality per person	$900,000
Nonfatal injury accidents	10,000
Property damage accident	1,800

From these the data aggregate cost of motor-vehicle accidents per death may be calculated.

Fatality per person	$900,000
Nonfatal injury accident $10,000 × 35	350,000
Property damage accident $1,800 × 240	432,000
Total	$1,682,000

The death rate on the highway in question has been 8 per 100,000,000 vehicle miles. A proposal to add a third lane is under consideration. It is estimated that the cost per mile will be \$1,500,000, the service life of the improvement will be 30 years, and the annual maintenance will be 3% of the first cost. The traffic density on the highway is 10,000 vehicles per day, and the interest rate is 7%. It is estimated that the death rate will decrease to 4 per 100,000,000 vehicle miles. Although other benefits will result from widening the highway, it is argued that the reduction in accidents is sufficient to justify the expenditure.

To verify the economic desirability of the widening project, the highway department performs the following calculations. The *AE(i)* benefit per mile to the public is

$$\frac{(8 - 4)(10{,}000)(365)(\$1{,}682{,}000)}{100{,}000{,}000} = \$245{,}572.$$

And the *AE(i)* cost per mile to the state is

$$\$1{,}500{,}000(\overset{A/P,7,30}{0.0806}) + \$1{,}500{,}000(0.03) = \$165{,}900.$$

Using Equation (9.2) results in a benefit-cost ratio of

$$BC(7) = \frac{\$245{,}572}{\$165{,}900} = 1.48.$$

Thus, widening the highway is justified based on the benefits to be derived from a reduction in accidents alone. Other benefits, such as the reduction in trip time, have not been included in the analysis and would increase the ratio.

Calculating the benefit-cost ratio from Equation (9.3) for this problem requires that the *AE(i)* operating costs be included in the numerator rather than in the denominator. This approach yields

$$BC'(7) = \frac{\$245{,}572 - \$45{,}000}{\$120{,}900} = \frac{\$200{,}572}{\$120{,}900} = 1.66.$$

The result indicates that a net savings of \$1.66 for each dollar invested will be realized from the widening project. Both ratios indicate that substantial benefits will occur from such an expenditure of funds.

It should be noted that for both of the benefit-cost ratio definitions, the ratios will equal 1 when the benefit and cost cash flows are evaluated at the IRR, i^*. Figure 9.1 illustrates, for the highway problem just discussed, the range of values of the two benefit-cost ratios as a function of the interest rate. Observe that the two curves intersect at a value of 1, where the interest rate of 13% is the IRR for this cash flow.

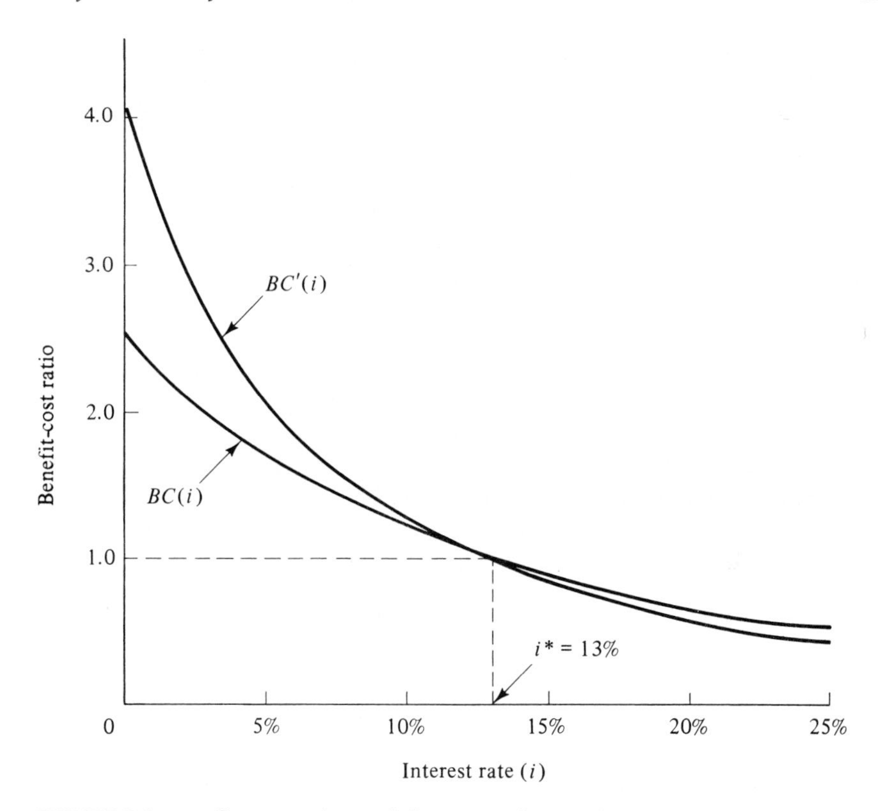

FIGURE 9.1 Benefit-cost ratios and the internal rate of return.

9.5.5 Benefit-Cost Analysis for Multiple Alternatives

The example of the previous paragraphs illustrated a situation in which the sponsoring agency had the simple choice of widening the highway or leaving it as is. Usually, however, a sponsoring agency finds that different benefit levels and different costs result in meeting a specific objective. When this is the case, the problem of interpreting the corresponding benefit-cost ratios presents itself. A hypothetical situation and the correct method of analysis is presented in the following paragraphs.

Suppose that four mutually exclusive alternatives have been identified for providing recreational facilities in a certain urban area. The $AE(i)$ benefits, $AE(i)$ costs, and benefit-cost ratios are given in Table 9.1. Inspection of the BC ratios might lead one to select Alternative $A2$ because its ratio is a maximum. Actually, this choice is *not* correct. The correct alternative can be selected by applying the principle of incremental analysis as described in Sections 7.3 and 7.4.

TABLE 9.1.
BENEFIT-COST RATIOS ON TOTAL INVESTMENT FOR FOUR ALTERNATIVES

Alternative	*AE(i)* Benefits	*AE(i)* Costs	*BC* Ratio
*A*1	$182,000	$91,500	1.99
*A*2	167,000	79,500	2.10
*A*3	115,000	78,500	1.46
*A*4	95,000	50,000	1.90

When using the benefit-cost ratio to compare mutually exclusive alternatives, the benefit-cost ratio must be applied incrementally.

In this instance, the additional increment of outlay is economically desirable if the incremental benefit realized exceeds the incremental outlay. Thus, when comparing mutually exclusive alternatives $A1$ and $A2$ the decision rule is

$$\begin{array}{ll} BC(i)_{A2-A1} > 1: & \text{accept } A2. \\ BC(i)_{A2-A1} \leq 1: & \text{reject } A2 \text{ and accept } A1. \end{array} \qquad (9.4)$$

When comparing alternatives on an incremental basis in Sections 7.4 and 7.5 we arranged the alternatives in order of increasing first-year cost. *However, for incremental benefit-cost analysis the alternatives must be arranged by increasing order of their denominators.* (Remember for the benefit-cost ratio, costs in the denominator and benefits in the numerator are always considered to be positive values.) Thus, the alternative with smallest denominator should be first, the alternative with the next smallest denominator second, and so forth. For the alternatives in Table 9.1 this ordering would be $A4$, $A3$, $A2$, and $A1$. This ordering will assure that the denominator of the increment being considered will always be positive, so that the decision rules just described will produce correct results.

If for some unusual situation the project revenues to the sponsor (−) should exceed the costs to the sponsor (+), the denominator value would then be negative. Other methods should be utilized in this case, since the decision rules for incremental benefit-cost analysis are applicable only when the incremental denominator is positive.

Applying these decision rules to the alternatives described in Table 9.1 indicates that Alternative $A1$ and *not* Alternative $A2$ is the more desirable. The calculations of the incremental benefit-cost ratios are summarized in Table 9.2.

TABLE 9.2.
INCREMENTAL BENEFIT-COST RATIOS

Alternative	Incremental Annual Benefits	Incremental Annual Costs	Incremental *BC* Ratio	Decision
$A4 - A0$	\$95,000	\$50,000	1.90	Accept *A*4
$A3 - A4$	20,000	28,500	0.70	Reject *A*3
$A2 - A4$	72,000	29,500	2.44	Accept *A*2
$A1 - A2$	15,000	12,000	1.25	Accept *A*1

If the Do Nothing alternative, *A*0, is to be considered, assume that the cash flow associated with that alternative is zero. This approach is described earlier in Section 7.3.3. When comparing an alternative to the *A*0 option, compute the incremental benefit-cost using this assumption and apply the decision rules just described.

The sequence of calculations required to produce the results in Table 9.2 are presented next. Since the Do Nothing alternative is considered to be feasible, the first incremental comparison is between the alternative wih the smallest denominator and *A*0. This comparison of *A*4 to *A*0 yields

$$BC(i)_{A4-A0} = BC(i)_{A4} = \frac{\$95,000}{\$50,000} = 1.90$$

because the cash flow for *A*0 is considered to be zero. Since this benefit-cost ratio is greater than 1, Alternative *A*4 is seen to be preferred to the Do Nothing alternative. Therefore, the Do Nothing alternative (the initial "current best" alternative) is rejected and Alternative *A*4 becomes the new "current best" alternative.

Next, it is necessary to determine whether the incremental benefits that would be realized if Alternative *A*3 were undertaken would justify the additional expenditure. Therefore, compare Alternative *A*3 to Alternative *A*4 as follows.

$$BC(i)_{A3-A4} = \frac{\$115,000 - \$95,000}{\$78,500 - \$50,000} = \frac{\$20,000}{\$28,500} = 0.70.$$

The incremental benefit-cost ratio is less than 1, and therefore Alternative *A*3 is rejected and Alternative *A*4 remains as the "current best" alternative.

Next, compare Alternative *A*2 to Alternative *A*4 as follows.

$$BC(i)_{A2-A4} = \frac{\$167,000 - \$95,000}{\$79,500 - \$50,000} = \frac{\$72,000}{\$29,500} = 2.44.$$

The incremental benefit-cost ratio in this instance exceeds 1, and therefore Alternative $A2$ is preferred to Alternative $A4$. Alternative $A2$ becomes the new "current best" alternative.

Alternative $A1$ is now compared to Alternative $A2$ as follows.

$$BC(i)_{A1-A2} = \frac{\$182{,}000 - \$167{,}000}{\$91{,}500 - \$79{,}500} = \frac{\$15{,}000}{\$12{,}000} = 1.25.$$

Since the incremental benefit-cost ratio for this comparison is greater than 1, Alternative $A1$ is preferred to Alternative $A2$. Alternative $A1$ becomes the "current best" alternative, and there are no more comparisons to be made. The alternative that should be selected is the current best alternative that remains after the final comparison. Therefore, Alternative $A1$ is the preferred of the four alternatives. Selection of this alternative will assure that the $AE(i)$ benefits less the $AE(i)$ costs are maximized and that its BC ratio is greater than 1. To demonstrate that the alternative chosen is the one that will maximize the equivalent benefits less the equivalent costs, Table 9.3 presents this net figure for each alternative.

Remember in Chapter 7 that, when comparing mutually exclusive alternatives, present worth (annual equivalent) on total investment, present worth (annual equivalent) on incremental investment, and rate of return on incremental investment are all consistent decision criteria. That incremental benefit-cost analysis will lead to the same conclusions as these decision criteria can be confirmed analytically. To show this, incremental analysis using the benefit-cost ratio will be compared to present worth on total investment.

Suppose that two mutually exclusive alternatives $A1$ and $A2$ are being considered for investment. Assume that the present worth on total investment is known for each alternative and that the following relationship exists:

$$PW(i)_{A2} > PW(i)_{A1}.$$

TABLE 9.3.
BENEFITS LESS COSTS FOR FOUR ALTERNATIVES

Alternative	Annual Equivalent Benefits	Annual Equivalent Costs	Net Improvement of General Welfare
$A1$	$182,000	$91,500	$90,500*
$A2$	167,000	79,500	87,500
$A3$	115,000	78,500	36,500
$A4$	95,000	50,000	45,000

**Alternative A1 provides the maximum net improvement.*

That is, the net present worth for $A2$ exceeds $A1$ and, therefore, $A2$ is economically more desirable than $A1$. Now define

$B(i)_j$ = present worth of benefits for alternative j;

$I(i)_j$ = present worth of investment for alternative j;

$C(i)_j$ = present worth of operating costs for alternative j.

It follows that the incremental benefit-cost ratio based on Equation (9.2) is

$$BC(i)_{A2-A1} = \frac{B(i)_{A2} - B(i)_{A1}}{I(i)_{A2} - I(i)_{A1} + [C(i)_{A2} - C(i)_{A1}]}.$$

This ratio should be greater than 1 if $A2$ is economically preferred to $A1$, as was initially assumed. Does it lead to the same conclusion as the present worth on total investment criterion?

If

$$BC(i)_{A2-A1} > 1$$

then

$$\frac{B(i)_{A2} - B(i)_{A1}}{I(i)_{A2} - I(i)_{A1} + [C(i)_{A2} - C(i)_{A1}]} > 1$$

which gives

$$B(i)_{A2} - B(i)_{A1} > I(i)_{A2} - I(i)_{A1} + C(i)_{A2} - C(i)_{A1}$$

and by transposing

$$B(i)_{A2} - I(i)_{A2} - C(i)_{A2} > B(i)_{A1} - I(i)_{A1} - C(i)_{A1}$$

which by definition is

$$PW(i)_{A2} > PW(i)_{A1}.$$

This same argument can be used for the benefit-cost ratio based on Equation (9.3.). Thus, it can be demonstrated that incremental analysis will again provide results consistent with the decision criteria presented in Sections 7.4–7.6.

9.6 IDENTIFYING BENEFITS, DISBENEFITS, AND COSTS

As indicated in Section 9.5, it is extremely important how the accounting for benefits and costs is accomplished. First, the traditional definition of the benefit-cost ratio requires that the net benefits to the user be placed in the numerator and the net costs to the sponsor in the denominator. To find the net benefits it is necessary to identify those consequences which are favorable and unfavorable to the user. These unfavorable benefits are usually referred to as *disbenefits.* When they are deducted from the positive effects to be realized by the user, the resulting figure represents the net "good" to be engendered by the project.

To determine the net cost to the sponsor it is necessary to identify and classify the outlays required and the revenues to be realized. These revenues or savings usually represent income generated from the sale of products or services that are developed from the project. These "costs" include both disbursements and receipts related to the project's initial investment and to its annual operation.

Presented below is an example of the classification of benefits and costs that would be related to the completion of a new toll road through a rural area in order to substantially shorten the distance between two large communities.

Benefits to Public

Reduced vehicle operating costs (excluding fuel tax)
Reduced commercial and noncommercial travel time
Increased safety
Increased accessibility between communities
Ease of driving
Appreciation of land values

Disbenefits to the Public

Land removed from agricultural production
Damages resulting from changes of water flow
Decreased movement of livestock across highway
Increased air pollution and litter

Costs to State

Construction costs
Maintenance costs
Administrative costs

Savings to State

Toll revenues

Increased taxes due to appreciated land and increased business activity

9.6.1 Types of Benefits

An important question in the classification of benefits is, "To what length should one go to trace all the consequences of a project?" Not only is the answer critical when attempting to quantify a project's contribution to the general welfare, but it can substantially affect the cost of undertaking the benefit-cost analysis. To distinguish between those benefits that can be directly attributable to the project and those that are less directly connected, benefits are generally classified as follows.

Primary benefits represent the value of the direct products or services realized from the activities for which the project was undertaken.

Secondary benefits represent the value of those additional products and services realized from the activities of or stimulated by the project.

Most public projects provide both primary and secondary benefits. Irrigation projects increase the crop yield (primary benefits) as well as the economic strength of the farming community (secondary benefit). A benefit-cost analysis should always consider the primary benefits and whenever appropriate should consider the secondary benefits. Inclusion of secondary benefits should be a function of their effect compared to that of the primary benefits and to the cost of determining them.

9.6.2 Valuation of Benefits

Public activities provide such a variety of benefits that it is impossible to always value benefits in monetary terms. What is important is that both benefits and costs be represented by measures that are most meaningful to those who are involved in project assessment.

A benefit-cost analysis not only compares the quantifiable consequences but also describes the irreducibles and nonquantifiable characteristics in whatever terms are feasible.

For certain benefits and costs the market price accurately reflects their true value. For other benefits there is a market price, but this price fails to realistically represent their actual value (e.g., products or services that are subsidized, price-supported, or artificially restrained from trade).

In addition, for some benefits there is no market value available but an economic value may be imputed. One approach for evaluating such benefits is to consider the least expensive means of achieving the same service. Another method is to infer what a user is willing to pay for a service by observing the amount he spends to take advantage of this service. This latter approach is frequently applied to ascertain the economic worth associated with recreation. Thus, to determine the value of recreation for a water resources project, an analysis is made of what the user spends to avail himself of the project's recreation opportunities.

Last, there are benefits and cost for which it would be impossible to assign economic values. If a benefit can be quantified in realistic and meaningful terms, then it should be quantified (e.g., number of trees, classified by height, as a measure of the aesthetics of a hardwood grove). For benefits for which suitable measures cannot be discovered, qualitative descriptions will suffice. However, it is important that all significant benefits and costs be included regardless of the degree to which they can be quantified.

9.6.3 Consideration of Taxes

Many public activities result in loss of taxes through the removal of property from tax rolls or by other means, as, for example, the exemption from sales taxes. In a nation in which free enterprise is a fundamental philosophy, the basis for comparison of the cost of carrying on activities is the cost for which they can be carried on by well-managed private enterprises. Therefore, it seems logical to take taxes into consideration in a benefit-cost analysis, particularly when the activities are competitive with private enterprise.

The federal government agreed to pay $300,000 annually for 50 years to each of the states of Arizona and Nevada in connection with the Hoover Dam project. These payments are to partially compensate for tax revenue that would accrue to these states if the project had been privately constructed and operated. These payments are a cost and were considered in the economic justification of the project.

Because governments do not pay taxes, it is possible to omit them from consideration in some cases. For example, when the government is comparing its proposals to each other, the net change to the general welfare is unchanged. (Many tax payments are a transfer of economic value from one group to another.) In highway studies it is common to exclude the fuel tax from the vehicular operating costs.[1] This exclusion then reduces this cost to the user by the amount of the fuel taxes.

9.6.4 Benefits and Costs for Multiple-Purpose Projects

As noted in Section 9.2.2, many public ventures are multiple-purpose projects. In particular, water-resource projects that provide a variety of functions, includ-

[1] *C. H. Oglesby and R. G. Hicks,* Highway Engineering, *4th ed. (New York: John Wiley & Sons, 1982).*

ing electric power, flood control, irrigation, navigation, and recreation, fit this category.

One difficulty that may arise when analyzing multipurpose projects is the assessment of the desirability of each of its functions. For example, the electric power generated by the stored water may be in competition with private power companies. Should turbines be included in the dam or should electric power be supplied from private sources? To answer such a question it is necessary to isolate the costs directly identifiable with power generation. The dam, which provides flood control, water for irrigation, and other benefits, is also an integral part of power generation. Thus, the cost of the dam must be *jointly* distributed among all the project functions. Unfortunately, the inability to allocate joint costs accurately is a fact of life. As a consequence, many procedures have been developed to assist in this allocation and none can be considered to be perfect.

The same problem exists when accounting for the benefits and disbenefits of multipurpose projects. Many of these benefits are inseparable from the project as a whole. For example, the project just discussed introduces a disbenefit representing the land removed from agricultural production when it is flooded by the reservoir. How should this disbenefit be apportioned among the various project functions? It becomes evident that the separate justification of integral units within a multiple-purpose project does not provide completely satisfactory conclusions. The following table shows some of the costs (joint and otherwise) and the benefits (separable and otherwise) that might be included in a benefit-cost analysis of the construction of a large multipurpose dam.

Function	*Benefits*	*Disbenefits*	*Costs*	*Savings*
Hydroelectric power	Increased availability	Land flooded	Investment and operating	Sales of power
Flood control	Reduced flood damage	Land flooded	Investment and operating	Flood damage costs avoided
Irrigation	Increased crop yield	Land flooded	Investment and operating	Water revenue
Navigation	Savings on shipping costs	Loss of rail-road traffic	Investment and operating	Vessel berth charges
Recreation	Increased accessibility	Destruction of scenic river	Investment and operating	Use charges

9.7 COST-EFFECTIVENESS ANALYSIS

Cost-effectiveness analysis had its origin in the economic evaluation of complex defense and space systems. Its predecessor, benefit-cost analysis, had its origin in the civilian sector of the economy and may be traced back to the Flood Control Act of 1936. Much of the philosophy and methodology of the cost-effectiveness approach was derived from benefit-cost analysis, and as a result there are many similarities in the techniques. The basic concepts inherent in cost-effectiveness analysis are now being applied to a broad range of problems in both the defense and the civil sectors of public activities.

In applying cost-effectiveness analysis to complex systems three requirements must be satisfied. First, the systems being evaluated must have common goals or purposes. The comparison of cargo aircraft with fighter aircraft would not be valid, but comparison with cargo ships would if both the aircraft and ships were to be utilized in military logistics. Second, alternate means for meeting the goal must exist. This is the case with cargo ships being compared with the cargo aircraft. Finally, the capability of bounding the problem must exist. The engineering details of the systems being evaluated must be available or estimated so that the cost and effectiveness of each system can be estimated.

9.7.1 The Cost-Effectiveness Approach

Certain steps constitute a standardized approach to cost-effectiveness evaluations.[2] These steps are useful in that they define a systematic methodology for the evaluation of complex systems in economic terms. The following paragraphs summarize these steps.

First, it is essential that the desired goal or goals of the system be defined. In the case of military logistics mentioned earlier, the goal may be to move a certain number of tons of men and supporting equipment from one point to another in a specified interval of time. This may be accomplished by a few relatively slow cargo ships or a number of fast cargo aircraft. Care must be exercised in this step to be sure that the goals will satisfy mission requirements. Each delivery system must have the capability of delivering a mix of men and equipment that will meet the requirements of the mission. Comparison of aircraft that can deliver only men against ships that can deliver both men and equipment would not be valid in a cost-effectiveness study.

Once mission requirements have been identified, alternate system concepts and designs must be developed. If only one system can be conceived, a cost-effectiveness evaluation cannot be used as a basis for selection. Also, selection must be made on the basis of an optimum configuration for each system. In Section 10.3.1 the minimum-cost pier spacing was found for two competing bridge designs before a choice was made. A similar comparable basis must be established for a ship cargo system and an aircraft cargo system.

System evaluation criteria must be established next for both the cost and the effectiveness aspects of the system under study. Ordinarily, less difficulty exists in establishing cost criteria than in establishing criteria for effectiveness. This does not mean that cost estimation is easy; it simply means that the classifications and basis for summarizing cost are more commonly understood.

[2] *A.D. Kazanowski, "A Standardized Approach to Cost-Effectiveness Evaluations," Chapter 7 in J. Morley English, ed.,* Cost Effectiveness *(New York: John Wiley & Sons, 1968).*

Among the categories of cost are those arising throughout the system life cycle, which include costs associated with research and development, engineering, testing, production, operation, and maintenance. The phrase *life-cycle costing* is often used in connection with cost determination for complex systems. Life-cycle cost determination for a specific system is normally on a present or equivalent annual basis. The Department of Defense has adopted the policy of procuring systems on the basis of life-cycle cost as opposed to first cost, as had been the practice in the past.

System evaluation criteria on the effectiveness side of a cost-effectiveness study are quite difficult to establish. Also, many systems have multiple purposes that complicate the problem further. Some general effectiveness categories are utility, merit, worth, benefit, and gain. These are difficult to quantify, and so such criteria as mobility, availability, maintainability, reliability, and others are normally used. Although precise quantitative measures are not available for all of them, these evaluation criteria are useful as a basis for describing system effectiveness.

The next step in a cost-effectiveness study is to select the fixed-cost or the fixed-effectiveness approach. In the *fixed-cost approach* the basis for selection is the amount of effectiveness obtained at a given cost; in the *fixed-effectiveness* approach it is the cost incurred to obtain a given level of effectiveness. When multiple alternatives that provide the same service are compared on the basis of cost, the fixed-effectiveness approach is being used. This was the case with the two competing bridge designs presented in Section 10.3.1.

Candidate systems in a cost-effectiveness study must be analyzed on the basis of their merits. This may be accomplished by ranking the systems in order of their ability to satisfy the most important criterion. For example, if the criterion in military logistics is the number of tons of men and equipment moved from one point to another in a specified interval of time, this criterion becomes the primary one. Other criteria, such as maintainability, would be ranked in a secondary position. Often this procedure will eliminate the least promising candidates. The remaining candidates can then be subjected to a detailed cost and effectiveness analysis. If the cost and the effectiveness for the top contender are both superior to the respective values for other candidates, the choice is obvious. If criteria values for the top two contenders are identical, or nearly identical, and no significant cost difference exists, either may be selected based on irreducibles. Finally, if system costs differ significantly, and effectiveness differs significantly, the selection must be made on the basis of intuition and judgment. This latter outcome is the most common in cost-effectiveness analysis directed to complex systems.

The final step in a cost-effectiveness study involves documentation of the purpose, assumptions, methodology, and conclusions. This is the communication step and it should not be treated lightly. No wise decision maker would base a major expenditure of capital on a blind trust of the analyst.

9.7.2 A Cost-Effectiveness Example

As an example of some aspects of cost-effectiveness analysis consider the goal of moving men and equipment from one point to another, as discussed earlier. Suppose that only the cargo aircraft mode and the ship mode are feasible. Also suppose that some design flexibility exists within each mode, so that the effectiveness in tons per day may be established through design effort.

Assume that the Department of Defense has convinced Congress that such a military logistic system should be developed and that Congress has authorized a research and development program for a system whose present life-cycle cost is not to exceed 1.2 billion dollars. The Department of Defense, in conjunction with a nonprofit research and engineering firm, has decided that three candidate systems should be conceived and costed. Table 9.4 shows the resulting present life-cycle cost and corresponding effectiveness in tons per day for each system at maximum utilization. These data are also exhibited in Figure 9.2.

The three candidate systems were studied because of the following logic. First, using the fixed-cost approach the study team projected the configuration of Aircraft System I and the Ship System and found that these systems would have effectivenesses of 1,620 and 1,410 tons per day, respectively. If the study had stopped here, the conclusion would be to spend the 1.2 billion dollars for Aircraft System I. Realizing, however, that the Congress and the Department of Defense would welcome estimates of effectiveness for other levels of expenditure, the study team chose to determine the cost for Aircraft System II, which would be designed to have an effectiveness equal to that of the Ship System costing 1.2 billion dollars. The cost of this system was estimated to be 1.0 billion dollars. This is an example of the fixed-effectiveness approach.

At this point a decision must be made between Aircraft System I and Aircraft System II. The Ship System has a lower effectiveness for the same cost as Aircraft System I and a higher cost for the same effectiveness as Aircraft System II and so will not be a candidate in the decision process. In choosing between Aircraft System I and Aircraft System II, the Congress in conjunction with the Department of Defense must decide if the increase in effectiveness of 210 tons per day is worth a present life-cycle cost of 0.2 billion dollars.

This example is quite simplified in its assumptions and analysis. Nothing has been said about how the study team was able to determine that a certain ef-

TABLE 9.4.
COST AND EFFECTIVENESS FOR THREE SYSTEMS

System	Present Cost in Billions of $	Effectiveness in Tons per Day
Aircraft I	1.2	1,620
Ship	1.2	1,410
Aircraft II	1.0	1,410

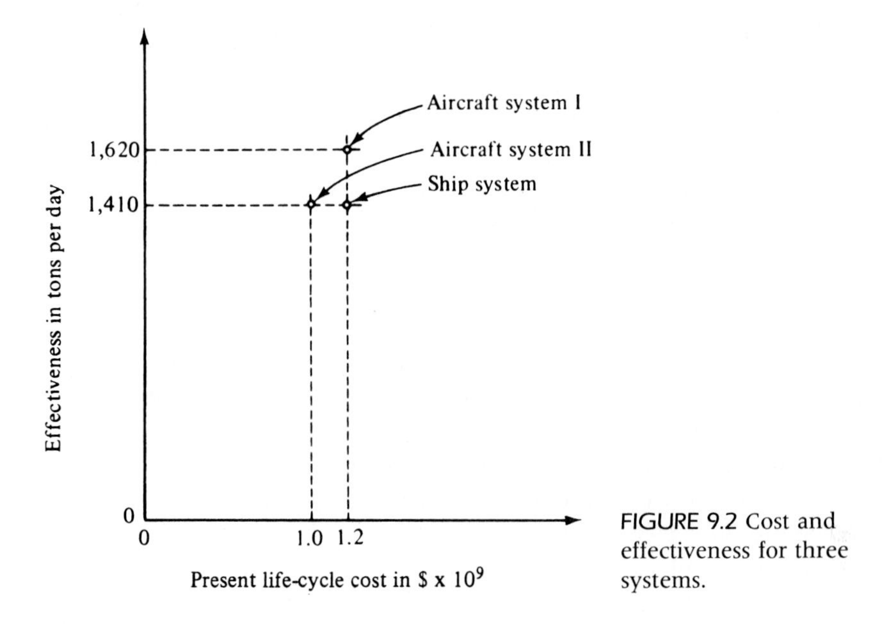

FIGURE 9.2 Cost and effectiveness for three systems.

fectiveness would result from a system with a present life-cycle cost of 1.2 billion dollars. Also, the example did not point out any secondary measures of effectiveness that would complicate the choice.

KEY POINTS

- A primary objective of government is to invest its resources so there is an increase in the general welfare of its citizens.
- The correct point of view for a project should include all the significant economic and noneconomic consequences of that project.
- The benefits and costs associated with a project should reflect the change in those items that would be expected with or without the project.
- The benefit-cost ratio for a government activity quantifies the economic benefits that are expected to accrue to the citizenry divided by the costs expected to be incurred by the activity.
- To assure that projects sponsored by government increase the general welfare, the benefits of each project are expected to exceed the costs of that project (that is, the benefit-cost ratio must exceed 1).
- Incremental analysis with the benefit-cost ratio must be employed when comparing mutually exclusive projects.

- The two benefit-cost ratios represented by Equations (9.2) and (9.3) will lead to the same conclusion when they are applied incrementally.

QUESTIONS AND PROBLEMS

1. Describe the meaning of the general welfare objective as it relates to engineering economy analysis applied to public activities.
2. Contrast the criteria for evaluation of private and public activities.
3. Describe the impediments to efficiency in governmental activities.
4. Outline the function of government and the nature of public activities.
5. Name the three basic methods for financing public activities.
6. What is the basic philosophy of government in the expenditure of tax funds?
7. What is the basic philosophy of government in the collection of taxes?
8. Name three types of issues used by the federal government to borrow from its citizenry.
9. Describe the difference between general obligation bonds and revenue bonds.
10. How do municipal bonds differ from bonds issued by the federal government?
11. Name the elements that should be considered in deciding on an interest rate to be used in the evaluation of public activities.
12. Show that project selection based upon the alternative incremental benefit-cost ratio $(B - C)/I$ leads to present-worth maximization.
13. Briefly describe the factors to be considered in identifying benefits, disbenefits, and costs in the evaluation of public activities.
14. List the factors that would have a significant effect on a city's decision to undertake a mass-transportation system. Indicate which factors could be quantified and which would be considered nonquantifiable.
15. List some noneconomic indicators that could be useful in evaluating whether or not public activities are in the interest of the general welfare.
16. Assume that a certain state is contemplating a highway development embracing the construction of 8,000,000 square yards of pavement. Vehicle registration in the state is

Passenger cars	1,000,000
Light trucks	200,000
Medium trucks	40,000
Heavy trucks	10,000

The pavements necessary to carry the vehicles are

Class of Vehicle	*Pavement Thickness (Inches)*	*Cost per Sq. Yard*
Passenger cars	5.5	$10.50
Light trucks	6.0	11.80
Medium trucks	6.5	12.50
Heavy trucks	7.0	13.00

On the assumption that paving costs should be distributed on the basis of the number of vehicles in each class and the incremental costs of paving required for each class of vehicle, what should be the taxes per vehicle for each vehicle class?

17. A state whose population is 4,600,000 has 13 state-supported colleges and universities with a total enrollment of 65,000 students and an annual budget for instructional programs approximating $162,500,000 per year.

- **a.** Most students enrolled in the colleges believe that prospective increased earnings is an important reason for attending. What groups within the state's population do you believe desire to support the program of higher education through taxation for this reason?
- **b.** What prospective annual increase in earnings must you achieve to recover the state's cost of your four years of education? Assume that the first increase will occur at the end of the second year of work, you work 45 years, the state provides $2,500 per student per year, and the interest rate is 10%.
- **c.** Indicate any benefits that may accrue to the state from its investment in education.

18. A toll road has been opened between two cities. The distance between the entrances of the two cities is 93 miles via the toll road and 104 miles via the shortest alternate free highway. Determine the economic advantage, if any, of using the toll road for the following conditions applicable to the operation of a light truck: toll cost, $4.50; driver cost, $12.50 per hour; average driving rate between entrances via toll road and free road respectively, 55 and 50 mph; estimated average cost of operating truck per mile via toll road, $0.20, via free road, $0.22.

19. Analysis of accidents in one state indicates that increasing the width of highways from 6 meters to 7.5 meters may decrease the annual accident rate from 125 per 100,000,000 to 72 per 100,000,000 vehicle-kilometers. Calculate the average daily number of vehicles that should use a highway to justify widening on the basis of the following estimates: average loss per accident, $1,500; per-kilometer cost of widening pavement 1.5 meters, $150,000; useful life of improvement, 25 years; annual maintenance, 3% of first cost; interest rate, 5%. Answer: 52,210 vehicles per day

20. A government agency has $5,000,000 available to spend on flood-control projects. Project *A* is expected to generate flood prevention benefits of $460,000 per year for an investment of $3,000,000 and annual maintenance expenses of $57,000. Project *B*, costing $4,000,000 with maintenance expenses of $81,500 per year, is expected to produce annual savings of $613,000 due to reduced flooding. Both projects are expected to last 40 years. For an interest rate of 10% compute the benefit-cost ratio for each project and then calculate the incremental benefit-cost ratio. Which approach provides the correct result?

21. Two mutually exclusive projects are being considered for investment. Project *A*1 requires an initial outlay of $115,000 with net receipts estimated to be $34,500 per year for the next 5 years. The initial outlay for Project *A*2 is $230,000, and net receipts have been estimated at $56,000 per year for the next 7 years. The interest rate is 13%, and there is no salvage value associated with either project. Using the benefit-cost analysis, which project would you select? Answer: Select A2

22. The Tennessee Valley Authority (TVA) wants to construct a 150-MW peaking power plant to meet the demand for additional electricity. Two alternatives are under consideration. One is to construct a combustion turbine plant, which is particularly well suited for peaking operation because of its low capital cost. Its most serious disadvantage is its poor thermal efficiency, which affects the fuel and operating costs. Also under consideration is a fuel-cell plant, which provides better thermal efficiency even though it is relatively more expensive to construct. Assume the fuel and operating costs are expected to increase at the rate of 6% per year, even though the amount of energy produced is expected to be the same each year. The fuel cost in any year *n* is represented by the following expression:

$$F_n = (C)(H)(U)\left(\frac{8{,}760 \text{ hours/year}}{10^6}\right)P_n$$

where

$$P_n = P_{n-1}(1 + 0.06).$$

The operating cost in any year *n* is given by

$$O_n = (C)(H)(U)\left(\frac{8{,}760 \text{ hours/year}}{10^6}\right)Q_n$$

where

$$Q_n = Q_{n-1}(1 + 0.06).$$

It would be 2 years before either plant would be in full commercial operation. The revenue generated each year is assumed to be the same and to exceed the total $AE(i)$ cost for either plant. Shown below are the economic and operating data associated with each type of generating plant. (Assume all construction costs occur when the plant commences full commercial operation.)

	Combustion Turbine	*Fuel-Cell Plant*
C = Plant size (kW)	150,000	150,000
H = Heat rate (Btu/kWh)	12,700	9,300
U = Plant utilization factor	15%	15%
N = Economic service life (years)	25	25
P_0 = Starting fuel cost ($\$/10^6$ Btu)	2.20	2.20
Q_0 = Starting maintenance cost ($\$/10^6$ Btu)	0.19	0.19
Construction cost ($/kW)	175	240
Other annual expenses (insurance, tax, depreciation, and maintenance cost) as a fixed % of construction costs	5.5%	5.5%
Salvage value as a fixed % of construction cost	−1%	−1%

a. For an interest rate of 12%, determine which alternative is more attractive to undertake ($AE(i)$ cost comparison over commercial operation period).

b. By using the incremental BC ratio $(B - C)/I$, what alternative would be chosen?

c. By using the incremental BC ratio $B/(I + C)$, what alternative would be chosen? Discuss any difficulty associated with applying the BC ratio to this problem.

d. For a lower plant utilization factor at 5%, repeat (a), (b), and (c).

23. An inland state is presently connected to a seaport by means of a railroad system. The annual goods transported is 500 million ton-miles. The average transport charge is 20 mills per ton-mile (1 mill = $0.001). Within the next 20 years, the transport is likely to increase by 25 million ton-miles per year. It is proposed that a river flowing from the state to the seaport be improved at a cost of $600,000,000. This will make the river navigable to barges and will reduce the transport cost to 5 mills per ton-mile. The project will be financed by 80% federal funds at no interest, and 20% raised by 9% bonds issued by the state at par with annual interest payments and maturity in 20 years. There would be some side effects of the changeover as follows: (1) The railroad will be bankrupt and sold for no salvage value. The right-of-way, presently worth about $40 million, will revert to the state. (2) 3,600 employees will be out of work. The state will have to pay them welfare checks of $560 per month. (3) The reduction in the income from the taxes on the railroad will be compensated by the taxes on the barges. The interest rate is 12%.

a. Show the cash flows from the state's point of view.

b. Using the general welfare viewpoint, compute the benefit-cost ratio based on the next 20 years of operation.

c. At what average rate of transport per year will the two alternatives be equal using the federal welfare viewpoint?

24. In a community the fire insurance premium rate is $0.57 per $100 of the insured amount. There are approximately 8,000 dwellings of an average valuation of $80,000 in the town. It is estimated that the insured amount represents 70% of the value of the dwellings. The town commissioners have been advised that the fire insurance premium rate will be reduced to $0.50 per $100 if the following improvements are made to reduce fire losses.

Improvement	*Cost*	*Life*
Increase capacity of trunk water lines from pumping station	$1,500,000	40 years
Increase capacity of pumps	$ 16,000	20 years
Add supply tank to increase pressure in remote sections of town	$ 60,000	40 years
Purchase additional fire truck and related equipment	$ 100,000	20 years
Add two firemen	$ 70,000 per year	

Operation costs of added improvements are expected to be offset by decreased pumping cost brought about by the enlargement of trunk water lines. The city is in a position to increase its bonded indebtedness and can sell bonds bearing 7% interest at par. Should the above improvements be made on the basis of prospective savings in insurance premiums paid by the homeowners? What is the benefit-cost ratio? Answer: BC (7) = 1.58

25. Two sections of a city are separated by a marsh area. It is proposed to connect the sections by a four-lane highway. Plan *A* consists of a 2.4-mile highway directly over the marsh by the use of earth fill. The initial cost will be $21,500,000 and the required annual maintenance will be $28,000. Plan *B* consists of improving a 4.2-mile road skirting the swamp. The initial cost will be $10,500,000 with an annual maintenance cost of $15,000. A traffic survey estimates the traffic density to be as follows.

Years after Construction	*Traffic Density in Vehicles per Hour*
1–5	150
6–10	800
11–20	2,200

The estimated average speed under these densities would be 55, 45, and 30 mph, respectively. The traffic consists of 80% noncommercial vehicles with an operating cost of $0.16 per mile and 20% commercial vehicles with an operating cost of $0.22 per mile and $12.50 per hour. If Plan *B* is accepted, the development of the property adjacent to the highway will result in an increase in tax revenue of $400,000 per year.

a. Compare the alternatives on the basis of benefit-cost for a 20-year period and 6% interest.

b. At what traffic density are the plans equivalent for an interest rate of 6%? Answer: 284 vehicles per hour

26. The federal government is planning a hydroelectric project for a river basin. In addition to the production of electric power, this project will provided flood control, irrigation, and recreation benefits. The estimated benefits and costs that are expected to be derived from the three alternatives under consideration are listed below.

	Alternatives		
	A	*B*	*C*
Initial cost	$25,000,000	$35,000,000	$50,000,000
Annual benefits and costs:			
Power sales	$1,000,000	$1,200,000	$1,800,000
Flood-control savings	250,000	350,000	500,000
Irrigation benefits	350,000	450,000	600,000
Recreation benefits	100,000	200,000	350,000
Operating and maintenance costs	200,000	250,000	350,000

The interest rate is 5%, and the life of each project is estimated at 50 years.

a. Using incremental benefit-cost analysis, determine which project should be selected.

b. Calculate the benefit-cost ratio for each alternative. Is the best alternative selected if the alternative with the maximum benefit-cost ratio is chosen?

c. If the interest rate is 8%, what alternative will be chosen?

27. A municipal government is considering the construction of a new county stadium in its city, and three alternatives have been selected for comparison:

*B*1: Consists of the construction of a new football stadium and the use of the existing stadium for baseball games.

*B*2: Considers the construction of a new multiple-purpose stadium for both football and baseball games.

*B*3: Requires the construction of a stadium for football and a stadium for baseball.

The total construction costs are $16 million, $23 million, and $30 million, respectively, for alternatives *B*1, *B*2, and *B*3. The construction of the football stadium for *B*3 will be financed with private money and is expected to cost $16 million. The relevant costs and benefits of these alternatives are shown below.

	Alternatives		
	*B*1	*B*2	*B*3
Annual benefits and costs			
Revenue to the city from ticket sales outside the city	$900,000	$1,000,000	$600,000
Image and other intangible benefits	300,000	500,000	300,000
Maintenance costs	300,000	350,000	400,000
Increase in hotel revenues from fans living outside the city	380,000	850,000	850,000
Loss in property values due to increased traffic	90,000	270,000	410,000
Expected life (years)	30	35	35

If a 6% interest rate is used and using benefit-cost analysis from the city's point of view, which of the three proposed alternatives should be selected? Answer: Alternative *B*2

28. The highway department is considering the construction of a bridge over a narrow inlet of a lake, thereby reducing the trip on this section of highway by 20 miles. Three options are being studied: (1) keep using the present highway, (2) build a bridge, (3) build the bridge and charge a toll. If the bridge is built, the present highway will continue to be maintained, but it is expected that traffic volume on it will be significantly reduced. As a result roadside businesses will suffer a loss in revenues. Listed below are the various costs and revenues associated with each option.

	Present Highway	*Bridge with Toll*	*Bridge*
Investment cost	$ 0	$46,000,000	$46,000,000
Annual operating and maintenance cost	160,000	80,000	20,000
Annual vehicle operating costs	3,300,000	100,000	100,000
Annual time saving	—	2,000,000	2,000,000
Annual cost of accidents	500,000	10,000	10,000
Annual revenue lost by roadside businesses	—	1,000,000	1,900,000
Annual toll revenues	—	1,200,000	—

If the bridge is expected to have a life of 40 years with no salvage value, compare these three options using benefit-cost analysis. In studies of this type the state uses an interest rate of 10% and a study period of 20 years.

29. The municipal authorities want to improve the city zoo. Two proposals are under consideration. Proposal *A*2 consists of a 4-acre expansion (minor building program), and Proposal *A*3 consists of a 10-acre expansion (major building program). The estimated costs and benefits are summarized in the following table.

	*A*1 *No Expansion*	*A*2 *4 Acre Expansion*	*A*3 *10 Acre Expansion*
Construction cost			
End of year 0	$ 0	$5,000,000	$4,000,000
End of year 1	0	0	6,000,000
Development cost of convenient access			
End of year 0	50,000	100,000	90,000
End of year 1	0	0	110,000
Dispossessed landowners' losses			
End of year 0	0	350,000	700,000
Annual benefits and costs			
Souvenir sales	30,000	85,000	170,000
Operating and maintenance cost	140,000	170,000	300,000
City business benefits	100,000	400,000	650,000
Recreational benefits to public	210,000	480,000	900,000
Expected life (years)	30	30	35

Based on a benefit-cost analysis with an interest rate of 9%, which of the three alternatives should be selected? The study period is 30 years. Assume there is no charge for admittance to the zoo.

30. A state highway department is considering the location of a new rural highway. Two possible locations are under consideration. For each location, 2-lane vs. 4-lane designs are being studied. It is estimated that 2 traffic lanes will be adequate for the next 10 years, but thereafter 4 traffic lanes will eventually be required. Adding 2 more lanes after 10 years will cost 1.8 times current 2-lane construction cost. The following data have been collected for the two locations. In comparing the economic desirability of the highway location, an interest rate of 6% is to be used. For an analysis period of 20 years with zero salvage value, determine which location should be selected by comparing the benefit-cost ratios.

	Location A		*Location B*	
Road Characteristics	*A1 2-Lane*	*A2 4-Lane*	*B1 2-Lane*	*B2 4-Lane*
Distance (miles	12	12	10.5	10.5
Total construction costs	$20,100,000	$31,500,000	$22,500,000	$42,000,000
Annual maintenance and operating costs per mile	$3,000	$3,600	$2,850	$3,300
Resurfacing cost per mile, end of 10th year	$156,000	$249,000	$156,000	$249,000

Traffic Characteristics	*Passenger Car*	*Commercial Vehicles*
Average daily traffic, (all routes) for the first year	3,500	500
Annual growth per year	160	35
Equivalent uniform speed	55 mph	50 mph
Value of travel time per vehicle-hour	$4.00	$12.50
Vehicle operating cost per mile	$0.16	$0.30

31. The parks and recreation agency in a state government is considering the development of one recreation area from land presently owned. Currently, two locations are under consideration, one in the mountains and the other on the coast. It is estimated that the mountain location will require an initial investment of $8,500,000 with annual maintenance costs of $120,000. Because of its excellent access it will generate an additional $100,000 in park-use fees with respect to the coastal location. The coastal location is expected to have an initial investment and annual maintenance costs of $5,000,000, and $80,000, respectively. Annual profits to park-operators (private companies allowed to operate within the park) are estimated to be $300,000 and $200,000 at the mountain and coastal location, respectively. Recreational benefits and general disadvantages associated with each project are estimated to be $700,000 and $40,000 at the mountain location, and $500,000 and $60,000 at the coastal location. Assuming that the park will be maintained indefinitely, use the *BC* method to determine which location should be selected. The interest rate is 10%. Answer: Mountain location

32. Four flood-control projects are being considered. In summary, the investment required and the annual benefits and costs resulting from these investments are

	Alternatives			
	*R*1	*R*2	*R*3	*R*4
Annual benefits and costs				
Operating and maintenance	$ 6,500,000	$ 8,250,000	$ 7,000,000	$ 5,900,000
Savings to the government	3,500,000	3,500,000	3,500,000	3,500,000
Benefits to the public	10,000,000	13,200,000	11,200,000	8,000,000
Disbenefits to the public	3,250,000	4,300,000	3,600,000	2,300,000
Initial investment	25,000,000	33,000,000	28,000,000	18,000,000
Expected life (years)	40	40	45	45

a. If an interest rate of 5% is used, determine the benefit-cost ratio for each alternative.

b. Determine which alternative should be chosen, using the interest rate given in part (a) and incremental benefit-cost ratio analysis. Assume that the Do Nothing alternative is not considered.

33. A suburban area has been annexed by a city, which henceforth is to supply water service to the area. The prospective growth of the suburb has been estimated, and on this basis five alternative plans for meeting the projected demand have been proposed. The pumping costs which are the same for each plan are as follows.

Years	*Pumping Cost per Year*
1–14	$282,500
15–25	$316,400
26–40	$350,300

Listed below are the current costs for the various sizes of pipe.

Pipe Diameter (Inches)	First Cost
10	$3,400,000
14	$4,500,000
18	$5,100,000

From the standpoint of water-carrying capacity, a 14-inch pipe is equivalent to two 10-inch pipes and an 18-inch pipe to three 10-inch pipes. On the basis of incremental benefit-cost analysis using constant dollars, compare over the next 40 years the following plans if the inflation-free rate is 9% and if used pipeline is valued in constant dollars at 15% of its original cost. The benefits are considered to be identical for each plan. At the end of 40 years all pipes in place will be sold.

Years from Now	Plan A	Plan B	Plan C	Plan D	Plan E
0	Install 10″	Install 10″	Install 14″	Insall 14″	Install 18″
14	Install 14″ Sell 10″	Install 10″	—	—	—
25	Install 18″ Sell 14″	Install 10″	Install 10″	Install 18″ Sell 14″	—

34. Five mutually exclusive public projects from a federal agency have been submitted for evaluation.

Project	Initial Investment	Annual Expenses	Annual Benefits
S1	$83,000,000	$16,430,000	$49,750,000
S2	89,000,000	18,690,000	50,789,000
S3	57,000,000	9,120,000	26,101,000
S4	86,000,000	17,200,000	51,885,000
S5	51,000,000	8,160,000	25,591,000

The life of all five projects is estimated to be 18 years. Compare the projects using incremental benefit-cost analysis for multiple alternatives and an interest rate of 9%. Which of the five projects should be funded?

35. The United States government is planning to develop advanced power-generation systems so that its coal and uranium resources can be utilized most effectively. The six cases listed below appear to be the most promising ways of using available resources and meeting the need for a clean environment.

	Base	I	II	III	IV	V	VI
		Cases					
Energy resources	Petroleum	Uranium	Coal	Coal	Coal	D_2O, Li	Solar
Power system	Present technology	Fast breeder-steam	MHD, fuel cell com-bined GT/ST*		Fuel cell	Fusion	Solar cells

*Comb. GT/ST, combined gas-turbine/steam-turbine.

Based upon the advanced power-generation concepts above, the following economic gains for each case (compared to present technology) are identified.

ECONOMIC BENEFITS OF ADVANCED POWER-GENERATION METHODS (PERIOD, 2000-2040)

	Base	Case I	Case II	Case III	Case IV	Case V	Case VI
	PW*	PW	PW	PW	PW	PW	PW
Total capital investment	98	114	99	110	121	117	92
Operating cost	212	149	190	158	132	132	196
Benefits	320	340	400	328	335	330	360

*Present worth at 7% in units of $\$10^9$.

For this study, the capital costs of power-generating units were based upon commercial operation after a sufficient R&D period. All present-worth calculations were made at an interest rate of 7%. If benefit-cost analysis is to be considered as a guide to decision making, which project has the highest funding preference? Assume that there is no Do Nothing alternative.

36. For the various alternatives described in Problem 35 consider the following. In the evaluation of advanced power systems the federal government believes that it is necessary to examine the contributions of each power system to the problem of environmental pollution. The level of emissions and water use can vary for the same power demand, depending upon the type of fuel utilized and the efficiency of power conversion. Thus, it is important to consider the cost of these systems and their environmental effects.

	Base	Case I	Case II	Case III	Case IV	Case V	Case VI
Cumulative							
Total air pollutants (10^9 lb)	4,530	3,080	3,249	3,249	3,080	1,945	1,945
Total water make-up (10^{12} gal)	249	254	194	194	161	253	215
Development cost* ($\$10^{10}$)	0	4.3	2.0	1.7	1.1	4.7	3.0

**Present worth at 7%.*

Show how you can use the cost-effectiveness approach to consider these various factors.

37. The federal government is presently considering a number of proposals for improving the speed of mail handling in large urban post offices. The measure of effectiveness to be used to evaluate these mail-handling systems is the volume of mail processed per day. The cost of purchasing and installing these various systems, their resulting savings, and their effectiveness are as follows.

System	*Initial Cost in Millions of $*	*Annual Savings in Millions of $*	*Effectiveness in Millions of Letters Processed per Day*
A	$1,200	$100	5
B	2,000	140	8
C	2,600	230	12
D	4,000	340	13
E	5,100	500	14

If the interest rate is 12% and the life of each system is estimated to be 10 years, plot the cost and effectiveness relationship for each proposal. Which alternatives can be dropped from further consideration? How would you decide between the remaining alternatives?

38. A firm is considering the purchase of an automated storage and retrieval system (AS/RS) for its major warehouse. Five AS/RS designs have been submitted by different manufacturers, and they have the following costs and material handling capacities.

Design	Initial Cost in Millions of $	Maintenance Cost ($ milions/year)	Salvage Value ($millions)	Expected Life (years)	Effectiveness Number of Operations/Week
J1	68	0.8	1.2	6	6,900
J2	95	1.3	2.0	7	10,600
J3	125	0.4	4.4	8	16,000
J4	120	2.7	12.0	7	18,000
J5	126	1.6	17.5	9	21,300

For an interest rate of 7%, plot the total cost versus the effectiveness relationship for each design. Determine graphically if any designs can be rejected. How should the best design be selected from the remaining candidates?

10

Break-Even and Optimization Analysis

In some situations encountered in engineering economy, the cost incurred by an alternative may be a function of a decision variable. When two or more alternatives are a function of the same variable, it may be desirable to find the value of the variable that will result in equal cost for the alternatives being considered. This value is known as the *break-even point.*

In other situations, a decision variable may lead to increasing and decreasing costs, the total of which is a minimum at a specific value of the variable. This value is known as the *optimum point* for the alternative being considered. When two or more alternatives depend upon the same variable, they may be compared equivalently after optimization.

10.1 BREAK-EVEN ANALYSIS, TWO ALTERNATIVES

When the cost of two alternatives is affected by a common independent decision variable, there may exist a value of the variable for which the two alternatives will incur equal cost. The costs of each alternative can be expressed as functions of the common variable and will be of the form

$$TC_{A1} = f_1(x) \quad \text{and} \quad TC_{A2} = f_2(x) \tag{10.1}$$

where

TC_{A1} = a specified total cost per time period, per project, or per item applicable to Alternative 1;

TC_{A2} = a specified total cost per time period, per project, or per item applicable to Alternative 2;

x = a common independent decision variable affecting Alternative 1 and Alternative 2.

Solution for the value of x resulting in equal cost for Alternatives 1 and 2 is accomplished by setting the total-cost functions equal: $TC_{A1} = TC_{A2}$. Therefore,

$$f_1(x) = f_2(x) \tag{10.2}$$

which may be solved for x. The resulting value of x yields equal cost for the alternatives being considered and is designated the break-even point.

10.1.1 An Equipment-Selection Example

Assume that a 20-hp pumping unit is needed to remove water from a tunnel. The number of hours that the equipment will operate per year depends upon the rainfall and is, therefore, uncertain. The pump unit will be needed for a period of 4 years.

Two alternatives are under consideration. Alternative 1 calls for the provision of a power supply and the purchase of an electric unit, at a total cost of \$1,800. The salvage value of this equipment at the end of the 4-year period is estimated at \$600. The cost of electric power per hour of operation is estimated at \$1.10, maintenance is estimated at \$360 per year, and the interest rate is 12%. No operating personnel will be needed, since the equipment is automatic. Let

TC_{A1} = annual equivalent total cost of Alternative 1;

$CR(i)_{A1}$ = annual equivalent cost of capital recovery with return

$$= (\$1{,}800 - \$600)(\overset{A/P,12,4}{0.3292}) + \$600(0.12) = \$467;$$

C_M = annual maintenance cost = \$360;

C_P = power cost per hour of operation = \$1.10;

H = number of hours of operation per year.

Then

$$TC_{A1} = CR(i)_{A1} + C_M + C_P H.$$

Alternative 2 calls for the purchase of a gasoline-powered pumping unit at a cost of $550. This equipment will have no salvage value at the end of the 4 year period. The cost of fuel and oil per hour of operation is estimated at $0.60, maintenance is estimated at $0.35 per hour of operation, and the cost of operator attention chargeable to the equipment when it runs is $1.40 per hour. Let

TC_{A2} = annual equivalent total cost of Alternative 2;

$CR(i)_{A2}$ = annual equivalent cost of capital recovery with return

$$= \$550(\overset{A/P,12,4}{0.3292}) = \$181;$$

C_F = hourly cost of fuel, maintenance, and operator attention

$$= \$0.60 + \$1.40 + \$0.35 = \$2.35;$$

H = number of hours of operation per year.

Then

$$TC_{A2} = CR(i)_{A2} + C_F H.$$

There is a value of H for which the two alternatives will incur equal cost. This value may be found by setting $TC_{A1} = TC_{A2}$ and solving for H as follows:

$$CR(i)_{A1} + C_M + C_P H = CR(i)_{A2} + C_F H$$

$$H = \frac{CR(i)_{A1} - [CR(i)_{A2} + C_M]}{C_P - C_F}.$$

Substituting the estimated values gives

$$H = \frac{\$181 - (\$467 + 360)}{\$1.10 - \$2.35} = 517 \text{ hours.}$$

The annual equivalent total cost can be shown to be equal for the two alternatives at the break-even point as follows:

$$TC_{A1} = TC_{A2}$$

$$CR(i)_{A1} + C_M + C_P H = CR(i)_{A2} + C_F H$$

or

$$\$467 + \$360 + 517(\$1.10) = \$181 + 517(\$2.35)$$

$$\$1,396 = \$1,396.$$

For the estimates given, the cost of the two alternatives is found to be equal for 517 hours of operation per year. The annual equivalent total cost for each alternative, as a function of the number of hours of operation per year, is shown graphically in Figure 10.1.

If the equipment is used less than 517 hours per year, selection of the gasoline powered unit is most economical; for more than 517 hours of operation, the electric unit is most economical. The difference in annual equivalent cost between the alternatives may be calculated for any number of hours of operation.

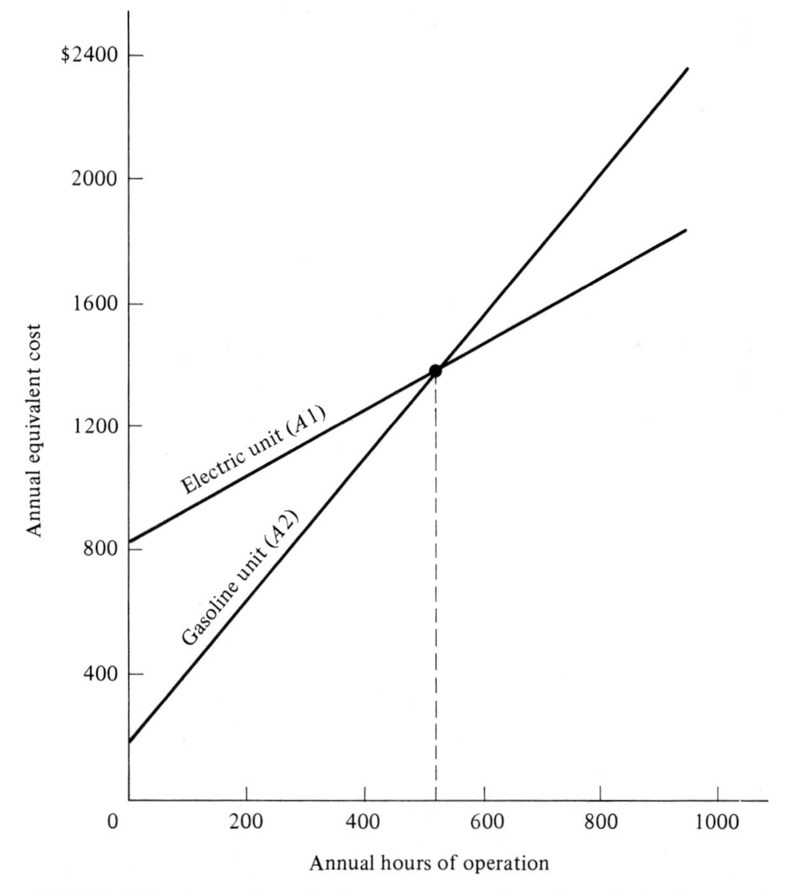

FIGURE 10.1. Annual equivalent cost as a function of the hours of operation per year.

For example, suppose the equipment is to be operated 300 hours per year. Then

$$TC_{A1-A2} = TC_{A1} - TC_{A2}$$

$$= CR(i)_{A1} + C_M + C_P H - [CR(i)_{A2} + C_F H].$$

Substituting the stated values gives

$$TC_{A1-A2} = \$467 + \$360 + 300(\$1.10) - \$181 - 300(\$2.35)$$

$$= \$271.$$

10.1.2 An Equipment-Replacement Example

As an application of break-even analysis to the replacement problem, suppose that a decision maker is interested in the *comparative use value* of Machine A (a defender) in comparison with Machine B (a challenger). Assume that Machine A will serve satisfactorily for 6 more years and be worth $2,000 for salvage at that time. Machine B is being offered for $24,000 and has an anticipated service of life of 10 years and an estimated salvage value at that time of $3,000. The attractiveness of the challenger is attributed to an annual saving in maintenance and operating cost from the $7,000 now being incurred, to $4,000 if the replacement is made. The MARR is 20%.

The break-even point, or comparative use value, of Machine A in comparison with Machine B is found by equating the annual equivalent costs, with x being the comparative use value found as follows:

$$(x - \$2{,}000)(\overset{A/P,20,6}{0.3007}) + \$2{,}000(0.20) + \$7{,}000$$

$$= (\$24{,}000 - \$3{,}000)(\overset{A/P,20,10}{0.2385}) + \$3{,}000(0.20) + \$4{,}000$$

Solving for x

$$0.3007x - \$601.40 + \$400 + \$7{,}000 = \$5{,}008.50 + \$600 + \$4{,}000$$

$$x = \$9{,}345.$$

Machine A should be replaced if it can be disposed of for more than $9,345 in the used-equipment market. Because of its current condition and the technology it embraces in comparison with a challenger, a defender will have a break-even use value which is helpful in making the replacement decision. Of course, the unequal life considerations discussed in Chapter 8 should be part of the final decision.

10.2 BREAK-EVEN ANALYSIS, MULTIPLE ALTERNATIVES

In the previous examples, break-even analysis was applied where only two alternatives confronted the decision maker. Break-even analysis can easily be extended to multiple alternatives.

As an example, consider an architectural engineering firm that has been asked to prepare preliminary plans for the construction of a one-story commercial building. To arrive at some quantitative basis for recommending a type of construction, the engineering firm develops fixed and variable cost data. These data, given below, are assumed to represent the cost of construction and operation for a building enclosing between 2,000 and 6,000 gross square feet.

Steel:

First cost per square foot	\$72
Annual maintenance	\$14,000
Annual HVAC per square foot	\$3.00
Estimated life in years	20
Estimated sale value is 80% of first cost.	

Concrete block:

First cost per square foot	\$76
Annual maintenance	\$9,000
Annual HVAC per square foot	\$3.40
Estimated life in years	20
Estimated sale value is 100% of first cost.	

Frame and brick:

First cost per square foot	\$81
Annual maintenance	\$6,000
Annual HVAC per square foot	\$3.90
Estimated life in years	20
Estimated sale value is 110% of first cost.	

The total cost for each type of construction will be a function of the number of gross square feet enclosed by the building, A. For an interest rate of 18%, the total cost for each alternative construction type is as follows:

Steel:

$$TC = \$72(A)(0.2)(\overset{A/P,18,20}{0.1868}) + \$72(A)(0.80)(0.18) + \$3.00(A) + \$14,000$$

$$= \$16.06(A) + \$14,000.$$

Concrete block:

$$TC = \$76(A)(0.0)(\overset{A/P,18,20}{0.1868}) + \$76(A)(1.00)(0.18) + \$3.40(A) + \$9{,}000$$

$$= \$17.08(A) + \$9{,}000.$$

Frame and brick:

$$TC = \$81(A)(-0.10)(\overset{A/P,18,20}{0.1868}) + \$81(A)(1.10)(0.18) + \$3.90(A) + \$6{,}000$$

$$= \$18.42(A) + \$6{,}000.$$

Solution for each of the break-even points may be done by considering the alternatives in pairs. For steel vs. concrete

$$\$16.06(A) + \$14{,}000 = \$17.08(A) + \$9{,}000$$

$$A = 4{,}900 \text{ ft}^2.$$

For steel vs. frame

$$\$16.06(A) + \$14{,}000 = \$18.42(A) + \$6{,}000$$

$$A = 3{,}390 \text{ ft}^2.$$

And, for concrete vs. frame

$$\$17.08(A) + \$9{,}000 = \$18.42(A) + \$6{,}000$$

$$A = 2{,}240 \text{ ft}^2.$$

These break-even points are shown graphically in Figure 10.2. Note from the graph that the break-even point between steel and frame is not active in the decision. It is dominated by two break-even points: one between frame and concrete and the other between concrete and steel.

Suppose that a client is considering a building enclosing 40 feet by 100 feet. In this case, the engineering firm would recommend construction from concrete and brick. However, if the client required a building of more than 4,900 square feet, it might be most economical to use steel and brick for the basic structure. For a building of less than 2,240 square feet, construction of frame and brick would probably be most economical.

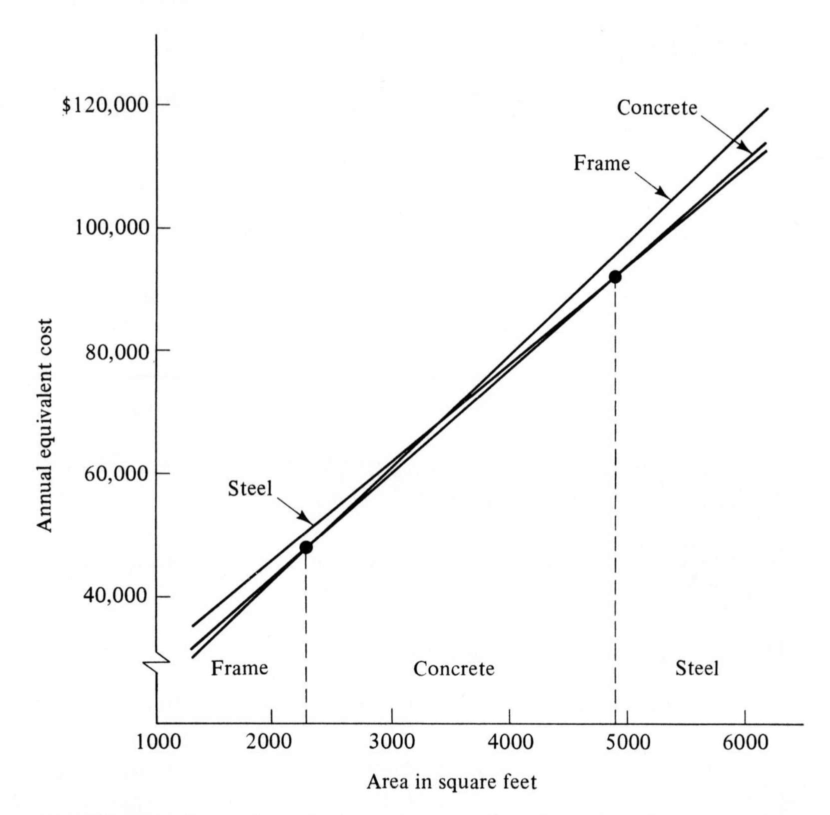

FIGURE 10.2. Annual equivalent cost as a function of enclosed area for three construction alternatives.

10.3 OPTIMIZATION ANALYSIS

An alternative may have two or more cost components that are modified differently by a common independent variable. Certain cost components may vary directly with an increase in the value of the variable while others may vary inversely. When the total cost of an alternative is a function of increasing and decreasing cost components, usually a value exists for the common variable that will result in a minimum cost for the alternative.

The general solution of the situation described may be demonstrated for an increasing-cost component and a decreasing-cost component as follows:

$$TC = Ax + \frac{B}{x} + C \tag{10.3}$$

where

TC = a specified total cost per time period, per project, etc.;

x = a common variable;

A, B, and C = constants.

Taking the first derivative, equating the result to zero, and solving for x gives

$$\frac{dTC}{dx} = A - \frac{B}{x^2} = 0$$

$$x = \sqrt{\frac{B}{A}}. \qquad (10.4)$$

The value for x found in this manner will be optimum and is, therefore, designated the minimum-cost point.

10.3.1 Bridge Design Optimization

As an example application of optimization analysis where one alternative is proposed, suppose that a double-walkway pedestrian bridge is to be constructed. The crossing is 600 feet and the superstructure design being considered will have a weight per foot of $W = 12(S) + 600$, where S is the span between piers. It is estimated that the superstructure will be erected at a cost of \$0.52 per pound. Piers and abutments (one abutment at each end) are estimated to cost \$135,000 each.

Before engaging in detail design, it is desirable to determine the optimum (minimum-cost) pier spacing. The general relationship is that of an inverse relationship between the cost of superstructure and the number of piers. As the number of piers increases, the cost of superstructure decreases. Conversely, as the number of piers decreases, the amount and cost of the superstructure increases. Pier cost for the bridge is directly proportional to the number required. Thus, this design situation involves increasing- and decreasing-cost components, the sum of which will be a minimum for a certain number of piers.

The total cost for the superstructure and piers, including the two abutments, as a function of the span between piers, is

$$TC = [12(S) + 600](\$0.52)(600) + \left(\frac{600}{S} + 1\right)\$135{,}000$$

$$= \$3{,}744\,S + \frac{\$81{,}000{,}000}{S} + \$322{,}200.$$

The minimum-cost span between piers can be found mathematically by differential calculus as follows:

$$\frac{dTC}{dS} = \$3{,}744 - \frac{\$81{,}000{,}000}{S^2} = 0$$

$$S = \sqrt{\frac{\$81{,}000{,}000}{\$3{,}744}} = 147 \text{ ft.}$$

Accordingly, four spans should be specified to minimize the total cost of the bridge. The number of piers would be specified as three in addition to the two required abutments.

Total cost for the bridge is tabulated as a function of the number of piers in Table 10.1. The minimum-cost design is also shown graphically in Figure 10.3. In each case, the minimum number of piers is three. Note the nature of the increasing- and decreasing-cost components.

TABLE 10.1.
BRIDGE COST AS A FUNCTION OF THE NUMBER OF PIERS

Piers	Cost of Piers	Abutment Cost	Superstructure Cost	Total Cost
1	$135,000	$270,000	$1,310,400	$1,715,400
2	270,000	270,000	936,000	1,476,000
3	405,000	270,000	748,800	1,423,800
4	540,000	270,000	636,480	1,446,480

10.3.2 Minimum-Cost Life

Maintenance and operating costs generally increase with an increase in the age of equipment. This rising trend is offset by a decrease in the annual equivalent cost of capital recovery with return. Thus, for some age, there will be a minimum total cost or a minimum-cost life.

If the time value of money is neglected, the average annual total cost of an asset with constantly increasing operating and maintenance cost is as follows:

$$TC = \frac{P}{n} + M + (n - 1)\frac{m}{2} \tag{10.5}$$

where

TC = average annual total cost;

P = first cost of the asset;

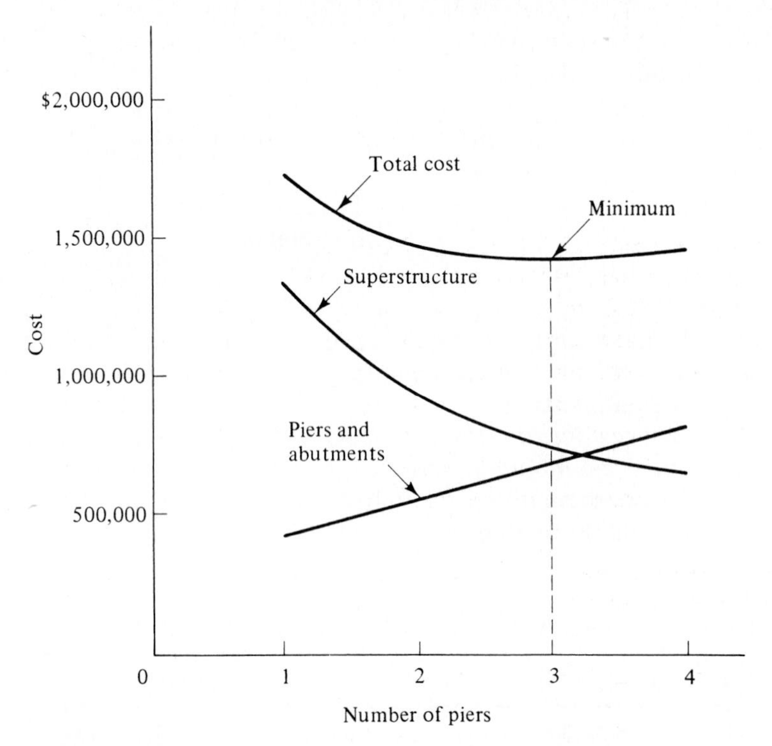

FIGURE 10.3. Bridge cost as a function of the number of piers.

M = annual constant portion of operating cost (equal to first-year operation cost of which maintenance is a part);

m = amount by which maintenance costs increase each year;

n = life of the asset in years.

The minimum-cost or optimum life can be found mathematically by differential calculus as follows:

$$\frac{dTC}{dn} = -\frac{P}{n^2} + \frac{m}{2} = 0$$

$$n = \sqrt{\frac{2P}{m}}. \tag{10.6}$$

As an example application, consider an asset with a first cost of $8,000, a salvage value of zero at any age, and with maintenance costs increasing by $1,000 after the first year of service. The minimum-cost life is

$$n = \sqrt{\frac{2(\$8{,}000)}{\$1{,}000}} = 4 \text{ years.}$$

This situation can also be studied by tabulating the cost components and the average total cost as shown in Table 10.2. The minimum total cost occurs at a life of four years and is $3,500. Figure 10.4. illustrates the nature of the increasing- and decreasing-cost components and the economic life of 4 years.

TABLE 10.2.
ECONOMIC HISTORY OF AN ASSET WITH CONSTANTLY INCREASING MAINTENANCE COSTS

End of Year	Maintenance Cost at End of Year	Cumulative Maintenance Cost	Average Maintenance Cost	Average Capital Cost	Average Total Cost
		ΣB	$C \div A$	$\$8{,}000 \div A$	$D + E$
A	*B*	*C*	*D*	*E*	*F*
1	$ 0	$ 0	$ 0	$8,000	$8,000
2	1,000	1,000	500	4,000	4,500
3	2,000	3,000	1,000	2,667	3,667
4	3,000	6,000	1,500	2,000	3,500
5	4,000	10,000	2,000	1,600	3,600
6	5,000	15,000	2,500	1,333	3,833

10.4 OPTIMIZATION ANALYSIS, MULTIPLE ALTERNATIVES

A classical example of optimization analysis is illustrated by the increasing- and decreasing-cost components involved in the choice of the cross-sectional area of an electrical conductor. Since resistance is inversely proportional to the size of the conductor, it is evident that the cost of power loss will decrease with increased conductor size. However, as the size of the conductor increases, an increased investment cost will be incurred. For a given conductor cross section, the sum of the two cost components will be a minimum.

As an illustration of the application of minimum-cost analysis where two design alternatives exist, consider the following situation. The daily electrical load to be transmitted in a power plant is 1,900 amperes for 365 days per year.

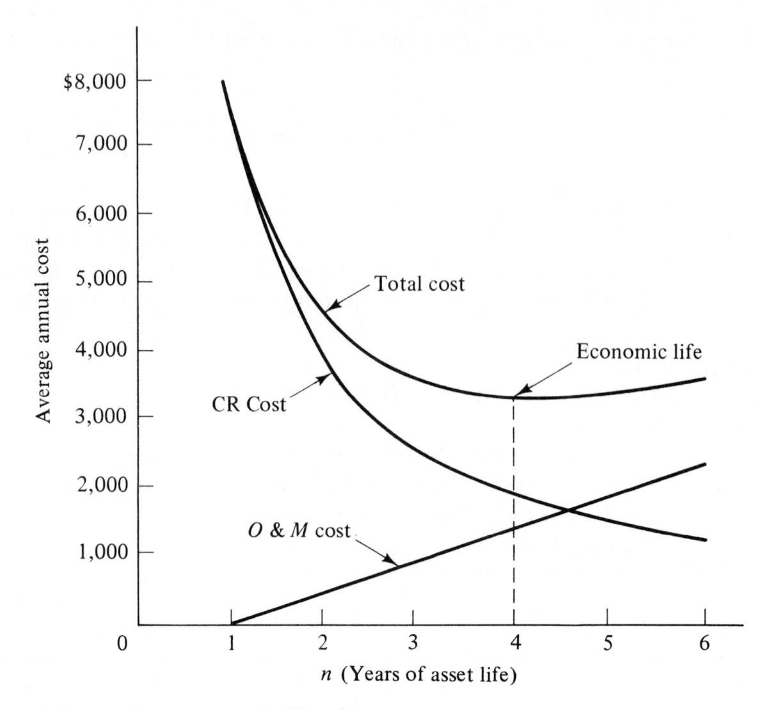

FIGURE 10.4. Economic life of an asset.

Two conductor materials are under consideration: copper and aluminum. Design data for the competing materials are given in Table 10.3.

The energy loss in kilowatthours in a conductor due to resistance is equal to I^2R times the number of hours divided by 1,000, where I is the current flow in amperes and R is the resistance in the conductor in ohms. The electrical resistance is inversely proportional to the area of the cross section. Lost energy is valued at \$0.052 per kilowatthour. The interest rate is 15%.

TABLE 10.3.
DESIGN DATA FOR TWO CONDUCTOR ALTERNATIVES

	Material	
Design Data	Copper	Aluminum
Length	120 ft	120 ft
Installed cost	\$200 + 1.10/lb	\$200 + 1.00/lb
Estimated life	10 years	10 years
Salvage value	\$0.62/lb	\$0.56/lb
Electrical resistance of conductor 120 ft for 1-in^2 cross section	0.000982 ohms	0.001498 ohms
Density of material	555 lb/ft^3	168 lb/ft^3

The I^2R loss cost in dollars per year for the copper alternative is

$$(1{,}900)^2 \frac{(0.000982)}{A} \frac{(24 \times 365)}{1{,}000} (\$0.052) = \frac{\$1{,}615}{A}.$$

Its weight in pounds is

$$\frac{(120)(12)(555)A}{1728} = 462.50A.$$

And the *CR* cost in dollars per year is

$$[\$200 + (\$1.10 - \$0.62)(462.50)A](\overset{A/P,15,10}{0.1993}) + \$0.62(462.50)A(0.15)$$

$$= \$39.86 + \$87.25A.$$

The annual equivalent total cost for copper is

$$TC = \$39.86 + \$87.25A + \frac{\$1{,}615}{A}.$$

Optimization to find the minimum-cost cross section may be done mathematically as follows:

$$\frac{dTC}{dA} = \$87.25 - \frac{\$1{,}615}{A^2} = 0$$

$$A = \sqrt{\frac{\$1{,}615}{\$87.25}} = 4.30 \text{ in}^2.$$

And the minimum annual equivalent total cost for copper is

$$TC = \$39.86 + \$87.25(4.30) + \frac{\$1{,}615}{4.30} = \$791.$$

This same analysis is now applied to the aluminum alternative. The I^2R loss cost in dollars per year is

$$(1{,}900)^2 \frac{(0.001498)}{A} \frac{(24 \times 365)}{1{,}000} (\$0.052) = \frac{\$2{,}463}{A}.$$

Its weight in pounds is

$$\frac{(120)(12)(168)A}{1{,}728} = 140A.$$

And the *CR* cost in dollars per year is

$$[\$200 + (\$1.00 - \$0.56)(140)A](\overset{A/P,15,10}{0.1993})$$

$$+ \$0.56(140)A(0.15)$$

$$= \$39.86 + \$24.03A.$$

The annual equivalent total cost for aluminum is

$$TC = \$39.86 + \$24.03A + \frac{\$2{,}463}{A}.$$

As before, optimization to find the minimum-cost cross section may be done mathematically as follows:

$$\frac{dTC}{dA} = \$24.03 - \frac{\$2{,}463}{A^2} = 0$$

$$A = \sqrt{\frac{\$2{,}463}{\$24.03}} = 10.12 \text{ in}^2.$$

And the minimum annual equivalent total cost for aluminum is

$$TC = \$39.86 + \$24.03(10.12) + \frac{\$2{,}463}{10.12} = \$526.$$

The calculations above can be used by designer to choose the least-cost alternative material and also to specify the optimum value for the design variable. For this example, aluminum should be chosen for the conductor and its cross-sectional area should be as close to 10.12 square inches as possible. Figure 10.5 illustrates the nature of this design decision.

The extension of optimization analysis to cases where there are more than two alternatives follows the same procedure. Suppose three plans for constructing a large earth-filled dam are under consideration by a prime contractor. Plan I will require a greater amount of labor costs than either of the other two plans, while Plan III requires the largest expenditure for construction equipment. The

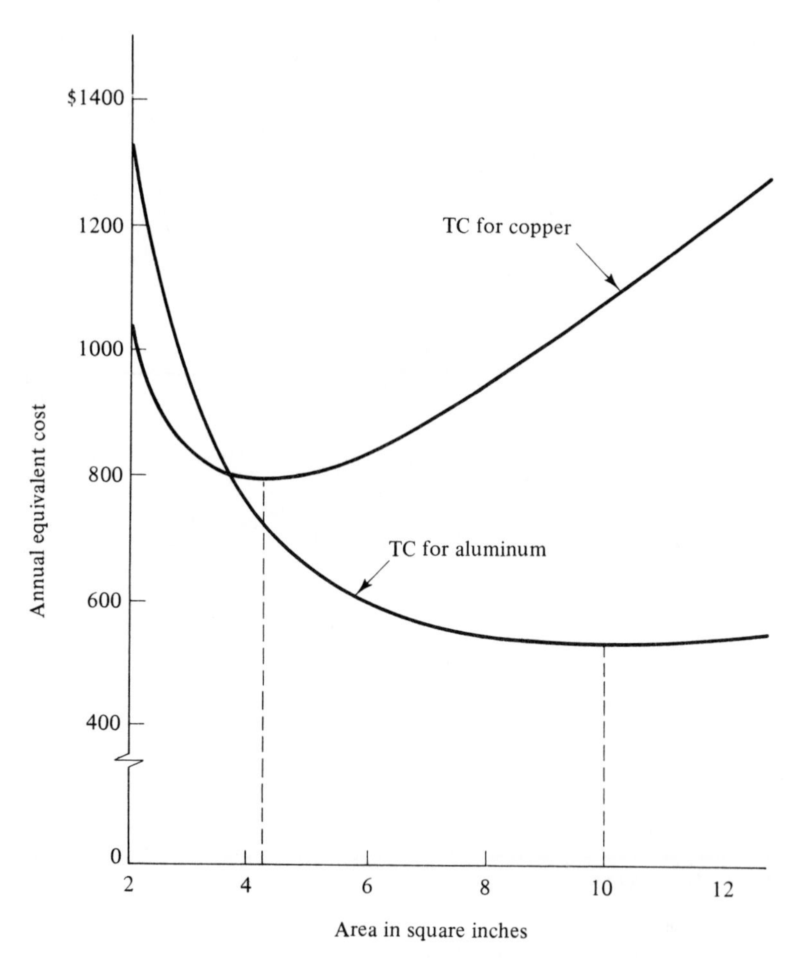

FIGURE 10.5. Aluminum vs. copper in the design of an electrical conductor.

total equivalent annual cost for each plan is described in Figure 10.6 as a function of the time required to complete the project.

From the annual-cost curves in Figure 10.6 it is seen that the overall minimum-cost plan ($12 million per year) would be Plan III if the job were completed in 3.8 years. If it were desired to complete the project in less than 3.3 years, the least expensive plan would be Plan I. The least-cost plan if the project were completed after 4.1 years would be Plan II. Any expected completion date between 3.3 years and 4.1 years would favor the selection of Plan III. This type of optimization analysis for multiple alternatives provides information that can be extremely helpful in the process of selecting the preferred alternative.

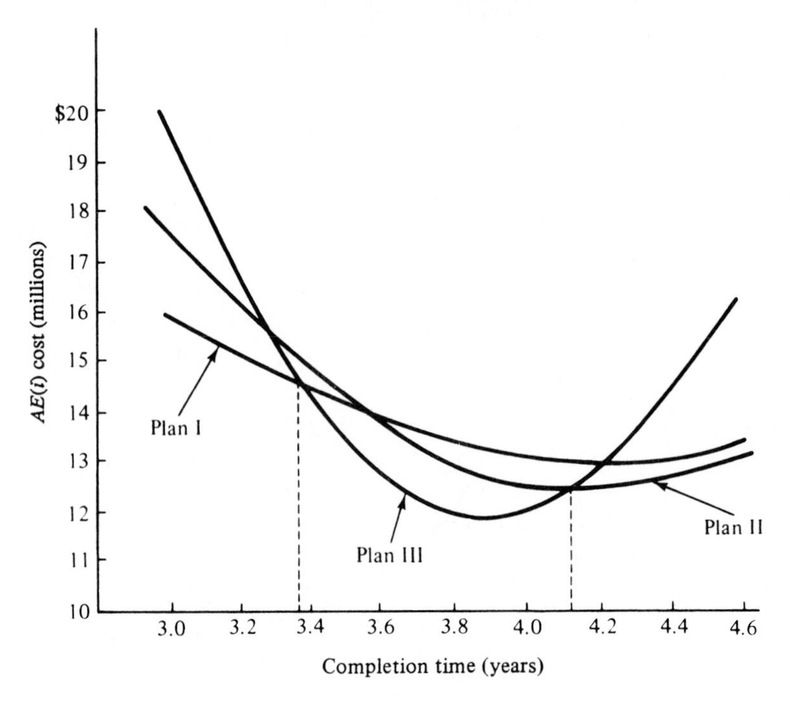

FIGURE 10.6. Minimum costs for three alternatives.

KEY POINTS

- In some evaluation situations, the economic measure for two or more alternatives will be a function of a common decision variable.
- Break-even and optimization analysis may be applied when a common desicion variable affects two or more alternatives simultaneously.
- When alternatives involve increasing- and decreasing-cost components, they may be compared equivalently only after optimization.
- Some common examples of break-even analysis are in selecting equipment, determining comparative use value, and deciding upon a type of construction.
- Some common examples of optimization analysis are in bridge design, conductor design, and selecting from among construction or other plans.

PROBLEMS

1. In considering the purchase of an automobile with air conditioning, it is estimated that when the air conditioning is off, the mileage will be 24.0 miles per gallon, and with air conditioning on, it will be 21.1 miles per gallon. The air conditioning will add $600 to the first cost of the automobile and will increase its trade-in value by $300 after a service life of 5 years. The air conditioning will be used 20% of the time and it will add $60 per year to the maintenance of the automobile. The benefit of having air conditioning when needed is considered to be worth $0.04 per mile. If gasoline costs $1.30 per gallon and if the interest rate is 8%, how many miles per year must be driven for the air conditioner to pay for itself?
2. Two brands of a protective coating are being considered. Brand *A* costs $15.80 per gallon. Past experience with Brand *A* has revealed that it will cover 350 square feet of surface per gallon, will give satisfactory service for 3 years, and can be applied at the rate of 70 square feet per hour. Brand *B*, which costs $20.40 per gallon, is estimated to cover 400 square feet of surface per gallon and can be applied at the rate of 80 square feet per hour. The wage rate of the worker is $17 per hour. If an interest rate of 10% is used, how long should Brand *B* last to provide service at equal cost with that provided by Brand *A*? Answer: 2.75 years
3. Machine *A* costs $2,000, has zero salvage value at any time, and labor cost per unit of product on it is $11.40. Machine *B* costs $36,000, has zero salvage value at any time, and labor cost per unit of product on it is $8.30. Neither machine can be used except to make the product in question. The interest rate is 16%. If the annual rate of production is 3,000 units, in how many years will the two machines break even in annual equivalent costs?
4. An electronic products manufacturer is considering two methods for producing a special circuit board. The board can be hand wired at an estimated cost of $2.40 per unit and with an annual equipment cost of $300. A printed equivalent of the required circuit can be produced with an investment of $9,000 in printed-circuit processing equipment, which will have an expected life of 9 years and a salvage value of $1,000. It is estimated that labor cost will be $0.52 per unit and that the processing equipment will cost $400 per year to maintain. If all other costs are assumed equal, and if the MARR is 18%, how many circuit boards must be produced each year for the two methods to break even in annual equivalent cost? Answer: 1,137
5. Three years ago an asset was placed in service with an anticipated life of 8 years and a salvage value then of $4,000. These estimates are still valid. A

challenger is being offered for $52,000 with an anticipated salvage value of $7,000 and a life of 8 years. Maintenance costs are estimated to be $8,000 less per year if the existing asset is replaced. If the MARR is 16%, find the break-even use value of the existing asset.

6. A certain area can be irrigated by piping water from a nearby river. Two competing installations are being considered, for which the following engineering and cost data apply:

	6-inch System	*8-inch System*
Size motor required	25 hp	10 hp
Energy cost per hour of operation	$ 1.25	$ 0.55
Cost of motor installed	$ 800	$ 500
Cost of pipe and fittings	$ 2,900	$3,700
Salvage value at end of 10 years	$ 360	$ 280

On the basis of a 10-year life with an interest rate of 12%, determine the number of hours of operation per year for which the two systems will break even.

7. A company may furnish a car for use by a service man for transportation between its properties, or the company may pay the service man for the use of his car at a rate of $0.25 per kilometer for this purpose. The following estimated data apply to company-furnished cars: A car costs $10,000 and has a life of 4 years and a trade-in value of $2,500 at the end of that time. Monthly storage cost for the car is $30 and the cost of fuel, tires, and maintenance is $0.12 per kilometer. How many kilometers must a service man travel annually by car for the cost of the two methods of providing transportation to be equal if the interest rate is 15%? Answer: 25,863

8. An engineering consulting firm can purchase a small electronic computer for $30,000. It is estimated that the life and salvage value of the computer will be 6 years and $4,000, respectively. Operating expenses are estimated to be $50 per day, and maintenance will be performed under contract for $3,000 per year. As an alternative, sufficient computer time can be rented at an average cost of $130 per day. If the interest rate is 12%, how many days per year must the computer be needed to justify its purchase? Answer: 123

9. The cost of constructing a motel, including materials, labor, taxes, and carrying charges during construction and other miscellaneous items, is $1,100,000. The land on which the motel is located cost $300,000. Furniture and furnishings cost $200,000. Working capital of 30 days' gross income at 100% capacity is required. The investment in the furniture and furnishings should be recovered in 7 years and the investment in the motel should be recovered in 25 years. The land on which the motel is built is considered not to depreciate in value. If the motel operates at 100% capacity, the gross annual income wll be $2,000 per day for 365 days. The motel has fixed operating expenses (exclusive of capital recovery and interest) amounting to $115,000 a year and a variable operating cost of $85,000 for 100% capacity. Assume that the variable cost varies directly with the level of operation. If interest is taken at 15% compounded annually, at what percent of capacity must the motel operate to break even?

10. A coffee processor has a weigher for filling cans with coffee. Because of its lack of sensitivity the weigher must be set so that the average filled can contains 1.005 kilograms in order to insure that no can contains less than 1 kilogram of coffee. The present weigher was purchased 10 years ago at a price of $1,500 and is expected to last an additional 5 years. A new weigher is being considered whose sensitivity is such that the average "overage" per kilogram can is 3.5 grams. This weigher will cost $6,400, on which $300 will be allowed for the present weigher. The salvage value of either weigher 10 years from now is considered to be negligible, and their operating costs are estimated to be equal. The value of the coffee is $10 per kilogram. What is the minimum number of 1-kilogram packages packed annually for which the purchase of the new weigher is justified, if the interest rate is 20%? Answer: 95,072

11. A contractor is offered his choice of either a gasoline, diesel, or butane engine to power a bulldozer he is to purchase. The gasoline engine will cost $2,000, will have an estimated maintenance cost of $200 per year, and will consume $7.20 worth of fuel per hour of operation. The diesel engine will cost $2,800, will cost an estimated $240 per year to maintain, and will consume $6.60 worth of fuel per hour. The butane engine will cost $3,300, will cost $315 per year to maintain, and will consume $5.80 worth of fuel per hour of operation. Since the salvage value of each engine will be identical, it may be neglected. All other costs associated with the three engines are equal and the interest rate is 15%. The service life of each engine is 5 years.

a. Plot the total annual cost of each engine as a function of the number of hours of operation per year.

b. Find the range of number of hours of operation for which it would be most economical to specify the gasoline engine; the diesel engine; the butane engine.

12. A 3-phase, 220-volt, squirrel-cage induction motor is to be used to drive a pump, which will fill a series of storage tanks. It is estimated that 300 horse power-hours will be required each day for 365 days per year. Motors having the following characteristics may be leased.

Size (hp)	Lase Rate (per year)	Operating Cost (per hp-h)
20	$ 928	$0.044
40	1,000	0.037
75	1,142	0.031
100	1,250	0.027
150	1,510	0.025

Plot the total yearly cost as a function of the size of the motor and select the size that will result in a minimum cost per year.

13. It has been found that the heat loss through the ceiling of a building is 0.13 Btu per hour per square foot of area per degree Fahrenheit. If the 2,500-square-foot ceiling is insulated, the heat loss in Btu per hour per degree temperature difference per square foot of area is taken as being equal to

$$\frac{1}{\dfrac{1}{0.13} + \dfrac{t}{0.27}}$$

where t is the thickness in inches. The in-place cost of insulation 4, 6, and 10 inches thick is $0.16, $0.29, and $0.42 per square foot, respectively. The building is heated to 75 degrees 3,000 hours per year by a gas furnace with an efficiency of 50%. The mean outside temperature is 45 degrees, and the natural gas used in the furnace costs $4.43 per 1,000 cubic feet and has a heating value of 2,000 Btu per cubic foot. What thickness of insulation, if any, should be used if the interest rate is 6% and the resale value of the building 6 years hence is enhanced $180 if insulation is added, regardless of the thickness?

14. An overpass is being considered for a certain railroad crossing. The superstructure design under consideration will be made of steel and will have a weight per foot depending upon the span between piers in accordance with $W = 32(S) + 1{,}850$. Piers will be made of concrete and will cost $215,000 each. The superstructure will be erected at a cost of $0.45 per pound. If the number of piers required is to be one less than the number of spans, find the number of piers that will result in a minimum total cost for piers and superstructure if $L = 1{,}275$ feet. Answer: 9 piers

15. An hourly electric load of 1,600 amperes is to be transmitted from a generator to a transformer in a certain power plant. A copper conductor 150 feet long can be installed for $160 + $0.64 per pound, will have an estimated life of 20 years, and can be salvaged for $0.50 per pound. Power loss from the conductor will be a function of the cross-sectional area and may be expressed as $25{,}875 \div A$ kilowatthours per year. Energy lost is valued at

$0.052 per kilowatthour; taxes, insurance, and maintenance are negligible; the interest rate is 15%. Copper weighs 555 pounds per cubic foot.

a. Plot the total annual cost of capital recovery and return plus power-loss cost for conductors for cross sections of 1, 2, 3, 4, and 5 square inches.

b. Find the optimum cross section mathematically and check the result against the minimum point found in part (a). Answer: 4.87 in^2

16. The maintenance cost of a certain machine is zero the first year and increases by $200 per year for each year thereafter. The machine costs $2,000 and has no salvage value at any time. Its annual operating cost is $1,000 per year. If the interest rate is 10%, what life will result in minimum equivalent annual cost? Solve by trial and error, showing yearly costs in tabular form.

17. A special-purpose machine is to be purchased at cost of $20,000. The following table shows the expected annual operating costs, maintenance costs, and salvage values for each year of service. If the rate of interest is 10%, what is the optimum economic life for this machine? Answer: 5 years

Year of Service	*Operation Cost for Year*	*Maintenance Cost for Year*	*Salvage Value at End of Year*
1	$2,000	$200	$10,000
2	3,000	300	9,000
3	4,000	400	8,000
4	5,000	500	7,000
5	6,000	600	6,000
6	7,000	700	5,000

18. Ethyl acetate is made from acetic acid and ethyl alcohol. Let x = pounds of acetic acid input, y = pounds of ethyl alcohol input, and z = pounds of ethyl acetate output. The relationship of output to input is

$$\frac{z^2}{(1.47x - z)(1.91y - z)} = 3.91.$$

a. Determine the output of ethyl acetate per pound of acetic acid, where the ratio of acetic acid to ethyl alcohol is 2, 1.5, 1, 0.67, and 0.50, and graph the result.

b. Graph the cost of material per pound of ethyl acetate for each of the ratios given and determine the ratio for which the material cost per pound of ethyl acetate is a minimum if acetic acid costs $0.80 per pound and ethyl alcohol $0.92 per pound.

19. The government is considering charging a polluter's fee on the amount of pollutants discharged into our rivers in order to significantly reduce the pollution of our waterways. The polluter's fee is to be based on the amount of pollutant remaining in the effluent being discharged. The indicator to be used to measure the level of pollution of industrial wastes is biochemical oxygen demand, BOD. A firm is presently discharging 4,000,000 pounds of BOD per year into a nearby river because it does not treat its waste products. The cost of installing and operating a waste-treatment system that reduces the BOD output to a percent of its present level is shown below.

Percent BOD Remaining	*Initial Investment*	*Operating Expenses*
5%	$250,000	$40,000
10%	100,000	25,000
15%	85,000	15,000
20%	75,000	10,000
25%	75,000	5,000

It is expected that the treatment system will last 10 years and have zero salvage value at that time. The firm's MARR is 15%.

a. If the polluter's fee is $0.10 per pound of BOD remaining, what percent of BOD remaining will minimize the firm's annual costs?

b. If the polluter's fee is $0.05 per pound of BOD remaining, what percent of BOD remaining will minimize the firm's costs?

c. Plot a graph of the optimum level of BOD remaining as a function of the polluter's fee. Let the polluter's fee range from $0.02 per pound of BOD to $0.20. If you were charged with deciding the amount of the polluter's fee for this firm, what amount would you choose?

20. An overpass bridge is to be constructed for a 1,200-foot crossing. Two girder designs have been proposed. The first will result in a superstructure weight per foot of $W_1 = 22(S) + 800$, where S is the span between piers. The second will result in a superstructure weight per foot of $W_2 = 20(S) + 1{,}000$. Regardless of the girder design chosen, the piers and two needed abutments will cost $220,000 each. The superstructure will be erected at a cost of $0.40 per pound. All other costs for the competing designs are the same. Choose the girder design that will result in a minimum-cost bridge and specify the optimum pier spacing. Answer: Second design with pier spacing of 166 ft

Part Four

ACCOUNTING, DEPRECIATION, AND INCOME TAXES

Although there is a vast difference in the fields of accounting and engineering economy, important relationships do exist. Accordingly, the fundamentals of accounting, cost accounting, and depreciation accounting are presented in this part of the text. Also presented herein is the role of accounting in economic analysis for the comparison of alternatives employing capital assets. In this application, charges from depreciation accounting are shown to have a significant effect on the after-tax cash flows and on the chosen evaluation measure. Other after-tax situations in engineering economic analysis are also presented. These include the evaluation of financing methods, investment credit and capital gains, inflationary effects, and replacement studies.

11 Accounting and Depreciation Accounting

The accounting system of an enterprise provides a medium for recording historical data arising from the essential activities employed in the production of goods and services. Accounting has the objective of providing summaries of the status of an enterprise in terms of assets and liabilities so that the condition of the enterprise may be judged at any point in time. Accounting records are thus important sources of data for engineering economy studies. In them the analyst seeking to quantify expected future differences in the worth and cost of alternative engineering proposals will find detailed quantitative data useful in estimating the future outcome of activities similar to those completed.

The accounting process also provides the depreciation charges that are essential for the calculation of income taxes. This chapter discusses basic accounting concepts and the methods of depreciation available to the taxpayer.

11.1 GENERAL ACCOUNTING

The purpose of accounting is to record the financial activity of an enterprise. The accounting system provides periodic reporting of this activity through the preparation of numerous financial statements and reports.

11.1.1 Basic Financial Statements

The primary purpose of the general accounting system is to make possible the periodic preparation of two basic financial statements for an enterprise. These are

1. A *balance sheet* setting out the assets, liabilities, and net worth of the enterprise at a stated date.
2. A *profit and loss statement* showing the revenues and expenses of the enterprise for a stated period.

Thus, the balance sheet presents a "snapshot" of the financial condition of an enterprise at a specified point in time. To reflect the level of transactions of the enterprise occurring between the time the previous balance sheet was prepared and the next one is prepared, the enterprise provides the profit and loss statement. Figure 11.1 depicts the relationship between the balance sheet and the profit and loss statement.

The accounts of an enterprise fall into five general classifications: assets, liabilities, net worth, revenue, and expense. Three of these—assets, liabilities, and net worth—serve to give the position of the enterprise at a certain date. The other two—revenue and expense—accumulate profit and loss information for a stated period, reflecting changes in the position of the enterprise at different points in time. Each of these five accounts is a summary of other accounts utilized as part of the total accounting system.

The *balance sheet* is prepared for the purpose of exhibiting the financial position of an enterprise at a specific point in time. It lists the assets, liabilities, and net worth of the enterprise as of a certain date. The monetary amounts recorded in the accounts must conform to the fundamental accounting equation

$$\text{assets} = \text{liabilities} + \text{net worth}.$$

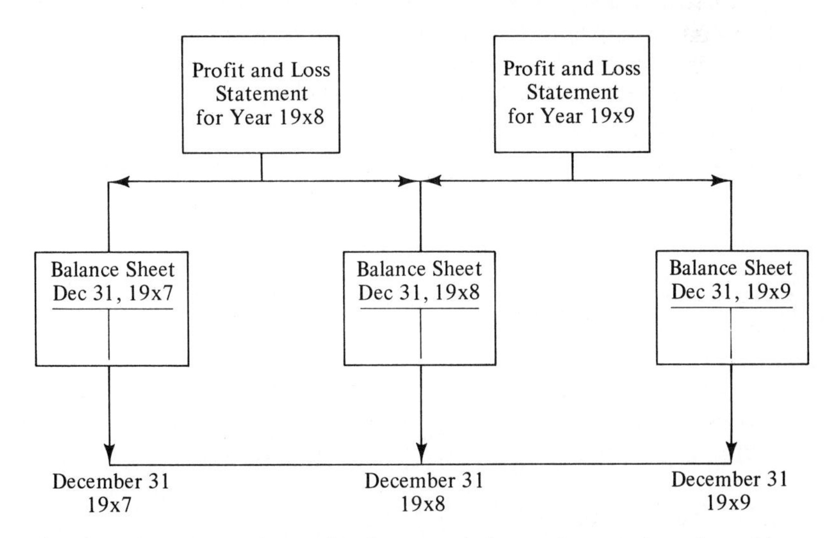

FIGURE 11.1. Time relationship between balance sheet and profit and loss statement.

For example, the balance sheet for Ace Company shows these major accounts as of December 31, 19xx.

ACE COMPANY BALANCE SHEET DECEMBER 31, 19xx

Assets		Liabilities	
Cash	$143,300	Notes payable	$22,000
Acounts receivable	7,000	Accounts payable	4,700
Raw materials	9,000	Accrued taxes	3,200
Work in process	17,000	Declared dividends	40,000
Finished goods	21,400		$69,900
Land	11,000		
Factory building	82,000	Net Worth	
Equipment	34,000	Capital stock	$200,000
Prepaid services	1,300	Profit for December	56,100
			256,100
	$326,000		$326,000

Balance sheets are normally drawn up annually, quarterly, monthly, or at other regular intervals. The change of a company's condition during the interval between balance sheets may be determined by comparing successive balance sheets.

Information relative to the changes of conditions during the interval between successive balance sheets is provided by a *profit and loss statement.* This statement is a summary of the income and expense for a stated period. For example, the profit and loss statement for the Ace Company shows the income, expense, and net profit for the month of December, 19xx.

ACE COMPANY PROFIT AND LOSS STATEMENT MONTH ENDING DECEMBER 31, 19xx

Gross income from sales		$251,200
Cost of goods sold		142,800
Net income from sales		$108,400
Operating expense:		
Rent	$11,700	
Salaries	28,200	
Depreciation	4,800	
Advertising	6,500	
Insurance	1,100	$52,300
Net profit from operations before taxes		$56,100
Federal and state taxes		12,300
Net profit from operations after taxes		$43,800

11.1.2 Profit as a Measure of Success for the Enterprise

Profit is the resultant of two components, one of which is the economy associated with income from the activity. It is obvious that some activities have greater profit potentialities than others. In fact, some activities can result only in loss.

When profit is a consideration, it is important that activities be evaluated with respect to their effect on profit. The first step in making a profit is to secure an income. But to acquire an income necessitates certain activities resulting in certain costs. Profit is, therefore, a resultant of activities that produce income and involve outlay, which may be expressed as

$$\text{profit} = \text{income} - \text{outlay}.$$

The total success of an organization is the summation of the successes of all the activities that it has undertaken. Also, the success of a major undertaking is the summation of the successes of the minor activities of which it is constituted. In Figure 11.2 each vertical block represents the income potentialities of a venture, the outlay incurred in seeking it, the oulay incurred in prosecuting it, and the net gain of carrying it on. It is apparent that the extent of the success of an activity depends upon its potentialities for income less the sum of the costs of funding it and carrying it on. In this conceptual scheme the various ventures are considered to be measured in a single commensurable term, such as money.

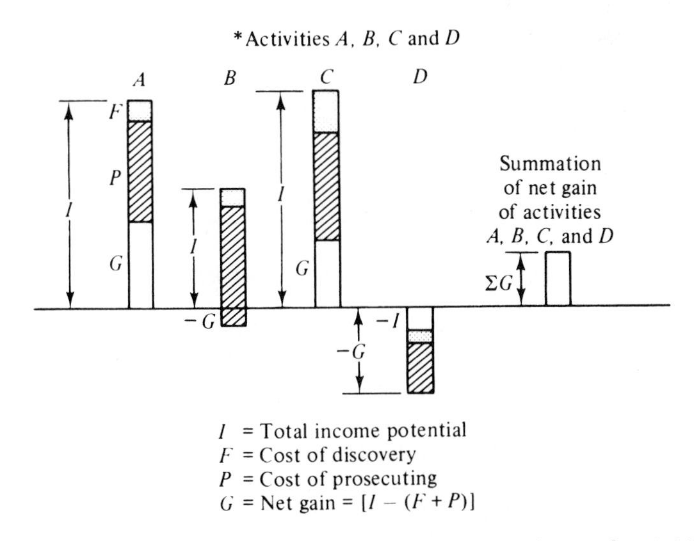

FIGURE 11.2. Illustration of the final outcome of several activities.

11.2 COST ACCOUNTING

Cost accounting is a branch of general accounting adapted to registering the costs for labor, material, and overhead on an item-by-item basis as a means of determining the cost of production. The final summary of this information is presented in the form of a cost of goods sold statement. It lists the costs of labor, material, and overhead applicable to all goods made and sold during a certain period. For example, the cost of goods sold statement for the Ace Company during the month of December, 19xx, is as follows.

ACE COMPANY COST OF GOODS SOLD STATEMENT DECEMBER 31, 19xx

Direct Material		
In process Dec. 1, 19xx	$ 3,400	
Applied during the month	$39,500	
Total	$42,900	
In process Dec. 31, 19xx	4,200	$ 38,700
Direct Labor		
In process Dec. 1, 19xx	$ 4,300	
Applied during the month	51,900	
Total	$56,200	
In process Dec. 31, 19xx	5,700	$ 50,500
Overhead		
In process Dec. 1, 19xx	$ 5,800	
Applied during the month	60,100	
Total	$65,900	
In process Dec. 31, 19xx	7,100	$ 58,800
Cost of Goods Made		$148,000
Finished goods Dec. 1, 19xx		16,200
Total		$164,200
Finished goods Dec. 31, 19xx		21,400
Cost of Goods Sold		$142,800

11.2.1 Product Flow and Cost of Goods Sold

In an enterprise where goods are manufactured, the costs of production are accounted for as the product's components physically move from the raw-material stage through partial completion, to their sale as finished goods. The cost of goods sold statement reflects summary data indicating the three categories of costs of material, labor, and overhead incurred in the process of production. These basic elements of cost are allocated to the product as it passes through each stage of production. The accumulation of these costs is reflected in the work-in-process inventory accounts at each intermediate stage of production.

When the product is completed, the total cost accumulated as work in process is transferred to the finished goods, inventory account. Once the product is sold, the finished goods cost for the product becomes the cost of the item sold. This process, illustrated in Figure 11.3, is basic to cost accounting in manufacturing.

The cost of goods sold statement is subsidiary to the profit and loss statement, since the "cost of goods sold" amount it develops is transferred to the profit and loss statement. (Note the cost of goods sold amount of $142,800 on the profit and loss statement.) Similarly, the profit and loss statement was shown to be subsidiary to the balance sheet, since the profit or loss amount developed thereon is transferred to the net worth section of the balance sheet.

The costs that are incurred to produce and sell an item of product are commonly classified as direct material, direct labor, manufacturing overhead, administrative cost, and selling cost. The first three are exhibited on the cost of goods sold statement, and they give rise to the cost of goods sold entry on the profit and loss statement. Administrative and selling costs appear on the profit and loss statement under operating expense, and they are subtracted from net income to arrive at the net profit amount as seen in Figure 11.4. Each of these cost classifications will be considered in the paragraphs that follow.

11.2.2 Direct Material

The material whose cost is directly charged to a product is termed *direct material.* Ordinarily, the costs of principal items of material required to make a product are charged to it as direct-material costs. Charges for direct material are made to the product at the time the material is issued, through the use of forms and pro-

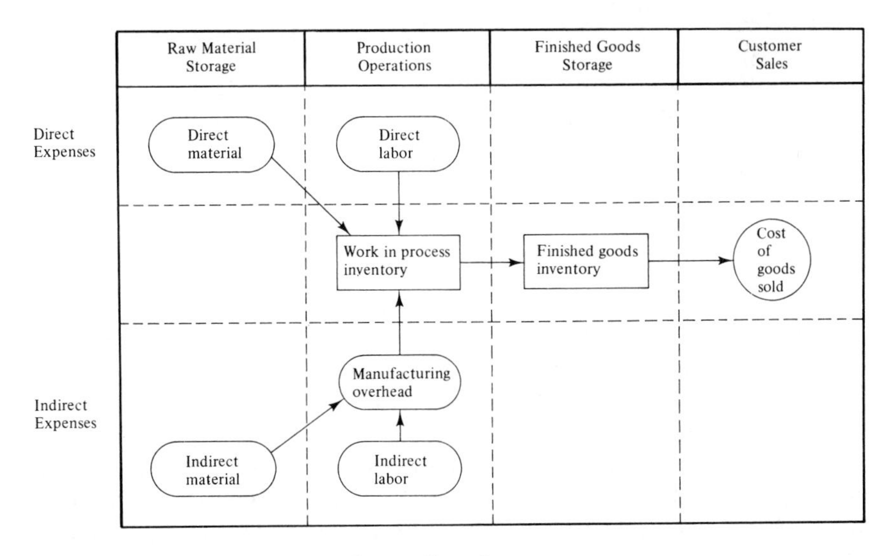

FIGURE 11.3. Product flow and cost allocation.

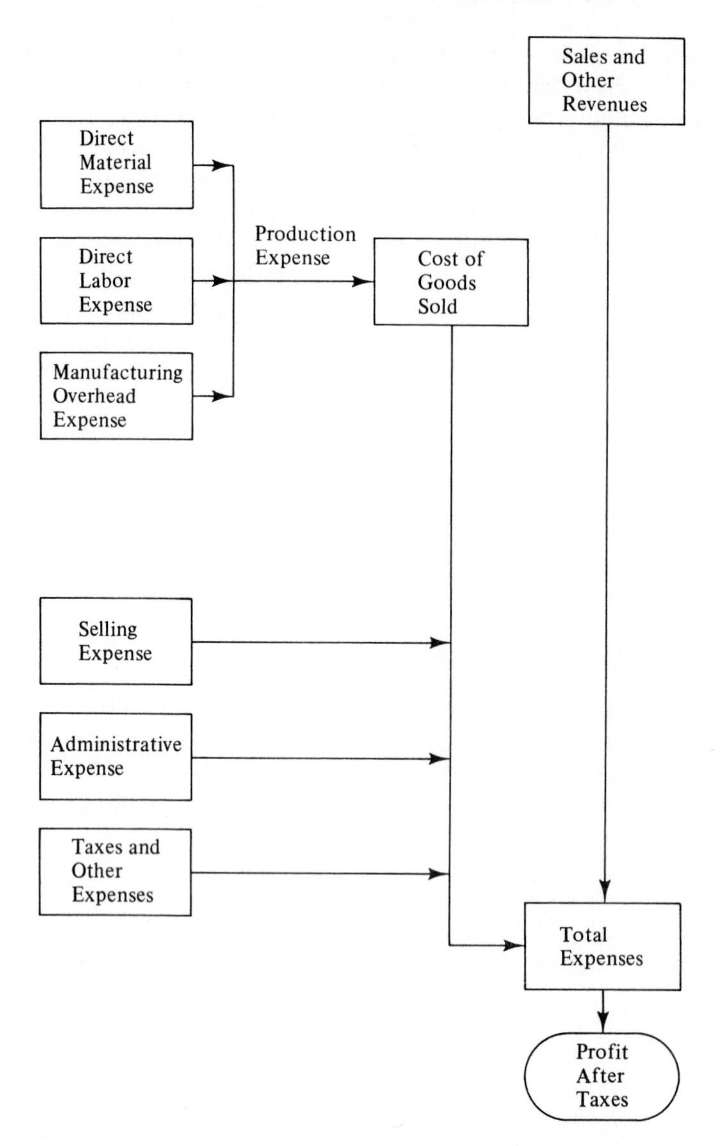

FIGURE 11.4. Relationship of income and expenses to profit after taxes.

cedures designed for that purpose. The sum of charges for materials that accumulate against the product during its passage through the plant constitutes the total direct-material cost.

In the manufacture of many products, small amounts of a number of items of material may be consumed which are not directly charged to the product. These items, referred to as *indirect material,* are charged to manufacturing overhead, as seen in Figure 11.3. They are not directly charged to the product on the premise that it would be impractical to trace their use to the product.

Although perhaps less subject to gross error than records of other elements of cost, records of direct-material costs should not be used in enginering economy studies without being questioned. Their accuracy in regard to quantity and price of material should be ascertained. Also, their applicability to the situation being considered should be established before they are used.

11.2.3 Direct Labor

Direct labor is labor whose cost is charged directly to the product. The source of this charge is time tickets or similar forms used to record the time and wages of workmen whose efforts are applied to a product during its journey through the production facility. Direct labor is usually measured by multiplying the hours of direct labor activity by the hourly wage rate. Other labor costs such as pensions, sick leave, vacations and other fringe benefits increase the actual labor costs; therefore, accounting practices sometimes include these costs as direct labor costs. More commonly, these fringe benefits are accounted for as part of manufacturing overhead.

Certain types of labor activity are not directly related to the products being produced, although they serve indirectly in the manufacturing process. Such items of labor costs are called *indirect labor,* and they become part of manufacturing overhead as shown in Figure 11.3. The labor of personnel engaged in engineering, testing, inspection, cleaning, material handling, supervision, and other support services are often charged in this manner.

As a result of either carelessness or a desire to conceal an undue amount of time spent on a job, some of the time applied to one job may be reported as being applied to another. Unless the allocation of labor costs to products is very closely controlled, records of labor costs charged to specific products are likely to be in error. Thus, direct-labor cost records should be carefully examined for accuracy and applicability to the situations under investigation before being used as data for engineering economy studies.

11.2.4 Manufacturing Overhead

Manufacturing overhead is also designated by such terms as factory overhead, shop expense, burden, and indirect costs. Manufacturing overhead costs embrace all expenses incurred in the production of products which are not directly charged to the products as direct material or direct labor.

Costs such as indirect materials and indirect labor are combined, as seen in Figure 11.3, with other costs that cannot be directly related to the product being manufactured. These other costs embrace charges for such things as research, software development, furniture and equipment, property taxes, insurance, interest, rent, depreciation (capital recovery), maintenance of buildings, and salaries of factory supervision, which are considered to be independent of volume of production.

The practice of applying overhead charges arises because cost accounting procedures would become prohibitively expensive if it was required that all items of cost be charged directly to the product. Thus, practicality motivates the use of overhead charges.

11.2.5 Production Cost

The *production cost* of a product is the sum of direct material, direct labor, and manufacturing overhead. It is these items that are summarized on the cost of goods sold statement. This cost classification separates the manufacturing cost from administrative and selling costs, as seen in Figure 11.4, thus giving an indication of production costs over time.

11.2.6 Administrative Expenses

Administrative expenses arise from expenditures for such items as salaries of executive, clerical, and technical personnel, office space, office supplies, depreciation of office equipment, travel, and fees for legal, technical, and auditing services that are necessary to direct the enterprise as a whole. These expenses are distinct from its production and selling activities. Expenses so incurred are often recorded on the basis of the cost of carrying on subdivisions of administrative activities deemed necessary to take appropriate action to improve the effectiveness of administration.

In most cases, it is not practical to relate administrative expenses directly to specific products. The usual practice is to allocate administrative expenses to the product as a percentage of the product's production expense. For example, if the annual administrative expenses and production expenses of a concern are estimated at $10,000 and $100,000, respectively, for a given year, 10% will be added to the production cost of products manufactured to absorb the administrative costs.

11.2.7 Selling Expense

The *selling expense* of a product arises from expenditures incurred in disposing of the products and services produced. This class of expense includes such items as salaries, commissions, office space, office supplies, rental and depreciation, operation of office equipment and automobiles, travel, market surveys, entertainment of customers, displays, and sales space.

Selling expenses may be allocated to various classes of products, sales territories, sales of individual salesmen, and so forth, as a means of improving the effectiveness of selling activities. In many cases it is considered adequate to allocate selling expense to products as a percentage of their production expense. For example, if the annual selling expense is estimated as $22,000 and the annual production expense is estimated at $110,000, 20% will be added to the production expense of products to obtain the cost of sales.

11.3 CLASSIFICATIONS OF DEPRECIATION

As physical assets lose value with the passage of time it is said that they *depreciate* in value.

With the possible exception of land and collectible items, this phenomenon is a characteristic of all physical assets. A common classification of the types of depreciation include (1) physical depreciation, (2) functional depreciation, and (3) accidents. The first two of these will be defined and explained below.

11.3.1 Physical Depreciation

Depreciation resulting in physical impairment of an asset is known as *physical depreciation*. Physical depreciation manifests itself in such tangible ways as the wearing of particles of metal from a bearing and the corrosion of the tubes in a heat exchanger. This type of depreciation results in the lowering of the ability of a physical asset to render its intended service. The primary causes of physical depreciation are

1. Deterioration due to action of the elements including the corrosion of pipe, the rotting of timbers, chemical decomposition, bacterial action, and so on. Deterioration is substantially independent of use.
2. Wear and tear from use that subjects the asset to abrasion, shock, vibration, impact, and the like. These forces are occasioned primarily by use and result in a loss of value over time.

11.3.2 Functional Depreciation

Functional depreciation results not from a deterioration in the asset's ability to serve its intended purpose, but from a change in the demand for the services it can render. The demand for the services of an asset may change because it is more profitable to use a more efficient unit, there is no longer work for the asset to do, or the work to be done exceeds the capacity of the asset. Depreciation resulting from a change in the need for the service of an asset may be the result of

1. Obsolescence resulting from the discovery of another asset that is sufficiently superior to make it uneconomical to continue using the original asset. Assets also become obsolete when they are no longer needed.
2. Inadequacy or the inability to meet the demand placed upon it. This situation arises from changes in demand not contemplated when the asset was acquired.

A disposition to replace machines when it becomes profitable to do so instead of when they are worn out has probably been an important factor in the rapid development of the United States. The sailing ship has given way to the steamship. In street transportation, the sequence of obsolescence has been horse cars, cable cars, electric cars, and automotive buses. One generation of computers is rapidly being replaced by the next. In power generation, nuclear plants are an alternative to fossil-fuel plants. In manufacturing, improvements in the arts of processing have resulted in widespread obsolescence and inadequacy of equipment. Each technological advance produces improvements that result in the obsolescence of existing assets.

11.4 ACCOUNTING FOR THE DEPRECIATION OF CAPITAL ASSETS

An asset such as a machine is a unit of capital. Such a unit of capital loses value over a period of time in which it is used in carrying on the productive activities of a business. This loss of value of an asset represents actual piecemeal consumption or expenditure of capital. For instance, a truck tire is a unit of capital. The particles of rubber that wear away with use are actually small physical units of capital consumed in the intended service of the tire. In a like manner, the wear of machine parts and the deterioration of structural elements are physical consumptions of capital. Expenditures of capital in this way are often difficult to observe and are usually difficult to evaluate in monetary terms, but they are nevertheless real.

An understanding of the concept of depreciation is complicated by the fact that there are two aspects to be considered: (1) the actual lessening in value of an asset with use and the passage of time, and (2) the accounting for this lessening in value.

> **Depreciation,** as an element of the basic accounting statements, views the cost of an asset as a prepaid expense that is to be charged against profits over some reasonable period of time.

Rather than charging the entire cost as an expense at the time the asset is purchased, the accountant attempts, in a systematic way, to spread the anticipated loss in value over the life of the asset. This concept of amortizing the cost of an asset so that the profit and loss statement is a more accurate reflection of capital consumption is basic to financial reporting and income tax calculation.

11.4.1 Value-Time Function and Book Value

In considering depreciation for accounting purposes, the pattern of the future value of an asset should be predicted. It is customary to assume that the value of an asset decreases yearly in accordance with one of several mathematical functions. Which function to select involves decisions as to the life of the asset, its salvage value, and the form of the mathematical function. Once a value-time function has been chosen, it is used to represent the value of the asset at any point during its life. A general value-time function is shown in Figure 11.5.

Accountants use the term *book value* to represent the undepreciated value of a firm's assets.

> **Book value** is the acquisition cost of an asset less its accumulated depreciation charges.

Thus, a function similar to Figure 11.5 can represent book value.

The book value at the end of any year is equal to the book value at the beginning of the year less the depreciation expense charged during the year. Table 11.1 presents the calculation of book value at the end of each year for an asset with a first cost of $12,000, an estimated life of 5 years, and a salvage value of zero with assumed depreciation charges.

In determining book value the following notation will be used. Let

P = first cost of the asset;

F = estimated salvage value;

B_t = book value at the end of year t;

D_t = depreciation charge during year t;

n = estimated life of the asset.

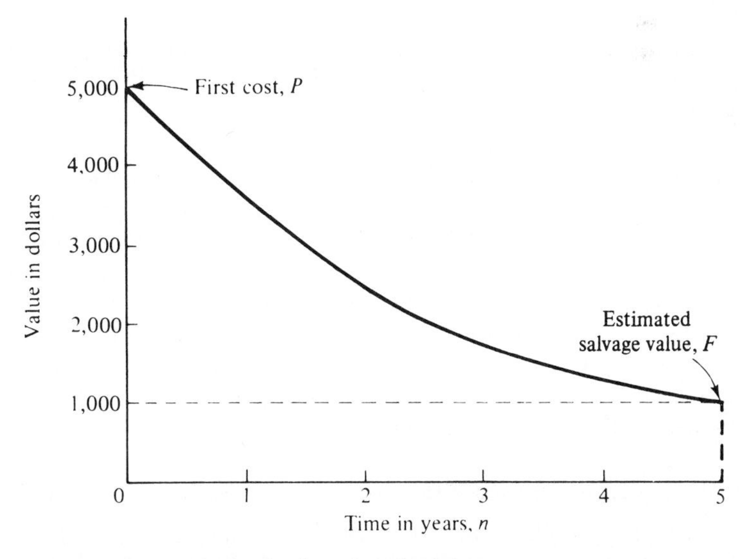

FIGURE 11.5. A general value-time function.

TABLE 11.1.
BOOK-VALUE CALCULATION

End of Year	Depreciation Charge During Year	Book Value at End of year
0	—	$12,000
1	$4,000	8,000
2	3,000	5,000
3	2,000	3,000
4	2,000	1,000
5	1,000	0

A general expression relating book value and the depreciation charge is given by

$$\boxed{B_t = B_{t-1} - D_t} \tag{11.1}$$

where $B_0 = P$. Another form of the book-value equation is:

$$B_t = P - \sum_{j=1}^{t} D_t. \tag{11.2}$$

The sections that follow are concerned with methods for determining D_t for $t = 1, 2, \ldots, n$.

11.5 DEPRECIATION METHODS PRIOR TO 1981

Because an asset's allowable depreciation depends on the tax law in effect at its time of acquisition, it is necessary to understand the peculiarities of the tax laws for various time periods in the recent past. Three of these periods include the time prior to 1981, the time from 1981 through 1986, and 1987 and later. Certain methods used prior to 1981 are important because they are the basis for the depreciation schedules of the latter periods.

Prior to 1981 the taxpayer had considerable latitude in the choice of depreciation method. The three most common methods were (1) straight-line depreciation, (2) declining-balance depreciation and (3) sum-of-the-years-digits depreciation.

Usually, the most important factor in figuring the annual depreciation charge for a pre-1981 asset is the estimate of its useful life. The useful life can be determined from the firm's experience with particular assets, or, if the estimator's experience is inadequate, the general experience of the industry can provide a basis for this estimate. As long as the lives utilized in depreciation calculations are reasonable and there is no clear and convincing basis for change, adjustments in these estimated lives are not required.

An alternative to determining useful life for pre-1981 assets, provided by the Internal Revenue Service (IRS), was the more liberal Class Life Asset Depreciation Range (CLADR) system. The CLADR system allowed taxpayers to take as a reasonable depreciation charge an amount based on any life selected within a range specified for designated classes of assets. Some examples of these asset classes and the ranges of lives permitted are presented in Table 11.2.

TABLE 11.2.
CLASSES OF ASSETS AND RANGES OF LIVES PERMITTED FOR THE CLASS LIFE ASSET DEPRECIATION RANGE (CLADR) SYSTEM*

Asset Description	Lower Limit	ADR Life	Upper Limit
Depreciable Assets Used in All Business Activities			
Aircraft	5	6	7
Automobiles, taxis	2.5	3	3.5
Railroad cars and locomotives	12	15	18
Vessels, barges, tugs	14.5	18	21.5
Depreciable Assets Used in the Following Activities			
Agriculture equipment	8	10	12
Computers	5	6	7
Construction	4	5	6
Electric production plant			
Hydraulic	40	50	60
Nuclear	16	20	24
Steam	22.5	28	33.5
Manufacturing			
Electrical equipment	9.5	12	14.5
Electronic products	6.5	8	9.5
Ferrous metals	14.5	18	21.5
Furniture	8	10	12
Motor vehicles	9.5	12	14.5
Paper pulps	13	16	19
Petroleum			
Drilling	5	6	7
Exploration	11	14	17
Refining and marketing	13	16	19
Rubber products	11	14	17
Textile mill products	11	14	17
Recreation and amusement	8	10	12

**Complete CLADR tables in Revenue Procedure 83–35 (IRS).*

CLADR was developed as a feature of the tax code prior to 1981. Nevertheless, this system continues to be used in all subsequent tax codes including the current one. As seen in Table 11.2, the midpoint of the asset depreciation range is denoted the ADR life for the property class. This ADR life is the basis on which certain property is assigned to the various property classes as defined by the depreciation systems in effect since 1981. How the ADR life is used is described in the property class definitions of Sections 11.6 and 11.7.

The Tax Reform Act of 1986 also assigned new ADR lives to certain property. These new ADR lives are listed below.

Property	*ADR Life*
Semiconductor manufacturing equipment	5 years
Computer-based central office switching equipment	9.5 years
Railroad track	10 years
Single-purpose agricultural and horticultural structures	15 years
Telephone distribution plant and comparable two-way communication equipment	24 years
Municipal wastewater treatment plants	25 years
Sewer pipes	50 years

11.5.1 Straight-Line Method of Depreciation

The straight-line depreciation model assumes that the value of an asset decreases at a constant rate. Thus, if an asset has a first cost of $5,000 and an estimated salvage value of $1,000, the total depreciation over its life will be $4,000. If the estimated life is 5 years, the depreciation per year will be $4,000 ÷ 5 = $800. This is equivalent to a depreciation rate of 1 ÷ 5 or 20% per year. For this example, the annual depreciation and book value for each year are given in Table 11.3.

General expressions for the calculation of depreciation and book value may be developed for the straight-line method. Table 11.4 shows the depreciation charge and book-value expressions for each year. Thus, the depreciation in any

TABLE 11.3.
THE STRAIGHT-LINE METHOD

End of Year t	Depreciation Charge During Year t	Book Value at End of Year t
0	—	$5,000
1	$800	4,200
2	800	3,400
3	800	2,600
4	800	1,800
5	800	1,000

TABLE 11.4.
GENERAL EXPRESSIONS FOR THE STRAIGHT-LINE METHOD

End of Year t	Depreciation Charge During Year t	Book Value at End of Year t
0	—	P
1	$\frac{P-F}{n}$	$P - \left(\frac{P-F}{n}\right)$
2	$\frac{P-F}{n}$	$P - 2\left(\frac{P-F}{n}\right)$
3	$\frac{P-F}{n}$	$P - 3\left(\frac{P-F}{n}\right)$
t	$\frac{P-F}{n}$	$P - t\left(\frac{P-F}{n}\right)$
n	$\frac{P-F}{n}$	$P - n\left(\frac{P-F}{n}\right)$

year is

$$D_t = \frac{P-F}{n}, \tag{11.3}$$

the book value is

$$B_t = P - t\left(\frac{P-F}{n}\right), \tag{11.4}$$

and the depreciation rate per year is $1/n$.

11.5.2 Declining-Balance Method of Depreciation

The declining-balance method of depreciation assumes that an asset decreases in value faster early rather than in the latter portion of its service life. By this method a fixed percentage is multiplied times the book value of the asset at the beginning of the year to determine the depreciation charge for that year. Thus, as the book value of the asset decreases through time, so does the size of the depreciation charge. For an asset with a \$5,000 first cost, a \$1,000 estimated salvage value, an estimated life of 5 years, and a depreciation rate of 30% per year the depreciation charge per year is shown in Table 11.5.

For a depreciation rate α the general relationship expressing the depreciation charge in any year for declining-balance depreciation is

$$D_t = \alpha \cdot B_{t-1}. \tag{11.5}$$

TABLE 11.5.
THE DECLINING-BALANCE METHOD

End of Year t	Depreciation Charge During Year t	Book Value at End of Year t
0	—	\$5,000
1	(0.30)(\$5,000) = \$1,500	3,500
2	(0.30)(3,500) = 1,050	2,450
3	(0.30)(2,450) = 735	1,715
4	(0.30)(1,715) = 515	1,200
5	(0.30)(1,200) = 360	840

From the general expression for book value shown in Section 11.41 it is seen that

$$B_t = B_{t-1} - D_t.$$

Therefore, for declining-balance depreciation

$$B_t = B_{t-1} - \alpha B_{t-1} = (1 - \alpha)B_{t-1}.$$

Using this recursive expression, we can determine the general expressions for the depreciation charge and the book value for any point in time. These calculations are shown in Table 11.6 for an asset with first cost P. Thus, the depreciation in any year is

$$D_t = \alpha(1 - \alpha)^{t-1}P \tag{11.6}$$

and the book value is

$$B_t = (1 - \alpha)^t P. \tag{11.7}$$

TABLE 11.6.
GENERAL EXPRESSIONS FOR THE DECLINING-BALANCE METHOD OF DEPRECIATION

End of Year t	Depreciation Charge During Year t	Book Value at End of Year t
0	—	P
1	$\alpha \times B_0 = \alpha(P)$	$(1 - \alpha)B_0 = (1 - \alpha)P$
2	$\alpha \times B_1 = \alpha(1 - \alpha)P$	$(1 - \alpha)B_1 = (1 - \alpha)^2 P$
3	$\alpha \times B_2 = \alpha(1 - \alpha)^2 P$	$(1 - \alpha)B_2 = (1 - \alpha)^3 P$
t	$\alpha \times B_{t-1} = \alpha(1 - \alpha)^{t-1} P$	$(1 - \alpha)B_{t-1} = (1 - \alpha)^t P$
n	$\alpha \times B_{n-1} = \alpha(1 - \alpha)^{n-1} P$	$(1 - \alpha)B_{n-1} = (1 - \alpha)^n P$

If the declining-balance method of depreciation is used for income-tax purposes, the maximum rate that may be used is double the straight-line rate that would be allowed for a particular asset or group of assets being depreciated. Thus, for an asset with an estimated life of n years the maximum rate that may be used with this method is $\alpha = 2/n$. Many firms and individuals choose to depreciate their assets using declining-balance depreciation with the maximum allowable rate. Such a method of depreciation is commonly referred to as the *double-declining balance* method of depreciation.

In Table 11.5 the book value at the end of year 5 is $840, which is less than the estimated salvage of $1,000. If the salvage value for this example had been zero, for this method of depreciation the book value would never reach zero, regardless of the time span over which the asset was depreciated. Thus, adjustments are necessary to recognize the differences between the estimated and calculated book value of the asset. In most situations these adjustments are made at the time of disposal of the asset, when acounting entries are made to account for the difference between the asset's actual value and its calculated book value.

11.5.3 Declining-Balance Switching to Straight-Line Depreciation

It is allowable under pre-1981 tax law to depreciate an asset over the early portion of its life using declining balance and then switching to straight-line depreciation for the remainder of the asset's life. The switch usually occurs when the next period's straight-line depreciation amount on the undepreciated balance of the asset exceeds the next period's declining-balance depreciation charge. The switch would occur at the beginning of the period t, where first

$$\text{declining-balance depreciation}_t < \begin{pmatrix}\text{straight-line depreciation} \\ \text{on undeprecidated balance}_t\end{pmatrix}$$

or

$$\alpha(B_{t-1}) < \frac{B_{t-1} - F}{n - (t - 1)}. \tag{11.8}$$

For an investment of $5,000 with $F = 0$, a 5-year life, and $\alpha = 0.3$, evaluate the above expression for

$$t = 1: \qquad \$1{,}500 > \frac{\$5{,}000 - \$0}{5} = \$1{,}000$$

$$t = 2: \qquad \$1{,}050 > \frac{\$3{,}500 - \$0}{4} = \$\ \ 875$$

TABLE 11.7.
DECLINING-BALANCE DEPRECIATION SWITCHING TO STRAIGHT-LINE DEPRECIATION

End of Year t	Depreciation Charges During Year t	Book Value at the End of Year t
0	—	$5,000
1	$1,500 (DB)	3,500
2	1,050 (DB)	2,450
3	817 (SL)	1,633
4	817 (SL)	816
5	816 (SL)	0

$$t = 3: \qquad \$735 < \frac{\$2{,}450 - \$0}{3} = \$\ 817.$$

The left-hand values for these comparisons are obtained from Table 11.5.

These calculations indicate a switch from declining-balance depreciation to straight-line depreciation at the beginning of the third period. Since declining-balance depreciation does not consider the salvage value in its calculation, the depreciation charges are the same for an assumed zero salvage as for the salvage of $1,000 used to develop Table 11.5. The depreciation charges and the accompanying book value for an asset having an original cost of $5,000 and a zero salvage are shown in Table 11.7.

11.6 ACCELERATED COST RECOVERY SYSTEM, ACRS (1981–1986)

The passage of the Economy Recovery Act of 1981 significantly changed the calculation of depreciation for assets in service between 1981 and 1987. An important feature of this tax law was the Accelerated Cost Recovery System, referred to as ACRS for the purposes of this text. This system signified a dramatic departure from previous depreciation methods by specifying that all eligible depreciable assets are to be assigned to one of only four separate classes of property. These four classes are identified by the number of years over which the property will be depreciated. The types of property associated with each of the four classes for tangible depreciable property are listed as follows:

3-year property includes cars and light-duty trucks. Also included is machinery and equipment used in research and experimentation and all other equipment having an ADR life of 4 years or less.

5-year property includes all personal property not included in any other class. Also included is most production equipment and public utility property with an ADR life between 5 and 18 years.

10-year property includes public utility property with an ADR life between 18 and 25 years and depreciable real property such as buildings and structural components with an ADR life less than or equal to 12.5 years.

15-year property includes depreciable real property with an ADR life greater than 12.5 years and public utility property with an ADR life greater than 25 years.

11.6.1 Prescribed Method, ACRS

Once the property is assigned to the proper class, the depreciation rates are prescribed by the IRS. The rates for each of the classes are presented in Table 11.8. These rates are based on 150% declining-balance depreciation switching to straight-line depreciation. (See Section 11.5.3.) The percentage (α) applied for the declining-balance portion of this method is determined by dividing 1.5 by the number of years in the recovery period for the property class. The first year's rate is $\alpha \div 2$, which assumes a half-year's use of the property during its initial year. For 15-year property $\alpha = 1.5/15$, so that in the first year the percentage is $1.50/15 \div 2 = 0.05$ or 5%. The second-year percentage is found from $\alpha(1 - 0.05) = 0.10(1 - 0.05) = 0.095$ rounded to 10%.

TABLE 11.8.
DEPRECIATION PERCENTAGE FOR ACRS CLASSES

Recovery Year is	3-Year	5-Year	10-Year	15-Year Public Utility
1	25	15	8	5
2	38	22	14	10
3	37	21	12	9
4		21	10	8
5		21	10	7
6			10	7
7			9	6
8			9	6
9			9	6
10			9	6
11				6
12				6
13				6
14				6
15				6

The percentage listed is applied to the unadjusted basis (original cost) of the asset. Thus, there is no consideration of possible future salvage value in the depreciation charges resulting from the use of ACRS. If, however, at the time the asset is retired, an actual salvage value is realized, any difference between this actual salvage amount and the book value determines the tax liability. Therefore, an asset classified as 5-year property having a $5,000 first cost and estimated salvage value of $1,000 will have the depreciation charges shown in Table 11.9.

Another consideration at the time of retirement for ACRS *is whether the property is disposed of during or before its last recovery year.* If it is, no depreciation deduction is allowed for the year of disposition. If the asset in Table 11.9 is disposed of in year 3, the only depreciation charges allowable are $750 and $1,100. The book value at the time of disposal will be calculated as follows:

$$B_3 = \$5,000 - \$750 - \$1,100 = \$3,150.$$

If the property is disposed of after the last recovery year, the property is fully depreciated and there is no additional depreciation. For 5-year property, the depreciation would be as shown in Table 11.9, and the half-year's depreciation applies only to the initial year.

11.6.2 Alternative Method, ACRS

The only other option under ACRS is the use of straight-line depreciation for the recovery periods shown in Table 11.10. These recovery periods are determined by the ACRS property classes, and the taxpayer can select one of three possible recovery periods. Since this option is a part of ACRS, any salvage values are ignored.

If a taxpayer elects the alternative method, it is mandatory that the half-year convention be applied. If the *property is fully depreciated before disposal,* one half of the full annual depreciation is charged to the initial year and the year following the recovery period. For example, for 5-year property with a $5,000

TABLE 11.9.
AN EXAMPLE OF ACRS DEPRECIATION

End of Year t	Depreciation Charges During Year t	Book Value at the End of Year t
0	—	$5,000
1	(0.15)($5,000) = $ 750	4,250
2	(0.22)(5,000) = 1,100	3,150
3	(0.21)(5,000) = 1,050	2,100
4	(0.21)(5,000) = 1,050	1,050
5	(0.21)(5,000) = 1,050	0

TABLE 11.10.
STRAIGHT-LINE RECOVERY PERIODS FOR ACRS

Type of Property	Optional Recovery Periods Available
3-year property	3, 5, or 12 years
5-year property	5, 12, or 25 years
10-year property	10, 25, or 35 years
15-year public utility property	15, 35, or 45 years

orginal cost and $1,000 estimated salvage, the depreciation charges are $500, $1,000, $1,000, $1,000, $1,000 and $500, if the disposal occurs after year 6. If *disposal occurs before the property is fully depreciated,* a half-year's depreciation is allowed for the year of disposal; thus, for the property just described, if disposal occurs during the third year, the depreciation charges are $500 and $1,000, with $500 allowed for the year of disposal.

11.7 MODIFIED ACCELERATED COST RECOVERY SYSTEM, MACRS (1987 AND LATER)

The Tax Reform Act of 1986 represents the most complete revision of the federal tax code in the last four decades. Although individual and corporate income tax remain as the primary sources of federal revenue, the tax base is broadened and the marginal tax rates are lowered. The result is a shifting of the tax burden onto corporations and away from individual taxpayers.

The basic structure of ACRS is retained by the Modified Accelerated Cost Recovery System of 1986, referred to here as MACRS. For this new system, there is an increase in the number of property classes and a change in the depreciation rates. For *personal property,* the 3-, 5-, 10-, and 15-year classes remain, with some realignment of class definitions. Two new classes, for 7- and 20-year property, are added. Depreciation rates for classes 3, 5, and 10 are increased by using 200% rather than 150% declining balance switching to straight-line. Also, for depreciation purposes, there is no distinction between new and used property, and salvage value is still ignored.

For *real property* (land and buildings) the 15-year and 18-19-year classes in ACRS have been substantially increased to 27.5 years for residential property and 31.5 years for nonresidential property. In addition, both of these types of property for MACRS must be depreciated by the straight-line method. The effect of lengthening the depreciation period and changing from the accelerated methods of ACRS leads to significantly lower depreciation amounts for real property.

11.7.1 Property Classes

MACRS allows for the recovery of depreciable property over designated recovery periods. Property is usually depreciable if its usefulness is exhausted over time and it is either

1. Property used in a trade or business
2. Property held for the production of income

The Internal Revenue Code of 1954 defines two major categories of depreciable property as Section 1245 property and Section 1250 property. Section 1245 property includes tangible and intangible *personal property.* All business property, other than structural components, contained in or attached to buildings or real property is tangible personal property. (Property temporarily attached to buildings with screws and bolts that can be removed is personal property.) Types of property that belong to this classification are machinery, vehicles used for business, and equipment. Copyrights and patents are considered to be intangible personal property.

Section 1250 property is *real property* (property that is permanently part of the land) such as buildings and their structural components and appurtenances. This classification of property includes buildings, parking facilities, bridges, and swimming pools. Structural components such as central air conditioning systems, plumbing, and wiring are considered real property. Land, although it may be defined as real property, is not depreciable.

To determine the depreciation schedule appropriate for depreciable property, MACRS has defined six recovery period classes for personal property and two classes for real property. The definitions of these classes for personal property follow, and the depreciation method applicable is given.

3-year property includes special material handling devices and special tools for manufacturing.
ADR life $\leq$ 4 years.
Depreciation method: 200% declining-balance switching to straight-line with half-year convention.

5-year property includes automobiles, light and heavy trucks, computers, copiers, semiconductor manufacturing equipment, qualified technological equipment, and equipment used in research.
4 $<$ ADR life $<$ 10 years.
Depreciation method: 200% declining-balance switching to straight-line with half-year convention.

7-year property includes property that is not assigned to another class, such as office furniture, fixtures, single-purpose agricultural structures, and railroad track.

10 year ≤ ADR years < 16.
Depreciation method: 200% Declining-balance switching to straight-line with half-year convention.

10-year property includes assets used in petroleum refining, in the manufacture of castings, forgings, and in the manufacture of tobacco products and certain food products.
16 years ≤ ADR life years < 20 years.
Depreciation method: 200% declining-balance switching to straight-line with half-year convention.

15-year property includes telephone distribution equipment and municipal water and sewage treatment plants.
20 years ≤ ADR life < 25 years.
Depreciation method: 150% declining-balance switching to straight-line with half-year convention.

20-year property includes vessels, barges and tugs, and municipal sewers.
25 years ≤ ADR life.
Depreciation method: 150% declining-balance switching to straight-line with half-year convention.

For real property the classes are described as follows.

Residential property includes apartment buildings and rental houses.
Depreciation method: Straight-line depreciation with half-year convention over 27.5 years.

Nonresidential property includes office buildings, warehouses, manufacturing facilities, refineries, mills, parking facilities, fences, and roads.
Depreciation method: Straight-line depreciation with half-year convention over 31.5 years.

11.7.2 Depreciation Schedules and Averaging Conventions

The depreciation method applicable to each of the eight property classes is listed along with each of the class descriptions just given. Although the IRS has not provided an official listing of the depreciation percentages for each of these classes, these percentages are presented in Table 11.11 for property assigned to the 3-, 5-, 7-, 10-, 15-, and 20-year classes, when using the half-year convention. Since the taxpayer is expected to calculate the appropriate depreciation rates, an example of these calculations is also presented.

The two *real* property classes are depreciated using straight-line depreciation over 27.5 and 31.5 years. Again the half-year convention is mandated; so there is a half year's depreciation for the initial year of service and year of dis-

TABLE 11.11.
DEPRECIATION PERCENTAGE FOR MACRS CLASSES

Recovery Year	3-Year Class (200% DB)	5-Year Class (200% DB)	7-Year Class (200% DB)	10-Year Class (200% DB)	15-Year Class (150% DB)	20-Year Class (150% DB)
1	33.33	20.00	14.29	10.00	5.00	3.75
2	44.44	32.00	24.49	18.00	9.50	7.22
3	14.81	19.20	17.49	14.40	8.55	6.68
4	7.41	11.52	12.49	11.52	7.70	6.18
5		11.52	8.92	9.22	6.93	5.71
6		5.76	8.92	7.37	6.23	5.28
7			8.92	6.55	5.90	4.89
8			4.46	6.55	5.90	4.52
9				6.55	5.90	4.46
10				6.55	5.90	4.46
11				3.28	5.90	4.46
12					5.90	4.46
13					5.90	4.46
14					5.90	4.46
15					5.90	4.46
16					2.95	4.46
17						4.46
18						4.46
19						4.46
20						4.46
21						2.23

posal. Since there is no salvage-value consideration, the formula that gives the depreciation rate for the n year recovery period is

$$D_t = (.5)\frac{1.0}{n} \qquad \text{for } t = 1 \text{ and } n + 1.$$

$$D_t = \frac{1.0}{n} \qquad \text{for } t = 2, 3, \ldots n.$$

MACRS treats all personal and real property as being placed in service and being disposed of in the middle of the year.[1] For the years when property is placed into service or disposed of the taxpayer charges one-half of the year's annual depreciation amount. This convention, coupled with the fact that salvage value is disregarded, leads to the percentages listed in Table 11.11 for each property class.

The percentages in Table 11.11 are based on declining balance depreciation switching to straight-line depreciation. This switch depends on whether the straight-line depreciation on the remaining book value exceeds the declining balance depreciation for the period being considered, as shown in Table 11.7. However, adjustments are made to account for the half-year convention and no salvage value. The resulting formulas for declining-balance and straight-line depreciation on the remaining balance are as follows. Let

n = recovery period for the property class;

D_t = depreciation charge for year t;

$B(\text{DB})_t$ = book value for declining balance at the end of year t;

$B(\text{DB})_0$ = original cost of property;

α = (200% or 150%) percentage $\div\ n$.

Declining-balance depreciation is

$$D(\text{DB})_t = (0.5)\alpha B(\text{DB})_0 \qquad \text{for } t = 1.$$

$$D(\text{DB})_t = \alpha B(\text{DB})_{t-1} \qquad \text{for } t > 1.$$

[1] *If the property placed in service during the last 3 months of the tax year exceeds 40% of the total property placed in service, all property placed in service during the year requires the use of a midquarter convention.*

Straight-line depreciation on the remaining declining-balance book value is

$$D(\mathrm{SL})_t = (0.5)B(\mathrm{DB})_0 \div n \qquad \text{for } t = 1.$$

$$D(\mathrm{SL})_t = B(\mathrm{DB})_{t-1} \div (n - t + 1.5) \qquad \text{for } t = 2, \ldots n.$$

$$D(\mathrm{SL})_{n+1} = \text{book value at } t = n. \qquad \text{for } t = n + 1$$

The switch to straight-line depreciation from declining-balance depreciation occurs for time t when

$$D(\mathrm{DB})_t < D(\mathrm{SL})_t$$

To better understand how the percentages are determined, an example based on the 5-year property is presented. In this example it is assumed that the initial book value of the property is 1.0; that is, $B_0 = 100\%$.

5-YEAR PROPERTY (200% DECLINING-BALANCE SWITCHING TO STRAIGHT-LINE WITH HALF-YEAR CONVENTION)

Declining-balance Percentages	Straight-line Percentages (Based on Remaining Book Value and Remaining Years)
$D(\mathrm{DB})_1 = \frac{1}{2}\left[\frac{200\%}{n}\right]B_0$	$D(\mathrm{SL})_1 = \frac{1}{2}\left(\frac{100.00}{5}\right) = 10.00$
$= \frac{1}{2}\left[\frac{2.00}{5}\right](100.00) = 20.00^*$	
$D(\mathrm{DB})_2 = \left[\frac{200\%}{n}\right]B_1$	$D(\mathrm{SL})_2 = \frac{(100.00 - 20.00)}{4.5} = 17.77$
$= \left[\frac{2.00}{5}\right](100.00 - 20.00) = 32.00^*$	
$D(\mathrm{DB})_3 = \left[\frac{200\%}{n}\right]B_2$	$D(\mathrm{SL})_3 = \frac{(80.00 - 32.00)}{3.5} = 13.71$
$= \left[\frac{2.00}{5}\right](80.00 - 32.00) = 19.20^*$	
$D(\mathrm{DB})_4 = \left[\frac{200\%}{n}\right]B_3$	$D(\mathrm{SL})_4 = \frac{(48.00 - 19.20)}{2.5} = 11.52$
$= \left[\frac{2.00}{5}\right](48.00 - 19.20) = 11.52^\dagger$	

Declining-balance Percentages	Straight-line Percentages (Based on Remaining Book Value and Remaining Years)
$D(\text{DB})_5 = \left[\frac{200\%}{n}\right]B_4$	$D(\text{SL})_5 = \left(\frac{28.80 - 11.52}{1.5}\right) = 11.52^*$
$= \left[\frac{2.00}{5}\right](28.00 - 11.52) = 6.91$	
	$D(\text{SL})_6 = (0.5)(11.5) = 5.76^*$

* *Depreciation percentages in Table 11.11.*

11.7.3 Alternative Method, MACRS

As with ACRS the taxpayer has the option of substituting straight-line depreciation for most property. If this option is selected, the half-year convention is used with straight-line depreciation, assuming no salvage value. The half-year convention is applicable for the year, the property is placed in service and its disposal year, as is done by the prescribed percentage depreciation under MACRS.

Generally, for personal property that is assigned a property class, the life to be used is the recovery period for that class. For personal property with no assignable class life, a 12-year recovery period is used. If the taxpayer chooses the alternative method, it must be applied to *all* property in that class that is placed in service during that tax year.

11.7.4 Effects of Time of Disposal on Depreciation Schedules

With the half-year convention now mandated by all MACRS depreciation methods, the time of disposal affects the depreciation schedules of MACRS and ACRS differently. For MACRS the year the property is placed in service receives one half of that year's annual depreciation amount, and the year of disposal is charged one half of its annual depreciation amount. As an example, assume an asset originally costing \$10,000 with a 3-year recovery period. Given in Table 11.12 are depreciation schedules applicable for three different times of disposal for the prescribed percentage and alternative methods under MACRS. The times of disposal shown are (a) before the asset is fully depreciated, (b) one year after the recovery period, and (c) after the asset is fully depreciated.

Replacement studies require the knowledge of the depreciation schedules in effect at the time of acquisition. Since an item may have been placed in service prior to 1987, it is necessary to know how the conventions used by ACRS

TABLE 11.12.
EFFECTS OF TIME OF DISPOSAL UNDER MACRS

End of Year	Prescribed Percentage Method: 200% Declining-Balance Switching to Straight-Line Depreciation			Alternative Method: Straight-Line Depreciation		
	(a)	(b)	(c)	(a)	(b)	(c)
1	$3,333	$3,333	$3,333	$1,667	$1,667	$1,667
2	2,222*	4,444	4,444	1,667*	3,333	3,333
3	—	1,481	1,481	—	3,333	3,333
4	—	741*	741	—	1,667*	1,667
5	—	—	—*			—*

* *Year of disposal.*

affect the depreciation schedules. For ACRS, the year of acquisition is charged one half-year's depreciation, where the effect of the half-year convention is reflected in the prescribed percentages. However, in the year of disposal, *no* depreciation is charged. Table 11.13 presents the application of the ACRS depreciation schedules to the example used in Table 11.12.

Examination of these two tables demonstrates how differently the same asset would be depreciated depending on the appropriate tax law and the time of disposal.

TABLE 11.13.
EFFECTS OF TIME OF DISPOSAL UNDER ACRS

End of Year	Prescribed Percentage Method: 150% Declining-Balance Switching to Straight-Line Depreciation			Alternative Method: Straight-Line Depreciation		
	(a)	(b)	(c)	(a)	(b)	(c)
1	$2,500	$2,500	$2,500	$1,667	$1,667	$1,667
2	—*	3,800	3,800	1,667*	3,333	3,333
3	—	3,700	3,700	—	3,333	3,333
4	—	—*	—	—	1,667*	1,667
5	—	—	—*	—	—	—*

* *Year of disposal.*

11.8 UNITS-OF-PRODUCTION DEPRECIATION

In some cases, it may not be advisable to assume that capital is recovered in accordance with a theoretical value-time model such as those considered previously. An alternative is to assume that depreciation occurs on the basis of the amount of work performed without regard to the duration of the asset's life. Thus, a trencher might be depreciated on the basis of pipeline trench completed. If the trencher has a first cost of $11,000 and a salvage value of $600, and if it is estimated that the trencher would dig 1,500,000 linear feet of pipeline trench over its life, the depreciation charge per foot of trench dug may be calculated as

$$\frac{\$11,000 - \$600}{1,500,000} = 0.006933 \text{ per foot.}$$

The undepreciated capital at the end of each year is a function of the number of feet of trench dug during the year. If, for instance, 300,000 feet of pipeline trench were dug during the first year, the undepreciated balance at the end of the year would be $11,000 − 300,000($0.006933) = $8,920.10. This analysis would be repeated at the end of each year.

11.9 DEPLETION

Depletion differs in theory from depreciation in that the latter results from use and the passage of time and the former from intentional, *piecemeal removal* of certain types of assets. Depletion refers to an activity that tends to exhaust a supply; the word literally means emptying. When natural resources are exploited in production, depletion indicates a lessening in value with the passage of time. Examples of depletion are the removal of coal from a mine, timber from a forest, stone from a quarry, and oil from a reservoir.

There is a difference in the manner in which the capital recovered through depletion and depreciation must be handled. In the case of depreciation, the asset involved usually may be replaced with a like asset, but in the case of depletion such replacement is usually not possible. In manufacturing, the amounts charged for depreciation are reinvested in new equipment to continue operation. In mining, however, the amounts charged to depletion cannot be used to replaced the ore deposit, and the venture may sell itself out of business. The depletion returned in such a case represents the funds necessary to acquire new properties, thus enabling the venture to continue.

To illustrate the two methods available for calculating depletion, an example of oil depletion is presented. Although the depletion allowance is generally repealed for oil and gas production, there are two basic exemptions that still apply. One of those exemptions is for small producers and royalty owners. In this example it is assumed that the depletion allowance is being calculated for a small producer. (A small producer has an average production that does not exceed 1,000 barrels per day.)

11.9.1 The Cost Method

This method of depletion is similar to the units-of-production method of depreciation discussed in Section 11.8. That is, the depletion charge is based on the amount of the resource that is consumed and the initial cost of the resource. Suppose a reservoir containing an estimated 1,000,000 barrels of oil required and initial investment of $7,000,000 to develop. For this reservoir the

$$\text{Unit depletion rate} = \frac{\$7{,}000{,}000}{1{,}000{,}000\ \text{bbl}} = \frac{\$7}{\text{bbl}}.$$

If 50,000 bbl of oil are produced from this reservoir during a year, the

$$\text{Depletion charge} = 50{,}000\ \text{bbl}(\$7/\text{bbl}) = \$350{,}000.$$

Since the estimates of the number of units of the resource remaining vary from year to year, the unit depletion rate is recalculated by dividing the unrecovered cost of the resource by the estimated units of the resource remaining.

11.9.2 The Percentage Method

For certain resources the United States income tax laws provide an optional method of calculating the depletion, sometimes referred to as the *percentage* method. Percentage depletion allows a fixed percentage of the gross income produced by the sale of the resource to be the depletion charge. Thus, over the life of an asset the total depletion charges may exceed the initial cost of the asset. However, it is required that for any period the depletion charge may not exceed 50% of the taxable income for that period computed without the depletion deduction.

A few examples of the fixed percentages defined by the tax statutes for certain natural resources follow.

Uranium and sulfur	22%
Oil and gas	
1. Regulated natural gas and gas sold under fixed contract	22%
2. Small producers and royalty owners	15%
Coal and salt	10%
Gravel	5%

Using the oil-reservoir example again, suppose that the price of oil is $20 per barrel and all other expenses associated with the operation of the oil reservoir are $200,000. To find the allowed depletion charge, the following calculations are required:

Gross depletion income (50,000 bbl @ $20/bbl)	$1,000,000
Depletion rate	15%
Percentage depletion charge	$ 150,000

Now it is necessary to determine if the deduction of $150,000 exceeds the maximum depletion charge allowed by the tax laws for percentage depletion. The following calculations specify the maximum depletion charge that is permitted:

Gross depletion income	$1,000,000
Less expenses	200,000
	$ 800,000
Deduction limitation	50%
Maximum depletion charge	$ 400,000

Since $150,000 is less than $400,000, the full depletion charge of $150,000 is allowable in this circumstance. For these properties to which the percentage depletion method applies, the taxpayer each year may determine the depletion deduction using both methods. The law allows the taxpayer to take whichever deduction is larger in that year. For the example, comparing the $150,000 amount to the $350,000 deduction permitted under the *cost* method indicates it would be advantageous to apply the *cost* method in this situation.

KEY POINTS

- Property can be depreciated by its owner if its usefulness is exhausted over time and it is used in business or is held for the production of income.
- The accountant's view of depreciation is that it is a periodic cost charged to an asset to reflect its loss in value over its lifetime.
- The book value of an asset at some point in time is its cost of acquisition less its accumulated depreciation to that point in time.
- Depreciation charges are essential in the calculation of income taxes.
- The selection of a depreciation schedule for tax purposes is based on the tax code in effect at the time an asset is acquired.

1. Assets acquired since 1987 are depreciated based on MACRS.
2. Assets acquired from 1981 through 1986 are depreciated based on ACRS.
3. Assets acquired before 1981 used the traditional depreciation methods of straight-line, declining balance, or sum-of-the-years'-digits.

- Once a depreciation method is selected for an asset, that method is in effect for the lifetime of the asset.
- Depletion is the gradual reduction of certain natural resources, and depletion deductions reflecting this loss of value are allowed by federal tax law.

QUESTIONS AND PROBLEMS

1. What is the relationship between the balance sheet and the profit and loss statement?
2. Describe the difference between general accounting activities and engineering economy studies.
3. Describe the difference in viewpoint in respect to time between a balance sheet and an economy study.
4. Explain the relationship between the cost of goods sold statement and the profit and loss statement.
5. Define the terms *direct labor* and *direct materials.*
6. What are the cost components of manufacturing overhead?
7. Define the costs elements that comprise the production cost of a unit of output.
8. What is the difference between the accountant's concept of depreciation and depreciation as it is commonly understood by the general public?
9. Describe the value-time function and name its essential components.
10. Does a depreciation charge represent an actual disbursement of funds? Explain.
11. Discuss the difference between physical and functional depreciation.
12. How does depreciation affect the profit and loss statement of a company?
13. Explain the difference between capital recovery with return as used by the engineering economist and the accountant's concept of depreciation. Under what conditions will the annual depreciation charge equal capital recovery with return?
14. What is the effect of inflation on depreciation charges?
15. How does an increase (or decrease) in depreciation expense affect the net profit shown on the income statement?

16. What is the effect on federal and state income taxes if the depreciation expense is increased (or decreased)?

17. An individual purchased a car for $15,000. He was offered amounts for his car in succeeding years as follows:

Year	*1*	*2*	*3*	*4*	*5*	*6*
Offer	$12,000	$9,500	$7,500	$5,500	$4,000	$3,000

On the basis of the offers, calculate the depreciation and the undepreciated value of the car for each year of its life. Is this the accountant's view of depreciation?

Assuming the amounts offered were based on no inflation since the car was purchased, what would have been the amounts offered if the annual inflation rate over the 6 years had been 7% per year? Does this effect change the depreciation to be charged based on the accountant's view of depreciation?

18. Using pre-1981 depreciation methods, calculate the book value at the end of each year using straight-line and double declining balance methods of depreciation for an asset with an initial cost of $50,000 and an estimated salvage value of $10,000 after 4 years. Present the results in tabular form.

19. An asset was purchased during 1980 for $4,800. It is being depreciated in accordance with the straight-line method for an estimated total life of 20 years and salvage value of $800. What is the difference in its current book value and the book value that would have resulted if declining-balance depreciation at a rate of 10% had been applied for 10 years?

20. A central air-conditioning unit was purchased in 1978 for $250,000 and had an expected life of 10 years. The salvage value for the unit at that time was expected to be $40,000. What will be the book value at the end of 7 years for (a) straight-line depreciation, and (b) double-declining-balance depreciation?

21. A pre-1981 asset had a first cost of $60,000 with a salvage of $5,000 after 11 years. If it is depreciated by the declining-balance method using a rate of 10%, what will be the

- **a.** depreciation charge in the tenth year? Answer: $2,325
- **b.** depreciation charge in the eighth year?
- **c.** book value in the eighth year? Answer: $25,828

22. A truck purchased before 1981 cost $46,000 and had an estimated useful life of 7 years and estimated salvage value of $4,000. How much depreciation should be allocated on the new truck in each year and what is the undepreciated value of the truck at the end of the fifth year, assuming the use of straight-line depreciation?

23. A machine acquired in 1979 at a cost of $32,000 had an estimated life of 5 years. Residual salvage value was estimated to be $3,500.

a. Determine the annual depreciation charges during the useful life of the machine using straight-line depreciation.

b. What depreciation rate makes the book value at the end of 5 years equal to the esimated salvage value if the declining-balance method of depreciation is used? Answer: $\alpha = 0.3576$

24. A drilling rig was purchased in 1980 for 1.8 million dollars. It has an estimated life of 10 years, and the salvage value at that time is expected to be $500,000. After 5 years what will be the total depreciation charged for (a) straight-line depreciation, and (b) double-declining balance depreciation?

25. A special-purpose machine was purchased in 1980 for $200,000 with an expected life of 12 years. If its salvage value is expected to be zero, what will be the depreciation charge in the first and fifth years and what will be the book value at the end of the third year for (a) straight-line depreciation, and (b) double-declining-balance depreciation?

26. A piece of earthmoving equipment was purchased in 1986 by a road construction firm at the cost of $120,000. It is estimated that after 6 years the equipment will be sold for $18,000. Using ACRS prescribed percentages for a 5-year recovery period, find the asset's book value for each year over its recovery period.

27. An oil refinery has added a new computer to control one of its refining units. Under ACRS for a 5-year recovery period, what would be the largest possible depreciation percentage for the fourth year after the placement of the computer in service? Use the ADR. Answer: 21%

28. An electric generating unit placed in service in 1985 by a public utility has an ADR life of 35 years. Show the difference between the percentage depreciation rates each year if the unit is depreciated (a) under ACRS using the prescribed percentages or (b) using straight-line for the shortest recovery period allowed. What happens to this difference if the longest recovery period is used with straight-line depreciation?

29. Equipment acquired 5 years ago by Firm *A* is purchased by Firm *B* for $1,900,000. The equipment is expected to have a service life of 20 more years. If its ADR life is 12 years, to what property class does the equipment belong under MACRS? What is the depreciation charge for the equipment 4 years after its purchase by Firm *B*? Answer: 7-year property; $237,310.

30. A company purchased 20 new automobiles at a cost of $20,000 each. After 4 years the cars were sold for $6,000 each. What were the depreciation charges over the life of these cars if the depreciation method under MACRS giving the largest write-off in year 1 was selected?

31. A supply of special tools for manufacturing was purchased in 1992 for $35,000. This 3-year property is expected to be sold during its fifth year of service for $2,000. Show the depreciation charges over the tools' lifetime using the prescribed percentages and the alternative method for MACRS.

32. A coal barge was acquired in 1992 for $900,000. It is expected that it will be sold for scrap 25 years from its purchase date for $35,000. Using MARCS, find the book value of the barge at the end of 1998 for the prescribed-percentage method and the alternative method.

33. An evaporation tower in a petroleum refinery was placed in service during 1990 at a cost of $1.2 million. This 10-year property is expected to be removed at a cost of $100,000 during 1999. Find the depreciation charges over the tower's lifetime using MACRS for the

a. prescribed-percentage method

b. alternative method

34. A 1990 asset had a first cost of $70,000 with recovery period of 7 years. The salvage value at that time is estimated to be $5,000. What is the total accumulated depreciation charged between 1990 and the end of 1993 under MACRS? Answer: $48,132

35. In 1988 a company purchased a numerically controlled machine for $850,000. It was estimated that it would have a salvage value of $100,000 and be classified as 5-year property. Find the depreciation charges over an 8-year period (assume the machine is sold at 7 years of age) using the prescribed-percentage method and the alternative method for MACRS.

36. A gas chromatograph has been purchased by a laboratory for use in medical research for $45,000. The expected service life is 8 years and the expected salvage value at that time is projected to be $8,000. Show each year's depreciation charge under MACRS over the recovery period for that type of property.

37. A personal computer purchased during 1989 had a first cost of $2,500 wih an estimated salvage value of $400 at the end of 1995. Compute the book value at the time of disposal of the computer if it was sold at the end of

a. 1989. Answer: $2,000

b. 1990.

c. 1991.

d. 1992.

e. 1993. Answer: $288

f. 1994.

g. 1995.

38. An automatic control mechanism is estimated to provide 3,000 hours of service during its life. The mechanism costs $5,200 and will have a salvage value of zero after 3,000 hours of use. What is the depreciation charge if the number of hours the mechanism is used per year is (a) 600, (b) 1,300? Answer: $2,253

39. An impact tester is utilized on a stand-by basis. It is estimated that a tester of this type, which costs $55,000, can sustain 2,000,000 test impacts over its lifetime. What is the depreciation charge this year if the number of impacts during the year was (a) 140,000, (b) 80,000?

40. Filters costing $1,000 per set are used to separate particles from hot gases being discharged into the air. These filters must be replaced every 2,500 hours. If the plant operates 24 hours per day, 30 days per month, what is the depreciation charge per month for these filters?

41. A gold mine that is expected to produce 250,000 ounces of gold is purchased for 30 million dollars. Gold can be sold for $350 per ounce. If 27,000 ounces are produced this year, what will be the depletion allowance for (a) cost depletion and (b) percentage depletion where the fixed percentage for gold is 22%? For what price of gold would it be advantageous to apply percentage depletion rather than cost depletion?

42. A sulfur-mining operation has a gross income of $500,000 in February from the production of sulfur. All expenses with the exception of depletion expenses for this month amount to $390,000. The percentage depletion rate for sulfur is 22%. What is the depletion allowance for this month?

43. A natural gas well produces 400,000 Mcf during this year. The well has been producing 3 years and at the beginning of this year the recoverable reserves were estimated at 8,000,000 (Mcf). The gas is being sold to a pipeline company under a fixed contract at the price of $7 per Mcf. What will be this year's percentage depletion if the annual expenses associated with this gas well are (a) $700,000, (b) $1,800,000? Answer: (b) $500,000

44. An oil reservoir is estimated to have 350,000 barrels of oil. The cost of purchasing the lease and well is $4,200,000. If crude oil is selling for $20 per barrel and 30,000 barrels of oil per year are produced, what is the amount of the depletion allowance for (a) cost depletion and (b) percentage depletion at a fixed percentage of 15%?

45. For each of the following sets of information select the depletion amount that will provide the largest allowable depletion charge for this year.

	Amount Produced This Year	*Gross Income*	*Expenses before Depletion*	*Unit Depletion Rate*	*Percentage Depletion Rate*
(a)	900,000 lb	$27,000,000	$14,000,000	$2.50/lb	10%
(b)	20,000 tons	$ 800,000	$ 500,000	$5.00/ton	22%
(c)	7,200 oz	$ 2,500,000	$ 1,500,000	$200/oz	22%
(d)	2,000,000 lb	$ 3,600,000	$ 2,600,000	$0.31/lb	15%
(e)	3,000,000 tons	$66,000,000	$40,000,000	$3.00/ton	10%

12

Income Taxes in Economic Analysis

Income-tax laws are the result of legislation over a period of time. Since they are man-made, they incorporate many diverse ideas, some of which appear to be in conflict. These laws are expressed through a number of provisions and rules intended to meet current conditions. The provisions and rules are not absolute in application, but rather are subject to interpretation when applied.

Income taxes are levied by the federal government as well as by many individual states. State income-tax laws will not be considered here because the principles involved are similar to those for federal tax laws, because state income rates are relatively small, and because there is a great diversity of state income-tax law provisions.

12.1 RELATION OF INCOME TAXES TO PROFIT

In most cases, the desirability of a venture is measured in terms of differences between income and cost, receipts and disbursements, or some other measure of profit. It is the specific function of economic analysis to determine future profit potential that may be expected from prospective engineering proposals being examined. But income taxes are levies on profit that result in a reduction of its magnitude.

In relation to engineering economy studies, income taxes are merely another class of expenditure which require special treatment. Such taxes must be taken into account along with other classes of costs in arriving at the profits accruing to sponsors of an undertaking.

Of the aggregate profit earned in the United States from such productive activities as manufacturing, construction, mining, lumbering, and others in which engineering analysis is important, the aggregate disbursement for income taxes will be approximately 30% to 40% of net income. Their importance becomes clear when it is realized that in some cases accumulated local, state and federal income taxes may result in disbursements of more than 50% of the profit to be expected from a venture.

12.2 FEDERAL INCOME TAX FOR CORPORATIONS

The term *corporation,* as used in income-tax law, is not limited to the artificial entity usually known as a corporation but may include joint stock associations or companies, some types of trusts, and some limited partnerships. In general, all business entities whose activities or purposes are the same as those of corporations organized for profit are taxed as such. This discussion will be based upon the tax requirements that apply to the usual business corporation.

In general, a corporation's income tax is a levy on its net earnings, the difference between the income derived from and the expense incurred in business activity, with some exceptions. This net income before taxes or taxable income, along with the following tax schedule is the basis for determining a corporation's federal tax obligation.

Corporation's Taxable Income	*Tax Rates*
$ 0 to $ 50,000	15%
$ 50,000 to $ 75,000	25%
$ 75,000 to $100,000	34%
$100,000 to $335,000	39%
Over $335,000	34%

For the first $50,000 of taxable income the marginal tax rate is 15%. The marginal rates for the next two $25,000 increments of taxable income are 25% and 34%. After reaching the $100,000 level an additional 5% tax is imposed until the tax saving realized due to the graduated rates for the first $100,000 is fully recovered. This tax saving of $11,570 represents the difference in taxes based on the graduated rates and a *flat* tax of 34% over the first $100,000 of taxable income.

$$[(34\%)(\$100,000) - (15\%)(\$50,000) - (25\%)(\$25,000) - (34\%)(\$25,000)] = \$11,750.$$

To recover this amount, the 5% tax must be applied to the next \$235,000 of taxable income (\$11,750/0.05). Therefore, the 39% tax rate is applicable for taxable income ranging from \$100,000 to \$335,000. Any taxable income over \$335,000 will be taxed at a marginal tax rate of 34% *and* a flat tax rate of 34%.

The total taxes a corporation would pay on taxable income less than \$335,000 are presented in Figure 12.1. The slope of each segment in Figure 12.1 represents the federal tax for that increment of taxable earnings. If taxable income were between \$75,000 and \$100,000, Figure 12.1 indicates that taxes of \$13,750 would be paid on the first \$75,000, with additional taxes of 34% on the portion of taxable income occurring between \$75,000 and \$100,000.

For most large or moderate size corporations taxable income will generally exceed \$335,000. The equation applicable for any corporation with taxable income greater than this amount is

$$T = TI \times 34\%, \qquad \text{TI} > \$335{,}000$$

where

T = total income tax;

TI = earnings before taxes (taxable income).

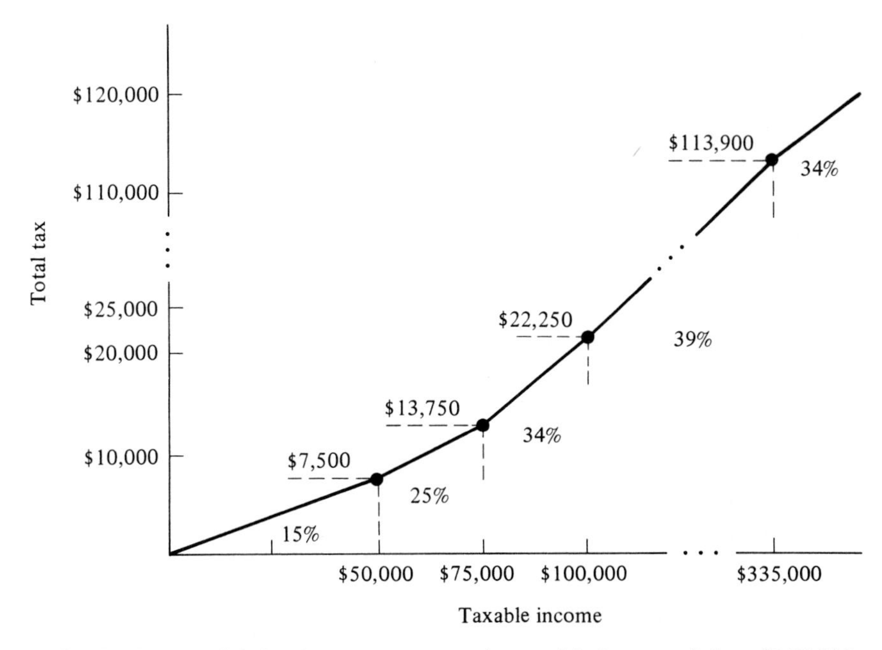

FIGURE 12.1. Total federal corporation tax for taxable incomes below \$335,000.

It is important to note that any additional taxable income earned, if the firm has already exceeded the $335,000, will be taxed at 34%. Thus the total tax rate and the marginal tax rate are the same above this amount.

The following is an illustration of the calculation of corporation taxes for a particular firm.

1.	Gross income. This item embraces gross sales less cost of sales, dividends received on stocks, interest received on loans and bonds, rents, royalties, gains and losses from capital or other property.		$780,000
2.	Deductions. This item embraces expense not deducted elsewhere, as, for example, in cost of sales. Includes compensation of officers, wages, and salaries, rent, repairs, bad debts, interest, taxes, contributions, losses by fire, storms, and theft, depreciation, depletion, advertising, contributions to employee benefit plans, and special deductions for partially exempt bond interest and partially exempt dividends.		600,000
3.	Taxable income.		$180,000
4.	First $50,000 of taxable income, $50,000 × 0.15	= $ 7,500	
5.	Next $25,000 of taxable income, $25,000 × 0.25	= 6,250	
6.	Next $25,000 of taxable income, $25,000 × 0.34	= 8,500	
7.	Taxable income in excess of $100,000, $80,000 × 0.39	= 31,200	
8.	Total tax (items 4, 5, 6, 7)	= $53,450	

To contrast federal income tax rates with those rates applicable to individuals see Tables 12.1 and 12.2. Notice that the tax rates and the ranges of values for which the various rates are applicable are quite different. In addition, the

TABLE 12.1.
TAX RATES FOR SINGLE TAXPAYERS (1993)

Taxable Income		Tax	
Over	But Not Over	Percent	Of the Amount Over
$ 0	$20,350	15%	$ 0
20,350	49,300	$3,052.50 + 28%	20,350
49,300		11,158.50 + 31%	49,300

TABLE 12.2.
TAX RATES FOR MARRIED TAXPAYERS FILING JOINT RETURNS OR SURVIVING SPOUSES (1993)

Taxable Income		Tax	
Over	But Not Over	Percent	Of the Amount Over
$ 0	$34,000	15%	$ 0
34,000	82,150	$5,100 + 28%	34,000
82,150		18,582 + 31%	82,150

ranges of values for corporate rates are not indexed for inflation, while those for individuals are indexed. The ranges that are indexed are adjusted each year, based on the previous year's inflation rate. In theory, the individual taxpayer will not be shifted to a higher tax rate because of inflationary effects on taxable income.

12.2.1 Capital-Gains Tax Rates for Corporations

Previous tax law distinguished between short- and long-term capital gains and losses based on the length of time the asset was owned. For corporations current tax law continues this distinction, but now both will be taxed as ordinary income at a maximum rate of 34%. Even though the tax rate for capital gains will be the same as for ordinary income, the tax code retains the statutory distinction between short- and long-term gains so that tax differentials could be reinstated later.

Suppose a firm has capital gains and losses as follows.

Capital gain	$45,000
Capital losses	15,000
Net capital gain	$30,000

If the firm has earned $90,000 gross income excluding any capital gains or losses and has itemized deductions of $25,000, its taxes are calculated as

Gross income (excluding capital gain)	$90,000
Itemized deductions	−25,000
Capital gain	30,000
Total taxable income	$95,000
Tax $13,750 + (0.34)($95,000 − $75,000)	$20,550

For this circumstance the $30,000 capital gain has $10,000 taxed at 25% and the remaining $20,000 portion taxed at 34%.

If the same firm earned $250,000 with $130,000 of deductions with the same $30,000 net capital gain, the taxes would be

Gross income (excluding capital gain)	$250,000	
Itemized deductions	−130,000	
Taxable income (excluding capital gain)	$120,000	
Taxes excluding capital gain		
$22,250 + (0.39)($120,000 − $100,000)		$30,050
Capital gain	$30,000	
Tax on capital gain (0.34)($30,000)		10,200
Total tax		$40,250

In this example the firm pays the 5% additional tax on a portion of its ordinary income while the capital gain is taxed at the maximum rate of 34%. Any firm with taxable income excluding capital gains that exceeds $335,000 will pay a total tax based on 34% of all taxable income including any capital gains. The tax on capital gains can be less than 34% if the taxable income excluding these gains falls below $75,000.

12.3 EFFECTIVE INCOME-TAX RATES

Economy studies involving income taxes can be simplified by the determination of applicable *effective income-tax rates.* An effective income-tax rate, as the term is used here, is a single rate which when multiplied by the taxable income of a venture under consideration will result in the income tax attributable to the venture. Effective income-tax rates are essentially incremental rates that are applicable over ranges of income.

The effective tax rate, t, of any increment of taxable income may be calculated as follows: Find the difference of the tax payable on incomes corresponding to the upper and lower limits of the increments and divide this difference by the amount of the increment. For example, suppose that a corporation with a taxable income of $45,000 wishes to find the effective tax rate on additional income of $15,000. The tax payable on $45,000 is equal to $45,000 × 0.15 = $6,750, and the tax payable on $60,000 is equal to $7,500 + ($60,000 − $50,000) × 0.25 = $10,000. The difference is $10,000 − $6,750 = $3,250. The average tax rate over the increment is

$$t = \frac{\$3{,}250}{\$60{,}000 - \$45{,}000} = \frac{\$3{,}250}{\$15{,}000} = 0.22 \text{ or } 22\%.$$

Therefore the effective tax on the additional $15,000 in taxable earnings is 22%.

For corporations with earnings in excess of \$335,000 the effective federal tax rate on any additional earnings will be at the incremental rate of 34%. However, because of state income taxes the total effective tax rate may be higher. Since state income taxes are deductible as expenses when determining federal taxes, the effective rate for both federal and state taxes can be calculated from the following expression:

$$\text{effective tax rate} = (\text{incremental federal rate})(1 - \text{incremental state rate}) + \text{incremental state rate}.$$

For example, if the incremental federal rate is 39% and the incremental state tax rate is 7%, then the effective incremental state and federal tax rate is $0.39(1.0 - 0.07) + 0.07 = 0.4327$, or 43.27%. Tax rates will be stated as effective tax rates unless indicated otherwise.

Once they are determined for a particular purpose, effective rates serve as a means for transmitting the thought and analyses that went into the determination in a single figure. By the use of effective rates the consideration of income taxes in economy studies is reduced to two distinct factors: (1) the determination of effective tax rates applicable to particular activities or particular classes of activities, and (2) the application of the effective tax rates in economy studies.

12.4 INTEREST AND INCOME TAXES

Interest *earned* from funds loaned by an individual or corporation is usually considered an income item when calculating income taxes. An important exception is the interest earned from municipal bonds, which are not taxed by the federal government. The economic importance of the tax treatment of interest earned can be observed from the following example.

Suppose an individual is planning to invest \$10,000 in one of two types of bonds. The first is a \$10,000, 12% corporate bond with interest payable annually, selling at par value. The other, also selling at par value, is a \$10,000 municipal bond with a 9% annual tax-free interest payment. Both bonds will mature in 7 years, and the individual's incremental tax rate is 33%. The after-tax rate of return on each of the bonds is determined from the after-tax cash flows in Figure 12.2. The cash flows yield after-tax IRR's of 8.04% for the corporate bond and 9% for the municipal bond. Thus, the after-tax interest earned determines the revenue that will actually be realized, and this is the value on which the decision should be based.

Interest *paid* for funds borrowed by an individual or a corporation to carry on a profession, trade, or business is deductible from income as the expense of

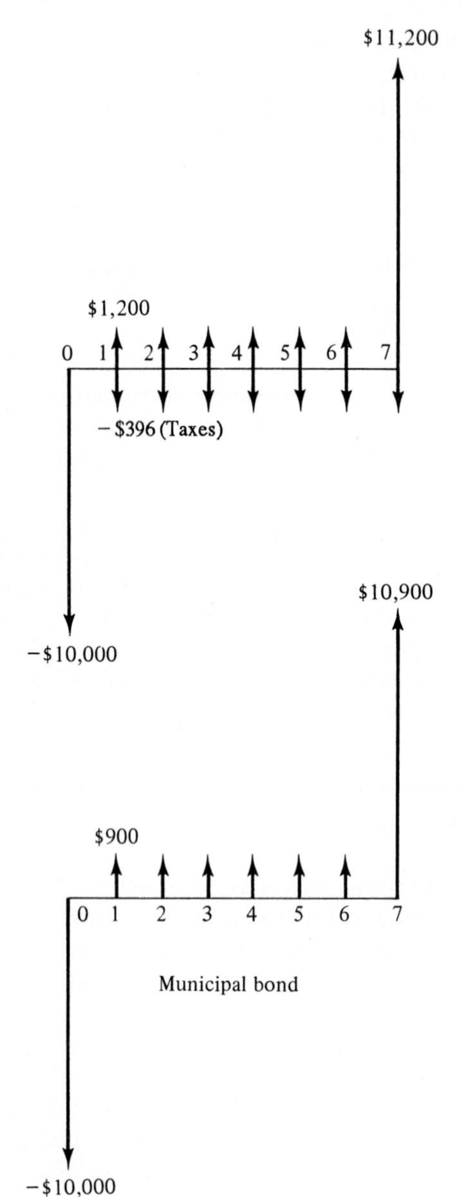

FIGURE 12.2. After-tax cash flow for two types of bonds.

carrying on such activity. In addition, interest paid on funds borrowed for home ownership is deductible from the adjusted income in computing an individual's income tax.

The federal government, by allowing the deduction of most interest payments as expense, is in effect lowering the cost of borrowing. For example, consider two corporations, Firm *A* and Firm *B*, that are essentially identical except that in Firm *A* no borrowed funds are used and in Firm *B* an average of $200,000 is borrowed at the rate of 12% during a year. Assume that taxable income before interest payments in both cases is $400,000. Shown below are the taxable income and the income taxes that each firm would be required to pay. The taxes are based on the corporate tax schedule given in Section 12.2.

	Firm A	*Firm B*
Taxable income before interest deduction	$400,000	$400,000
Interest expense	0	24,000
Taxable income	$400,000	$376,000
Taxes (effective tax rate, 0.34)	$136,000	$127,840

The difference in taxes to be paid represents an annual saving in taxes of $136,000 − 127,840 = $8,160 to be realized if the $200,000 is borrowed. With the interest costs on the loan being $24,000 and a tax saving of $8,160, Firm *B* is effectively paying only $24,000 − $8,160 = $15,840 to borrow the money. This is equivalent to an interest rate of 7.92% = (1 − 0.34)12%.

When analyzing the effect of an investment alternative it is important to know how the alternative is to be financed. *If the firm has a capital structure that includes significant borrowing or if a special borrowing is directly associated with the investment, the after-tax consequences of the associated interest payments must be considered.*

No matter what the form of the repayment schedule, it is always necessary to identify the interest payments separately from the repayment of the borrowed principal. Although both principal payments and interest payments affect the final cash flow, only the interest payments affect the tax cash flows. The inclusion of both these effects in the analysis of alternatives is demonstrated in Section 12.10.

12.5 DEPRECIATION AND INCOME TAXES

Since the amount of taxes to be paid during any one year is dependent upon deductions made for depreciation, the latter is a matter of consideration for the Internal Revenue Service of the United States Treasury Department and state taxing agencies. Directives are issued by governmental agencies as guides to the taxpayers in properly handling depreciation for tax purposes.

The taxpayer who incurs an economic loss with the loss in value of an asset over time is entitled to the depreciation deduction. Ordinarily, this is the taxpayer that owns the asset. If a taxpayer owns only a portion of a depreciable asset, then only that portion of the depreciation deductions can be claimed.

12.5.1 Effect of Method of Depreciation on Income Taxes

Prior to 1981 a taxpayer had considerable choice in the method of depreciation he could use for tax computation. With the adoption of the Accelerated Cost Recovery System (ACRS) for 1981 and the subsequent modification of that system for 1987, MACRS, and later, the options available to the taxpayer are substantially reduced. Under MACRS the only options are to use the prescribed-depreciation schedule or straight-line depreciation as discussed in Section 11.7.

If the effective tax rate remains constant and the operating expense is constant over the life of the equipment, the depreciation method used will not alter the total of the taxes payable over the life of the equipment. But methods providing for high depreciation and consequent low taxes in the first years of life will be of advantage to the taxpayer because of the time value of money.

For illustration of the comparative effect of the straight-line method and the MACRS prescribed-percentage method, consider the following example. Assume that a taxpayer has just installed an asset whose first cost is $1,000, whose

TABLE 12.3.
ALTERNATIVE A

Income Taxes for Straight-Line Method of Depreciation and 10-Year Life

Year End *A*	First Cost *B*	Income Before Depr. and Income Tax *C*	Annual Depr. *D*	Income Less Depr. (taxable income) $C - D$ *E*	Income Tax Rate *F*	Income Tax $E \times F$ *G*
0	$1,000					
1		$ 200	$ 50	$150	0.4	$ 60.00
2		200	100	100	0.4	40.00
3		200	100	100	0.4	40.00
4		200	100	100	0.4	40.00
5		200	100	100	0.4	40.00
6		200	100	100	0.4	40.00
7		200	100	100	0.4	40.00
8		200	100	100	0.4	40.00
9		200	100	100	0.4	40.00
10		200	100	100	0.4	40.00
11		200	50	150	0.4	60.00
		$2,200	$1,000			$480.00

$$\text{Present worth of income taxes} = \sum_{t=1}^{11} [\text{Col. } G \times (\overset{P/F,15,t}{\qquad})] = \$231$$

property class is 10 years, and whose salvage value is nil. The asset is estimated to have a constant operating income before depreciation and income taxes of $200 per year. The taxpayer estimates the applicable effective income tax rate for the life of the machine to be 40%. He considers money to be worth 15% and wishes to compare the effect of the straight-line method and the MACRS prescribed-percentage method. Let the first method be represented by Alternative *A* and the second method by Alternative *B*. These alternatives are summarized in Tables 12.3 and 12.4.

The total income tax paid during the life of the asset is equal to $480 with either alternative, but the present worths of the taxes paid differ. For Alternative *A*, the present worth of taxes paid as of the beginning of year 1 is equal to $231. The corresponding figure for Alternative *B* is $199. The difference in favor of the prescribed-percentage method is $32.

By examining the book value of an asset over its life for the various methods of depreciation, one can easily see which methods produce larger deprecia-

TABLE 12.4.
ALTERNATIVE B

Income Taxes for MACRS Prescribed-Percentage Method of Depreciation and 10-Year Recovery Period

Year End *A*	First Cost *B*	Income Before Depr. and Income Tax *C*	Annual Depr. *D*	Income Less Depr. (taxable income) $C - D$ *E*	Income Tax Rate *F*	Income Tax $E \times F$ *G*
0	$1,000					
1		$ 200	$ 100	$100	0.4	$ 40.00
2		200	180	20	0.4	8.00
3		200	144	56	0.4	22.40
4		200	115	85	0.4	34.00
5		200	92	108	0.4	43.20
6		200	73	127	0.4	50.80
7		200	66	134	0.4	53.60
8		200	66	134	0.4	53.60
9		200	66	134	0.4	53.60
10		200	66	134	0.4	53.60
11		200	32	168	0.4	67.20
		$2,200	$1,000			$480.00

$$\text{Present worth of income taxes} = \sum_{t=1}^{11} [\text{Col. } G \times (\overset{P/F,15,t}{\quad\quad})] = \$199$$

tion deductions in the early portion of the asset's life. The time-value functions for three of the depreciation methods discussed in Chapter 11 are presented in Figure 12.3. Since the book value of an asset is the first cost of the asset less accumulated depreciation charges, the curves with the lowest values in their early life are those which are most favorable to the taxpayer.

From Figure 12.3 it is seen the MACRS has a faster depreciation write-off than straight-line depreciation. Thus, prior to 1981 for tax purposes double-declining-balance depreciation found wide acceptance, since it effectively postponed the payment of taxes to a later date. Such a postponement was favorable to the taxpayer because money has a time value, and a dollar paid in taxes at the present is worth more than a dollar paid in taxes at some future date.

Both ACRS and the MACRS depreciation methods are based on declining balance early in the asset's life with a switch to straight-line depreciation later. *Therefore, MACRS depreciation can be considered to be an accelerated method that will postpone the payment of taxes.* It has been noted that under the current MACRS system there is an optional or an alternative depreciation system based on straight-line depreciation. This alternative system would not be classified as using an accelerated depreciation method.

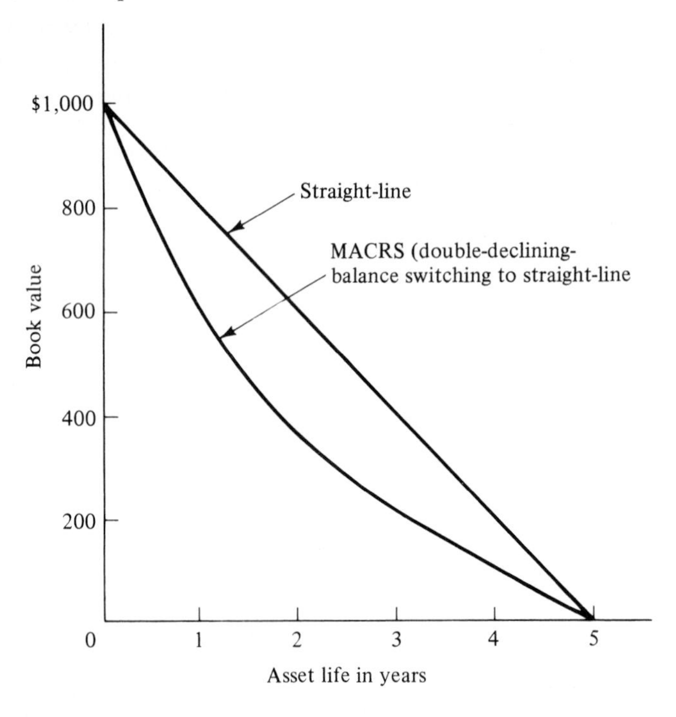

FIGURE 12.3. Book value computed by three methods of depreciation.

The depreciation schedule applicable to an acquired asset depends on the tax law in effect at the time of purchase.

Assets purchased before 1981 cannot use the ACRS schedule required for 1981 through 1986, or its modified version applicable to 1987 and beyond. The significant advantage of ACRS over the methods used prior to 1981 was the substantially shorter period over which an asset could be depreciated. While MACRS has generally increased the time of depreciation compared to ACRS, the current system still allows for faster recovery of depreciation than the pre-1981 methods.

12.5.2 Effect of Recovery Period on Income Taxes

There is often little connection between the life of an asset that will be realized and that which a taxpayer may use for tax purposes. If the applicable effective tax rate remains constant through the realized life of an asset, the use of a shorter recovery period for tax purposes will usually be favorable to a taxpayer. The reason is that use of short lives for tax purposes results in relatively high annual depreciation, low annual taxable income, and, consequently, low annual income taxes during the early years of an asset's life. Even though there will be correspondingly higher annual income taxes during the later years of an asset's life, the present worth of all income taxes during the asset's life will be less.

Suppose the conditions of Alternative *A* in Table 12.3 had permitted the use of a five-year property class. For this property class, the time span over which the depreciation charges are incurred in considerably shorter (6 years) than for the example (11 years). Applying straight-line depreciation with the half-year convention for this shorter depreciation period over the same 11-year period yields the following annual taxes.

Year-End	0	1	2	3	4	5	6	7	8	9	10	11
Annual taxes		$40	0	0	0	0	$40	$80	$80	$80	$80	$80

The present worth of these taxes at 15% is equal to

$$\$40(\overset{P/F,15,1}{0.8696}) + \$40(\overset{P/F,5,6}{0.4323}) + \$80(\overset{P/A,15,5}{3.3522})(\overset{P/F,15,6}{0.4323}) = \$168.$$

The difference in favor of the shorter life is \$231 − \$168, a present worth saving of \$63. This example demonstrates the important economic advantage of charging depreciation or other expense items as early as possible over the life of an asset.

12.6 GAIN OR LOSS ON DISPOSITION OF DEPRECIABLE ASSETS

When a depreciable asset used in business is exchanged or sold for an amount greater or smaller than its book value, this gain or loss on disposal has an important effect on income taxes. At the time of disposal there is an accounting entry that identifies whether a gain or loss has occurred. For the current tax law this classification of gains and losses on disposal continues. If an asset is sold for an amount equal to its book value at the time of disposal, no taxes are incurred on the amount received.

Figure 12.4 presents examples of these different situations for an asset having a book value of $2,880 at the time of disposal. For Case 1, the excess of salvage value over book value ($5,000 − $2,880) represents a capital gain of $2,120. Case 2 indicates a large capital gain of $12,000 − $2,880, since the asset is sold for an amount greater than its original cost. Case 3 illustrates a capital loss of $2,880 − $1,000. If the asset's salvage value equals $2,880, the book value, no taxes are calculated on that cash receipt.

Because capital gains and losses are taxed as ordinary income under current tax law, the tax considerations at the time of disposal may be greatly simplified. Net capital gains or losses are combined with the firm's other income and the applicable tax rate is applied (see Section 12.2.1).

A firm with a taxable income of $500,000 excluding capital gains will pay $34,000 in federal taxes on a $100,000 capital gain (0.34 × $100,000). In fact, the firm's total taxes are 0.34 × $600,000 or $204,000.

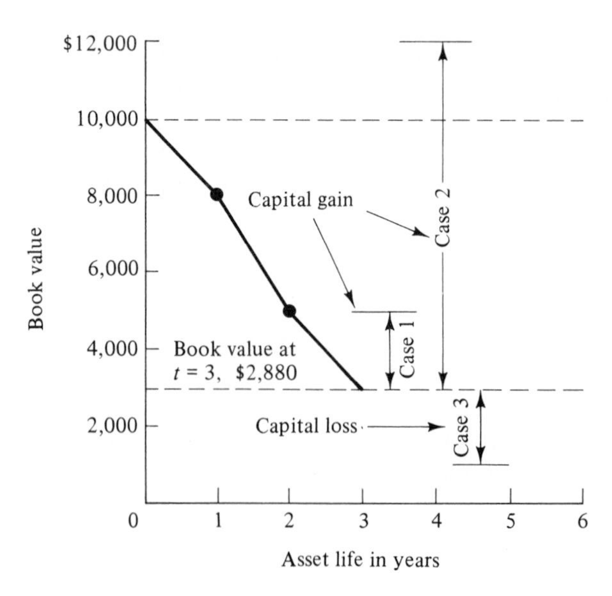

FIGURE 12.4. Capital gains and losses on disposal.

Unfortunately, the situation will not always be so straightforward. If the firm pays federal, state, and local income taxes, the combined tax rate applicable to ordinary income may differ from the rate applicable to capital gains. Usually state and local governments *do not* tax capital gains, so the only rate applied is the ordinary federal tax rate with a maximum at 34%.

12.7 TAX CREDITS

Tax credits are amounts determined by the tax code that reduce the tax liability of an individual or corporation. These credits are made available to taxpayers for a variety of purposes. Some credits reduce the tax liability of the elderly, the disabled, and individuals with low incomes. Others are intended to stimulate investment in particular areas of the economy. Credits of this type include the investment credit, credit for research expenses, energy credits, and fuel-production credits. Because of the multitude of rules associated with the use of various tax credits, the taxpayer should depend on expert tax advice for specifics on the use of tax credits.

12.7.1 The Investment Credit

To encourage the use of new technology and equipment, the federal government has in the past provided tax incentives that favor the acquisition of capital assets. The investment credit was an allowed amount that directly reduced a taxpayer's liability, and it was determined by the dollar amount of investment for "qualified property."[1]

The Tax Reform Act of 1986 repealed the regular investment credit for property placed in service after 1986. However, for a few special situations and types of property the investment credit remains in effect. In addition, regular investment credits that are being carried forward can continue to be claimed although in reduced amounts. Therefore, it is important to understand how tax credits are reflected in an after-tax analysis.

In general, tax credits represent a dollar of reduction in federal taxes owed for each dollar of tax credit. Tax credits are unlike depreciation or other expense items, which reduce tax owed by only a fraction of the expense amount. Thus tax credits can be considered a direct saving in the amount of the tax credit. The regular investment credit prior to 1987 was allowed only for the year during which the property was placed in service. This one-time credit effectively reduced the purchase price of the property by the amount of the allowable credit.

For most of the common types of capital investment being currently undertaken, the regular investment credit no longer applies. Rather than focus on

[1] *Prentice Hall Federal Tax Handbook (Englewood Cliffs, N.J.: Prentice Hall, Inc.).*

certain special types of tax credit or special tax situations, all problems and examples in this book assume that the tax credit is *not* applicable unless specifically stated otherwise.

12.8 DEPLETION AND INCOME TAXES

Natural resources such as minerals, oil, gas, timber, and certain others are found in deposits or stands of definite quantities. They are exploited by removal. As quantities are removed to produce income, the amount remaining in a deposit or stand is reduced.

In arriving at taxable income, depletion is deducted from gross income. As seen in Section 11.9 the unit method for determining the amount of depletion for a given year is based on the total cost of the property, the total number of units in the property, and the number of units sold during the year in question. This may be expressed by the equation

$$\text{depletion for year} = \frac{\text{cost of property}}{\text{total units in property}} \times \text{units sold during year.}$$

Oil, gas, and mineral properties are also depleted for tax purposes on the basis of a percentage of gross income, provided that the amount allowed for depletion is not greater than 50% of the taxable income of the property before depletion allowances. In general, percentage depletion will be elected if it results in less tax than does depletion based on cost.

Typical of percentage depletion allowances for mineral and similar resources are the following:

Sulphur, uranium, asbestos, bauxite, graphite, mica, antimony, bismuth, cadmium, cobalt, lead, manganese, nickel, tin, tungsten, vanadium, zinc	22%
Gas wells (only certain gas wells)	22%
Domestic gold, silver, oil shale, copper, and oil (small producers) iron ore	15%
Various clays, diatomaceous earth, dolomite, feldspar, and metal mines if not in 22% group	14%
Coal, lignite, sodium chloride	10%
Brick and tile clay	$7\frac{1}{2}$%
Gravel, peat, pumice, and sand	5%

Timber may not be depleted on a percentage basis. In general, timber is depleted for tax purposes on the basis of its cost at a specific date, the total number of units of timber on the property, and the units of timber removed during the tax year. Special rules are applicable for determining cost and for revising the number of units on the property from time to time.

12.8.1 Effect of Depletion Method on Income Taxes

Most resources subject to depletion may be depleted on either a cost basis or a percentage basis in computing income taxes. Consider an example of a mineral deposit estimated to contain 60,000 units of ore and purchased at a cost of $24,000 including equipment. The depletion rate is $24,000 ÷ 60,000, or $0.40 per unit of ore.

This property is also subject to percentage depletion at a rate of 22% per year, but not in excess of 50% of the net taxable income of the property before depletion. The effective tax rate for this enterprise is 30%. An analysis of this situation is presented in Table 12.5.

By taking advantage of the two methods of depletion, the income taxes on the property above total $8,880. If only cost depletion had been used, income taxes in successive years would have been $3,600, $2,880, $1,560, $1,440, and $120 for a total of $9,600.

TABLE 12.5.
APPLICATION OF COST AND PERCENTAGE DEPLETION TO A MINERAL DEPOSIT

Year *A*	Units Produced *B*	Gross Income (*B* × $2) *C*	Operating Cost *D*	Net Income Before Depletion and Income Tax *E*	50% of Net Income (*E* × 0.5) *F*	Allowable Cost Depletion at $0.40 per Unit *H*	Allowable Percentage Depletion at 22% (*C* × 0.22) *I*	Taxable Income (*E* less *F*, *H*, or *I*) *J*	Income Tax (Tax Rate of 0.3)(*J* × 0.3) *K*
1	20,000	$40,000	$20,000	$20,000	$10,000	$8,000	$8,800‡	11,200	3,360
2	16,000	32,000	16,000	16,000	8,000	6,400	7,040‡	8,960	2.688
3	12,000	24,000	14,000	10,000	5,000*	4,800	5,280	5,000	1,500
4	8,000	16,000	8,000	8,000	4,000	3,200	3,960‡	4,040	1,212
5	4,000	8,000	6,000	2,000	1,000	1,600†	1,760	400	120
									$8,880

**Deduct this amount for depletion, as it is the maximum percentage depletion allowed, Col F, and because it is greater than cost depletion, Col. H.*

†Deduct this amount for depletion because cost depletion, Col. H, is greater than the allowable percentage depletion, Col. I.

‡Deduct this amount for depletion because it is full percentage-depletion allowance, Col. I, and it is greater than allowable cost depletion, Col. H.

12.9 CALCULATING AFTER-TAX CASH FLOWS

The basic philosophy of the comparison of alternatives on an after-tax basis requires that the resulting tax and other effects on the flrm be evaluated with and without the alternatives.

This approach forces each alternative to be judged on its particular merits, as best as they can be determined. To quantify the after-tax effects of a single alternative it is necessary to convert the before-tax cash flow into an after-tax cash flow. The most common method of accomplishing this conversion is to use the tabular method presented in Table 12.6. The example is based on the following information:

First cost:	\$60,000
Salvage value:	None
Annual income:	\$50,000
Annual operating costs:	\$24,000
Depreciation method:	Straight-line (3-year recovery period, half-year convention)
Estimated life:	4 years
Effective tax rate:	40%

To assure that the tabular calculation procedure is general, positive values are assigned to revenues while costs are assigned negative values. Notice that in Column *B* the net annual income is found by adding the annual receipts (+) to the annual costs (−), giving net receipts of \$50,000 − \$24,000 or \$26,000. The initial outlay of the alternative is shown occurring at the present (the end of year 0). If there had been a salvage value at the time of disposal, it would have been included as a cash flow at the end of year 4. This before-tax cash flow represents the net cash flow for all activities except taxes.

Column *C* presents the depreciation charges as negative amounts. These amounts are costs, but they are *not* cash flows. Their purpose is to allow for the calculation of taxes, which *are* cash flows. Although the recovery period is 3 years, there are depreciation charges over 4 years because of the half-year convention.

The taxable income for each year in Column *D* is calculated by adding the amounts in Column *B* (excluding the first cost) to the amounts in Column *C*. The investment of funds has no effect on taxable income, since these are funds on which taxes were paid at the time they were earned. Therefore, an alternative's first cost is not considered in the calculation of taxable income.

The taxes for each year shown in Column *E* are found by multiplying the tax rate of −0.40 times the taxable income. Since the taxable income is a positive value in this example, the taxes are negative amounts, indicating that they are disbursements or taxes paid. If the taxes were positive, this would represent a situation where taxes are being saved. That is, less taxes are being paid with the alternative than without it.

The last step is to add the taxes in Column *E*. which are cash flows, to the before-tax cash flows in Column *B*. The intermediate step dealing with depreciation is only to account for its tax effects. Since depreciation is not an actual disbursement of cash, it must not be included in the sum of the other cash flows when finding the after-tax cash flow. If the tax cash flows associated with an alternative are known, the use of Columns *C* and *D* is not necessary.

The tabular approach is intended to focus on the individual merits of the alternative. By observing the firm's profit and loss with or without the alternative, any change associated with the alternative is represented in Table 12.6. To observe this fact, suppose that for year 2 the firm has the following profit and loss statement with and without the alternative.

PROFIT AND LOSS (YEAR 2)

	Without Alternative	With Alternative	Net Change
Gross income	$600,000	$650,000	$50,000*
Expenses:			
Operating expenses	300,000	324,000	24,000*
Depreciation	140,000	160,000	20,000
Taxable income	$160,000	$166,000	$ 6,000
Taxes	64,000	66,400	2,400*
After-tax profit	$ 96,000	$ 99,600	

**Actual cash flows associated with alternative*

TABLE 12.6.
TABULAR CALCULATION OF AFTER-TAX CASH FLOW

End of Year *A*	Before-Tax Cash Flow *B*	Depreciation *C*	Taxable Income $B^* + C$ *D*	Taxes $-0.40 \times D$ *E*	After-Tax Cash Flow $B + E$ *F*
0	−$60,000				−$60,000
1	26,000	−$10,000	$16,000	−$6,400	19,600
2	26,000	− 20,000	6,000	− 2,400	23,600
3	26,000	− 20,000	6,000	− 2,400	23,600
4	26,000	− 10,000	16,000	− 6,400	19,600

**Consider only those revenues and costs that affect taxable income. (The initial cost of investment has no effect on taxable income.)*

Notice that the change in taxes above corresponds to the taxes in Table 12.6 for year 2. Also, the net change in gross income ($50,000) and operating expenses ($24,000) is properly identified in Column *B* as a net receipt of $26,000. From a cash flow point of view the after-tax profit from the profit and loss statement will not agree with the after-tax cash flow in Table 12.6. Cash flow analysis recognizes depreciation as a cost only in its effect on taxes. Profit and loss analysis includes depreciation as an expense item to be deducted from revenue. Therefore any results based on profit and loss analysis are expected to differ from those results produced by cash flow analysis.

Since the after-tax cash flow from Table 12.6 represents the incremental cash flow compared to Do Nothing, the incremental after-tax IRR can be determined from the following expression.

$$0 = -\$60{,}000 + \$19{,}600(\overset{P/A,i^*,4}{\quad}) + \$4{,}000(\overset{P/A,i^*,2}{\quad})(\overset{P/F,i^*,1}{\quad})$$

$$i^* = 16.3\% \text{ (after-tax).}$$

This return compares to the incremental before-tax IRR found from

$$0 = -\$60{,}000 + \$26{,}000(\overset{P/A,i^*,4}{\quad})$$

where

$$i = 26.3\% \text{ (before-tax).}$$

To decide whether this alternative meets the profit objectives of the firm, however, the incremental after-tax rate of return must be compared to the after-tax MARR stipulated by the firm.

In many instances in the evaluation of economic alternatives only the costs of the alternatives are considered. Most often this occurs where alternatives that are intended to provide the same service are to be compared. Since the benefits to be derived are equal for each alternative, it is common practice to eliminate the estimation of associated revenues to reduce the evaluation effort.

It might appear that ignoring revenues for alternatives would prevent the comparison of these alternatives on an after-tax basis. Fortunately, this is not the case. It can be seen that the direct application of the procedures presented in Table 12.6 will permit the incorporation of tax effects in the comparison of cash flows having equal revenue streams. By following the sign convention just used, (+) revenues, (−) costs, the tabular method will produce tax-adjusted cash flows that can be directly compared as long as (1) their revenues are assumed to be equal, and (2) the firm is realizing a profit on its other activities. To see how the procedure operates, examine the cost-only cash flow in Table 12.7.

For this example, MACRS straight-line depreciation is used. The estimated salvage for the investment is $5,000, and the effective tax rate is 42%. Since no

TABLE 12.7.
TAX EFFECTS FOR A COST CASH FLOW

End of Year A	Before-Tax Cash Flow B	Depreciation Charges** C	Taxable Income B + C D	Taxes (Savings) −0.42 × D E	After-Tax Cash Flow B + E F
0	−$30,000				−$30,000
1	− 15,000	−$ 5,000	−$20,000	$ 8,400	− 6,600
2	− 15,000	− 10,000	− 25,000	10,500	− 4,500
3	− 15,000	− 10,000	− 25,000	10,500	− 4,500
4	− 15,000	− 5,000	− 20,000	8,400	− 6,600
5	− 15,000	0	− 15,000	6,300	− 8,700
5	5,000*	0	5,000	−2,100	2,900

* *Salvage value.*
** *Depreciation charges ignore salvage value in MACRS.*

annual operating revenues are being considered, the taxable income appears as $20,000 in costs for year 1, $25,000 in costs for year 2, etc. That is, the effect of this project in year 1 is to reduce the firm's taxes by $8,400 per year (if revenues aren't considered). Assuming that the firm has sufficient earnings from other activities to offset these costs, there will be a "savings" in taxes of $8,400. By following the sign convention utilized in the previous tables, the taxes in Column *E* are positive, indicating that these tax amounts would be avoided yearly if the costs in Columns *B* and *C* were incurred.

The after-tax cash flow is found by adding these tax savings to the before-tax cash flow shown in Column *B*. By comparing this after-tax cash flow with a similarly calculated after-tax cash flow for a competing alternative, the alternative with the minimum equivalent after-tax cost can be identified.

Because of the tabular format used to solve for after-tax cash flows, there is widespread use of computer spreadsheets to solve the types of problems. Use of spreadsheets such as Lotus 1-2-3®, Excel®, Quatro®, and SuperCalc® is encouraged.

12.10 EVALUATION OF FINANCING METHODS USING AFTER-TAX CASH FLOWS

There are numerous financial arrangements by which assets may be acquired. In many cases the asset is purchased outright with funds that have been accumulated through the retention of earnings. For this situation the asset is owned by the firm, and any tax benefits associated with property ownership are available to the firm.

Another means of ownership is to finance the first cost of the asset partially or totally through the use of borrowed funds. A firm for which 40% of its capital is debt can consider each acquisition as being financed with 40% borrowed funds and 60% retained earnings. In other instances it will be possible to identify a specific loan and its repayment schedule with a particular acquisition. Many times a variety of borrowing arrangements are provided by the dealer selling the asset. Again ownership resides with the purchaser, although the provider of credit may have a legal claim on the asset in case the purchaser fails to make the specified loan payments.

By leasing an asset, the lessee has use and possession of the asset without having ownership. In this type of financing the asset's owner, the lessor, provides the lessee with the asset for a periodic rental charge. Here the user of the asset does not need the funds required to purchase the asset, and any concern associated with the disposal of the asset is avoided. On the other hand, any advantages stemming from asset ownership, such as depreciation write-offs and appreciation in value, cannot be realized by the lessee.

To understand the evaluation of alternatives with regard to how they are financed, let us look at an example dealing with the purchase of a single used asset for research purposes. The acquisition of this asset is analyzed for three general methods of financing:

1. Purchase with retained earnings
2. Borrowing
3. Leasing

The pertinent information regarding this asset is as follows.

First cost:	$30,000
Salvage value:	None
Annual income:	Not shown
Annual operating costs:	$6,000
Depreciation:	Straight-line (recovery period, 3 years, half-year convention)
Estimated life:	5 years
Effective tax rate:	50%
MARR (after-tax):	10%

For this example the same asset is being considered with regard to various means of acquisition, so there is no need to quantify the revenues, since they will be identical for each alternative. Also, since the asset is a used one with the estimated salvage equaling the book value at retirement, capital-gains tax is not considered.

12.10.1 PURCHASE WITH RETAINED EARNINGS

If an asset is purchased with funds that have been previously earned, the analysis follows the tabular format shown in Table 12.8. Notice that the recovery period for which there are depreciation charges does not need to equal the estimated service life of the asset. Also, the calculation of taxable income in Column *D* is the year-by-year sum of Column *B* and Column *C* with one exception. At the end of year 0, there is no effect on taxable income as a consequence of investing funds on which taxes have been previously collected. The $PW(i)$ of this after-tax cash flow at the after-tax MARR of 10% is

$$PW(10) = -\$30{,}000 - \$500(\overset{P/F,10,1}{0.9091}) + \$2{,}000(\overset{P/F,10,2}{0.8265})$$

$$+ \$2{,}000(\overset{P/F,10,3}{0.7513}) - \$500(\overset{P/F,10,4}{0.6830}) - \$3{,}000(\overset{P/F,10,5}{0.6209})$$

$$= -\$29{,}503.$$

12.10.2 Purchase with Borrowed Funds

There are numerous arrangements by which assets may be purchased with borrowed funds. To demonstrate the method for analyzing the effect of borrowing on after-tax cash flows, two different borrowing arrangements are presented. The repayment schedule for the first example assumes a simple-interest loan with the borrowing rate at 12% per year. Thus, $30,000 is borrowed initially, with only the interest of $3,600 being repaid annually for 5 years. Then at the end of year 5 the principal of the loan is repaid in its entirety. The tabular calculation of the after-tax cash flow for this loan arrangement is presented in Table 12.9. Because only the interest payments affect taxes, the principal payments

TABLE 12.8.
PURCHASE OF AN ASSET WITH RETAINED EARNINGS

End of Year *A*	Before-Tax Cash Flow *B*	Depreciation Charges *C*	Taxable Income $B + C$ *D*	Taxes (Savings) $-0.5 \times D$ *E*	After-Tax Cash Flow $B + E$ *F*
0	−$30,000				−$30,000
1	− 6,000	−$ 5,000	−$11,000	$5,500	− 500
2	− 6,000	− 10,000	− 16,000	8,000	− 2,000
3	− 6,000	− 10,000	− 16,000	8,000	− 2,000
4	− 6,000	− 5,000	− 11,000	5,500	− 500
5	− 6,000	0	− 6,000	3,000	− 3,000

TABLE 12.9.
PURCHASE OF AN ASSET WITH A SIMPLE INTEREST LOAN AT 12%

End of Year	Before-Tax Cash Flow	Loan Cash Flow		Depreciation Charges	Taxable Income	Taxes (Savings)	After-Tax Cash Flow
		Principal	Interest		$B + D + E$	$-0.5 \times F$	$B + C + D + G$
A	B	C	D	E	F	G	H
0	−$30,000	$30,000					$ 0
1	− 6,000		−$3,600	−$ 5,000	−$14,600	$7,300	− 2,300
2	− 6,000		− 3,600	− 10,000	− 19,600	9,800	200
3	− 6,000		− 3,600	− 10,000	− 19,600	9,800	200
4	− 6,000		− 3,600	− 5,000	− 14,600	7,300	− 2,300
5	− 6,000	− 30,000	− 3,600	0	− 9,600	4,800	− 34,800

TABLE 12.10.
PURCHASE OF AN ASSET WITH A COMPOUND-INTEREST LOAN AT 12%

End of Year	Before-Tax Cash Flow	Loan Cash Flow		Depreciation Charges	Taxable Income	Taxes (Savings)	After-Tax Cash Flow
		Principal	Interest		$B + D + E$	$-0.5 \times F$	$B + C + D + G$
A	B	C	D	E	F	G	H
0	−$30,000	$30,000					$ 0
1	− 6,000	− 4,722	−$3,600	−$ 5,000	−$14,600	$7,300	− 7,022
2	− 6,000	− 5,289	− 3,033	− 10,000	− 19,033	9,517	− 4,805
3	− 6,000	− 5,924	− 2,398	− 10,000	− 18,398	9,199	− 5,123
4	− 6,000	− 6,634	− 1,688	− 5,000	− 12,688	6,344	− 7,978
5	− 6,000	− 7,431	− 891	0	− 6,891	3,446	− 10,876

and the interest payments are separated in Columns C and D. The taxable income (Column F) is found by adding operating expenses (Column B), interest (Column D), and depreciation (Column E). Once the taxes are determined (Column G), they are added to Columns B, C, and D giving the after-tax cash flow (Column H).

For the second loan arrangement, \$30,000 is borrowed at 12% to be repaid in equal annual payments of

$$\$8{,}322 = (\overset{A/P,12,5}{0.2774}).$$

The tabular calculation of after-tax cash flows for a loan of this type is given in Table 12.10. The calculation of Columns C and D is based on Equations (4.3) and (4.4) developed in Section 4.7 dealing with loans.

Calculating the $PW(i)$ of the after-tax cash flows for the two loan arrangements presented gives the following results.

Simple-interest loan

$$PW(10) = -\$2{,}300(\overset{P/F,10,1}{0.9091}) + \$200(\overset{P/F,10,2}{0.8265}) + \$200(\overset{P/F,10,3}{0.7513})$$

$$-\$2{,}300(\overset{P/F,10,4}{0.6830}) - \$34{,}800(\overset{P/F,10,5}{0.6209})$$

$$= -\$24{,}954$$

Compound-interest loan

$$PW(10) = -\$7{,}022(\overset{P/F,10,1}{0.9091}) - \$4{,}805(\overset{P/F,10,2}{0.8265}) - \$5{,}123(\overset{P/F,10,3}{0.7513})$$

$$-\$7{,}978(\overset{P/F,10,4}{0.6830}) - \$10{,}876(\overset{P/F,10,5}{0.6209})$$

$$= -\$26{,}407$$

We see that the equivalent costs are lower for the alternative incorporating simple-interest loan than for the one using compound interest. Since all features of the alternatives are identical with the exception of the different loan schedules, it is concluded that with an after-tax MARR of 10% the simple-interest loan is the least expensive of the loan arrangements.

Comparing the after-tax cash flows for the three alternatives just examined gives

Puchase with retained earnings

$$PW(10) = -\$29{,}503$$

Purchase with simple-interest borrowing

$$PW(10) = -\$24{,}954$$

Purchase with compound-interest borrowing

$$PW(10) = -\$26{,}407$$

For these three alternatives the most expensive option is to purchase the asset with retained earnings. It might appear that borrowing at 12% would not be wise if the borrowing cost is greater than the opportunity cost (after-tax MARR = 10%). However, a closer examination reveals that the borrowing cost of 12% is *before taxes* and the MARR is stated on an *after-tax* basis. With the effective tax rate at 50% the after-tax cost of borrowing will be approximately 6%. (See Section 12.4.) An evaluation of the after-tax cash flows in Tables 12.8 and 12.10 at an after-tax MARR of 6% yields

$$PW(6) = -\$30{,}000 - \$500\overset{P/F,6,1}{(0.9434)} + \$2{,}000\overset{P/F,6,2}{(0.8900)} + \$2{,}000\overset{P/F,6,3}{(0.8396)}$$

$$-\$500\overset{P/F,6,4}{(0.7921)} - \$3{,}000\overset{P/F,6,5}{(0.7473)}$$

$$= -\$29{,}650$$

and

$$PW(6) = -\$7{,}022\overset{P/F,6,1}{(0.9434)} - \$4{,}805\overset{P/F,6,2}{(0.8900)} - \$5{,}123\overset{P/F,6,3}{(0.8396)}$$

$$-\$7{,}978\overset{P/F,6,4}{(0.7921)} - \$10{,}876\overset{P/F,4,5}{(0.7473)}$$

$$= -\$29{,}650.$$

Thus, evaluated at an after-tax rate of 6%, the borrowing option at compound interest is equal to the option that relies on retained earnings for financing. If these two options are evaluated at an after-tax MARR less than the after-tax cost of borrowing, then borrowing will be the less desirable option.

12.10.3 Leasing

When an asset is leased, there will be no depreciation charges, since the asset is not owned by the lessee. Neither will any loan arrangement be associated with the acquisition of this asset. Nevertheless, substantial tax savings will accrue to the lessee, since the lease charges are deductible as ordinary expense items. Suppose, for this example, that the asset can be leased for an annual charge of

TABLE 12.11.
ASSET OBTAINED BY LEASING

	Before-Tax Cash Flow					
End of Year A	Operating Expenses B	Lease Charges C	Depreciation Charges D	Taxable Income $B + C$ E	Taxes (Savings) $-0.5 \times E$ F	After-Tax Cash Flow $B + C + F$ G
0						0
1	−$6,000	−$8,000	$0	−$14,000	$7,000	−$7,000
2	− 6,000	− 8,000	0	− 14,000	7,000	− 7,000
3	− 6,000	− 8,000	0	− 14,000	7,000	− 7,000
4	− 6,000	− 8,000	0	− 14,000	7,000	− 7,000
5	− 6,000	− 8,000	0	− 14,000	7,000	− 7,000

$8,000. In this case the lease charges exclude any operating expenses, since these expenses will be borne by the lessee. Table 12.11 presents the tabular calculation of the after-tax cash flow for this case.

The $PW(i)$ calculation of the after-tax cash flow in Table 12.11 yields

$$PW(10) = -\$7,000(\overset{P/A,10,5}{3.7908}) = -\$26,535.$$

The $PW(i)$ amounts at an after-tax rate of 10% for each for the four financing options are summarized as follows.

Purchase with retained earnings:	$PW(10) = -\$29,503$
Purchase with simple-interest loan:	$PW(10) = -\$24,954$
Purchase with compound-interest loan:	$PW(10) = -\$26,407$
Lease:	$PW(10) = -\$26,535$

Clearly, the simple-interest loan provides the least expensive method for acquiring the asset. More important, the range of costs associated with the various financing options are presented. With this information it is now possible to weigh the economic trade-offs if one of the more costly options happens to be preferred. This presentation of the range of equivalent costs associated with the various investment alternatives is a primary goal of any economy study intended to assist those responsible for making the final selection.

12.11 INVESTMENT CREDIT AND CAPITAL GAINS IN AFTER-TAX CASH FLOWS

The discussions of Sections 12.6 and 12.7 indicate that for most assets purchased during 1987 or later, the regular investment credit is not allowed and there is no differential federal tax for capital gains. However, for assets purchased prior to

1987, the investment credit was a significant economic consideration when assessing either the final or prospective profitability of an investment. *For such investments the investment credit must be assessed according to the tax law in effect at time of purchase, while the capital-gain effect is determined by relevant tax law at the time of disposal.*

Suppose, for example, a construction firm purchased and later sold a large piece of earthmoving equipment. This equipment was purchased in 1986 and its service life is 6 years. Since this is 5-year recovery property, the investment credit is 8% of the original cost of the equipment. Based on the data that follow, the firm desires to know the after-tax IRR earned over its lifetime by the investment in this equipment:

First cost:	\$200,000
Salvage value:	\$40,000
Net annual income:	\$35,000 for first year, increasing by \$5,000 each year thereafter
Depreciation method:	ACRS prescribed percentage for 5-year recovery period (Table 11.8)
Service life:	6 years
Effective tax rate:	50%
MARR (after-tax):	10%

The above data are in terms of actual dollars received, which reflects any inflation over the 6-year time span. The firm uses an after-tax MARR of 10% to determine if equipment investments are in accord with the firm's profit objectives. The calculation of the equipment's after-tax cash flow is illustrated in Table 12.12. It is assumed that this equipment was purchased with retained earnings.

The depreciation charges are calculated by applying the ACRS depreciation percentages that apply for 1986 acquisitions. These percentages are given in Table 11.8. Note that, although the investment credit is taken, the equipment is still fully depreciated. For a 1986 acquisition, if the 8% investment credit is applied, the original cost becomes the depreciation basis. Thus depreciation percentages are multiplied by the original cost of the equipment to determine the annual depreciation charges; for example first-year depreciation = (0.15)\$200,000 = \$30,000.

Notice the effect of the 8% investment credit in reducing the before-tax outlay of \$200,000 to \$184,000. Although this credit may be taken at the end of the tax year, the convention of associating it at the time of the capital investment is used here. It is assumed that the firm had other tax liabilities at the time of investment, so the full savings due to the credit were realized. The same assumption is applied at the end of year 2, when the depreciation charges exceeded the net income. Again, the tax saving of \$2,000 is shown as a positive value.

The advantage of being able to take the investment credit as compared to not taking it can be assessed by calculating the IRR on the after-tax cash flow in

TABLE 12.12.
EFFECT OF INVESTMENT CREDIT AND SALVAGE VALUE ON AFTER-TAX CASH FLOW

End of Year A	Before-Tax Cash Flow B	Depreciation Charges C	Taxable Income B + C D	Taxes (Savings) −0.5 × D E	After-Tax Cash Flow B + E F
0	−$200,000			$16,000**	$184,000
1	35,000	−$30,000	$ 5,000	− 2,500	32,500
2	40,000	− 44,000	− 4,000	2,000	42,000
3	45,000	− 42,000	3,000	− 1,500	43,500
4	50,000	− 42,000	8,000	− 4,000	46,000
5	55,000	− 42,000	13,000	− 6,500	48,500
6	60,000	0	60,000	− 30,000	30,000
6	40,000*		40,000†	− 13,600	26,400

*Salvage value.
**Investment credit (8% of orginal cost, no adjustment of depreciation base).
†Capital gain (taxed at 34%).

Table 12.12, changing only the initial investment from $184,000 to $200,000. This calculation gives after-tax IRR's of 11.1% and 8.5%, respectively. If the equipment had not qualified for the investment credit, it would not have earned a level of return considered adequate by the firm's MARR of 10%.

Thus prior to 1987 the taxpayer had a significant tax advantage that is not available for later capital investments. The 1986 Tax Reform Act's substitution of lower corporate tax income rates for the investment credit will have a profound effect on those industries that are capital intensive.

Since the capital gain on the investment in Table 12.12 is realized after 1987, the tax law in effect at the time of disposal determines the tax liability. The current tax law indicates that capital gains are taxed at the ordinary corporate rate but limited to a maximum tax rate of 34%. In this example, the 50% rate applied to taxable income includes federal, state, and local taxes. Usually state and local governments do not impose capital-gain taxes, so the capital-gain tax used here is the maximum federal rate of 34%.

12.12 INFLATION AND ITS EFFECTS ON AFTER-TAX FLOWS

When dealing with after-tax cash flows, it is important to realize that inflation affects some elements of the calculation differently than other elements. In particular, depreciation charges are not affected by inflation, because they represent the recovery of an asset's orginal cost. Normally, repayment schedules associated with loans are stated in actual dollars, taxes paid are in actual dollars, while in many instances the revenues, operating and maintenance costs, and future salvage values are estimated in terms of constant dollars.

When calculating after-tax cash flows from before-tax cash flows use actual-dollar analysis.

The application of this principle is demonstrated in Table 12.12, where the cash flows associated with purchase and subsequent sale of the earthmoving equipment are presented. The after-tax cash flows of the venture are evaluated based on all the data being stated in terms of actual dollars. Other previous examples of after-tax cash flow analysis in Tables 12.8, 12.9, and 12.10 also assume that all the cash flows are provided on an actual dollar basis. If any of the cash flows are stated in terms of constant dollars, they should be converted to actual dollars as discussed in Section 5.3; then the analysis proceeds as usual. Once the after-tax cash flows in actual dollars have been determined, it is a straightforward procedure to convert them to their constant-dollar equivalent as shown in Table 12.13.

To illustrate the consideration of inflation in an after-tax analysis, suppose the investment being reviewed is the earthmoving equipment analyzed in Table 12.12. The value of the inflation index for each year is presented in Column *B* of Table 12.13. The base year for the index is the beginning of the investment. Column *D* is the result of converting the actual-dollar cash flows to constant-dollar cash flows. Now it is possible to present the results on both an actual- and constant-dollar basis. This option can provide powerful insights regarding the

TABLE 12.13.
CONSTANT-DOLLAR AFTER-TAX CASH FLOW

End of Year *A*	Inflation Index *B*	Actual-Dollar After-Tax Cash Flow *C*	Constant-Dollar After-Tax Cash Flow *D*
0	100.0	−$184,000	−$184,000
1	105.0	$32,500	30,952
2	113.4	$42,000	37,037
3	115.7	$43,500	37,597
4	122.6	$46,000	37,520
5	128.7	$48,500	37,685
6	137.7	$30,000	21,786
6	137.7	$26,400	19,172
i^*		11.1%	5.4%

important economic features of an investment. Some examples of these insights follow.

Once the cash flows have been transformed to either actual dollars or constant dollars, then the IRR's can be determined in either domain, as demonstrated in Section 5.3.3. The IRR for the actual-dollar cash flow is 11.1%, while that for the constant-dollar cash flow is 5.4%. Thus, the after-tax return for this investment in terms of constant dollars is relatively low. If the firm knowingly selected their MARR at 10% so that it reflects inflationary effects, then this investment is acceptable (11.1% > 10%). However, if the firm desired to earn 10% after taxes with the effects of inflation removed, then this project has failed to meet the firm's objective (5.4% < 10%).

Direct examination of the after-tax cash flows in terms of actual and constant dollars provides additional information. It is observed that, although the revenues in actual dollars were increasing for years 2 through 5, inflation has kept the real gains (constant dollars) almost the same for that period. In other words, over those four years, the revenues generated by the project have not significantly increased purchasing power, which indicates that there is little improvement in the living standard provided by these receipts. This ability to measure the real improvements in standard of living is critical to any sound investment-evaluation technique.

12.13 AFTER-TAX ANALYSIS OF REPLACEMENTS

The comparison of replacement alternatives requires the simultaneous consideration of numerous tax effects. The decision to keep an asset that is presently owned (defender) not only incurs a cost (outsider viewpoint), but it saves the payment of any capital-gains taxes that might result. Knowledge of the depreciation schedule for the defender is necessary to determine the tax liability if the asset is sold immediately or at a later date.

The decision to acquire a new asset also has significant tax implications, as there is the possibility of investment credits for special circumstances. In addition, the depreciation schedule must be known so that the tax effects of any future disposal of the asset can be determined.

Knowing when the defender was acquired is also necessary, because the tax law at that time will determine the depreciation schedule applicable. Acquisitions prior to 1981 may use the depreciation methods discussed in Section 11.5. If the asset was purchased after 1980 but before 1987, the ACRS system described in Section 11.6 will be applied. For acquisitions after 1986 the MACRS system presented in Section 11.7 is applicable.

The following example presents the comparisons of two replacement alternatives (defender and challenger) under consideration at a time after 1987. The firm is expected to have taxable income in excess of $500,000 with a combined federal, state, and local income tax rate of 50%. Since capital gains are taxed only by the federal government, the tax rate applied to capital gains is 34%. This rate is the federal rate applicable to ordinary income for this firm's level of taxable income.

The defender and challenger represent the retention of an existing machine or its replacement with a new machine that has the same production output. Machine *A* was purchased 16 years ago at a cost of $26,000. At the time of purchase, it was estimated to have a 20-year service life with no salvage value. This machine is being depreciated over 20 years utilizing straight-line depreciation for pre-1981 acquisitions. The depreciation charges are $1,300 per year, since the depreciation rate of 5% is being applied to its original cost of $26,000.

Because Machine *A*, the defender, is a general-purpose machine, its resale value is expected to remain high relative to its first cost. However, it is also expected that operating expenses will increase substantially each year, owing to increased maintenance and inflation.

Machine *B*, the challenger, is a special-purpose machine presently costing $30,000. Because of its special-purpose nature, it is expected that machine's future salvage will decrease rapidly, stabilizing at the end of its expected life. The pertinent data relating to Machines *A* and *B* are presented in Table 12.14.

Using the outsider viewpoint, we will compare alternatives on an $AE(i)$ basis. The Method 2 (Section 7.8.2) approach is selected because this comparison assumes no knowledge of future challengers. Thus the economic life for each alternative must be found as described in Section 8.5.2. Tables 12.15 and 12.16 illustrate the calculations necessary to find the after-tax $AE(i)$ cost for Machine *A*, with a life of 2 years, and Machine *B*, with a life of 6 years.

Note in Table 12.15 that the outsider viewpoint for Machine *A* requires the $15,000 of present salvage value to be shown as an opportunity cost. (See Section 8.2.) In addition, the taxes that will be *paid* if the machine is *sold* are shown as a tax *savings* of $3,332 if the machine is *kept*. Thus, the capital-gains tax saving if the machine were kept would be

$$(\$15{,}000 - \$5{,}200)(0.34) = \$3{,}332.$$

In addition, capital-gains tax at the end of year 2 must be included in the analysis. If the machine is sold at the end of year 2, the capital-gains tax is

$$(\$12{,}000 - \$2{,}600)(0.34) = \$3{,}196.$$

The after-tax $AE(i)$ for a 2-year life for Machine *A* is calculated from the expression

$$AE(12)_{n=2} = [-\$11{,}668 - \$1{,}850(\overset{P/F,12,1}{0.8929}) + (\$8{,}804 - \$2{,}850)(\overset{P/F,12,2}{0.7972})](\overset{A/P,12,2}{0.5917})$$

$$= -\$5{,}073 \text{ per year.}$$

TABLE 12.14.
DATA FOR MACHINES *A* AND *B*

	Machine *A*				Machine *B*			
End of Year	Operating Costs	Depreciation	Book Value*	Salvage Value* If Sold in Year Indicated	Operating Costs	Depreciation†	Book Value††	Salvage Value** If Sold in Year Indicated
0			$5,200	$15,000			$30,000	
1	−$ 5,000	−$1,300	3,900	12,000	−$1,500	−$3,000	—	
2	− 7,000	− 1,300	2,600	12,000	− 1,500	− 6,000	24,000	$20,000
3	− 9,000	− 1,300	1,300	12,000	− 1,500	− 6,000	18,000	16,000
4	− 11,000	− 1,300	0	12,000	− 1,500	− 6,000	12,000	12,000
5					− 1,500	− 6,000	6,000	9,000
6					− 1,500	− 3,000	0	9,000
7							0	9,000

* *Value incurred at the end of the year indicated with the cash flow at the end of the year.*

** *Value incurred during the year indicated with the cash flow at the end of the year [e.g., at retirement at age 2, Machine B has a salvage value of $16,000; at the end of year 3, it has capital gains ($2,000 loss) cash flow].*

† *Depreciation schedule if Machine B sold during year 6 or 7 at age 5 or 6, respectively.*

†† *Because of the half-year convention, the book value at the end of the year is based on disposal during the year (e.g., for retirement at age 2, Machine B will have incurred depreciation charges of $3,000, $6,000, and $3,000 over 3 years).*

TABLE 12.15.
AFTER-TAX CASH FLOW FOR MACHINE *A* WITH A 2-YEAR LIFE

End of Year *A*	Before-Tax Cash Flow *B*	Depreciation Charges *C*	Taxable Income *D*	Taxes $-0.5 \times D$ *E*	Capital Gains Taxes *F*	After-Tax Cash Flow $B + E + F$ *G*
0	−$15,000		−($15,000 − $5,200)		$3,332	−$11,668
1	− 5,000	−$1,300	− 6,300	$3,150		− 1,850
2	− 7,000	− 1,300	− 8,300	4,150		− 2,850
2	12,000*		(12,000 − 2,600)		− 3,196	8,804

* *Salvage value.*

TABLE 12.16.
AFTER-TAX CASH FLOW FOR MACHINE *B* WITH A 6-YEAR LIFE

End of Year *A*	Before-Tax Cash Flow *B*	Depreciation Charges *C*	Taxable Income *D*	Taxes $-0.5 \times D$ *E*	Capital Gain Taxes *F*	After-Tax Cash Flow $B + E + F$ *G*
0	−$30,000					−$30,000
1	− 1,500	−$3,000	− $4,500	$2,250		750
2	− 1,500	− 6,000	− 7,500	3,750		2,250
3	− 1,500	− 6,000	− 7,500	3,750		2,250
4	− 1,500	− 6,000	− 7,500	3,750		2,250
5	− 1,500	− 6,000	− 7,500	3,750		2,250
6	− 1,500	− 3,000	− 4,500	2,250		750
7	9,000*		(9,000 − 0)		−3,060	5,940

* *Salvage value.*

The 12% represents the firm's after-tax MARR, which is considered to include the effects of inflation.

For Machine *B* the after-tax cash flow analysis for a 6-year life is presented in Table 12.16. Because this alternative is purchased after 1987, this investment does not qualify for the regular investment credit. Although the service life for this machine is expected to be 6 years, the depreciation recovery period is 5 years. The alternative MACRS depreciation system applicable to this post-1987 acquisition is based on straight-line depreciation. The depreciation amount for a full year of ownership is the original cost divided by the number of years in the recovery period (e.g., \$30,000 ÷ 5 years = \$6,000). Because MACRS mandates that the half-year convention be used, the depreciation amounts chargeable for the year of acquisition and the year of disposal are one-half of a full year's depreciation (\$6,000 × 0.5 = \$3,000).

The book values listed in Table 12.14 are provided so that the book value at the time of disposal can be used to determine any capital gain or loss. For Machine *A*, it is assumed that the half-year convention is *not* in effect and that Machine *A* is purchased at the beginning of the period and sold at the end of the period. If Machine *A* is sold when it is 2 years old, its book value is \$2,600 at the end of year 2.

Implementing the half-year convention for Machine *B* assumes purchase and disposal at the midpoint of the year. If Machine *B* is kept only for 2 full years, the depreciation amounts are \$3,000 for year 1, \$6,000 for year 2, and \$3,000 for year 3. However, the \$1,500 of operating expenses are considered to occur only at the end of the first 2 years. If Machine *B* is retained for 6 full years, the \$1,500 in operating expenses and the depreciation amounts are those shown in Table 12.16 (where disposal is assumed to occur during the 7th year). In addition, for calculation purposes, it is assumed that the cash flows associated with salvage value and any capital gains will be assigned to the end of the year of disposal, although the sale occurs during the year.

To determine the capital gain it is necessary to know the book value and the actual salvage value received at the time of disposal. For Machine *B*, Table 12.14 presents these two amounts at the *end* of the year during which the sale occurs. For a 2-year-old machine, the book value is \$18,000 and the actual salvage value is \$16,000. From these two amounts there is a capital loss of \$2,000, which provides a tax saving of (0.34)(\$2,000) or \$680. Thus, at the end of year 3, the after-tax cash flow consists of \$1,500 (the tax saving on depreciation of \$3,000) and \$16,680 (the salvage value plus the capital gains tax saving). The after tax $AE(i)$ values of utilizing Machine *B* for 2 years and 6 years are

$$AE(12)_{n=2} = [-\$30{,}000 + 750(\overset{P/F,12,1}{0.8929}) + \$2{,}250(\overset{P/F,12,2}{0.7972}) + (\$1{,}500 + \$16{,}680)(\overset{P/F,12,3}{0.7118})](\overset{A/P,12,2}{0.5917})$$

$$= -\$8{,}637 \text{ per year}$$

TABLE 12.17.
AFTER-TAX ECONOMIC LIVES COSTS FOR MACHINES *A* AND *B*

	Annual Equivalent Costs	
Life	Machine *A*	Machine *B*
1	−$5,672	−$12,440
2	− 5,073*	− 8,637
3	− 5,155	− 7,227
4	− 5,393	− 6,296
5	—	− 5,398
6	—	− 4,903*

* *Economic life.*

and

$$AE(12)_{n=6} = [-\$30{,}000 + \$750(\overset{P/A,12,6}{4.1114}) + (\$2{,}250 - \$750)(\overset{P/A,12,4}{3.0374})(\overset{P/F,12,1}{0.8929}) + \$5{,}940(\overset{P/F,12,7}{0.4524})](\overset{A/P,12,6}{0.2432})$$

$$= -\$4{,}903 \text{ per year.}$$

To find the economic life for each alternative, the after-tax $AE(i)$ costs for Machine *A* must be calculated for a possible life of 1, 2, 3, and 4 years. The calculation for 2 years is shown in Table 12.15. For Machine *B*, the after-tax $AE(i)$ costs must be found for a possible life of 1, 2, 3, 4, 5, and 6 years. Table 12.16 presents this calculation for a 6-year life. A summary of these after-tax $AE(i)$ costs is given in Table 12.17.

An examination of the results in Table 12.17 indicates that the economic life for Machine *A* is 2 years and for Machine *B*, 6 years. With the assumption of no knowledge of future challengers, Machines *A* and *B* are compared over the lives most favorable to each. Selecting the shortest economic life of the two machines as the study period allows the direct comparison of the annual equivalent amounts in Table 12.17. (See Section 8.5.2.) This comparison favors the purchase of Machine *B*, assuming it will be utilized over its full service life of 6 years. The advantage of Machine *B* over Machine *A* is $5,073 − $4,903 = $170 per year for 2 years.

KEY POINTS

- Sound economic analyses must include consideration of the effects of taxes.

- To judge an investment option on its merits, it is necessary to identify the tax effects associated with that option by examining the effects on the firm with and without the option.
- Corporations and individuals are affected by taxes, although their tax rates are significantly different.
- Items such as interest, depreciation, and other expenses of doing business are tax deductible. For a tax rate of 40%, each dollar of deductible expenses saves 40 cents of taxes if the enterprise is profitable.
- Accelerated depreciation methods are preferred if time-value-of-money considerations are paramount. Also, shorter recovery periods are preferred to longer ones.
- Gains or losses on an asset at the time of disposal are represented by the difference between the actual value received and the asset's book value. The taxes paid on these gains or losses are determined by the tax law in effect at the time of disposal.
- After-tax cash flow comparisons may be made for alternatives described by receipt and disbursement cash flows or for alternatives showing only their cost cash flows.
- Methods of financing assets have important effects on after-tax cash flows.
- When the effects of inflation are being considered, the conversion of before-tax cash flows to after-tax cash flows should be accomplished using actual-dollar analysis.
- The tax effects of the defender's current market value and book value must be considered in after-tax replacement analyses.

PROBLEMS

1. A single individual working for a large corporation has the following taxable incomes. Determine the total taxes owed and the highest tax rate paid by this individual.

a. $21,000. **e.** $ 70,000.
b. $35,000. **f.** $100,000.
c. $40,000. **g.** $120,000.
d. $50,000. Answer: $11,375.50 **h.** $160,000. Answer: $45,475.50

2. A family files a joint income tax return. If the family's taxable income is as follows, what is their total tax obligation and what is the maximum tax rate they are paying?

a. $30,000.
b. $45,000.
c. $50,000.
d. $80,000. Answer: $17,980
e. $100,000.
f. $130,000.
g. $170,000.
h. $200,000. Answer: $55,115.50

3. A corporation is considering a study to develop a new process to replace a present method of manufacturing electronic components. If successful, the new process is estimated to result in a saving of $300,000 in fulfilling a contract that will be completed during the current tax year. If not successful, no savings will be realized. The study is estimated to cost $25,000 during the same tax year. The corporation taxable income will be $5,000,000 before the investment.

a. If no savings are achieved, what will be the increase in income after taxes?

b. If the study is unsuccessful, what will be the decrease in income after taxes?

4. A corporation estimates its taxable income for next year at $1,000,000. It is considering an activity for next year that it estimates will result in an additional taxable income of $100,000. Calculate the effective income tax rate applicable

a. To the present estimated $1,000,000 taxable income.

b. To the $100,000 estimated increment of income.

c. If the profit to the new venture is $100,000 before income taxes, what will it be after income taxes are paid?

5. The same conditions exist as in Problem 4, except that a new activity has been undertaken that, instead of resulting in an increase of $100,000 in taxable income, as estimated, has resulted in a decrease in the taxable income of $50,000. Calculate

a. Net income after income taxes on a taxable income of $1,000,000. Answer: $660,000

b. Net income after income taxes if the new venture results in a loss of $50,000. Answer: $627,000

c. The loss in income after income taxes that would be caused by the $50,000 loss. Answer: $33,000

6. A small firm is considering an expense of $6,000 for a marketing program. It is estimated that the increase in total income from the marketing program in the coming year will be $14,000. Currently, the taxable income for the firm is $80,000 per year.

a. What will be the increase in the net income after taxes if the marketing program is successful?

b. What will be the decrease in the net income after taxes if the marketing program is unsuccessful (i.e., no income)?

7. A corporation has a taxable income of $100,000 during a year. It is considering a venture that will result in an additional income of $20,000 during the next year.
   a. What is the effective income-tax rate if earnings remain at $100,000? Answer: 22.25%
   b. What is the effective income-tax rate if the new venture is undertaken and turns out as estimated? Answer: 25.04%
   c. What is the effective income-tax rate applicable to the increment of income due to the new venture? Answer: 39%
8. Corporation *A* has total assets of $1,000,000, uses no borrowed funds, and has a taxable income of $150,000.
   a. How much will it have for payment of dividends after income taxes?
   b. If it is capitalized at $1,000,000, what would be the percent return it earned for its stockholders?

   Corporation *B* is identical with *A* except that it is capitalized at $800,000, uses $200,000 of funds borrowed at 9%, and its taxable income before payment of interest on borrowed funds is $150,000.
   c. How much will it have for payment of dividends after income taxes?
   d. What percent return did it earn for its stockholders?
9. An asset with an estimated life of 10 years and a salvage value of $5,000 was purchased by a firm for $60,000 in 1980. The firm's effective tax rate for state and federal taxes is 40% and the minimum attractive rate of return is 10%. The gross income from the asset before depreciation and taxes was estimated to be $8,500 per year. Calculate the present worth at 1980 of the taxes for the straight-line and double-declining-balance methods of depreciation. Assume that the firm is profitable in its other activities.
10. The applicable effective tax rate of a firm is 0.45. It purchased a machine prior to 1980 for $2,000 whose life and salvage value for tax purposes are 4 years and zero, respectively. The estimated annual income from the machine is $1,200 per year before income taxes. Compare the present worth of the income taxes that will be payable if the straight-line and double-declining-balance methods are used, and if the interest rate is 15%. Answer: $899; $874
11. A corporation purchases an asset for $60,000 with an estimated life of 5 years and a salvage value of zero. The gross income per year will be $20,000 before depreciation and taxes. If the tax rate applicable to this activity is 40% and if the interest rate is 12%, calculate the present worth of the income taxes if the straight-line method of depreciation is used; if the double-declining-balance method of depreciation is used; if the ACRS method of depreciation is used.
12. In 1979 a company purchased a $60,000 asset with an estimated life of 4 years and expected salvage value of $30,000. If after 4 years the asset is

sold for $20,000, what are the capital gains or losses for the straight-line depreciation method and the double-declining-balance depreciation method in the fourth year? Assume all accounting adjustments between actual and estimated salvage value occur at the time the asset is sold.

13. The income tax of a corporation is 46%. Five months ago it purchased a warehouse for $44,000. It has just received an offer to sell the warehouse for $50,000 for a short-term gain of $6,000. Under current tax law is there any tax advantage in waiting 3 more months to sell the warehouse?

14. A contractor purchased $400,000 worth of equipment to build a dam. For tax purposes, he was permitted to depreciate the equipment by the straight-line method under MACRS for 7-year property. After the completion of the dam, the contractor sold the 2 year old equipment for $300,000. (The equipment was purchased during the first year and sold during the third year.) The contractor had taxable income of $70,000. There were no short-term capital losses. The contractor's after-tax MARR is 15%.

a. How much capital gains tax would the contractor pay as a result of selling his equipment?

b. What was his total income tax including capital gains tax for the third year, and what was his total income after income taxes?

c. Suppose the contractor had spent an additional $4,000 per year over the three years on extra maintenance. If this additional maintenance had increased the selling price of the equipment by $15,000 during the third year, would these additional expenditures justify the additional gain on disposal on an after-tax basis? Make a $PW(i)$ comparison.

d. How much was his total income after income taxes for the 2-year period increased by his maintenance policy if his estimates were correct?

15. An oil lease is purchased for $1,800,000 by an independent operator, and it is estimated that there are 400,000 barrels of oil on this lease. It costs $60,000 per year exclusive of depletion charges to recover the oil, which can be sold for $21 per barrel. The effective tax rate is 40%, and 40,000 barrels of oil are produced each year. If the interest rate is 12% and the lease will produce for 10 more years, find the present worth of the after-tax cash flow for this lease for (a) unit depletion, and (b) percentage depletion.

16. A mineral deposit discovered by a mining company is estimated to contain 100,000 tons of ore. The company has made an initial investment of $5,000, 000 to recover the ore, which sells for $210 per ton. The company has a MARR of 15% and an effective tax rate of 40%. The fixed-percentage depletion rate for this mineral is 14%. If the ore is produced and sold at a rate of 20,000 tons per year and the operating expenses exclusive of depletion expenses are $1,000,000 per year, find the present worth of the after-tax cash flow for this company (a) when unit depletion is used, and (b) when percentage depletion is used.

17. A prospector acquired a mine containing 300,000 tons of tungsten ore for $600,000. Annual operating costs of the mine are $132,000, and 40,000 tons of ore are being sold at the rate of $14.50 per ton. The prospector wishes to compare the results of using cost depletion and percentage depletion (22%). Determine the income tax for each method, assuming that the applicable income tax rate is 0.35. Answer: $128,800; $112,140

18. An investment undertaken before 1981 is described by the following tabulation.

	Year		
	1	2	3
Gross income at end of year	$ 8,000	$16,000	$20,000
Investment at beginning of year	14,000	10,000	0
Operating cost	2,000	3,000	4,000
Depreciation charge	8,000	8,000	8,000

The company is profitable in its other activities. Thus, any depreciation not used to offset income for this proposal can be used to reduce taxes on the income from the other activities. Find the IRR earned by this proposal (a) before taxes, and (b) after taxes. The effective income-tax rate is 40%.

19. An investment proposal has the following estimated cash flow before taxes.

End of year	0	1	2	3	4
Cash flow	−$50,000	$30,000	$10,000	$5,000	$2,000 + $20,000* (*salvage value)

The effective tax rate is 40%, the tax rate on capital gains is 34%, and the interest rate is 10%. Assume that the company considering this proposal is profitable in its other activities. Find the after-tax cash flows and the present worth of those cash flows for (a) the straight-line method of depreciation, (b) the double-declining-balance method of depreciation, and (c) MACRS depreciation for a 3-year recovery period. Assume the depreciation base for parts (a), and (b) is determined by pre-1981 tax law.

20. Consider the following two investment alternatives.

	Alternative A	Alternative B
Initial investment	$10,000	$5,000
Service life	5 years	5 years
Salvage value	0	0
Depreciation method	MACRS	MACRS (Alternative method)
Recovery period	3 years	3 years

Estimated operating costs and revenues are as follows:

		End of Year				
		1	2	3	4	5
Alternative A	Operating cost	$10,000	$10,500	$12,000	$14,000	$16.000
	Revenue	17,000	16,900	17,800	19,200	21,400
Alternative B	Operating cost	$1,200	$1,000	$1,500	$1,300	$1,200
	Revenue	6,400	6,200	6,700	6,500	6,400

For an after-tax rate of interest of 6%, which alternative is more attractive to undertake? (Effective tax rate is 40%.)

21. Two types of machine tools are available for performing a particular job in a certain manufacturing firm. The estimated costs and salvage values are summarized in the following table.

	Machine A	Machine B
First cost	$30,000	$40,000
O&M costs/year	$3,000	$2,500
Service life	6 years	8 years
Salvage value	$6,000	$4,000
Depreciation method	MACRS (Alternative method)	MACRS
Recovery period	5 years	5 years

The effective state and federal tax rate is 50%, and the capital-gains rate is 34%. Make your comparison after income taxes, using a MARR of 10%. Answer: $AE(10)_A = -\$5{,}382$; $AE(10)_B = -\$5{,}620$

22. A $20,000 investment in machinery is under consideration. The project is expected to have a life of 6 years and no salvage value. The estimated annual income from the project is $10,000 with annual operating expenses of $4,000. The investment will be depreciated by the MACRS alternative method based on a 5-year recovery period. If a 40% income tax rate is applied with a MARR of 5%, compute the present worth on the proposed investment's after-tax cash flow under the following financial policies:

a. The investment is provided from the firm's retained earnings.

b. The initial investment is borrowed at 10% with repayment of interest at the end of each period and repayment of the loan principal at the end of 5 years.

c. Repeat parts (a) and (b) for a MARR of 9%.

23. An asset is being considered whose first cost, life, recovery period, salvage value, and annual operating expenses, respectively, are estimated at $15,000, 12 years, 10 years, zero, and $800. The asset, classified as 10-year property, will be depreciated by the MACRS alternative method of depreciation. The effective income tax is 40%. Determine the $AE(i)$ cost at 12% for the after-tax cash flow if

a. The initial investment is made from retained earnings.

b. The initial investment is borrowed at 8% with repayment of principal and interest in 10 equal annual amounts.

c. The initial investment is borrowed at 8% with repayment of interest at the end of each period and repayment of the loan principal at the end of 10 years.

24. A small firm is considering the acquisition of a speical-purpose machine for its research lab at a cost of $15,000. It is estimated that the operation of the new unit will cost $1,000 per year, that it will produce annual revenues of $6,000 over its economic life of 5 years, and that it will have a zero salvage value in the future. The machine will be depreciated by MACRS using a recovery period of 3 years. If the effective income tax rate is 50%, determine the $PW(i)$ at an after-tax rate of 15% for the after-tax cash flow if:

a. The initial investment is made from retained earnings.

b. The initial investment is borrowed at 10% with repayment of principal and interest in three equal amounts.

c. The asset is leased at a cost of $4,000 per year, paid at the end of each year. (Operating costs are not included in the lease payments.)

25. A computerized control system has been proposed for a chemical plant. If the new system is implemented, it will have an initial cost of $130,000 and will produce net savings of $60,000 per year over its life of 10 years. It is expected to have zero salvage value at the end of that time. Two financing alternatives are under consideration. The first is to obtain credit from the supplier; this calls for a 40% down payment and a 7% loan for the balance, which should be paid in equal amounts over a 5-year period. The second alternative is a 12% bank loan with the total amount to be paid in equal amounts over a 4-year period. If MACRS depreciation is used over a 7-year recovery period and the effective income tax rate is 42%, which financing alternative is most desirable? The company uses an after-tax MARR of 10%.

26. A bio-engineering firm is considering the acquisition of a new electronic microscope. The relevant information for the two models under study is given below.

	Model I	Model II
Initial cost	\$80,000	\$60,000
Annual operating costs	\$20,000	\$24,000
Economic life	5 years	5 years
Final salvage value	\$30,000	\$10,000

Assume that each model is 3-year MACRS recovery property and depreciation is based on prescribed percentages. Any capital gains should be included in the analysis. With an effective income tax rate of 50%, a before-tax MARR of 25%, and an after-tax MARR of 14%, find the alternative that should be selected if the analysis is made

a. Before taxes. Answer: $PW_I = -\$123{,}955$; $PW_{II} = -\$121{,}266$

b. After taxes. Answer: $PW_I = -\$72{,}921$; $PW_{II} = -\$74{,}424$

27. A car rental company is studying two proposals to buy 100 new automobiles. These two proposals are summarized below.

	Proposal I	*Proposal II*
Initial cost	\$800,000	\$880,000
Annual operating cost	\$420,000	\$330,000
Salvage value after 3 years	\$270,000	\$350,000
Answer: *PW*(15)	\$911,738	\$832,219 Select II

The company's policy is to keep the automobiles for 3 years. The automobiles are depreciated as MACRS 5-year property using prescribed percentages. Assuming that the effective income tax rate is 50% and the after-tax MARR is 15%, which proposal should be selected? Capital gains are to be considered and are taxed at 34%.

28. A company manufacturing semiconductors is considering the installation of an automatic control system, which will produce estimated savings in its production process equivalent to \$35,000 per year. Two different suppliers are under consideration. If the equipment is purchased from Supplier *A*, it will have an initial cost of \$50,000 and estimated operating costs of \$21,000. Its estimated salvage value at the end of its economic life of 7 years will be \$5,000. If the equipment is purchased from Supplier *B*, its initial cost and its estimated operating costs, economic life, and final salvage value will be $40,000, $25,000, 6 years, and $9,000, respectively. Assume that the control system is depreciated as MACRS 5-year property using prescribed percentages and the effective income tax is 40%. Capital gains are to be considered and they are taxed at 34%. If the after-tax MARR is 15%, which supplier should be selected?

29. A firm has an opportunity to expand one of its production facilities at a cost of $200,000. The asset has an expected economic life of 12 years, after which it is not expected to have any salvage value. MACRS prescribed-percentage depreciation with a recovery period of 10 years will be used for tax purposes for the new asset. The annual operating cost is estimated at $5,000. If the asset is purchased, the firm is capable of financing the acquisition entirely with an 8%, 10-year unsecured term loan. The loan is payable over this period in equal annual payments, which include both principal and interest. The income-tax rate applicable for this firm is 40%. Determine the after-tax $AE(i)$ cost at 14% for this investment.

30. The XYZ company is considering investing in a new computer. The computer costs $240,000 and is estimated to have a 7-year economic life, with operating costs of $45,000 per year and a $30,000 salvage value at the end of that time. If the computer is purchased, it will be financed with a 12% loan payable over 3 years in equal amounts. The same computer can be leased for an annual payment of $60,000 to be paid at the beginning of each year. Operating costs for the leased computer are expected to be $38,000 per year. If depreciation is based on the MACRS prescribed-percentage method, the effective income-tax rate is 40%, and the computer is considered to be 5-year recovery property, which alternative should be selected? The MARR = 10%, and capital gains are taxed at 34%.

31. A firm is considering buying a new machine that costs $40,000. The machine has an economic life of 6 years, an estimated salvage value of $4,000, and annual operating disbursements of $3,000. The machine is expected to generate a revenue of $15,000 annually and will be depreciated by MACRS prescribed percentages using a recovery period of 3 years. Assuming that the effective income tax rate is 40% on taxable income of $400,000 and considering capital gains, compute the $AE(i)$ at 10% for the after-tax cash flow under the following financing:

a. The initial investment is financed from retained earnings.

b. The initial investment is borrowed at 12% with repayment of principal and interest in six equal annual amounts.

c. The initial investment is borrowed at 14% with repayment of interest at the end of each period and repayment of the loan principal at the end of 6 years.

32. A telephone company is faced with the prospect of having to replace one of its central switching systems, which is estimated to have zero salvage value. The new computerized switching system will cost $2,100,000. The system is estimated to have a 15-year useful life with no salvage value and would be depreciated on a straight-line basis as 7-year property using the alternative method of MACRS. The system will result in a net annual savings of $300,000.

a. Find the $AE(i)$ on the project before taxes for a MARR of 20%.

b. Find the $AE(i)$ on the project after taxes if the effective tax rate is 40% and the after-tax MARR is 10%.

c. Repeat (b) if the above estimates are in constant dollars and the inflation rate is 6%.

33. A new machine costing $130,000 was installed in a certain manufacturing plant at the beginning of 1978. Its useful life was 6 years with a salvage value of $10,000. Annual net operating income before tax was $40,000. The machine was depreciated by the straight-line method over 6 years. A 10% tax credit was applicable to this asset. Assume that the firm was profitable in its other activities and that the effective tax rate is 40%.

a. Determine the $AE(i)$ profit earned at 10% based on the after-tax cash flow.

b. If the dollar amounts above are in constant dollars with the base year at the beginning of 1978, find the after-tax cash flow in actual dollars. Use the data in Table 5.1 to represent price changes for the years spanning this investment. Using an after-tax MARR of 10% find the $AE(i)$ of the after-tax cash flow. Use a capital gains tax rate of 20%.

34. A company purchased an asset at the beginning of 1987 for $100,000. It was depreciated on the basis of a 10-year recovery period using the MACRS prescribed-percentage method. However, the asset was sold during the sixth year for $20,000. The company financed $60,000 of the purchase price by a 5-year loan costing 12% per year. The loan was to be repaid in equal annual payments beginning one year after the date of purchase. The asset earned $15,000 per year during the 6 years it was in service. Consider any capital gain or loss applicable to this problem. Assume the effective income-tax rate for ordinary income is 40%. This rate includes both federal and state taxes.

Assume the company is profitable in its other activities. Disposal occurs during 1992. Also assume that this firm has other capital gains and losses, so that any gains or losses on this asset are taken at the time of disposal. The company's earnings without this asset are approximately $1,000,000 per year over the asset's lifetime.

If the *above* dollar values are all in terms of 1986 dollars, and the price index values based on 1982 dollars are given below, find the constant-dollar present worth in terms of 1986 dollars of the after-tax cash flow. The after-tax MARR is 9% (this is a market-determined rate).

End of Year	1986	1987	1988	1989	1990	1991	1992
Price Index	126	131	133	139	146	156	170

35. A company purchased an asset in 1980 that cost $500,000. The asset was depreciated using straight-line depreciation over 6 years based on an estimated salvage value of $80,000. In fact, the asset was used for 10 years when it was disposed of for $120,000. One-half of the original cost was borrowed at 10% interest, to be repaid with equal annual payments over the first 5 years. Any capital gain or losses were taxed at 34%. The asset earned $130,000 per year over its 10 year life, and an investment tax credit of 10% was taken at the time of acquisition. If, over the life of this asset, ordinary income was taxed at 50%

a. Find the after-tax rate of return earned on this investment. Answer: 24%

b. If the above values are in constant dollars and the net earnings were experiencing inflation at 6% per year while the salvage value was inflating at 4% per year, find the actual-dollar IRR realized from this investment. Answer: 32%

36. Owing to an increase in business, a firm is considering the replacement of an existing heater, which will be able to supply only 50% of the required heat-treating capacity. The existing heater was purchased 10 years ago at a significant discount for $4,500 and was estimated to have a 15-year useful life with $1,500 salvage value at the end of that time. Currently, it has a net salvage value of $3,600 and expected economic life, future salvage value, and annual operating expenses of 5 years, $600, and $4,000, respectively.

A new heater of the same capacity costs $4,200. It is estimated that this heater will have an economic life of 7 years with zero salvage value at the end of that time and operating costs of $3,500 per year.

As an alternative to the above, a new heater having the capacity of the two small heaters together can be purchased for $9,000. This unit is expected to have annual operating costs of $5,500 and a salvage value of $1,000 at the end of its economic life of 7 years.

Assume that the straight-line method of depreciation has been used for the existing heater, based on pre-1981 tax law, and that new units will be depreciated as 5-year recovery property using the MACRS prescribed percentages. The effective tax rate is 40% and the after-tax MARR is 15%. Net capital losses are expected to be offset by current capital gains, being taxed at 34%.

If the study period used for the heaters is selected to be 5 years and the future replacements are unknown, define all relevant alternatives and select the most desirable one.

37. A firm has an automatic chemical mixer that has been in use for 3 years since 1988. The mixer originally cost $36,000 and was estimated to have a $6,000 salvage value at the end of its 15-year economic life. Now, the present mixer can be sold for $15,000, or, if retained, it is estimated that it could be used for an additional 11 years, with annual operating costs of $10,000 and a $3,000 salvage value at the end of that time.

Owing to increased demand, if the old mixer is retained now, a new mixer with equal capacity will be purchased at a cost of $30,000. This new unit is expected to have annual operating costs, economic life, and salvage value of $8,000, 12 years, and $6,000, respectively.

If the old mixer is sold now, a new unit with the required total capacity can be purchased for $55,000. This full-capacity mixer is expected to have operating costs of $16,000 per year and a salvage value of $8,000 at the end of its 12-year economic life.

Assume that the MACRS alternative method for 10-year recovery property has been used for the old mixer. The new units are 10-year recovery property and will be depreciated according to MACRS prescribed percentages. The effective income-tax rate is 40%; net capital losses are offset by capital gains, which are taxed at 34%. Assuming future replacements are unknown and for an after-tax MARR of 12%, which alternative should be selected?

38. A packaging company is planning to replace its existing material handling system (MHS). The MHS that it now owns cost $135,000 when purchased 7 years ago and was estimated to have a 12-year useful life with a $3,000 salvage value at the end of that time. Currently, the MHS has a net salvage value of $40,000; and if it is not sold, it is expected to have operating costs of $24,000 per year and a salvage value decreasing at a rate of $8,000 per year from its present level.

A new MHS can be purchased for $85,000. This new system is expected to have operating costs of $17,000, an economic life of 7 years, and a salvage value of $10,000 at the end of that time.

For depreciation purposes, assume that the straight-line method has been used for the old MHS based on pre-1981 tax law, and that the new MHS is 5-year recovery property. The company uses the MACRS prescribed-percentage depreciation. Any short-term losses will be balanced with long-term gains, being taxed at the effective tax rate. If the effective tax rate is 40% and assuming no knowledge of future replacements, answer the following for a before-tax MARR of 20% and an after-tax MARR of 10%.

a. Compute the before-tax and after-tax cash flows for each alternative.

b. Which alternative should be accepted if the analysis is made before taxes; after taxes? Answer: (after taxes) Replace existing MHS.

Part Five

ESTIMATES, RISK, AND UNCERTAINTY

Since decision making is applicable only to the future, it should be based on anticipated outcomes. These outcomes must be estimated. In this part of the text the elements to be estimated are enumerated and methods for estimating their likely values are presented. Techniques are then discussed for dealing with inaccurate estimates. Among these are heuristic approaches, sensitivity analysis, and Monte Carlo analysis. The final chapters then present decision making under risk and decision making under uncertainty. Taken together, the four chapters in Part Five provide the essentials for understanding the degree to which engineering economy can be successful in leading a decision maker to the best course of action.

13

Estimating Economic Elements

Estimating in the physical environment approximates certainty in most applications. Examples are the pressure that a confined gas will develop under a given temperature, the current flowing in a conductor as a function of the voltage and resistance, and the velocity of a falling body at a given point in time. Much less is known with certainty about the economic environment with which the engineering process is concerned. Economic laws depend upon the behavior of people, whereas physical laws depend upon well-ordered cause-and-effect relationships.

Creative engineering activity searches for activities with a profit potential that is high in relation to the risk involved. This search necessitates the estimation of pertinent facets of the anticipated economic outcome. In this chapter, attention is directed to the methods and techniques for estimating economic elements.

13.1 THE ELEMENTS TO BE ESTIMATED

The success of an activity as a whole may be estimated. Thus, if a firm is considering the organization of a construction department, it might directly estimate that the department will yield a return equivalent to a certain percent per year on the amount invested. This return will be a resultant of a number of prospective receipts and disbursements, which may be classified as income, operating expenses, depreciation, interest, and taxes. It is a rare individual who can combine accurately four complex items to obtain their resultant without resort to formal analysis, even when the items are clearly known. Thus, for best results in estimating the final outcome of an undertaking, it will almost always be advantageous to begin with detailed estimates of income and costs as they may be expected to originate in the future. The detailed estimates are then combined to obtain their results.

Success of a venture in terms of economy is determined by considering the relationship between the input and output of the venture through time, with the time value of money taken into consideration. But anticipated success can be imaginary if certain elements are omitted. Thus, an important task in an economy study is the delineation of inputs and outputs associated with the proposed activity.

13.1.1 Outputs

Structures, processes, systems, and activities are normally proposed in response to a need or requirement. Therefore, outputs should be considered first and in conjunction with the need. Benefit, worth, effectiveness, and other terms are used to describe outputs in relationship to needs.

The outputs of commercial organizations and governmental agencies are endless in variety. Commercial outputs are differentiated from governmental outputs by the fact that it is usually possible to evaluate the former accurately, but not the latter. A commercial organization offers its products to the public. Each item of output is evaluated by its purchaser at the point of exchange. Thus, the monetary values of past and present outputs of commercial concerns are accurately known item by item.

Since engineering is concerned with the future, it is concerned with future output. Generally, information on two subjects is needed to come to a sound conclusion. One of these is the physical output that may be expected from a certain input. This is a matter for engineering analysis. The second is a measure of output that may be expressed in terms of monetary income.

Monetary income is dependent upon two factors: (1) the volume of output; (2) the monetary value of the output per unit. The determination of each of these items for the future must of necessity be based upon estimates. Market surveys and similar techniques are widely used for estimating the future output of commercial concerns. In the case of large-scale systems or projects the monetary income to the contractor is determined through contractual agreement.

Under some circumstances—for example, if the income is presented by the saving resulting from an improvement in a process for manufacturing a product made at a constant rate—an estimate is easily made. But estimating income for new products with reasonable accuracy may be very difficult. Extensive market surveys and even trial sales campaigns over experimental areas may be necessary to determine volume. When work is done on contract, as is the case with much construction work the necessity for estimating income is eliminated. Under these circumstances the income to be received is known in advance with certainty from the terms of the contract.

Internal intermediate outputs of commercial organizations are determined with great difficulty and are usually estimated by judgment. For example, the value of the contribution of an engineer, a maintenance technician, or a supervisor to a final output is rarely known with reasonable accuracy to either the employee or this superior. Similarly, it is difficult to determine the value of the contributions of most intermediate activities to the final result.

The outputs of many governmental activities are distributed without regard to the amount of taxes paid by the recipient. When there is no evaluation at the point of exchange, it appears utterly impossible to evaluate many governmental activities. John Doe may recognize the desirability of the Department of Defense, the Forest Service, or Public Health activities, but he will find it impossible to demonstrate their worth in monetary terms. However, some government outputs, particularly those which are localized, such as highways, drainage, irrigation, and power projects, may be fairly accurately evaluated in monetary terms by calculating the reduction in cost or the increase in income they bring about for the user.

13.1.2 Input

Any activity that is undertaken requires an input of thought, effort, material, and other elements for its performance. In a purposeful activity, an input of some value is surrendered in the hope of securing an output of greater value. The terms *input* and *output* as used here have the same meanings as when they are used to designate, for instance, the number of heat units that are supplied to a turbine and the number of energy units it contributes for a defined purpose.

A most important item of input is services of people, for which salaries and wages are paid. In a commercial organization the total input of human services, as measured by the cost for a given period of time, is ordinarily reflected quite accurately. The input of human services may be classified under the headings of direct labor, indirect labor, and investigation and research. Of these, direct labor is the only item whose amount is known with reasonable accuracy and whose identity is preserved until it becomes a part of output.

Input devoted to investigation and research is particularly hard to relate to particular units of output. Much research is conducted with no particular specified goal in mind, and much of it results in no appreciable benefit that can be associated with a particular output. Expenditures for people engaged in research and development may be made for some period of time before this type of service has a concrete effect upon output. Successful research of the past may continue to affect output for a long time in the future. The fact that expenditures for research do not parallel in time the benefits provided makes it very difficult to relate them to an organization's output.

Indirect labor, supervision, and management have characteristics falling between those of direct labor and those of investigation and research. The input of indirect labor and, to a lesser extent, that of supervision parallel the output fairly closely in time, but their effects can ordinarily be identified only with broad classes of output items. Management is associated with the operations of an organization as a whole. Its important function of seeking out desirable opportunities is similar in character to research. Input in the form of management effort is difficult to associate with output in relation either to time or to classes of product.

As difficult as it may be to associate certain inputs with a final measurable output, such as of product sold on the market, input can ordinarily be identified closely with intermediate ends that may or may not be measurable in concrete terms. For example, the costs of the input of human effort assigned to the engineering department, the legal department, the labor relations department, and the production department are reflected with a high degree of accuracy by the payrolls of each. However, the worth of the output of the personnel assigned to these departments may—and usually does—defy even reasonably accurate measurement.

A second major category of input is that of material. Many items of material are acquired to meet the objectives of commercial and governmental enterprise. For convenience, material items may be classified as direct material, indirect material, equipment, land, and buildings.

Inputs of direct material are directly allocated to final and measurable outputs. The measure of material items of input is their purchase price plus costs for purchasing, storage, and the like. This class of input is subject to reasonably accurate measurement and may be quite definitely related to final output, which in the case of commercial organizations is easily measurable.

Indirect material and energy inputs are measurable in much the same way and with essentially the same accuracy as are direct material and power inputs. One of the important functions of accounting is to allocate this class of input in concrete terms to items of output or classes of output. This may ordinarily be done with reasonable accuracy.

An input in the form of an item of equipment requires that an immediate expenditure be made, but its contribution to output takes place piecemeal over a period in the future which may vary from a short time to many years, depending upon the use life of the equipment. Inputs of equipment are accurately measurable and can often be accurately allocated to definite output items except in amount. The latter limitation is imposed by the fact that the number and kinds of output to which any equipment may contribute are often not known until years after many units of the product have been distributed. The function of depreciation accounting is to allocate equipment inputs to outputs.

Inputs of land and buildings are treated in essentially the same manner as inputs of equipment. They are somewhat more difficult to allocate to output because of their longer life and because a single item, such as a building, may contribute simultaneously to a great many output items. Allocation is made with the aid of depreciation and cost accounting techniques and practices.

Allocation of inputs of indirect materials, equipment, buildings, and land rests finally upon estimates or judgments. Although this fact is often obscured by the complexities of and the necessary reliance upon accounting practices for day-to-day operations, it should not be lost sight of when economy studies are to be made.

Capital in the form of money is an essential input, although it must ordinarily be exchanged for producer goods in order to make a contribution to output. Interest on money used is usually considered to be a cost and so may be considered an input. Its allocation to output will necessarily be related to the allocation of human effort, services, material, and equipment in which money has been invested.

Taxes are essentially the purchase of governmental service required by private enterprise. Since business activity cannot be carried on without the payment of taxes, they comprise a necessary input. There are many types of taxes, such as ad vaolerem, excise, sales, and income. The amounts may be precisely known and, therefore, are accurately called inputs. However, it is often difficult to allocate taxes to outputs, especially in the case of income taxes, which are levied only after a profit is derived.

13.2 COST ESTIMATING METHODS

A *cost estimate* is an opinion based upon analysis and judgment of the cost of a product, system, structure, or service. This opinion may be arrived at in either a formal or an informal manner by several methods, all of which assume that experience is a good basis for predicting the future. In many cases the relationship between past experience and future outcome is fairly direct and obvious; in other cases it is unclear, because the proposed product or service differs in some significant way from its predecessors. The challenge is to project from the known to the unknown by using experience with existing items. The techniques used for cost estimating range from intuition at one extreme to detailed mathematical analysis at the other.

13.2.1 Estimating by Engineering Procedures

Estimating by engineering procedures can be described as an examination of separate segments at a low level of detail. The engineering estimator begins with a set of drawings and specifies each engineering task, equipment and tool need, and material requirement. Costs are assigned to each element at the lowest level of detail, and these are then combined into a total for the product or project.

Time standards for construction or production operations exist for many common tasks. These are usually developed by industrial engineers and constitute the minimum time required to complete a given task with normal worker skill and tools. Standards are best applied in engineering estimating procedures when a long, stable production run of identical items is contemplated. They are normally not useful in estimating for complex systems in which one of a kind is to be fabricated. For example, production runs of advanced military and space systems are normally short, with both design configurations and production processes continuing to evolve rapidly over time.

Engineering estimating procedures require more hours of effort and data than are likely to be available in the development of some systems or projects. One large aerospace firm judges that the engineering approach in estimating the cost of an airframe would require thousands of estimates. The cost of an effort such as this makes other estimating methods attractive, especially if the alternative method gives comparable results.

Combining thousands of detailed estimates into an overall estimate can lead to an erroneous result, for the whole often turns out to be greater than the sum of its parts. The engineering estimator works from sketches, blueprints, or descriptions for some items that have not been completely designed. He can assign costs only to work that he knows about. The effect of a low estimate may be compounded because detail estimating is attempted on only a portion of the labor-hours. A number of construction or production labor elements, such as planning, rework, coordination, and testing, are usually factored in as a percentage of the detail estimates. Other cost elements, such as maintenance, inspection, and production control, are factored in as a percentage of the production labor required. Thus, small errors in detailed estimates can result in large errors in the total cost estimate.

Another source of error in estimates made by the engineering method is the significant variability that occurs in the fabrication of successive units. Production runs of like models may be of limited length and often are subject to design changes. In the case of defense systems, production rates vary frequently and unexpectedly. The proportion of new components may be significant from model year to model year as the manufacturer seeks to adapt a product to market demand. Sometimes the effect of these forces can be represented best by mathematical or statistical functions that describe technological progress and learning.

13.2.2 Estimating by Analogy

When a firm is venturing into a new area, estimating by analogy can be very effective. For example, aircraft companies bidding on missile programs in the 1950s drew analogies between aircraft and missiles as a basis for estimating. Appropriate adjustments were made for differences in size, number of engines, and performance. This is an instance of estimating by analogy at the macrolevel.

Estimating by analogy may also occur at the microlevel. The direct labor-hours required to make a part may be estimated by referring to that required on similar jobs. The basis for the estimate is the similarity that exists between the known item and the proposed part. Some estimators with backgrounds as mechanics, tool makers, or supervisors are able to estimate the required times very closely. They are often consulted when an estimate is needed quickly.

The cost of direct labor is often estimated in relation to the cost of direct material. These relationships are known with reasonable accuracy for different kinds of activities. For example, the cost of labor to lay 1,000 bricks is approximately equal to the cost of the bricks and necessary mortar.

At all levels of aggregation, much estimating is performed by analogy. For example, Project *A* required 12,000 direct labor-hours and 5,000 equipment-hours. Given the similarities and differences between Project *A* and proposed Project *B*, the direct labor-hours and equipment-hours might be estimated to be 8,000 and 3,200, respectively. By applying current labor- and equipment-hour rates, and an applicable overhead rate, a total project cost can be estimated. Or, with Project *A* the estimator may find elements that are analogous to elements in Project *B*. From this the cost of Project *B* may be estimated. In this instance, analogy becomes part of the engineering method of estimating.

A major disadvantage of estimating by analogy is the high degree of judgment required. Considerable experience and expertise are required to identify and deal with appropriate analogies and to make adjustments for perceived differences. However, because the cost of estimating by analogy is low, it can be used as a check on other methods. Often it is the only method that can be used because the product or service is in a preliminary stage of development.

13.2.3 Statistical Estimating Method

The statistical method of cost estimating may utilize statistical techniques ranging from simple graphical curve fitting to complex multiple correlation analysis. In either case the objective is to find a functional relationship between changes in costs and the factor or factors upon which cost depends, such as output rate, lot size, weight, and so forth.

Figure 13.1 illustrates a simple curve representing the cost per horsepower as a function of the horsepower for a class of electric motors. The data points were taken from the manufacturer's list price and provide a statistical base for

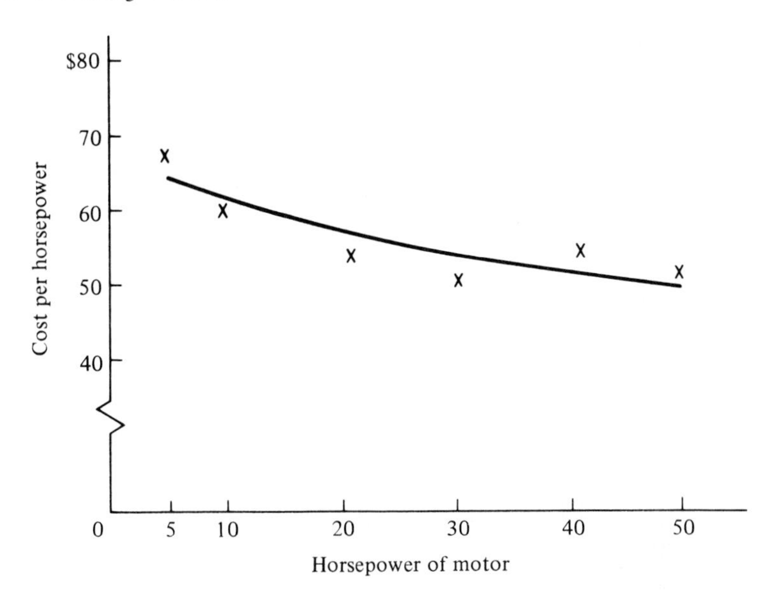

FIGURE 13.1. Cost of electric motors per horsepower based on manufacturer's list price.

the curve. Estimates of the cost of motors of any horsepower within the range illustrated can be made from the curve. However, it would be risky to attempt an estimate much beyond the range for which the cost data are available.

Although statistical cost-estimating techniques are preferred in most situations, there are cases in which engineering methods or estimating by analogy are required because the data for a systematic historical base do not exist. The product may utilize some new advanced manufacturing method, thus invalidating data from previous items as a statistical base.

There will always be situations in which analogy or engineering methods are required, but the statistical approach is useful and sufficient in long-range planning. Total cost may be estimated directly as a function of power output, weight, square feet, volume, and the like. Industrywide data or maintenance experience and energy consumed can be treated statistically and added to statistically estimated costs associated with the item itself.

Statistical estimating techniques will vary according to the purpose of the study and the information available. In a conceptual study, it is desirable to have a procedure that gives the total expected cost of the product or system. Allowances for contingencies are to compensate for unforeseen changes that will have to be made. Later, as the product or system moves closer to final design, it is desirable to have a procedure that will yield estimates of its component parts. Additional engineering effort can then be applied to reduce the cost of those components which are high contributors to overall cost.

In planning, industrywide labor and overhead rates can be obtained from statistical publications and used to give a rough cost estimate for a given item. As the item nears the time for bidding in its design cycle, data that are specific for a particular contractor and particular location can be used. As more is known, more specific statistical data can be used, as is the case for a product currently in production. In this state of product development, accounting records can be used to produce a relatively precise cost estimate.

13.3 ADJUSTMENT OF COST DATA

Data must be consistent and comparable if they are to be useful in an estimating procedure. Inconsistency is often inherent in cost data because there are differences in definitions and in production quantity, missing cost elements, inflation, and so forth. In this section some approaches found useful in coping with this inherent inconsistency are presented.

13.3.1 Cost Data Categories

Different accounting practices make it necessary to adjust basic cost data. Companies record their costs in different ways; often they are required to report costs to governmental agencies in categories that differ from those used internally. These categories also change from time to time. Because of these definitional differences, the first step in cost estimating should be to adjust all data to the definition being used.

Consider the following example: A manufacturing firm has a planning department that is responsible for both production planning and planning for plant expansion. Production planning costs are those associated with scheduling, sequencing, machine loading, and so on for a specific product. Plant-expansion planning leads to costs associated with adapting the manufacturing facility to the production demands created by the numerous products in process and the products anticipated to be in process.

Planning costs in the firm may be charged several ways. Some may be allocated to specific product lines, such as those anticipated with production planning. Plant planning costs may be charged to overhead, or they may be prorated to product lines on the basis of some formula. Also, all costs arising from the department could be charged to overhead and then apportioned back to the product lines, depending upon the labor- or machine-hours utilized by the product. Therefore, a precise definition applied consistently over time is required so that the planning-cost element will have value in estimating for a new product line.

Consistency is also needed in defining physical and performance characteristics. The weight of an item depends upon what is included. Gross weight, empty weight, and airframe weight apply to aircraft, with each term differing in exact meaning. Speed can be defined in many ways: maximum speed, cruising speed, and so on. Differences such as these can lead to erroneous cost estimates in which the estimate is statistically derived as a function of a physical characteristic. When cost data are collected from several sources, an understanding of the definitions of physical and performance characteristics is an important as an understanding of the cost elements.

Another matter requiring clear definition is that of nonrecurring and recurring costs. Recurring costs are a function of the number of units produced, but nonrecurring costs are not. If research and development work on new products is charged off as an expense against current production, it does not appear against the new product. In this case, the cost of initiating a new product would be understated and the cost of existing products overstated. Separation of nonrecurring and recurring costs should be made by means of a downward adjustment of the production costs for existing products and the establishment of an account to collect research and development costs for new products.

In addition to the first cost needed to construct or produce a new structure or system, there are recurring costs of a considerable magnitude associated with operations, maintenance, and disposal. Power, fuel, lubricants, spare parts, operating supplies, operator training, maintenance labor, and logistics support are some of these recurring costs. The life of many complex systems, when multiplied by recurring costs such as these, leads to an aggregate expenditure that may be large when compared with the first cost. Limited cost-estimating effort is often applied to the category of costs associated with operations and maintenance. This may be unfortunate, for the life-cycle cost of the structure or system is the only correct basis for judging its worth.

13.3.2 Price-Level Changes

The price of goods and services and of the labor, material, and energy required in their production changes over time. Down through history, inflationary pressures in the economic system have acted to increase the price of most items. These increases have been very significant in recent years. Prices have decreased for only very brief periods.

Figure 13.2 shows the change in average weekly earnings of production workers in manufacturing over the last two decades. The weekly wage rate has increased by a factor of about 5 from 1962 to 1992. Thus, if the labor-cost component of an automobile was $1,000 in 1962, it would be almost $5,000 in 1992. Fortunately, increased productivity has kept the labor cost of the automobile at about the same percentage over this 30-year period.

Adjustments in cost data are often made by means of an index constructed from data in which one year is selected as the base. The value of that year is expressed as 100, and the other years in the series are then expressed as a percentage of this base. Hourly earnings for any year for production workers could be

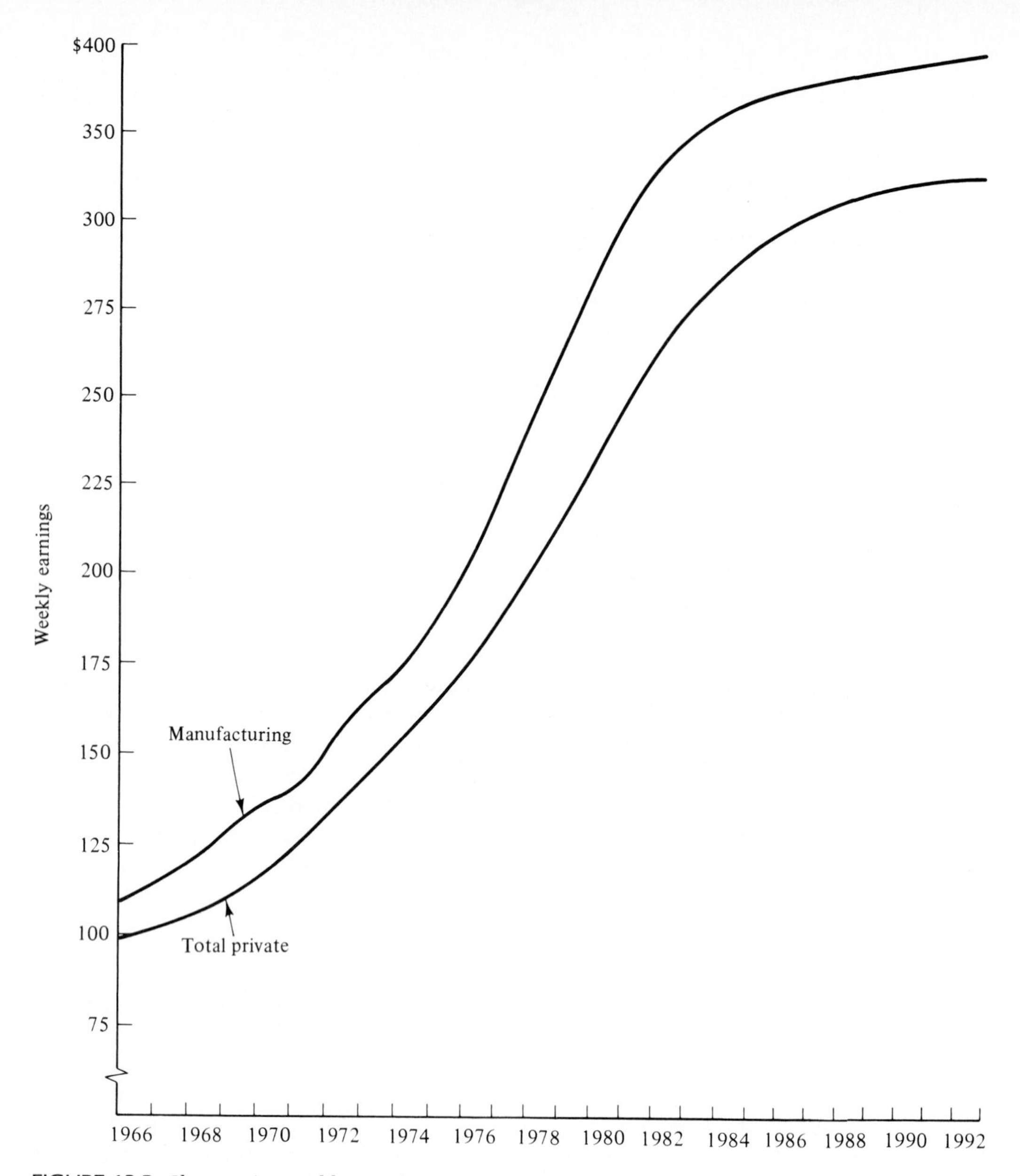

FIGURE 13.2. Changes in weekly earnings.

converted to an index by using any of the years as a base. Table 13.1 uses both 1967 and 1977 as a base.

The information needed to construct a labor index such as is given in Table 13.1 is available in the Bureau of Labor Statistics' monthly publication entitled *Employment and Earnings.* Indexes for labor categories can be developed from this source, as shown in Table 13.2. Average hourly earnings are also a function of location as shown in Table 13.3. The differences are quite significant.

TABLE 13.1.
AVERAGE HOURLY EARNINGS INDEX

Year	Average Hourly Earnings	Index with 1967 as Base Year	Index with 1977 as Base Year
1967	2.68	100	51
1968	2.85	106	54
1969	3.04	113	58
1970	3.23	121	62
1971	3.45	129	66
1972	3.70	138	70
1973	3.94	147	75
1974	4.24	158	81
1975	4.53	169	86
1976	4.86	181	93
1977	5.25	196	100
1978	5.69	212	108
1979	6.16	230	117
1980	6.66	249	127
1981	7.25	271	138
1982	7.68	287	146
1983	8.02	299	153
1984	8.32	310	158
1985	8.57	320	163
1986	8.76	327	167
1987	8.98	335	171
1988	9.28	346	177
1989	9.66	360	184
1990	10.02	374	191
1991	10.34	386	197

Figure 13.3 gives the cost index for industrial buildings as computed and published by the Austin Company. This index is based on prices in major industrial areas for a 116,760-square-foot flat-roofed, steel-framed structure and an 8,325-square-foot office building. Estimates include site work, fluorescent lighting with uniform illumination, mechanical ventilation, office air conditioning, sprinklers for fire protection, toilet facilities, process services, sanitary plumbing, and storm drainage. This index is an example of how the costs for labor, materials, and so on can be combined into one composite index.

The adjustment of costs for yearly price increases based on indexes is not always easy. While the average labor rate may increase by 6% in a given year, the labor rate of a particular firm may be either more or less. Also, indexes may

TABLE 13.2.
AVERAGE HOURLY EARNINGS INDEX FOR PRODUCTION LABOR CATEGORIES (1967 = 100)

Year	Electric and Electronic Equipment	Transportation Equipment	Chemicals and Allied Products	Petroleum and Coal Products	Miscellaneous Manufacturing Industries
1967	100	100	100	100	100
1968	106	107	105	105	106
1969	111	113	112	112	113
1970	119	118	119	120	120
1971	125	129	128	128	126
1972	130	140	137	139	132
1973	136	150	145	147	140
1974	146	161	157	159	150
1975	163	176	174	181	162
1976	172	192	191	201	172
1977	187	212	207	219	186
1978	204	230	226	241	200
1979	223	248	245	261	214
1980	252	272	268	282	232
1981	276	302	294	318	254
1982	298	323	321	348	273
1983	315	339	341	371	290
1984	328	355	357	375	300
1985	346	369	373	393	311
1986	360	372	386	396	321
1987	372	376	399	407	330
1988	385	386	410	418	340
1989	398	398	422	430	353
1990	412	412	438	446	366
1991	425	425	452	457	378

not be available for specific material items or for certain purchased parts. A third problem arises when expenditures are made over a number of years for a project of long duration. In this latter case, the costs in early years will require less adjustment than those in later years.

Whenever price-level changes are contemplated, one should consider the fact that increasing productivity tends to offset increased labor costs. The increase in productivity is not uniform from firm to firm or even within a specific

TABLE 13.3.
AVERAGE HOURLY EARNINGS OF PRODUCTION WORKERS BY LOCATION IN 1991

State	Average Hourly Earnings
Alabama	$ 9.61
California	11.76
Florida	9.23
Kentucky	10.82
Maryland	11.82
Massachusetts	11.73
Michigan	14.18
Nevada	10.83
New Mexico	9.26
New York	11.26
Pennsylvania	11.35
Texas	10.75
Washington	12.90
Wisconsin	11.37
Wyoming	10.89

company. However, it is a well-known fact of economic life that the upper limit on wage increases is set by the attainable productivity increase. Wage increases that exceed increases in productivity tend to depress profits and/or increase prices, which, in turn, adds to inflation.

13.3.3 Adjustments Because of Learning

Learning takes place within an individual or within an organization as a function of the number of units produced. It is commonly accepted that the amount of time required to complete a given task or unit of product will be less each time the task is undertaken. The unit time will decrease at a decreasing rate, and this time reduction will follow a predictable pattern.

The empirical evidence supporting the concept of a *learning curve* was first noted in the aircraft industry. The reduction in direct labor-hours required to build an aircraft was observed and found to be predictable. Since then the learning curve has found applications in other industries as a means for adjusting costs for items produced beyond the first one.

Most learning curves are based upon the assumption that the direct labor-hours needed to complete a unit of product will decrease by a constant percent-

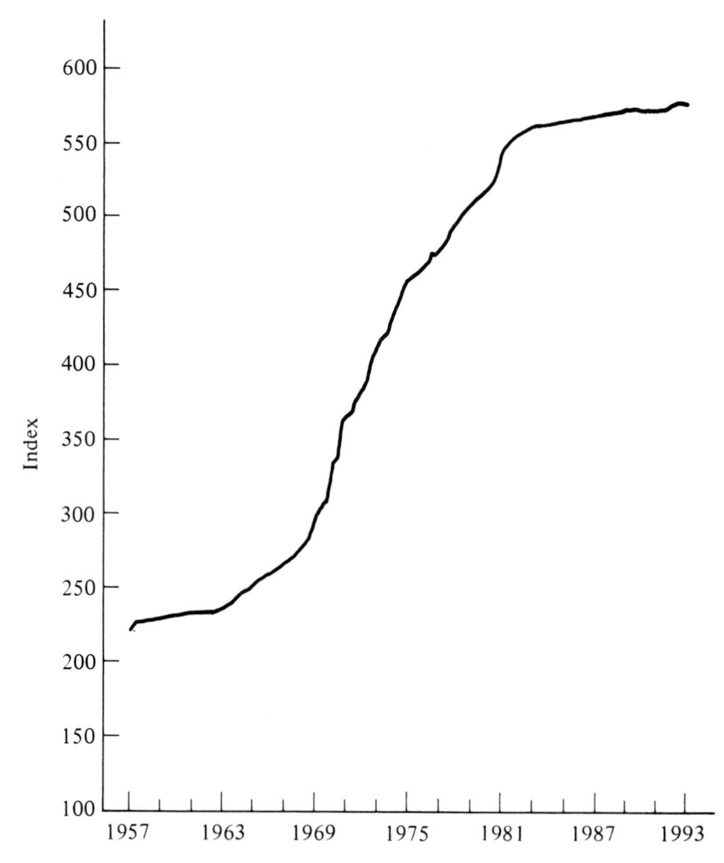

FIGURE 13.3. Industrial building cost index (1926 = 100). (Courtesy of Austin Company)

age each time the production quantity is doubled. A typical rate of improvement in the aircraft industry is 20% between doubled quantities. This establishes an 80% learning function and means that the direct labor-hours needed to build the second aircraft will be 80% of the hours required to build the first aircraft. The fourth aircraft will require 80% of the hours that the second require, the eighth aircraft will require 80% of the fourth, and so on. This relationship is given in Table 13.4.

An analytical expression for the learning curve may be developed from the above assumptions. Let

x = the unit number;

Y_x = the number of direct labor-hours required to produce the xth unit;

K = the number of direct labor-hours required to produce the first unit;

ϕ = the slope parameter of the learning curve.

TABLE 13.4.
UNIT, CUMULATIVE, AND CUMULATIVE AVERAGE DIRECT LABOR-HOURS FOR AN 80% FUNCTION WITH UNIT 1 AT 100 HOURS

Unit Number	Unit Direct Labor-Hours	Cumulative Direct Labor-Hours	Cumulative Average Direct Labor-Hours
1	100.00	100.00	100.00
2	80.00	180.00	90.00
4	64.00	314.21	78.55
8	51.20	534.59	66.82
16	40.96	892.01	55.75
32	32.77	1,467.86	45.87
64	26.21	2,392.45	37.38

From the assumption of a constant percentage reduction in direct labor-hours for doubled production units,

$$Y_x = K\phi^0 \qquad \text{where } x = 2^0 = 1$$

$$Y_x = K\phi^1 \qquad \text{where } x = 2^1 = 2$$

$$Y_x = K\phi^2 \qquad \text{where } x = 2^2 = 4$$

$$Y_x = K\phi^3 \qquad \text{where } x = 2^3 = 8.$$

Therefore,

$$Y_x = K\phi^d \qquad \text{where } x = 2^d.$$

Taking the common logarithm gives

$$\log Y_x = \log K + d \log \phi$$

where

$$\log x = d \log 2.$$

Solving for d gives

$$d = \frac{\log Y_x - \log K}{\log \phi} \quad \text{and} \quad d = \frac{\log x}{\log 2}$$

from which

$$\frac{\log Y_x - \log K}{\log \phi} = \frac{\log x}{\log 2}$$

$$\log Y_x - \log K = \frac{\log x(\log \phi)}{\log 2}.$$

Let

$$n = \frac{\log \phi}{\log 2}.$$

Therefore,

$$\log Y_x - \log K = n \log x.$$

Taking the antilog of both sides gives

$$\frac{Y_x}{K} = x^n$$

$$\boxed{Y_x = Kx^n.} \qquad (13.1)$$

Application of the above equation can be illustrated by reference to the example of an 80% progress function with unit 1 at 100 direct labor-hours. Solving for Y_8, we see that the number of direct labor-hours required to build the eighth unit gives

$$Y_8 = 100(8)^{\log 0.8/\log 2}$$

$$= 100(8)^{-0.322}$$

$$= \frac{100}{1.9535} = 51.2.$$

The information from the learning curve can be extended to cost estimates for labor by multiplying by the labor rate that applies. In doing this, the analyst must take into consideration that the subsequent units may be completed months or years after the initial units. Adjustments for labor-rate increases might have to be made along with the adjustment for learning that is inherent in the application of the learning curve.

13.3.4 Adjustments for the Degree of Perfection

Figure 13.4 shows the approximate general relationship between tolerance in inches and the cost of production. The specification of a tolerance of 0.001 inch would increase the cost over a specification of 0.003 inches by

$$\frac{32 - 12}{12} = 1.67, \text{ or } 167\%.$$

Adjustments in production cost data can be made for future items that exhibit either more or less perfection from relationships like that exhibited in Figure 13.4. However, perfection is often a false ideal unless it is taken to mean that which is most appropriate.

To secure a desirable result even in the physical world, compromises from perfection must often be made. For example, increasing the compression ratio of an internal-combustion engine, ideally, will increase its efficiency but at the

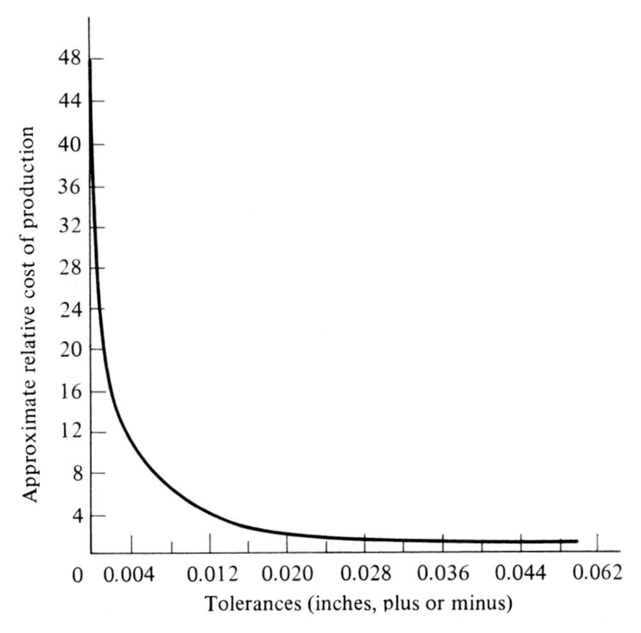

FIGURE 13.4. General cost relationship of various degrees of accuracy.

same time necessitate use of fuel of higher antiknock characteristics. In order to produce a smooth-running engine, a compromise between an ideal compression ratio and characteristics of available fuel will have to be made. Cost estimating for an appropriate degree of perfection suggests that consideration be given to the determination of the degree of perfection that, if specified, will be beyond the capability of the process to which it applies.

13.4 COST-ESTIMATING RELATIONSHIPS

Estimates of a result will usually be more accurate if they are based upon estimates of the factors having a bearing on the result than if the result is estimated directly. For example, in estimating the volume of a room it will usually prove more accurate to estimate the several dimensions of the room and calculate the volume than to estimate the volume directly. Similar reasoning applies to estimates of cost and other economic elements.

Statistical estimating methods often lead to mathematically fitted functions called *cost-estimating relationships* (CERs).

> A **cost estimating relationship** is a functional model that mathematically describes the cost of a structure, system, or service as a function of one or more independent variables.

Of course, there must be a logical or theoretical relationship of the variables to cost, a statistical significance of the contribution of the variables, and independence among the variables.

13.4.1 Determining Item Cost Involving Learning

As one example of the development of a cost-estimating relationship consider the production of a lot of N units. The cumulative number of direct labor-hours required to produce N units may be expressed as

$$T_N = Y_1 + Y_2 + \cdots + Y_N = \sum_{x=1}^{N} Y_x \quad (13.2)$$

where Y_x is the number of direct labor-hours required to produce the xth unit given by Equation (13.1).

An approximation for the cumulative direct labor-hours is given by

$$T_N \cong \int_0^N Y_x \, dx$$

$$\cong \int_0^N Kx^n \, dx$$

$$\boxed{\cong K\left(\frac{N^{n+1}}{n+1}\right).} \qquad (13.3)$$

Dividing by N gives an approximation for the cumulative average number of direct labor-hours expressed as

$$\boxed{V_N \cong \frac{1}{n+1}(KN^n).} \qquad (13.4)$$

Item cost per unit may be expressed in terms of the direct-labor cost, the direct-material cost, and the overhead cost. Let

LR = direct-labor hourly rate;

DM = direct-material cost per unit;

OH = overhead rate expressed as a decimal fraction of the direct-labor hourly rate.

Item cost per unit, C_i, can be expressed as

$$C_i = V_N(LR) + DM + V_N(LR)(OH). \qquad (13.5)$$

Or, by substituting for V_N,

$$\boxed{C_i = \frac{KN^n}{n+1}(LR)(1 + OH) + DM.} \qquad (13.6)$$

As an example of the application of this cost-estimating relationship, consider a situation in which 8 units are to be produced. The direct-labor rate is \$9 per hour, the direct-material cost per unit is \$600, and the overhead rate is 0.90.

It is estimated that the first unit will require 100 direct labor-hours and that an 80% learning curve is applicable. The item cost per unit is then estimated to be

$$C_i = \left[\frac{100(8)^{-0.322}}{-0.322 + 1}\right](\$9)(1.90) + \$600$$

$$= \left[\frac{51.19}{0.678}\right](\$9)(1.90) + \$600$$

$$= \$1{,}291 + \$600 = \$1{,}891.$$

This cost-estimating relationship gives the cost per unit as a function of the direct-labor cost, the direct-material cost, and the overhead cost. It also adjusts the direct labor-hours for learning in accordance with an estimated rate.

13.4.2 Power Law and Sizing Model

Equipment cost estimating can often be accomplished by using the power law and sizing model. Equipment to which this cost estimating relationship applies must be similar in type and vary only in size. The economies of scale in terms of size are expressed in the relationship

$$\boxed{C = C_r\left(\frac{Q_c}{Q_r}\right)^m} \qquad (13.7)$$

where

C = cost for design size Q_c;

C_r = known cost for reference size Q_r;

Q_c = design size;

Q_r = reference design size;

m = correlating exponent, $0 < m < 1$.

If $m = 1$, a linear relationship exists and the economies of scale do not apply. For most equipment m will be approximately 0.5, and for chemical processing equipment it is approximately 0.6.

As an example of the application of the power law and sizing model, assume that a 200-gallon reactor with glass lining and jacket cost \$9,500 in 1990. An estimate is required for a 300-gallon reactor to be purchased and installed in 1996. The price index in 1990 was 180 and is anticipated to be 210 in 1996. An estimate of the correlating exponent for this type of equipment is 0.50. Therefore, the estimated cost in 1990 dollars is

$$C_{1990} = \$9{,}500\left(\frac{300}{200}\right)^{0.5} = \$11{,}635$$

and the cost in 1996 dollars is

$$C_{1996} = \$11{,}635\left(\frac{210}{180}\right) = \$13{,}574.$$

This cost-estimating relationship relates a known cost to a future cost by means of an exponential relationship. Determination of the exponent is essential to the estimation process.

13.4.3 Other Estimating Relationships

The widespread use of estimating relationships speaks of their value in a wide range of situations. These relationships are available in the form of equations, sets of curves, nomograms, and tables.

In the aircraft and defense industry it is common to express the cost of airframe, power plant, avionics, and so forth as a function of weight, impulse, speed, etc. It is normally the next generation of aircraft or missile that is of interest. The estimated cost is determined by an extrapolation from known values for a sample group to an unknown proposed vehicle. This information is then used in cost-effectiveness studies and in budget requests.

Pollution-control equipment such as sewage plants can be estimated from functions that relate the investment cost to the capacity in gallons per day. A municipality considering a plant with primary and secondary treatment capability may refer to estimating a plant with primary and secondary treatment capability may refer to estimating relationships based upon plants already built. The estimator then would make two adjustments, one for the difference in construction cost due to the passage of time since the relationship was derived, the other for cost differences based on location.

Cost-estimating relationships are probably most plentiful for estimating the cost of buildings. These relationships are normally based upon square feet and/or volume. They take into consideration costs due to labor differentials for location. An individual beginning discussions about the construction of a new home will almost always begin thinking in terms of the cost per square foot.

Engineering costs for complex projects are available as a function of the installed cost of the structure or system being designed. These costs are expressed as a percentage of the installed cost of such projects as office buildings and laboratories, power plants, water systems, and chemical plants. In each case the engineering cost is a decreasing percentage as the installed project cost increases.

13.5 ESTIMATING AND ALLOCATING INDIRECT COSTS

Indirect or overhead costs embrace costs of material and labor not charged directly to product and fixed costs. Fixed costs embrace charges for such things as taxes, insurance, interest, rental, depreciation and maintenance of buildings, furniture, and equipment, and salaries of factory supervision, which are considered to be independent of volume of production. The practice of determining and allocating indirect cost is adopted to avoid the expense of charging all items of cost directly to the product being produced.

There are three common methods of allocating factory indirect charges to the product being produced: the direct-labor-cost method, the direct-labor-hour method, and the direct-material-cost method. Each of these bases for allocation will be illustrated by reference to a hypothetical manufacturing concern, which will serve to exhibit some aspects of indirect cost determination and allocation.

13.5.1 Indirect Cost-Estimation Example

As an example of indirect cost estimation, consider PLASCO Company, a small plastic manufacturing concern with plant facilities as shown in Figure 13.5. Land for the plant cost \$18,000 and the plant itself, constructed four years ago, cost \$9,000. Two-thirds of each of these items, or \$12,000 and \$60,000, are attributed to the production department. Initial cost of Machine X and Machine Y was \$36,000 and \$60,000, respectively. In addition, other assets of the firm are factory furniture with a first cost of \$8,000, small tools and dies which cost \$8,800, and stores and stock inventory with a current value of \$12,000.

The factory salaries and wages during the current year are estimated to be as follows:

Foreman F supervises factory operations	\$24,000
Handyman H moves material, takes care of stock and stores, and does janitor work	11,000
Total indirect labor	35,000
Worker W_1 operates Machine X, \$9/h × 2,000 h/y	18,000
Worker W_2 operates Machine Y, \$7/h × 2,000 h/y	14,000
	\$32,000

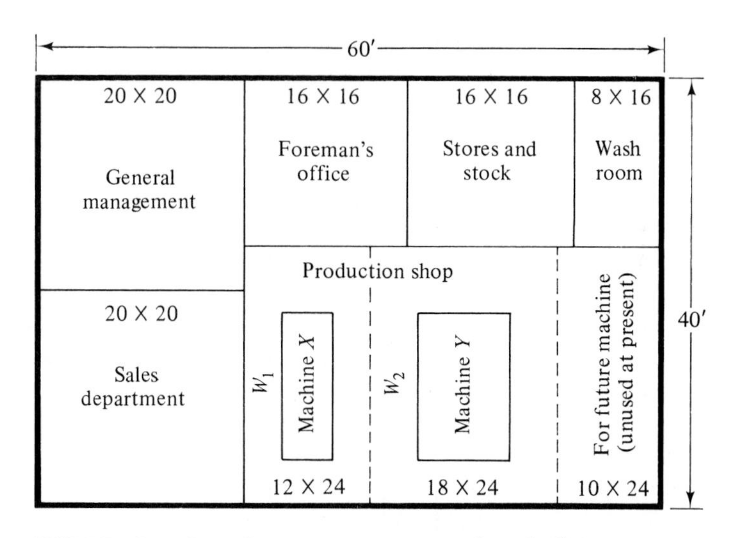

FIGURE 13.5. Plan of PLASCO Company plant facilities.

The supplies to be used in the factory during the fifth year are estimated as follows:

Office and general supplies	\$1,600
Water (estimated as $\frac{3}{4}$ of bill for entire building)	175
Lighting current (estimated as $\frac{2}{3}$ of bill for entire building)	900
Heating fuel (estimated as $\frac{2}{3}$ of bill for entire building)	1,410
Electric power (\$660 for Machine *X* and \$1,000 for Machine *Y*)	1,660
Maintenance supplies (\$580 for Machine *X* and \$800 for Machine *Y*)	1,380
	\$7,125

PLASCO Company makes three products *L*, *M*, and *N*. Estimated output, material cost, direct labor-hours, and machine-hours are given in Table 13.5.

The estimation of PLASCO Company indirect cost rates during the current year may be summarized as follows:

A. *Overhead Items for Building*

Depreciation, insurance, maintenance on building	\$ 8,800
Taxes on building and land	2,000
Interest on present value of building and land	8,400
Water, light, and fuel for factory	3,700
Total	\$22,900

TABLE 13.5.
ESTIMATED ACTIVITY DURING CURRENT YEAR

				Direct Labor-Hours				Machine Hours			
	Esti-			Worker		Worker		Machine		Machine	
Pro-	mated	Material Cost		W_1		W_2		X		Y	
duct	output	Each	Total	Each	Total	Each	Total	Each	Total	Each	Total
L	100,000	$0.10	$10,000	0.01	1,000	—	—	0.01	1,000	—	
M	140,000	0.08	11,200	—	—	0.01	1,400	—	—	0.01	1,400
N	80,000	0.12	9,600	0.0125	1,000	0.0075	600	0.0125	1,000	0.0075	600
			$30,800		2,000		2,000		2,000		2,000

B. *Miscellaneous Items of Overhead*	
Depreciation, taxes, insurance, and maintenance of factory furniture	$ 1,320
Interest on present value of factory furniture	740
Depreciation, taxes, insurance, and interest on present value of small tools	710
Taxes, insurance, and interest on stores and stock inventory	1,750
Office and general supplies	1,600
Total	$ 6,120
C. *Indirect Labor and Labor Overhead*	
Salaries of indirect labor of F and H	$35,000
Payroll taxes	7,800
Total	$42,800
D. *Machine X Overhead Items*	
Depreciation, taxes, insurance, and maintenance on Machine X	$6,100
Interest on present value of Machine X	2,800
Supplies for Machine X	580
Power for Machine X	660
Total	$10,140
E. *Machine Y Overhead Items*	
Depreciation, taxes, insurance, and maintenance on Machine Y	$ 9,400
Interet on present value of Machine Y	4,300
Supplies for Machine Y	800
Power for Machine Y	1,000
Total	15,500
Grand total, all factory overhead items	$97,460

On the basis of the information given, overhead allocation rates may be calculated as follows:

$$\text{direct-labor-cost rate} = \frac{\text{total factory overhead}}{\text{total direct-labor wages}}$$

$$= \frac{\$97{,}460}{\$32{,}000} = 3.05.$$

$$\text{direct-labor-hour rate} = \frac{\text{total factory overhead}}{\text{total hours of direct labor}}$$

$$= \frac{\$97{,}460}{4{,}000} = \$24.37.$$

$$\text{direct-material-cost rate} = \frac{\text{total factory overhead}}{\text{total direct-material cost}}$$

$$= \frac{\$97{,}460}{\$30{,}800} = 3.16.$$

The unit cost of Products *L*, *M*, and *N* may now be determined by each of the three methods of allocating factory overhead.

The Factory Cost of Product L

Direct-labor-cost method:	
Direct material	\$0.100
Direct labor, 0.01 × \$9	0.090
Overhead, \$0.090 × 3.05	0.275
	\$0.465
Direct-labor-hour method:	
Direct material	\$0.100
Direct labor, 0.01 × \$9	0.090
Overhead, 0.01 × \$24.37	0.244
	\$0.434
Direct-material-cost method:	
Direct material	\$0.100
Direct labor, 0.01 × \$9	0.090
Overhead, \$0.10 × 3.16	0.316
	\$0.506

The Factory Cost of Product M

Direct-labor-cost method:	
Direct material	\$0.080
Direct labor, 0.01 × \$7	0.070
Overhead, \$0.070 × 3.05	0.214
	\$0.364
Direct-labor-hour method:	
Direct material	\$0.080
Direct labor, 0.01 × \$7	0.070
Overhead, 0.01 × \$24.37	0.244
	\$0.394
Direct-material-cost method:	
Direct material	\$0.080
Direct labor, 0.01 × \$7	0.070
Overhead, \$0.08 × 3.16	0.253
	\$0.403

The Factory Cost of Product N

Direct-labor-cost method:	
Direct material	\$0.120
Direct labor, 0.0125 × \$9 + 0.0075 × \$7	0.165
Overhead, \$0.165 × 3.05	0.503
	\$0.788
Direct-labor-hour method:	
Direct material	\$0.120
Direct labor (calculated as above)	0.165
Overhead, 0.02 × \$24.37	0.487
	\$0.772
Direct-material-cost method:	
Direct material	\$0.120
Direct labor (calculated as above)	0.165
Overhead, \$0.12 × 3.16	0.379
	\$0.664

The cost of sales for the PLASCO Company is obtained by adding administrative and selling costs to the factory cost. Continuing with the example of the PLASCO Company, suppose that after careful analysis of expenditures, annual administrative costs have been estimated at \$54,000 and annual selling costs at \$30,200.

Annual direct-material costs, direct-labor costs, and factory overhead costs for PLASCO have been estimated previously as \$30,800, \$32,000, and \$97,460, respectively. Therefore, the annual estimated cost of sales of PLASCO Company for the current year may be summarized as follows:

Estimated annual direct-material cost	\$30,880		
Estimated annual direct-labor cost	32,000		
Estimated annual factory overhead cost	97,460		
Estimated annual factory cost		\$160,260	
Estimated annual administrative cost		54,000	
Estimated annual production cost			\$214,260
Estimated annual selling cost			30,200
Estimated annual cost of sales			\$244,460

13.5.2 Difficulties Associated with Overhead Allocations

An examination of the factory costs obtained for Product *L* in the PLASCO Company example with the three methods of factory overhead employed shows considerable variation. Each of the three costs that make up the total is subject to estimating error. Direct-material costs may not be accurate because of variations in pricing, charging a product with more material than is actually used, and the use of estimates. Similarly, for the same reasons, direct-labor costs as charged will usually be inaccurate to some extent. However, with reasonably good control and accounting procedures, direct-material costs and direct-labor costs are generally reliable.

If attention is directed to factory overhead costs allocated to Product *L*, it will be observed that amounts allocated by the several methods range from $0.434 per unit to 0.506 per unit. The ratio of these amounts is approximately 1.17. The fact that the amounts shown for the several methods differ is evidence that significant differences can result from the choice of a method of overhead allocation.

In actual practice the item of factory overhead is a summation of a great number and variety of costs. It is therefore not surprising that the use of a single, simple method will not allocate factory overhead costs to specific products with precision. Although a particular method may be generally quite satisfactory, wide variations may result in some specific situations.

For example, consider the direct-labor-hour method as it applies to Product *L* and Product *M*. The assumption is that $24.37 will be incurred for each hour of direct labor, regardless of equipment used. Thus, the amount of overhead allocated to Product *L* and Product *M* is identical, even though it is apparent that the cost of operating the respective machines on which these products are processed is quite different.

If the direct-material-cost method is used, it is clear that overhead allocations to products will depend upon the unit cost as well as upon the quantity of materials used. Suppose, for example, that in the manufacture of tables a certain model may be made from either pine or mahogany. Processing might be identical, but the amount charged for overhead might be several times as much for mahogany as for pine because of the difference in the unit cost of the materials used.

13.5.3 Effect of Changes in Level of Activity

In the determination of rates for the allocation of overhead, the activity of the PLASCO Company for the fifth year was estimated in terms of Products *L*, *M*, and *N*. This estimate served as a basis for determining annual material cost, annual direct-labor cost, and annual direct labor-hours. These items then became the denominator of the several allocation rates.

The numerator of the allocation rates was the estimated factory overhead, totaling $97,460. This numerator quantity will remain relatively constant for changes in activity, as an examination of the items of which it is composed will reveal.

For this reason the several rates for allocating factory overhead will vary in some generally inverse proportion with activity. Thus, if the actual activity is less than the estimated activity, the overhead rate charged will be less than the amount necessary to absorb the estimated total overhead. The reverse is also true. When the under- or overabsorbed balance of overhead becomes known at the end of the year, it is usually charged to profit and loss, or surplus.

In engineering economy, the effect of activity on overhead charges and overhead rates is an important consideration. The total overhead charges of the PLASCO Company would remain relatively constant over a range of activity represented, for example, by 1,200 to 2,000 hours of activity for each of the two machines. Thus, after the total overhead has been allocated, the incremental cost of producing additional units of product will consist only of direct-material and direct-labor costs.

13.6 ACCOUNTING DATA IN ESTIMATING

Since accounting data are the basis for many engineering economy studies, caution should be exercised in their use. An understanding of the relevance of accounting data is essential for its proper utilization. Two examples of the pertinence of cost accounting data are presented in this section.

13.6.1 Cost Data Must be Applicable

It is a common error to infer that a reduction in labor costs will result in a proportionate decrease in overhead costs, particularly if overhead is allocated on a labor-cost basis. In one instance a company was manufacturing an oil-field specialty. An analysis revealed per-item costs as follows:

Direct labor	\$16.20
Direct material	6.05
Factory overhead, \$16.20 × 2.30	37.26
Factory cost per item	\$59.51

The factory cost of \$59.51 was slightly less than the price of the item in question. The first suggestion was to cease making the article. But after further analysis it became clear that the overhead of \$37.26 would not be saved if the item was discontinued. The overhead rate used was based on heavy equipment required for most of the work in the department and on hourly earnings of workmen who averaged \$17.50 per hour. For the job in question only a light drill press and hand tools were used. Little actual reduction in cost would have resulted from not using them in the manufacture of the article.

It has previously been calculated that items of overhead of the PLASCO Company for the building are equal to \$9.54 per square foot of factory floor space. Figure 13.5 reveals a currently unused 10 feet by 24 feet space for future machines. The annual cost equivalent of this space is

$$10 \text{ ft} \times 24 \text{ ft} \times \$9.54/\text{ft}^2 = \$2{,}290.$$

This item need not be included as an item of cost attributable to a machine that may be purchased to occupy the space in question, since no actual additional cost will arise should the space be occupied. The $2,290 item has been entered as an overhead charge to be allocated to products made on the basis of one of several overhead rates. The addition of a new machine will probably result in changes in the overhead allocation rate used, but it will not result in a change in the overhead item pertaining to the building.

13.6.2 Average Costs are Inadequate for Specific Analysis

An important function of cost accounting, if not the primary one, is to provide data for decisions relative to the reduction of production costs and the increase of profit from sales. Cost data that are believed to be accurate may lead to costly errors in decisions. Cost data that give true average values and are adequate for overall analysis may be inadequate for specific detailed analyses. Thus, cost data must be carefully scrutinized and their accuracy established before they can be used with confidence in engineering economy studies.

In Table 13.6 actual and estimated cost data relative to the cost of three products have been tabulated. The actual production costs of Products *A*, *B*, and *C* are $9, $10, and $11, respectively, but because of inaccuracies in overhead costs are believed to be $10 for each.

Note that even though the average of a number of costs may be correct, there is no assurance that this average is a good indication of the cost of individual products. For this reason the accuracy of each cost should be ascertained before it is used in an economic analysis.

If, for example, the selling price of the products is based upon their believed production cost, Product *A* will be overpriced and Product *C* will be underpriced. Buyers may be expected to shun Product *A* and to buy large quantities of Product *C*. This may lead to a serious unexplained loss of profit. Average values of cost data are of little value in making decisions relative to specific products.

TABLE 13.6.
ACTUAL AND ESTIMATED COST DATA

Product	Direct Labor and Material Costs	Overhead Costs, Actual	Overhead Costs, Believed to Be	Production Cost, Actual	Production Cost, Believed to Be
A	$6.50	$2.50	$3.50	$ 9.00	$10.00
B	7.00	3.00	3.00	10.00	10.00
C	7.50	3.50	2.50	11.00	10.00
Average	7.00	3.00	3.00	10.00	10.00

13.7 JUDGMENT IN ESTIMATING

The need for judgment in the estimating process is often discussed. Although this need is self-evident, one of the problems in the past has been too much reliance on judgment and too little on quantitative approaches.

Because people are involved in applying judgment, there arises the problem of personal bias. A person's assigned role or position seems to influence his forecasts. Thus, a tendency toward low estimates appears among those whose interests they serve, principally proponents of a new structure, project, or product. Similarly, estimates by people whose interests are served by caution are likely to be higher than they would otherwise.

The primary use of judgment should be to decide whether or not an estimating relationship can be used for an advanced system. Second, judgment must determine what adjustments will be necessary to take into account the effect of a technology that is not present in the sample. Judgment is also required to decide whether or not the results obtained from an estimating relationship are reasonable in comparison with the past cost of similar items. Any estimate that implies that new, higher-performance equipment can be produced for less than the cost of existing units should be subject to critical examination. This will be true for computer hardware and electronic items but is rarely true in other areas.

When a proposed project contains considerable uncertainty, it is often wise to hold capital equipment investment to a minimum until outcomes become clearer, even though such a decision may result in higher maintenance and operation costs for the present system. Such action amounts to a decision to incur higher costs temporarily in order to reserve the privilege of making a second decision when the situation becomes clearer.

The accuracy of estimates with respect to events in the future is, at least to some extent, inversely proportional to the span of time between the estimate and the event. It is often appropriate to incur expense for the privilege of deferring the decision until better estimates can be made.

KEY POINTS

- Estimating in the economic environment cannot be done with certainty, since economic laws depend upon the behavior of people.
- For best results in estimating the final outcome of an undertaking, it will almost always be advantageous to begin with detailed estimates of income and costs.
- To come to a sound conclusion about an engineering alternative, both physical elements (input and output) and measures of these elements in economic terms are needed.

- Cost estimating can be done by engineering procedures, by analogy, or by statistical methods depending upon the phase of the project life cycle.
- Data must be made consistent and comparable if they are to be useful in an estimating procedure.
- Cost-estimating relationships are functions mathematically fitted to empirical data which can be used during conceptual or preliminary design.
- The allocation of indirect cost is an uncertain process which contributes significantly to the lack of pecision in the estimation of the cost of products, systems, structures, and services.
- Accounting data may be used in estimating economic elements, but its relevance to the proposed undertaking must be examined carefully.
- Although the need for judgment in the estimating process is self-evident, too much reliance has been placed on judgment and not enough on quantitative approaches.

QUESTIONS AND PROBLEMS

1. Why must engineering economy studies rely heavily upon estimates?
2. Explain why an estimate of a result will probably be more accurate if it is based upon estimates of the factors that have a bearing on the result than if the result is estimated directly.
3. Contrast outputs expected from commercial activities with those from governmental activities.
4. List the common items of input that are utilized in the production of goods and services.
5. Briefly describe the advantages and disadvantages of the three cost-estimating methods.
6. Use the data in Table 13.1 to compute a new set of indexes with 1980 as a base year.
7. Plot the data in Table 13.1 and estimate the average hourly earnings in 1994 by extrapolation.
8. If the total cost for manufacturing labor was $628,000 in 1978, estimate the total cost in 1988 using the information in Figure 13.2.
9. Suppose that 15,000 hours are required to build a certain system, for which the rate of learning for subsequent systems will be 15% on doubled quantities.
   a. Calculate the labor-hours to build the second, fourth, eighth, and 16th systems, and plot these points.

b. Estimate the labor-hours required for the 10th and 20th systems and compare this estimate to the result obtained from the formula.

10. Use the information in Problem 9 to compute the cumulative labor-hours for a group of 16 systems by both the exact and the approximate methods.

11. Estimate the cumulative average number of direct labor-hours for a product requiring 4 hours for the first unit, if 16 units are to be produced with an improvement of 20% between doubled production quantities.

12. Estimate the item cost per unit for a production lot of 1,000 units if the first unit requires 8 labor-hours, the 70% learning curve applies, the labor rate is $9.10 per hour, the direct-material cost is $40 per unit, and the overhead rate is 1.80.

13. A heat exchanger cost $7,500 in 1983 and must be replaced soon with a larger unit. The present unit has an effective area of 250 feet and its replacement should have an area of 350 feet. Replacement is anticipated in 1993 when the price index is estimated to be 170 with 1983 as the base year. Estimate the cost of the unit in 1993 if the correlating coefficient for this type of equipment is 0.6.

14. The manufacturing costs of Products X and Y are believed to be $10 per unit. On the basis of this estimate and a desired profit of 10%, the selling price is set at $11 per unit.

a. What is the profit if 800 units of Product X and 2,000 units of Product Y are sold?

b. If the actual manufacturing costs of Products X and Y are $9, and $11, respectively, what is the actual profit?

15. A small factory is divided into four departments for accounting purposes. The direct-labor and direct-material expenditures for a given year are as follows:

Department	*Direct Labor-Hours*	*Direct-Labor Cost*	*Direct-Material Cost*
A	900	$14,500	$19,000
B	945	14,900	6,800
C	1,050	15,050	11,200
D	1,335	13,900	15,000

Distribute an annual overhead charge of $42,000 to Departments *A*, *B*, *C*, and *D* on the basis of direct labor-hours, direct-labor cost, and direct-material cost.

16. A manufacturer makes 8,500,000 memory units per year. An assembly operation is performed on each unit, for which a standard piece rate of $1.05 per hundred pieces is paid. The standard cycle time per piece is 0.624 minute. Overhead costs associated with the operation are estimated at

$0.36 per operator-hour. If the standard cycle time can be reduced by 0.01 minute and if one-half of the overhead cost per operator-hour is saved for each hour by which total operator-hours are reduced, what will be the maximum amount that can be spent to bring about the 0.01-minute reduction in cycle time? Assume that the improvement will be in effect for one year, that piece rates will be reduced in proportion to the reduction in cycle time, and that the average operator's time per piece is equal to the standard time per piece in either case.

17. An automobile parts manufacturer produces batteries and distributor assemblies in his electrical products department. It is believed that the cost of manufacturing the batteries and the distributor assemblies is $18.40 and $23.20 per unit, respectively. These costs were derived on the basis of an equal distribution of overhead charges. A study of the firm's cost structure reveals that overhead would be more equitably distributed if overhead charges against the distributor assembly were 50% more than those against the battery. This conclusion was reached after careful consideration of the nature and source of a $220,000 annual overhead expenditure for these products.

   a. Calculate the unit production cost applicable to the battery and to the distributor assembly if the annual production is made up of 24,000 batteries and 16,000 distributor assemblies.

   b. What is the annual profit if the selling prices are $25.80 and $32 for the battery and the distributor assembly, respectively?

18. What precautions should be exercised in utilizing accounting data in engineering economic analysis?

19. Describe a situation in which a person's role or position has influenced his or her estimate of an undertaking.

14

Estimates and Decision Making

The future outcome of a venture can be predicted accurately if sufficient knowledge is available. Considerable emphasis is placed on securing appropriate data and using them carefully to arrive at estimates that are representative of future outcomes to the greatest possible extent. When all facts about alternative courses of action are known in accurate quantitative terms, the relative merit of each may be expressed in terms of a single decision number. Decision making in this case is simple.

Decisions must be made even though quantitative considerations are based upon estimates that are subject to error. It must be remembered that the final calculated amounts embody the errors in the estimated quantities. Qualitative knowledge must be used to fill gaps in what is known about an undertaking. This chapter presents several methods for recognizing and compensating for errors in estimates.

14.1 AN EXAMPLE DECISION BASED UPON ESTIMATES

An example will be used to illustrate some aspects of estimating, treating estimated data, and arriving at an economic decision. To simplify the discussion, the example presented in this section will be used in the next two sections and income taxes will not be considered.

The purchase of a machine for producing a microcircuit in a new manner is considered likely to result in a saving. It is known with absolute certainty that the machine will cost $80,000 installed. All other factors pertinent to the decision are unknown and must be estimated.

14.1.1 Income Estimate

From a study of available data and the result of judgment, it has been estimated that a total of 6,000 units of the circuit will be made during the next 6 years. The number to be made each year is not known; but, since it is believed that production will be fairly well distributed over the 6-year period, it is believed that the annual production should be taken as 1,000 units. A detailed consideration of materials used, time studies of the methods employed, wage rates, and the like, has yielded an estimated saving of $55 per unit, exclusive of the costs incident to the operation of the machine if the machine is used. Combining the estimated production and the estimated unit saving results in an estimated saving (income) of 1,000 × $55 = $55,000 per year.

14.1.2 Capital Recovery with Return Estimate

The machine is a single-purpose machine and no use is seen for it except in processing the product under consideration. Its service life has been taken to be 6 years to coincide with the estimated production period of 6 years. It is believed that the salvage value of the machine will be offset by the cost of removal at retirement. Thus, the estimated net receipts at retirement will be zero.

Interest is considered to be an expense in this evaluation, and the rate has been estimated at 14%. The next step is to combine the estimates of first cost, service life, and salvage value to determine the estimated annual capital recovery with a return. For the first cost of $80,000, a service life of 6 years, a salvage value of zero, and an interest rate of 14% the resultant estimate of the annual cost of capital recovery and return from Equation (6.10) will be

$$(\$80{,}000 - \$0)(\overset{A/P,\,14,6}{0.2572}) + \$0(0.14) = \$20{,}573.$$

14.1.3 Operation and Maintenance Costs Estimate

Operation and maintenance costs will ordinarily be made up of several items such as power, supplies, spare parts, and labor. For simplicity, consider that the operation and maintenance expenses of the equipment in this example consist of four items. Each is estimated on the basis of the number of units of product to be processed per year as follows:

Power	$ 1,600
Maintenance labor	3,400
Operating and maintenance supplies	2,200
Operating labor	21,400
	$28,600

The estimated net annual profit for the project may be summarized as follows:

Estimated total annual income		$55,000
Estimated annual capital recovery with return	$20,573	
Estimated annual operation and maintenance cost	$28,600	
Estimated total annual cost	$49,173	
Estimated net annual profit		$ 5,827

This final statement means that the project will result in an equivalent annual profit of $5,827 per year for a period of 6 years if the several estimates prove to be accurate.

The resultant equivalent annual profit is itself an estimate, and experience teaches that the most certain characteristic of estimates is that they nearly always prove to be inaccurate, sometimes in small degree and often in large degree. Once the best possible estimates have been made, however, whether they eventually prove to be good or bad, they remain the most objective basis upon which to base decisions. It should be realized that decision making can never be an entirely objective process.

14.2 ALLOWANCES FOR ERRORS IN ESTIMATES

The better the estimate, the less allowance need be made for error. Allowances for variances do not, of course, make up for a deficiency of knowledge in the sense that they correct errors. The allowances presented in this section are merely *heuristic* methods of eliminating some consequences of error.

As an example of the effect of an allowance in an economic undertaking, consider the following illustration. A contractor has estimated the cost of a project on which he has been asked to bid at $100,000. If he undertakes the job, he wishes to profit by 10% or $10,000. How shall he make allowance for errors in his cost estimate? If he makes an allowance of 10% for errors in his estimates, comparing to the very low factor of safety of 1.10, his estimated cost becomes $110,000. To allow for his profit margin, he will have to enter a bid of $121,000. But the higher his bid, the less the chance that he will be the successful bidder. If he is not the successful bidder, his allowance for errors in his estimate may have served to insure not only that he did not profit from the venture, but that he was left with a loss equal to the cost of making of bid. This illustration serves to emphasize the necessity for considering the cost of making allowances for errors in estimates.

14.2.1 Allowing for Error in Estimates by High Interest Rates

A policy common to many business firms is to require that prospective undertakings be justified on the basis of a high minimum acceptable rate of return, say 30%. One basis for this practice is that there are so many opportunities which will result in a return of 30% or more that those yielding less can be ignored. But since this is a much greater return than most concerns make on the average, the high rate of return represents an allowance for error. It is hoped that if undertaking of ventures is limited to those that promise a high rate of return, none or few will be undertaken that will result in a loss.

Returning to the example of microcircuit production given previously, suppose that the estimated income and the estimated cost of carrying on the venture when the interest rate is taken at 30% are as follows:

Estimated total annual income		$55,000
Estimated annual capital recovery with return,		
$80,000(\overset{A/P,30,6}{0.3784})$	$30,272	
Estimated annual operating and maintenance cost	$28,600	
Estimated total annual cost	$58,872	
Estimated net annual profit		−$3,872

If the calculated loss based on the high interest rate is the deciding factor, the venture will not be undertaken. Though the estimates as given above, except for the interest rate of 30%, might have been correct, the venture would have been rejected because of the arbitrary high interest rate taken, even though the resulting return would be almost 30%.

Suppose that the total operating costs had been estimated as above but that annual income had been estimated at $60,000. On the basis of a policy to accept ventures promising a return of 30% on investment, the venture would be accepted. But if it turned out that the annual income was, say, only $50,000, the venture would result in loss regardless of the calculated income with the high interest rate. In other words, an allowance for error embodied in a high rate of return does not prevent a loss that stems from incorrect estimates if a venture is undertaken that will, in fact, result in loss.

14.2.2 Allowing for Error in Estimates by Rapid Payout

The effect of allowing for error in estimates by rapid payout is essentially the same as that of using high interest rates for the same purpose. For example, assume that a policy exists that equipment purchases must be based upon a 3-year payout period when the interest rate is taken at 14%.

Returning to the example above, suppose that the estimated income and the estimated cost of carrying on the venture when a 3-year payout period are adopted is as follows:

Estimated annual income (estimated for 6 years but taken as 3 years to conform to policy)		$55,000
Estimated annual capital recovery with return,		
$\$80{,}000(\overset{A/P,\,14,3}{0.4307})$	$34,459	
Estimated annual operating and maintenance cost	$28,600	
Estimated total annual cost	$63,059	
Estimated net annual profit		−$ 8,059

Under these conditions the venture would not have been undertaken. The effect of choosing conservative values for the components making up an estimate is to improve the certainty of a favorable result if the outcome results in values that are more favorable than those chosen.

14.3 CONSIDERING A RANGE OF ESTIMATES

A plan considered to have some merit for the treatment of estimates is to make a least favorable estimate, a fair estimate, and a most favorable estimate of each situation. The *fair estimate* is the estimate that appears most reasonable to the estimator after a diligent search for and a careful analysis of data. This estimate might also be termed the most likely estimate.

The *least favorable estimate* is the estimate that results when each item of data is given the least favorable interpretation that the estimator feels may reasonably be realized. The least favorable estimate is definitely not the very worst that could happen. This is a difficult estimate to make. Each element of each item should be considered independently insofar as this is possible. The least favorable estimate should definitely not be determined from the fair estimate by multiplying the latter by a factor.

The *most favorable estimate* is the estimate that results when each item of data is given the most favorable interpretation that the estimator feels may reasonably be realized. Comments similar to those made in reference to the least favorable estimate, but of reverse effect, apply to the most favorable estimate. The use of the three estimates will be illustrated by application to the example of previous sections.

Items Estimated	*Least Favorable Estimate*	*Fair Estimate*	*Most Favorable Estimate*
Annual number of units	900	1,000	1,100
Savings per unit	$50	$55	$60
Annual savings	$45,000	$55,000	$66,000
Period of annual savings, n	4	6	8

Capital recovery with return,			
$\$80{,}000(\overset{A/P,\,14,n}{\quad})$	$27,456	$20,573	$17,248
Operating and mainteance cost:			
Item *a*	$2,000	$1,600	$1,200
Item *b*	4,500	3,400	2,300
Item *c*	3,300	2,200	1,150
Item *d*	24,600	21,400	18,200
Estimated total of capital recovery, return, and operating items	$61,856	$49,173	$40,098
Estimated net annual saving in prospect for *n* years	−$16,856	$5,827	$25,902

An important feature of the least favorable, fair, and most favorable estimate plan of comparision is that it provides for bringing additional information to bear upon the situation under consideration. Additional information results from the estimator's analysis and judgment in answering two questions relative to each item: "What is the least favorable value that this item may reasonably be expected to have?" and the reverse, "What is the most favorable value that this item may reasonably be expected to have?"

Judgment should be made item by item, for a summation of judgments can be expected to be more accurate than a single judgment of the whole. A second advantage of the three-estimate approach is that it reveals the consequences of deviations from the fair or most likely estimate. Even though the calculated consequences are themselves estimated, they show what is in prospect for different sets of conditions. It will be found that the small deviations in the direction of unfavorableness may have disastrous consequences in some situations. In others even a considerable deviation may not result in serious consequences.

The result above can be put on other bases for comparison. The present-worth basis has merit in this instance because of the variation in the number of years embraced by the above three estimates. The estimated present worth of savings is as follows:

Items Estimated	*Least Favorable Estimate*	*Fair Estimate*	*Most Favorable Estimate*
Net annual saving	−$16,856	$ 5,827	$ 25,902
Period of annual savings, *n*	4	6	8
Present worth of saving$(\overset{P/A,\,14,n}{\quad})$	−$49,113	$22,659	$120,157

Another viewpoint may be useful—for example, the net estimated savings per unit of product. This is computed as follows:

Items Estimated	Least Favorable Estimate	Fair Estimate	Most Favorable Estimate
Net annual saving	−$16,856	$5,827	$25,902
Annual number of units	900	1,000	1,100
Net savings per unit of product	−$18.73	$5.83	$23.55

Many estimators find it an aid to judgment to have several bases on which to compare a single situation. Since the cost of making extra calculations is usually insignificant in comparison with the worth of even a small improvement in decision, the practice should be followed by all who feel they benefit from the additional information. But there are limits beyond which further calculations can serve no useful purpose. This occurs when calculations are made but are beyond the scope of the data used. In the example above it would seem that no useful purpose would be served, for instance, by averaging the results calculated from the least favorable, fair, and most favorable estimates.

14.4 SENSITIVITY ANALYSIS

The crude techniques of the previous sections are limited in their usefulness. Decision makers are typically interested in the full range of possible outcomes that would result from variances in estimates. Sensitivity analysis permits a determination of how sensitive final results are to changes in the values of the estimates.

14.4.1 Sensitivity Involving a Single Alternative

As a simple example illustrating the concept of sensitivity analysis, assume that a present investment of $1,000 is anticipated to return $2,000 6 years hence. The expected rate of return on this investment may be found using

$$F = P(\overset{F/P,i,n}{\quad})$$

$$F = P(1 + i)^n$$

from which

$$i = \sqrt[n]{\frac{F}{P}} - 1.$$

For the stated example, the expected rate of return is

$$i = \sqrt[6]{\frac{\$2{,}000}{\$1{,}000}} - 1$$

$$= 1.1225 - 1$$

$$= 12.25\%.$$

The anticipated return of $2,000 may be more or less, which will make the expected rate of return more or less than 12.25%. Table 14.1 gives a range of possible values for F and rates of return that result. Also shown is the percentage change in i related to the percentage change in F. From this it can be concluded that the rate of return is quite sensitive to errors in the estimate of F, with the sensitivity being greater for underestimates of F than for overestimates. Figure 14.1 illustrates this situation for arbitrarily chosen maximum and minimum values of F.

In the previous example, only one estimate was varied as a basis for examining the impact on a measure of investment worth. As an example of sensitivity analysis involving a single alternative and multiple estimates, consider the cash flow illutrated in Figure 14.2.

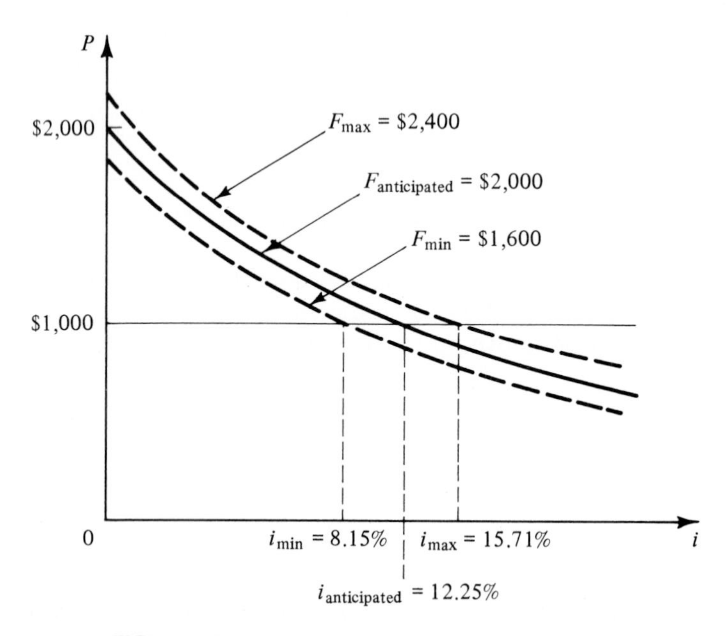

FIGURE 14.1. Sensitivity of i to F, given P and n.

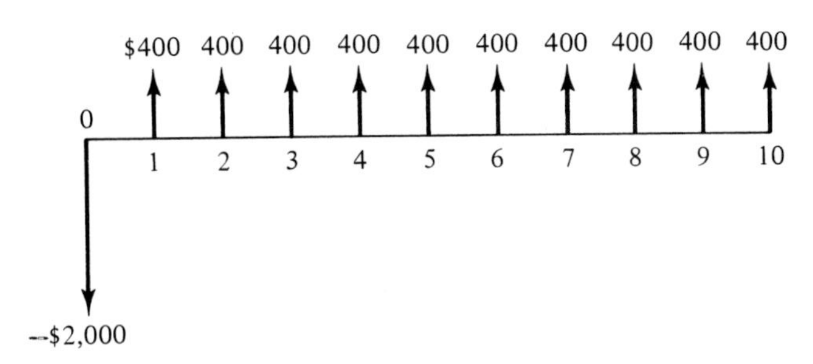

FIGURE 14.2. Cash flow for sensitivity-analysis example.

At an interest rate of 10%, the present worth of the cash flow is

$$PW(10) = -\$2{,}000 + \$400(\overset{P/A,\,10,10}{6.1446}) = \$458.$$

This result is valid only if all parameter estimates are correct. However, it is likely that errors will be made in estimating one or more of these parameters. To illustrate their individual impacts on present worth, assume that A, i, and n can vary between +50% and −50% of their anticipated or estimated values.

TABLE 14.1.
SENSITIVITY OF i TO F, GIVEN P AND N

F	i%	Percent Change in F	Percent Change in i
$1,600	8.15	−20	−33.5
1,700	9.25	−15	−24.5
1,800	10.29	−10	−16.0
1,900	11.29	− 5	− 7.8
2,000	12.25	0	0
2,100	13.16	5	7.4
2,200	14.04	10	14.2
2,300	14.89	15	21.6
2,400	15.71	20	28.2

If the estimate of the interest rate is incorrect and the estimates of A and n are assumed to be correct, the present worth of this investment can be expressed as

$$PW = -\$2{,}000 + \$400(\overset{P/A,i,10}{\quad\quad})$$

A ± 50% variation in i has an effect on PW as shown in Table 14.2. Likewise, the effects of ± 50% changes in A and n on PW are shown in Tables 14.3 and 14.4.

TABLE 14.2.
SENSITIVITY OF *PW* TO *i*, GIVEN *A* AND *n*

i%	*PW*
5	$1089
6	944
7	809
8	684
9	567
10	458
11	356
12	260
13	171
14	86
15	8

Sensitivity of present worth to changes in *A*, *n*, and *i* may be illustrated graphically as shown in Figure 14.3. Present worth is plotted versus the percentage error in the estimate of *A*, *n*, and *i*, from the information in Table 14.2, 14.3, and 14.4.

Figure 14.3 can now be used to reach conclusions about the desirability of pursuing the alternative under study. First, the *PW* is positive for all interest rates less than 15% (50% above the estimated value of 10%). *PW* is positive for investment durations exceeding approximately 7.5 years (25% below the estimated value of 10 years). Finally, *PW* is positive for annual returns exceeding about $325 (18.75% below the estimated value of $400).

TABLE 14.3.
SENSITIVITY OF *PW* TO *A*, GIVEN *i* AND *n*

A	*PW*
$200	−$771
240	−525
280	−278
320	−34
360	212
400	458
440	704
480	949
520	1195
560	1441
600	1687

TABLE 14.4.
SENSITIVITY OF *PW* TO *n*, GIVEN *i* AND *A*

n	PW
5	−$484
6	−258
7	−53
8	134
9	304
10	458
11	598
12	725
13	841
14	947
15	1042

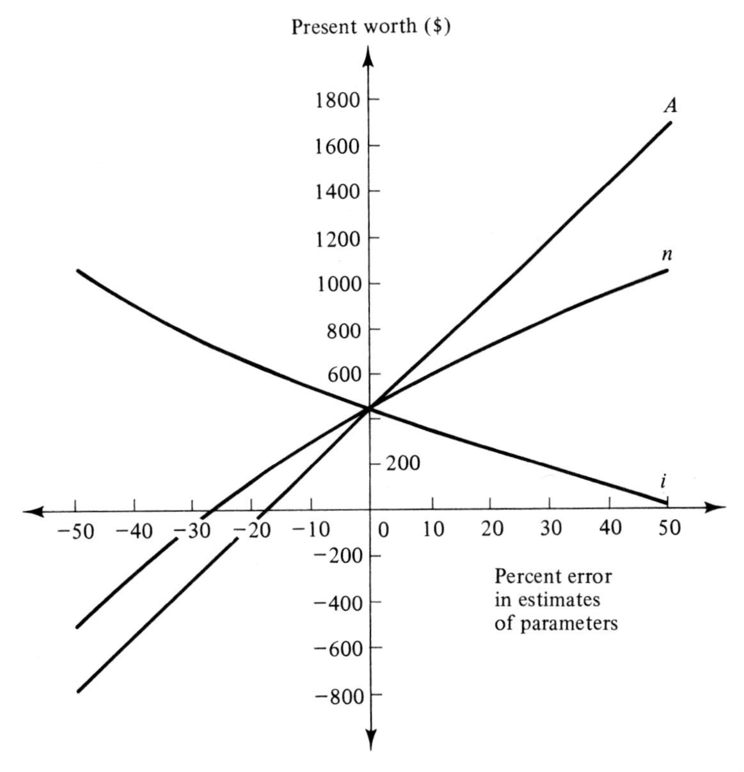

FIGURE 14.3. Sensitivity of present worth to changes in A, n, and i.

Sensitivity analysis simultaneously considering multiple estimates can give insight into planned courses of action by the decision maker. The same analysis can also be performed to determine the sensitivity of annual worth or other criteria which the decision maker may specify.

14.4.2 Sensitivity Involving Multiple Alternatives

As a more complex example of sensitivity analysis, the variation in the annual equivalent amount for three investment alternatives to changes in their gradient, the interest rate used, and their expected service lives will be examined. The annual equivalent profit expected from each of these alternatives is the basis for judging their economic desirability.

Alternative *A* requires an initial investment of \$1,000 followed by receipts that are a decreasing-gradient series, as shown in Figure 14.4. The annual equivalent profit for Alternative *A* is expressed as

$$AE(i)_A = -\$1{,}000(\overset{A/P,i,n}{\quad}) + \$1{,}000 - G(\overset{A/G,i,n}{\quad}).$$

Alternative *B* produces \$1,300 per year for *n* years from an initial investment of \$4,000, as shown in Figure 14.2. This alternative's receipts are an equal-payment series, and its annual equivalent profit is expressed as

$$AE(i)_B = -\$4{,}000(\overset{A/P,i,n}{\quad}) + \$1{,}300.$$

Alternative *C* requires an investment of \$5,000, which will produce revenues that are expected to increase uniformly over the investment's life, as shown in Figure 14.2. The equivalent annual profit for this alternative is expressed as

$$AE(i)_C = -\$5{,}000(\overset{A/P,i,n}{\quad}) + \$1{,}000 + G(\overset{A/G,i,n}{\quad}).$$

Consider the situation in which it is expected that the lives for each of the three alternatives will be 10 years and that the appropriate interest rate is 15%. What are the effects on profit for these three alternatives when the values of *G* range from \$0 to \$200? From the expressions for the three alternatives' annual equivalent profit, it is observed that Alternatives *A* and *C* are directly affected by changes in *G*, while Alternative *B* remains unaffected. These observations are clearly evident in Figure 14.5.

If, on the other hand, it is believed that the greatest variances will be associated with the estimated interest rate to be used, the changes in the annual equivalent profit as a function of interest rate can be investigated. Suppose that the gradient amount is \$100 for Alternatives *A* and *C* and that the expected life

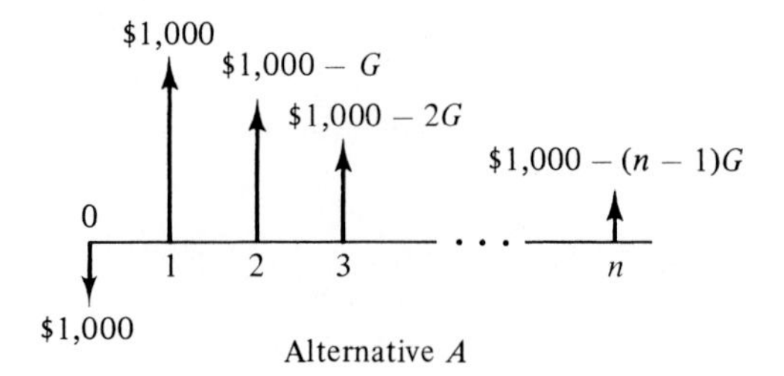

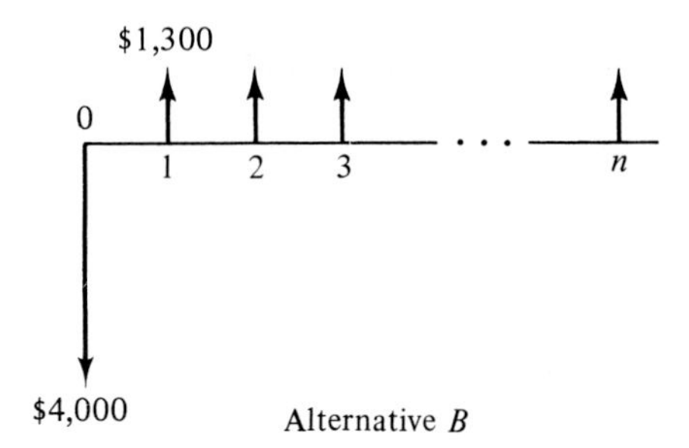

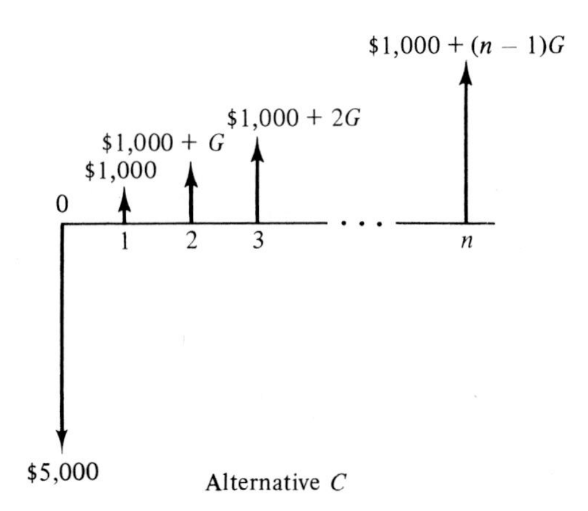

FIGURE 14.4. Cash flows for three investment alternatives.

for each of the three alternatives is 10 years. If these two parameters are held fixed, the changes in equivalent annual profit can be computed for interest rates ranging from 0% to 30%. The results of such an analysis are presented in Figure 14.6.

By holding the gradient and the interest-rate values constant, it is possible to study how variations in the estimated lives of these alternatives affect their equivalent annual profit. For this analysis the gradient amount was set at $100

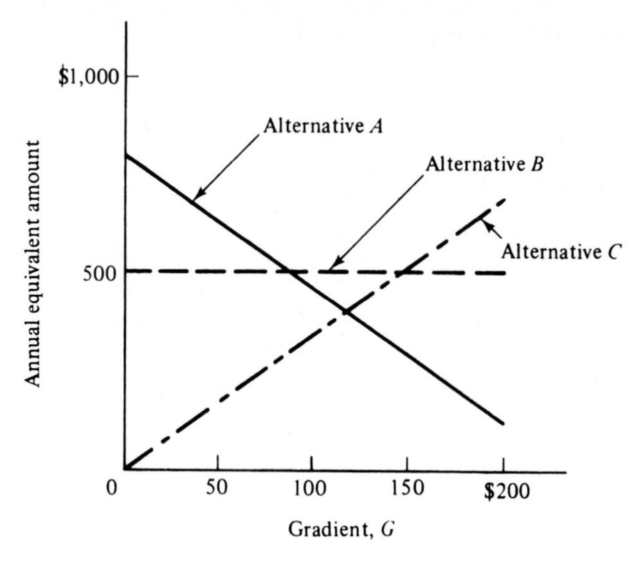

FIGURE 14.5. Sensitivity of the annual equivalent amount to the gradient.

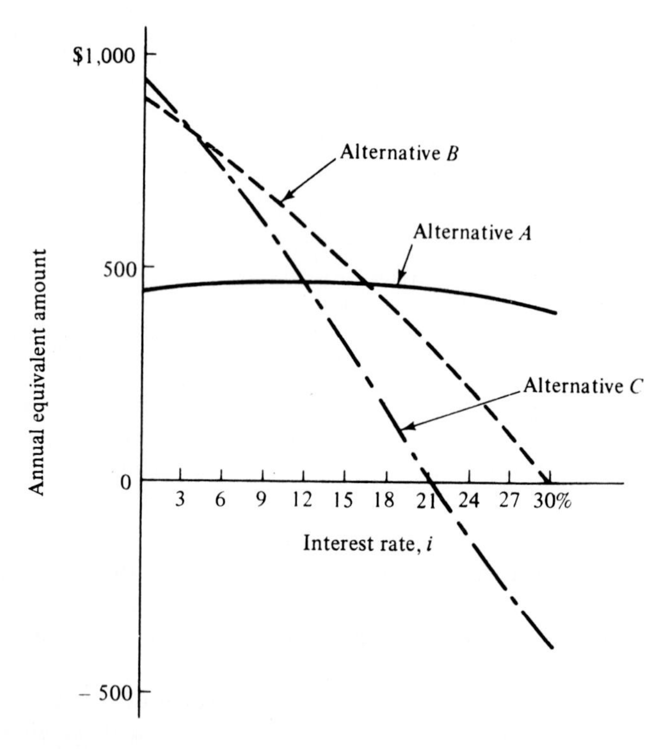

FIGURE 14.6. Sensitivity of the annual equivalent amount to the interest rate.

not included

and the interest rate was assigned a value of 15%. Figure 14.7 indicates the sensitivity of these three alternatives to various estimates of their lives.

An examination of Figure 14.5 reveals that the profitability of Alternatives *A* and *C* is significantly affected by the gradient amount that is estimated. Obviously, Alternative *B's* profitability is unaffected by such an estimate. It is also seen that for gradient amounts ranging from $0 to $200 all the alternatives will produce some positive contribution to profit. Thus, with regard to the estimate of the gradient amount, there seems to be little chance that financial losses will occur from any of these undertakings.

However, when comparing these three alternatives, it is possible to determine for a particular estimate which alternative would be preferred. From Figure 14.5 we see that Alternative *A* is preferred to the others as long as the esti-

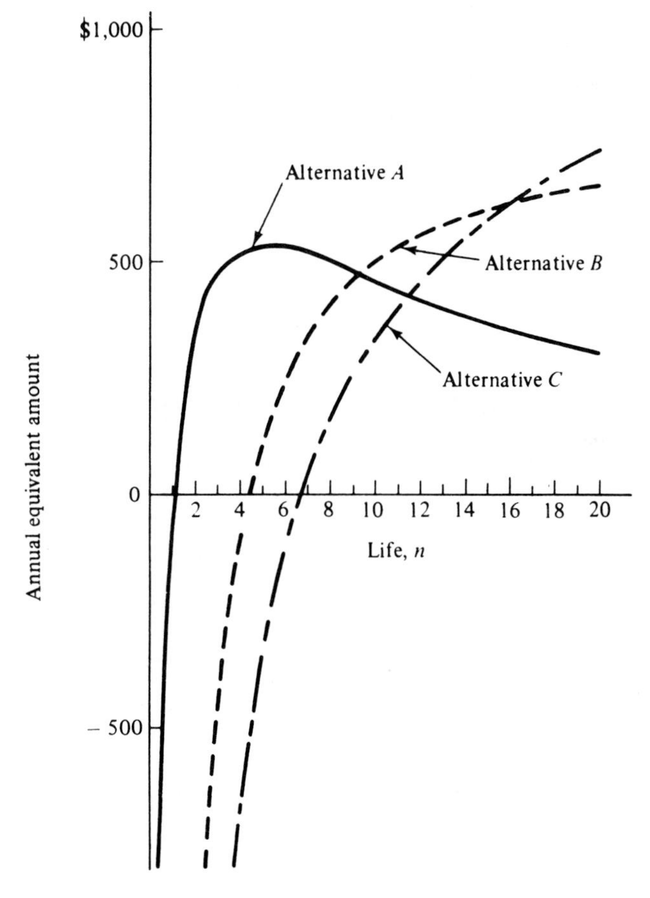

FIGURE 14.7. Sensitivity of the annual equivalent amount to the alternative's life.

mated gradient amount is less than \$88, while Alternative B is preferred for the range of values from \$88 to \$147. For values estimated over \$147, Alternative C is preferred. Use of this information provides insight regarding the sensitivity of these alternatives to the estimates of the gradient amount.

Similar interpretations can be made from Figures 14.6 and 14.7. For example, from Figure 14.6 we see that for the range of interest rates considered Alternative A will always be profitable. However, if the interest rate exceeds 21%, Alternative C will become unprofitable. Alternative B becomes unprofitable if the interest rate exceeds 30%. From Figure 14.6 it is possible to specify the range of interest rates for which one alternative is more profitable than the others. These ranges are as given in Table 14.5. Thus, if it is believed that the interest rate that should be utilized in this analysis would be in the vicinity of 10%, Alternative B is the obvious choice. Relatively large deviations from this 10% rate would not change the decision, and therefore the decision can be used with confidence.

The effect of various estimates of the alternatives' lives on profitability is presented in Figure 14.7. Here it is observed that Alternatives A, B, and C must have lives that exceed 2, 5, and 7 years, respectively, in order to assure a profit. In addition, the range of lives for which one alternative dominates the others can be easily determined. These ranges are given in Table 14.6.

Unless Alternative C is expected to have a life that exceeds 15 years, it should not be adopted. Thus, if it is anticipated that these undertakings would span approximately 10 years, Alternative B is favored. If the actual time span ex-

TABLE 14.5.
PREFERRED ALTERNATIVES FOR INTEREST-RATE RANGES

Preferred Alternative	Range of Interest Rates
Alternative C	$0 \leq i \leq 4\%$
Alternative B	$14\% \leq i \leq 16\%$
Alternative A	$16\% \leq i \leq 30\%$

TABLE 14.6.
PREFERRED ALTERNATIVES FOR LIFE RANGES

Preferred Alternative	Range of Life
Alternative A	$0 \leq n \leq 9$
Alternative B	$9 \leq n \leq 15$
Alternative C	$15 \leq n \leq 20$

ceeds the estimated life of 10 years by a substantial amount, Alternative *B* is still preferred. However, if the actual life realized is only slightly less than the 10 years estimated, Alternative *A* will be preferred to Alternative *B*. The impact of overestimating or underestimating the actual lives for these investments becomes evident. This type of information can provide the decision maker with a better understanding of the effect of estimates of future outcomes.

KEY POINTS

- Decisions must be made even though they are based on future outcomes, the estimates of which are subject to error.
- The final calculated evaluation measures embody errors in the estimated inputs and outputs, and so methods for treating these errors are needed.
- Heuristic methods for allowing for errors in estimates include requiring a high interest rate and rapid payout.
- Another heuristic method for treating errors in estimates is to allow estimates to vary over a range from least favorable to most favorable.
- The full range of possible outcomes that would result from variances in estimates can be studied through sensitivity analysis.
- Sensitivity analysis for multiple alternatives can lead to decision reversal.

QUESTIONS AND PROBLEMS

1. Relate the "factor of safety" concept in engineering design to allowances for estimating error in engineering economy studies.

2. Additional equipment must be purchased for use in a warehouse to meet an increased workload. It is estimated that the equipment would handle an additional 4,000 tons of goods per year, for which the estimated income would be $8 per ton. The equipment will cost $30,000, its service life is estimated to be 10 years, and the estimated annual maintenance and operation costs are $15,000. If the interest rate is 12%, compute the estimated net annual profit. Answer: $11,690

3. Would the equipment in Problem 2 be purchased if the additional workload per year were estimated to be 3,000 tons?

4. Would the equipment in Problem 3 be purchased if the minimum acceptable rate of return were 20%?

5. Suppose that the following most favorable estimates apply to the equipment in Problem 2:

Additional tonnage handled	5,500
Income per ton	$ 11
Service life in years	12
Annual maintenance and operation costs	$12,000

What is the annual net profit for this most favorable estimate? Answer: $43,658

6. Compute the least favorable, fair, and most favorable estimates of the cost per mile for an automobile you would personally use from estimates of the least favorable, fair, and most favorable first costs, service life, salvage value, interest rate, maintenance and operating costs, and miles driven per year. Your estimates should be those which you might personally experience.

7. A special metal-working machine was purchased for $75,000. It was estimated that the machine would result in a saving on production cost of $14,000 per year for 20 years. With a zero salvage value at the end of 20 years, what is the anticipated rate of return? It now appears that the machine will soon become inadequate. If it is sold after 6 years of use for $20,000, what is the change in the anticipated rate of return resulting from the incorrect service-life estimate?

8. A project is estimated in the face of a MARR = 10% as follows:

	Optimistic	*Most likely*	*Pessimistic*
Investment	$45,000	$50,000	$55,000
Service life	6 years	5 years	4 years
Salvage value	15,000	12,000	10,000
Net annual return	16,000	13,000	10,000

a. Find the annual equivalent amount for each possible outcome.

b. Tabulate the annual equivalent amount for all combinations of the estimates for service life and net annual return.

9. A machine design department is attempting to determine the annual equivalent cost that would be experienced by a machine currently being designed. It is estimated that it will cost $72,000 to design and build the machine. A service life of 7 years with a salvage value of $7,000 is anticipated, but these estimates are subject to error. A service life of 6 years with a salvage value of $8,000 and a service life of 8 years with a salvage value of $6,000 are also possible. If the interest rate is 10%, analyze the sensitivity of the annual equivalent cost to changes from the anticipated values.

10. A testing function can be performed by an automated diagnostic computer which will cost $100,000. Net annual savings in technician time are estimated to be $36,000. This equipment will serve for 6 years, but must be reprogrammed at the end of 3 years of operation at a cost of $40,000. Find the internal rate of return anticipated and then calculate the sensitivity of the IRR to the reprogramming cost.

11. The best estimates for a certain investment opportunity are as follows:

Initial investment	$105,000
Estimated life	8 years
Salvage value	22,000
Net annual return	34,000

If the MARR is 12%, calculate and graph (over a range of ± 40%) the sensitivity of the annual equivalent worth to (a) the interest rate, (b) the net annual return, and (c) the salvage value.

12. Three investment alternatives are under consideration with the following economic parameters and where $a, b, = 1, 2, \ldots, n - 1$.

Investment Alternative	Initial Investment	Annual Income	Service Life	Salvage Value	Interest Rate
A	X	$700	n	x	i
B	Y	$2,000 − aG	n	0	i
C	$3,000	$500 + bG	n	y	i

Analyze the sensitivity of the present worth if

a. G ranges from $50 to $200 with other parameters fixed at $X = \$2{,}000$, $Y = \$1{,}500$, $n = 10$, $x = \$200$, $y = \$300$, and $i = 10\%$.

b. n ranges from 0 to 20 years with other parameters fixed as in (a) with $G = \$100$.

c. i ranges from 0% to 25% with other parameters fixed as in (a) with $G = \$100$.

13. Analyze the sensitivity of the annual equivalent cost for the data given in Problem 12 if

a. X and Y vary between \$1,000 and \$3,000 with x = \$200, y = \$300, n = 10, i = 10%, and G = \$100.

b. x and y vary between \$0 and \$300 with X = \$2,000, Y = \$1,500, n = 10, i = 15%, and G = \$100.

15
Decision Making Involving Risk

There is usually little assurance that predicted outcomes will coincide with actual outcomes. The economic elements upon which a course of action depends may vary from their estimated values because of chance causes. Not only are the estimates of costs problematical but, in addition, the anticipated worth of most ventures is known only with a degree of assurance. This lack of certainty about the future makes economic decision making one of the most challenging tasks faced by individuals, industry, and government.

This chapter begins with an introduction to probability theory as a quantitative tool for dealing with risk in decision making. The techniques for using probability concepts assume that probabilities can be ascribed to future events. Decisions based on this assumption are classified as decisions under risk.

15.1 PROBABILITY CONCEPTS AND THEORY

To formally incorporate risk and uncertainty about future events into a decision process it is appropriate to utilize probability theory. Probability theory consists of an extensive body of knowledge concerned with the quantitative treatment of uncertainty. Since probability theory is well developed and rigorously defined, it is appropriate to apply it to decision problems involving chance causes.

By using probability theory it is possible to uniquely define events so that no ambiguities exist and so that each statement made within the theory is explicit and clearly understood. Probability theory allows uncertainty to be represented by a number, so that the uncertainty of different events can be directly compared. In addition, the structure of probability theory prevents the introduction of extraneous notions without the decision maker's full knowledge.

15.1.1 Some Probability Concepts

The probability that an event will occur may be expressed by a number that represents the likelihood of the occurrence. This likelihood may be determined by examining all available evidence related to the occurrence of the event. Thus, probability can be viewed as a state of mind, since it represents one's belief about the likelihood of an event's occurring.

To illustrate, suppose you were presented with a normal-looking coin and asked to guess the probability of its landing with heads up after it is tossed. Since you have never tossed this particular coin, you might answer that the probability of heads occurring is approximately one-half. This answer is based upon all your past experience of tossing two-sided coins.

Now suppose you were allowed to toss the coin in question 100 times and heads occurred 60% of the time. Your belief that this coin would yield heads 50% of the time might now be somewhat shaken. If you tossed the coin one million times and heads occurred 60% of the time, you would be convinced that this coin was not "fair" and that the likelihood of heads occurring was approximately 60%. Because your knowledge about the frequency of heads for this coin has increased, your feeling about the likelihood of getting a heads from a toss of this coin has been substantially modified. Thus, your information leads you to believe that the probability of heads occurring when *this* coin is tossed is approximately 0.6. Accordingly, any decision involving a wager on the basis of a toss of this coin will certainly be affected by your feeling about the likelihood of heads occurring. Since our beliefs about the occurrence of future events are the bases for making investment decisions, the concept of probability as a state of mind is quite useful in the analysis of economic alternatives.

The idea that probabilities can be subjective is disputed by some. However, because the usefulness of this concept for decision problems is well established, subjective probabilities have become an integral part of economic decision making. To take advantage of existing probability theory it will be assumed that subjective probabilities obey the laws of classical probability theory.

15.1.2 Probability Axioms

Three probability axioms are derived from classical probability theory. These are

1. For any event A, the probability of A is $P(A) \geq 0$.
2. $P(S)$-1, where S is the set of all possible events or outcomes under consideration.
3. If $AB = \emptyset$, then $P(A + B) = P(A) + P(B)$. That is, for two mutually exclusive events, the probability that event A or B occurs is equal to the sum of their probabilities.

The conditional probability of event A occurring, given that event B has occurred, is described as the probability of A given B,

$$P(A \mid B) = \frac{P(AB)}{P(B)} \tag{15.1}$$

where $P(AB)$ equals the probability of both event A and event B occurring.

When events A and B are independent,

$$P(AB) = P(A)P(B). \tag{15.2}$$

Thus, the probability of the event (A and B) equals the probability of event A multiplied by the probability of event B. When events A and B are independent, the conditional probability of event A occurring given that event B has occurred equals the probability that event A occurs, expressed as

$$P(A \mid B) = \frac{P(AB)}{P(B)} = \frac{P(A)P(B)}{P(B)} = P(A). \tag{15.3}$$

15.1.3 Probability Functions

A *random variable* is a function that assigns a value to each event included in the set of all possible events. For instance, if a coin is to be tossed twice, a random variable describing the number of heads occurring can have the values 0, 1, or 2. When a random variable is discrete, as in the coin-tossing example, a *probability function* is used to describe the probability of the random variable being equal to a particular value.

If the probability of a coin showing heads is considered to be 0.5, the probability function for the random variable "number of heads" for two tosses of the coin is found in the following manner. The possible outcomes of the two tosses can be described by the events H_1, heads on first toss; H_2, heads on second toss; T_1, tails on first toss; and T_2, tails on second toss. The possible outcomes of tossing the coin twice are shown in Column A of Table 15.1. The probability of each outcome is shown in Column B of Table 15.1. These probabilities are easy to cal-

TABLE 15.1.
RANDOM VARIABLE "NUMBER OF HEADS" FOR TWO TOSSES OF A COIN

Possible Outcomes A	Probability of Occurrence B	Number of Heads C
$H_1 H_2$	$P(H_1)P(H_2) = (0.5)(0.5) = 0.25$	2
$H_1 T_2$	$P(H_1)P(T_2) = (0.5)(0.5) = 0.25$	1
$T_1 H_2$	$P(T_1)P(H_2) = (0.5)(0.5) = 0.25$	1
$T_1 T_2$	$P(T_1)P(T_2) = (0.5)(0.5) = 0.25$	0

culate, since the outcome on the first toss is considered to be independent of the outcome on the second toss. Thus, the probabilities in Column *B* consist of the probability of the given event occurring on the first toss multiplied by the probability of the given event occurring on the second toss. The value of the random variable "number of heads" is shown in Column *C*.

The probability function can be directly described from the information in Table 15.1. The probability that the random variable "number of heads" has the value 2 is 0.25. The probability of having no heads occurring of two tosses is also 0.25. Since the events $H_1 T_2$ and $H_1 T_2$ are mutually exclusive, it is neccessary to sum the probabilities of those events to determine the probability that the random variable "number of heads" is equal to 1. Thus, the probability of getting one head in two tosses of the coin is 0.5.

The probability function describing the probability of there being a particular number of heads after two tosses of a coin is shown in Figure 15.1. For any probability function the sum of the probabilities for all possible outcomes must total to one and the probability of each possible outcome must be greater than or equal to zero and less than or equal to one. Thus, for the random variable x

$$\sum_x P(x) = 1 \tag{15.4}$$

where $0 \leq P(x) \leq 1$ and where $\sum_x$ indicates summing over all possible values of x.

When a random variable is continuous, a probability function is used to relate the probability of an event to a value or range of values for the random variable. A probability function for a random variable x is shown in Figure 15.2.

The probability of an event occurring is described by the area under the probability function for those values of x included in the event. For example, in Figure 15.2 the dark area under the curve for $1 \leq x \leq 3$ represents the probability that the random variable x will occur within that range. In order to satisfy the

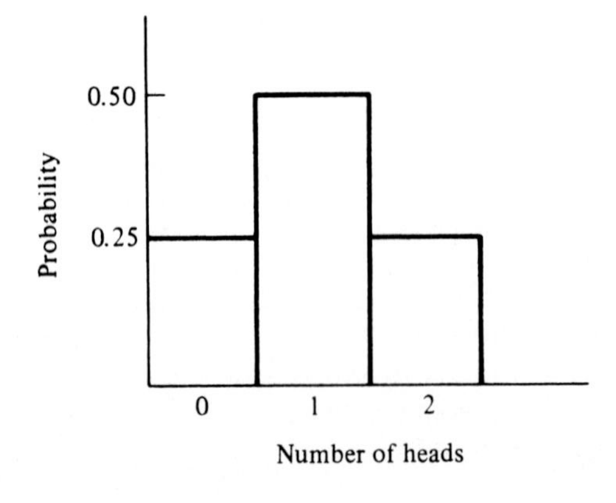

FIGURE 15.1. Probability function for the "number of heads" from the toss of two coins.

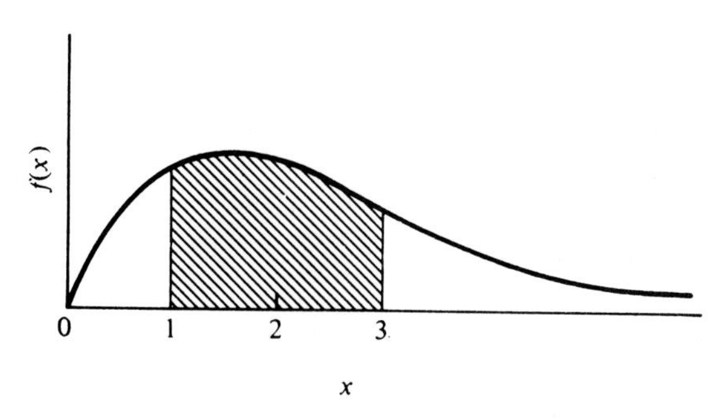

FIGURE 15.2. A continuous probability function.

axioms of probability, a probability function must have the following property:

$$\int_{-\infty}^{\infty} f(x)\, dx = 1. \tag{15.5}$$

with $0 \le f(x) < \infty$. The probability of an event in the range a to b is defined as

$$P(a \le x \le b) = \int_{a}^{b} f(x)\, dx. \tag{15.6}$$

15.1.4 Mean and Variance

Two parameters of a probability function help characterize the function. The first is the *mean* or *expected value* of the random variable. For a discrete random variable x, the expected value is defined as

$$E(x) = \sum_{x} xP(x). \tag{15.7}$$

For a continuous random variable x, the expected value is defined as

$$E(x) = \int_{-\infty}^{\infty} xf(x)\, dx. \tag{15.8}$$

The mean of a probability distribution is a measure of central tendency or concentration of mass for the distribution.

The second parameter is a measure of the spread of the probability distribution and is called the *variance*. The variance for any random variable x is defined as

$$\mathrm{Var}(x) = E\{[x - E(x)]^2\} = E(x^2) - [E(x)]^2. \tag{15.9}$$

For discrete random variables

$$E(x^2) = \sum_{x} x^2 P(x) \tag{15.10}$$

and

$$[E(x)]^2 = \left[\sum_x xP(x)\right]^2. \tag{15.11}$$

For continuous random variables

$$E(x^2) = \int_{-\infty}^{\infty} x^2 f(x)\, dx \tag{15.12}$$

and

$$[E(x)]^2 = \left[\int_{-\infty}^{\infty} xf(x)\, dx\right]^2. \tag{15.13}$$

Calculation of the mean and variance for the probability function in Figure 15.1 for the random variable "number of heads" N yields

$$E(N) = \sum_N NP(N) = 0(0.25) + 1(0.50) + 2(0.25) = 1$$

$$\text{Var}(N) = E(N^2) - [E(N)]^2 = 0^2(0.25) + 1^2(0.50) + 2^2(0.25) - 1^2 = 0.5.$$

15.2 EXPECTED-VALUE DECISION MAKING

If probability distributions are used to describe economic elements, the expected value of cost or profit can provide a reasonable basis for comparing alternatives. The expected profit or cost of a proposal reflects the long-term outcome that would be realized if the investment were repeated a large number of times with its probability distribution unchanged. Thus, when large numbers of investments are made, it may be reasonable to make decisions based upon the average or long-term effects of each proposal. Of course, it is necessary to recognize the limitations of using the expected value as a basis for comparison on unique or unusual projects where long-term effects are less meaningful.

Because most industries and governments are generally long-lived, the expected value as a basis for comparison seems to be a sensible method for evaluating investment alternatives under risk. The long-term objectives of such organizations may include the maximization of expected profits or the minimization of expected costs. To include the effect of the time value of money where risk is involved, all that is required is to state expected profits or costs as expected present worths, expected annual equivalents, or expected future worths.

15.2.1 Designing against Flood Damage

As a first example of decision making using expected values, suppose that a community has a water-treatment facility located on the floodplain of a river. The construction of a levee to protect the facility during periods of flooding is under consideration. Data on the costs of construction and expected flood damages are shown in Table 15.2.

Using historical records that describe the maximum height reached by the river during each of the last 50 years, the frequencies shown in Column *B* of Table 15.2 are ascertained. From these frequencies are calculated the probabilities that the river will reach a particular level in any one year. The probability for each height is determined by dividing the number of years for which each particular height was the maximum by 50, the total number of years.

The damages expected if the river exceeds the height of the levee are shown in Column *D*. It is observed that they increase in relation to the amount by which the flood crest exceeds the levee height. If the flood crest is 15 feet and the levee is 10 feet, the anticipated damages will be $100,000, whereas a flood crest of 20 feet for a levee 10 feet high would create damages of $150,000.

The estimated costs of constructing levees of various heights are shown in Column *E*. The community considers 12% to be the minimum attractive rate of return, and it is anticipated that after 15 years the treatment plant will be relocated away from the floodplain. The community wants to select the alternative that minimizes its total expected costs. Since the probabilities are defined as likelihood of a particular flood level in any one year, the expected equivalent annual costs are an appropriate choice for the basis for comparison.

TABLE 15.2.
PROBABILITY AND COST INFORMATION FOR DETERMINING OPTIMUM LEVEE HEIGHT

Feet (x) *A*	Number of Years River Maximum Level Was x Feet above Normal *B*	Probability of River Being x Feet above Normal *C*	Loss if River Level is x Feet above Levee *D*	Initial Cost of Building Levee x Feet High *E*
0	24	0.48	$ 0	$ 0
5	12	0.24	100,000	100,000
10	8	0.16	150,000	210,000
15	3	0.06	200,000	330,000
20	2	0.04	300,000	450,000
25	1	0.02	400,000	550,000
	50	1.00		

For example, the calculations required for each levee height for two of the alternatives are as follows:

5-Foot Levee

$$\text{Annual investment cost } \$100{,}000(\overset{A/P,12,15}{0.1468}) = \$14{,}682$$

$$\text{Expected annual damage} = (0.16)(\$100{,}000) + (0.06)(\$150{,}000) + (0.04)(\$200{,}000) + (0.02)(\$300{,}000) = 39{,}000$$

$$\text{Total expected annual equivalent cost} \quad \$53{,}682$$

10-Foot Levee

$$\text{Annual investment cost} = \$210{,}000(\overset{A/P,12,15}{0.1468}) = \$30{,}828$$

$$\text{Expected annual damage} = (0.06)(\$100{,}000) + (0.04)(\$150{,}000) + (0.02)(\$200{,}000) = 16{,}000$$

$$\text{Total expected annual equivalent cost} \quad \$46{,}828$$

The costs associated with all alternative levee heights are summarized in Table 15.3. The levee height that minimizes the total expected annual cost is 10 feet. A lower levee would not provide enough protection to offset the reduced construction costs. A levee higher than 10 feet requires more investment without providing proportionate savings from expected flood damage. The use of expected value in determining the cost of flood damage is reasonable in this case, since the 15-year period under consideration allows time for long-term effects to appear.

15.2.2 Introducing a New Product

Suppose a firm is planning to introduce a new product that is similar to an existing product. After a detailed study by the marketing and production departments, estimates are made of possible future cash flows as related to varying market conditions. The estimates shown in Table 15.4 indicate that this new product will produce cash flow *A* if demand decreases, cash flow *B* if demand remains constant, and cash flow *C* if demand increases. Based upon past experi-

TABLE 15.3.
SUMMARY OF ANNUAL CONSTRUCTION AND FLOOD DAMAGE COSTS

Levee Height (Feet)	Annual Investment Cost	Expected Annual Damage	Total Expected Annual Cost
0	$ 0	$80,000	$80,000
5	14,682	39,000	53,682
10	30,828	16,000	46,828
15	48,450	7,000	55,450
20	66,069	2,000	68,069
25	80,751	0	80,751

TABLE 15.4.
PROBABILITY OF OCCURRENCE FOR NEW-PRODUCT PROPOSAL CASH FLOWS

	Probability of Occurrence for Cash Flow		
	A	B	C
Year	$P(A) = 0.1$	$P(B) = 0.3$	$P(C) = 0.6$
0	−$30,000	−$30,000	−$30,000
1	11,000	11,000	4,000
2	10,000	11,000	7,000
3	9,000	11,000	10,000
4	8,000	11,000	13,000

ence and projections of future economic activity, it is believed that the probabilities of decreasing, constant, and increasing demand will be 0.1, 0.3, and 0.6, respectively.

This firm generally uses the present-worth criterion to make such decisions, and the minimum attractive rate of return is 10%. By calculating the present worth for each level of demand, the firm develops a probability function for the present-worth amount. Because this firm makes numerous decisions where risk is involved, it uses the expected value of the present-worth amount to determine the acceptability of such ventures. From Table 15.4 the expected present worth may be calculated using Equation (15.7) as follows:

$$
\begin{aligned}
E[PW(10)] &= (0.1)PW(10)_A + (0.3)PW(10)_B + (0.6)PW(10)_C \\
&= (0.1)\Big[-\$30{,}000 + \$11{,}000(\overset{P/A,10,4}{3.170}) \\
&\qquad -\$1{,}000(\overset{A/G,10,4}{1.3812})(\overset{P/A,10,4}{3.170})\Big] \\
&\quad +(0.3)\Big[-\$30{,}000 + \$11{,}000(\overset{P/A\,10,4}{3.170})\Big] \\
&\quad +(0.6)\Big[-\$30{,}000 + \$4{,}000(\overset{P/A,10,4}{3.170}) \\
&\qquad + \$3{,}000(\overset{A/G,10,4}{1.382})(\overset{P/A,10,4}{3.170})\Big] \\
&= (0.1)(\$492) + (0.3)(\$4{,}870) + (0.6)(-\$4{,}185) \\
&= -\$1{,}001.
\end{aligned}
$$

The expected value for the present worth of this proposal is negative, and the venture should be rejected. If the expected value had been positive, the new product could have been introduced.

15.3 EXPECTATION VARIANCE IN DECISION MAKING

In many decision situations it is desirable not only to know the expected value of the basis for comparison, but to also have a measure of the dispersion of its probability distribution. The variance of a probability distribution provides such a measure, and its value in decision making will be illustrated by examples.

15.3.1 Overall Outcome Uncertain

Suppose that a firm has a set of four mutually exclusive alternatives from which one is to be selected. The probability functions describing the likelihood of occurrence of the present-worth amounts for each alternative are given in Table 15.5. For example, the probability that Alternative $A2$ will realize a cash flow that has a present worth equal to \$60,000 is 0.4, while the probability that it will have a \$110,000 present worth is 0.2. The expected present worth and the variance of each probability distribution are given in Table 15.5. These values are calculated by using the definitions in Section 15.2. For example, the expected present worth for Alternative $A2$ using Equation (15.7) is

$$E(PW_{A2}) = (0.1)(-\$40{,}000) + (0.2)(\$10{,}000) + (0.4)\$60{,}000)$$

$$+(0.2)(\$110{,}000) + (0.1)(\$160{,}000) = \$60{,}000.$$

TABLE 15.5.
PROBABILITY DISTRIBUTIONS OF PRESENT-WORTH AMOUNTS FOR FOUR ALTERNATIVES

	Present Worth					Expected Present Worth	Variance
	−\$40,000	\$10,000	\$60,000	\$110,000	\$160,000		
Alternatives	*A*	*B*	*C*	*D*	*E*	*F*	*G*
*A*1	0.2	0.2	0.2	0.2	0.2	\$60,000	$\$5 \times 10^9$
*A*2	0.1	0.2	0.4	0.2	0.1	60,000	3×10^9
*A*3	0.0	0.4	0.3	0.2	0.1	60,000	2.5×10^9
*A*4	0.1	0.2	0.3	0.3	0.1	65,000	3.85×10^9

And the variance using Equation (15.9) is

$$\begin{aligned}\text{Var}(PW_{A2}) &= E(PW_{A2}^2) - [E(PW_{A2})]^2 \\ &= (0.1)(-\$40{,}000)^2 + (0.2)(\$10{,}000)^2 + (0.4)(\$60{,}000)^2 \\ &\quad +(0.2)(\$110{,}000)^2 + (0.1)(\$160{,}000)^2 - (\$60{,}000)^2 \\ &= \$3{,}000{,}000{,}000.\end{aligned}$$

An examination of the expected present-worth values in Table 15.5 indicates that there is relatively little difference among the alternatives. However, an examination of the distribution of possible present worth for each alternative gives additional insight into the desirability of each. First, it is important to determine the probability that the present worth of each alternative will be less than zero. This probability represents the likelihood of the investment yielding a rate of return less than the MARR. From Table 15.5, Column *A*

$$P(PW_{A1} \leq 0) = 0.2$$

$$P(PW_{A2} \leq 0) = 0.1$$

$$P(PW_{A3} \leq 0) = 0.0$$

$$P(PW_{A4} \leq 0) = 0.1.$$

Since it is desirable to minimize these probabilities, it appears that Alternative *A*3 is most desirable.

The second important consideration is the variance of these probability distributions. Because the variance represents the dispersion of the distribution, it is usually desirable to try to minimize it, since the smaller the variance the less the variability or uncertainty associated with the random variable. Alternative *A*3 has the minimum variance, as shown in Table 15.5.

It is clear that Alternative *A*3 is more desirable than *A*1 or *A*2, since their expected present worths are the same, and *A*3 is preferred on the basis of the two criteria discussed. However, the decision is not so obvious when comparing Alternatives *A*3 and *A*4. Alternative *A*4 has a larger expected present worth, but it is not as desirable as *A*3 on the basis of minimum variance and minimum chance of the present worth being less than zero. In cases such as these the decision maker must weigh the importance of each factor and decide if he would prefer more variability in the possible outcomes in order to achieve a higher expected value or less chance of the present worth being negative. If the relative importance of these three factors could be quantified, a single basis for comparison could be developed for each alternative.

15.3.2 Individual Elements Uncertain

As another example of expectation-variance decision making, suppose a defense contractor wishes to enter a bid on a project that needs special instrumentation. Two alternatives are being considered. The first has a high first cost and low operation and maintenance (O&M) costs, but the second has a low first cost and high O&M costs. Although the first cost of each instrumentation alternative is known with certainty, the annual O&M costs are uncertain, as is the contract duration.

Table 15.6 gives the probabilities associated with the O&M costs for each instrumentation alternative. Alternative *A* has a first cost of $70,000 because of its high degree of automation. Alternative *B* requires considerable operator attention and has a first cost of only $20,000. Neither instrumentation alternative will have any salvage value at the end of the contract period. The interest rate is 10%.

The contract duration is uncertain and is estimated to be either 1 year, 2 years, or 3 years. Table 15.7 gives the probability distribution describing this uncertainty, and Table 15.8 gives the capital recovery cost as a function of the contract duration for each alternative.

In this example, two discrete random elements must be combined to yield the expected cost and the cost variance. These two elements, the O&M cost and the contract duration, have no effect on each other and are therefore independent. Let

x_i = annual O&M cost, where i is the outcome;

y_j = annual capital-recovery cost, where j is the contract duration;

Px_i = probability that the outcome is x_i;

Py_j = probability that the outcome is y_i;

AE_x = annual equivalent cost for alternative x, where x is Alternative A or B.

TABLE 15.6.
PROBABILITIES FOR OPERATION AND MAINTENANCE COST

Alternative *A*		Alternative *B*	
O&M Cost	Probability	O&M Cost	Probability
$2,000	1/6	$12,000	1/6
$3,000	1/2	$25,000	1/3
$5,000	1/3	$30,000	1/3
		$40,000	1/6

TABLE 15.7.
CONTRACT DURATION PROBABILITIES

Contract Duration in Years, n	Probability the Duration is n
1	0.25
2	0.50
3	0.25

TABLE 15.8.
CAPITAL-RECOVERY COSTS

Contract Duration in Years, n	Capital-Recovery Cost	
	Alternative A $\$70{,}000(\overset{A/P,10,n}{\quad})$	Alternative B $\$20{,}000(\overset{A/P,10,n}{\quad})$
1	$77,000	$22,000
2	40,334	11,524
3	28,147	8,042

The annual equivalent cost will include the capital-recovery and the O&M costs. To determine the expected value and the variance for each alternative, each possible outcome must be considered. For independent events

$$E[AE_x] \sum_i \sum_j (x_i + y_j) Px_i Py_j \tag{15.14}$$

$$\text{Var}[AE_x] = E[(x_i + y_i)^2] - E[AE_x]^2 \tag{15.15}$$

where

$$E[(x_i + y_j)^2 = \sum_i \sum_j (x_i + y_j)^2 Px_i Px_j$$

For the example under consideration the mean for Alternative A is

$$\begin{aligned} E[AE_A] &= (2{,}000 + 77{,}000)\tfrac{1}{6} \cdot \tfrac{1}{4} + (2{,}000 + 40{,}334)\tfrac{1}{6} \cdot \tfrac{1}{2} \\ &\quad + (2{,}000 + 28{,}147)\tfrac{1}{6} \cdot \tfrac{1}{4} + (3{,}000 + 77{,}000)\tfrac{1}{2} \cdot \tfrac{1}{4} \\ &\quad + (3{,}000 + 40{,}334)\tfrac{1}{2} \cdot \tfrac{1}{2} + (3{,}000 + 28{,}147)\tfrac{1}{2} \cdot \tfrac{1}{4} \\ &\quad + (5{,}000 + 77{,}000)\tfrac{1}{3} \cdot \tfrac{1}{4} + (5{,}000 + 40{,}334)\tfrac{1}{3} \cdot \tfrac{1}{2} \\ &\quad + (5{,}000 + 28{,}147)\tfrac{1}{3} \cdot \tfrac{1}{4} \\ &= \$49{,}954 \end{aligned}$$

The variance for Alternative A is found as follows:

$$\begin{aligned} E[(x_i + y_j)^2] &= (2{,}000 + 7{,}000)^2 \tfrac{1}{6} \cdot \tfrac{1}{4} + (2{,}000 + 40{,}334)^2 \tfrac{1}{6} \cdot \tfrac{1}{2} \\ &\quad + (2{,}000 + 28{,}147)^2 \tfrac{1}{6} \cdot \tfrac{1}{4} + (3{,}000 + 77{,}000)^2 \tfrac{1}{2} \cdot \tfrac{1}{4} \\ &\quad + (3{,}000 + 40{,}344)^2 \tfrac{1}{2} \cdot \tfrac{1}{2} + (3{,}000 + 28{,}147)^2 \tfrac{1}{2} \cdot \tfrac{1}{4} \\ &\quad + (5{,}000 + 77{,}000)^2 \tfrac{1}{3} \cdot \tfrac{1}{4} + (5{,}000 + 40{,}334)^2 \tfrac{1}{3} \cdot \tfrac{1}{2} \\ &\quad + (5{,}000 + 28{,}147)^2 \tfrac{1}{3} \cdot \tfrac{1}{4} \\ &= \$2{,}832{,}405{,}430 \end{aligned}$$

$$\text{Var}[AE_A] = \$2{,}832{,}405{,}430 - (\$49{,}954)^2 = \$337{,}000{,}000.$$

The mean and variance for Alternative B are found to be

$$E[AE_B] = \$39{,}773$$

$$\text{Var}[AE_B] = \$89{,}400{,}000.$$

From the above it is observed that the annual equivalent cost for Alternative B is less than that for Alternative A. Likewise, the variance is lower for Alternative B, making it the preferred choice by the defense contractor.

15.4 MONTE CARLO ANALYSIS

Monte Carlo is the name given to a class of simulation approaches to decision making in which probability distributions describe certain economic elements. In many of these cases, an analytical solution is not possible because of the way in which the probabilities must be manipulated. In other cases, the Monte Carlo approach is preferred because of the level of detail that it provides.

Decision situations to which Monte Carlo analysis may be applied are characterized by empirical or theoretical distributions. The Monte Carlo approach utilizes these distributions to generate random outcomes. These outcomes are then combined in accordance with the economic analysis technique being applied to find the distribution of the present worth, the annual equivalent cost distribution, and so on.

It is necessary to generate values at random from the distributions representing problem parameters. There are many ways of doing this, including mechanical, mathematical, digital computer, and others. In this section, the defense-contractor example of Section 15.3 will be used to illustrate the mechanical approach for discrete distributions.

The distribution for contract duration in Table 15.7 is recognized to be the same as illustrated in Figure 15.1. Since this is the case, values can be generated for contract duration by tossing two coins. The method for generating a sequence of simulated contract durations is mechanical in nature.

A mechanical method can also be used to generate simulated O&M costs for each alternative. For Alternative *A* one die can be tossed, with certain outcomes representing certain cost occurrences. This same approach can be used for Alternative *B*. Table 15.9 summarizes the three mechanical means for generating the data pertinent to the instrumentation choice described in this example.

The process for generating simulated contract durations and O&M costs for the alternatives presented in Table 15.9 can now be applied with the techniques of engineering economy over several trials. This is shown in Table 15.10 for 100 trials with a summary at 10 trials. After 10 trials the annual equivalent cost for Alternative *A* is $48,630. For Alternative *B* the annual equivalent cost is $41,623 after 10 trials. The annual equivalent cost difference is $7,007, as shown in the last column of Table 15.10.

Although Alternative *B* appears to be best, it must be recognized that only 10 trials lead to this conclusion. It is entirely possible for a larger sample to yield different results. Consider the behavior of the mean annual equivalent cost for Alternative *A* as the number of trials increases to 100, as shown in Figure 15.3. Note how the average annual equivalent cost fluctuates early in the simulation and then stabilizes as the number of trials increases. It may be concluded that at

TABLE 15.9.
GENERATION OF SIMULATED VALUES

Simulated Value	Possible Outcomes	Probability of Outcome	Simulation Technique	Assignment of Outcome
Contract	1	1/4	Tossing	*HH*
Duration,	2	1/2	two	*HT* or *TH*
n	3	1/4	coins	*TT*
O&M	$ 2,000	1/6	Tossing	1
Costs for	$ 3,000	1/2	one	2, 3, or 4
Alternative *A*	$ 5,000	1/3	die	5 or 6
O&M	$12,000	1/6	Tossing	1
Costs for	$25,000	1/3	one	2 or 3
Alternative *B*	$30,000	1/3	die	4 or 5
	$40,000	1/6		6

TABLE 15.10.
MONTE CARLO COMPARISON OF TWO ALTERNATIVES

Trial Number	Outcome of Coin Tossing	Contract Duration n	(a) Capital Recovery Cost, Alternative A; $\$7,000(^{A/P,10,n})$	(b) Capital Recovery Cost, Alternative B; $\$20,000(^{A/P,10,n})$	Outcome of Die Toss Alternative A	(c) Annual O & M Costs Alternative A	(d) = (a) + (c) Annual Equivalent Cost, Alternative A	Outcome of Die Toss, Alternative B	(e) Annual O & M Costs, Alternative B	(f) = (b) + (e) Annual Equivalent Cost, Alternative B	(d) − (f) Difference in Annual Equivalent Cost, Alternative A − Alternative B
1	*HH*	1	$77,000	$22,000	2	$3,000	$80,000	3	$25,000	$47,000	$33,000
2	*HT*	2	40,334	11,524	4	3,000	43,334	3	25,000	36,524	6,810
3	*TH*	2	40,334	11,524	1	2,000	42,334	6	40,000	51,524	−9,190
4	*TH*	2	40,334	11,524	5	5,000	45,334	2	25,000	36,524	8,810
5	*TT*	3	28,147	8,042	2	3,000	31,147	1	12,000	20,042	11,105
6	*HH*	1	77,000	22,000	3	3,000	80,000	5	30,000	52,000	28,000
7	*TH*	2	40,334	11,524	6	5,000	45,334	4	30,000	41,524	3,810
8	*HT*	2	40,334	11,524	1	2,000	42,334	5	30,000	41,524	810
9	*TT*	3	28,147	8,042	3	3,000	31,147	6	40,000	48,042	−16,895
10	*TH*	2	40,334	11,524	6	5,000	45,334	4	30,000	41,524	3,810
.	.	.	.	.	.	.	486,300	.	.	$416,228	$70,070
.	.	.	.	.	.	.	Mean AEC =	.	.	Mean AEC =	Mean Diff. =
.	.	.	.	.	.	.	48,630	.	.	$41,623	$7,007
							.			.	.
							.			.	.
							.			.	.
100	*HT*	2	40,334	11,524	2	3,000	43,334	4	30,000	41,524	1,810

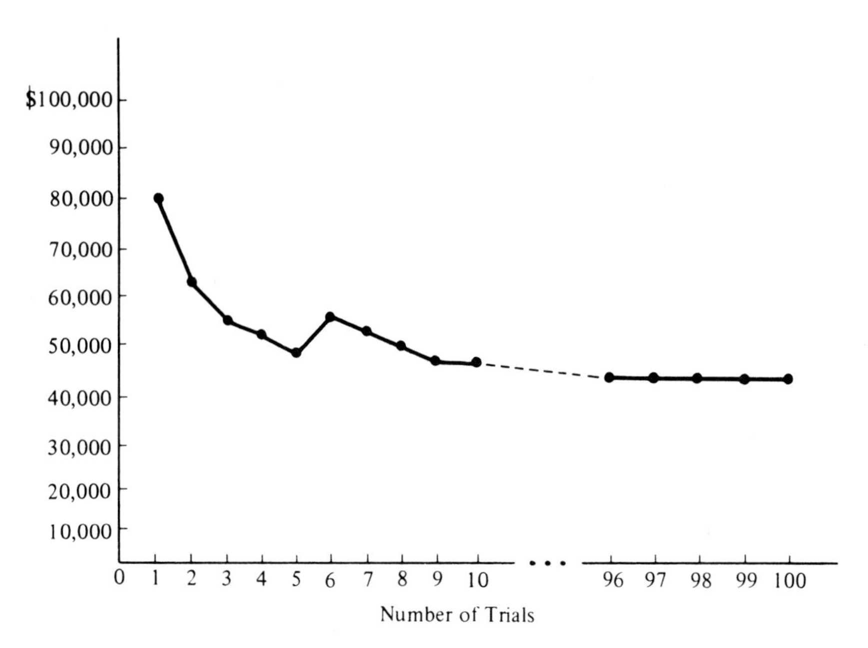

FIGURE 15.3. Convergence of the annual equivalent cost with increasing trials.

least 100 trials are required in order to obtain sufficiently stabilized results upon which to base a comparison.

The simulated data from Table 15.10 can be used to develop additional information about the cost of Alternative *A* and Alternative *B*. Tables 15.11 and 15.12 give the possible annual equivalent cost values for each alternative together with the frequency of occurrence for each.

The mean and variance for the annual equivalent cost under each alternative can be estimated from the frequencies given in Table 15.11 and 15.12. For Alternative *A* the mean is found from Equation (15.7) as

$$0.04(\$30{,}147) + 0.13(\$31{,}147) + \cdots + 0.08(\$82{,}000) = \$50{,}229.$$

And the variance is found from Equation (15.9) as

$$0.04(\$30{,}147)^2 + 0.13(\$31{,}147)^2 + \cdots + 0.08(\$82{,}000)^2 - (\$50{,}229)^2$$

$$= \$346{,}802{,}000.$$

For Alternative *B* the mean is

$$0.04(\$20{,}042) + 0.09(\$23{,}524) + \cdots + 0.05(\$62{,}000) = \$40{,}158.$$

And the variance is

$$0.04(\$20{,}042)^2 \quad 0.13(\$23{,}524)^2 + \cdots + 0.05(\$62{,}000)^2 - (\$40{,}158)^2$$

$$= \$101,\ 232{,}300.$$

TABLE 15.11.
FREQUENCY DISTRIBUTION OF THE ANNUAL EQUIVALENT COST FOR ALTERNATIVE A

Annual Equivalent Cost	Frequency in 100 Trials	Probability
$30,147	4	0.04
31,147	13	0.13
33,147	9	0.09
42,334	8	0.08
43,334	22	0.22
45,334	18	0.18
79,000	5	0.05
80,000	13	0.13
82,000	8	0.08
	100	1.00

TABLE 15.12.
FREQUENCY DISTRIBUTION OF THE ANNUAL EQUIVALENT COST FOR ALTERNATIVE B

Annual Equivalent Cost	Frequency in 100 Trials	Probability
$20,042	4	0.04
23,524	9	0.09
33,042	9	0.09
34,000	4	0.04
36,524	17	0.17
38,042	8	0.08
41,524	17	0.17
47,000	7	0.07
48,042	4	0.04
51,524	8	0.08
52,000	8	0.08
62,000	5	0.05
	100	1.00

The annual equivalent cost difference between Alternative *A* and Alternative *B* can now be estimated more precisely. It is $50,229 − $40,158 = $10,071. This compares with a difference of $7,007 after 10 trials. Thus, Alternative *B* is favored on the basis of the expected-value approach.

It should be noted that Alternative *B* also leads to a smaller variance in the annual equivalent cost than does Alternative *A*. Thus, from an expectation-variance viewpoint Alternative *B* is clearly superior to Alternative *A*. It is unlikely that this conclusion would be altered by pursuing the Monte Carlo analysis beyond 100 trials.

15.5 DECISION TREES IN DECISION MAKING

In many situations, future decisions are affected by actions taken at the present. Too often, decisions are made without consideration of their long-term effects. As a result, decisions that initially appear sound may place the decision maker in an unfavorable position with respect to future possibilities. For decisions where the consideration of sequences of decisions is important (and where probabilities of future events are known) the use of *decision trees* or *decision-flow diagrams* for analysis is often an effective approach.

The application of decision trees to investment analysis is illustrated by the following marketing problem. Suppose a firm is planning to produce a product that has not been marketed previously. Because this product is somewhat different from existing ones, it will be necessary to construct a separate production facility for its manufacture.

Based on information supplied by the firm's marketing group, it is believed that the demand for this new product will persist over the next eight years and the demand for it will be high. If, however, the product is popular only initially, it is anticipated that the demand will be high for the first two years followed by a low demand for the remaining six years. If the product is not popular, then the demand is expected to be low for the full eight years. Thus, the demand for this product is expected to follow one of the three patterns shown in Table 15.13.

TABLE 15.13.
THREE DEMAND PATTERNS

Sales	Demand		
	Period 1	Period 2	Designation
Popular	High	High	(H_1, H_2)
Initially popular	High	Low	(H_1, L_2)
Not popular	Low	Low	(L_1, L_2)

The first two years are designated the first period; the remaining six years, the second period.

The first alternative is to build an automated plant that would serve for the full eight years of the product's life. The second alternative is to build a conventional plant and, after two years of marketing the product, decide whether to automate the plant. The decision problem is which alternative to select.

15.5.1 Structuring the Decision Tree

A decision tree displaying the alternatives and the possible chance events (demand) that can occur is illustrated in Figure 15.4, along with the forecasts of receipts and disbursements. Beginning at the left and moving to the right, the decision tree spans 8 years, the life of the product. The nodes of the tree from which the tree's branches emanate are either decision nodes, □, or chance nodes, ○. The branches that emanate from a decision node represent alternative courses of actions about which the decision maker must make a choice; those leaving the chance nodes represent chance events. The occurrence of a chance event is considered to be a random variable over which the decision maker has no control. Chance events are usually controlled by exogenous forces such as weather, sun spots, or the marketplace. In this example, the chance events represent the amount of demand for the product over the next eight years.

The first node represents the choice between building an automated plant or a conventional plant. If an automated plant is built, the next concern relates to whether there is a high or a low demand for the first two years. If an automated plant is built and demand is high for the first two years, the next chance node represents the possibility of a high or low demand occurring for the remaining six years. Thus, one sequence of branches, beginning at the left and ending at the right, represents one possible sequence of events that can result from actions by the decision maker and by "nature." Note that if the conventional plant is built, the decision about whether to automate or not is made after two years only if there is a high demand. If the demand is low in the first two years, it is anticipated that demand will also be low for the remaining six years; so the decision not to automate is obvious, and no decision node is required.

Once the structure of the decision tree is defined, the next task is to determine the costs and revenues that are associated with each of the decision alternatives and the possible chance outcomes. These amounts should then be entered on the appropriate branches of the tree. In this example the costs of plant construction and the net profits expected from sale of the product for the various market conditions are shown in Figure 15.4. Thus, the investment required to build the automated plant is $6,000,000, and the net profit for the first two years will the $1,200,000 per year if the automated plant is built. If it is built and there is a high demand in the first two-year period, the net profit will be $2,848,000 per year for the remaining six years, given a high demand during this period. The cost to build a conventional plant is $4,000,000, and the cost of

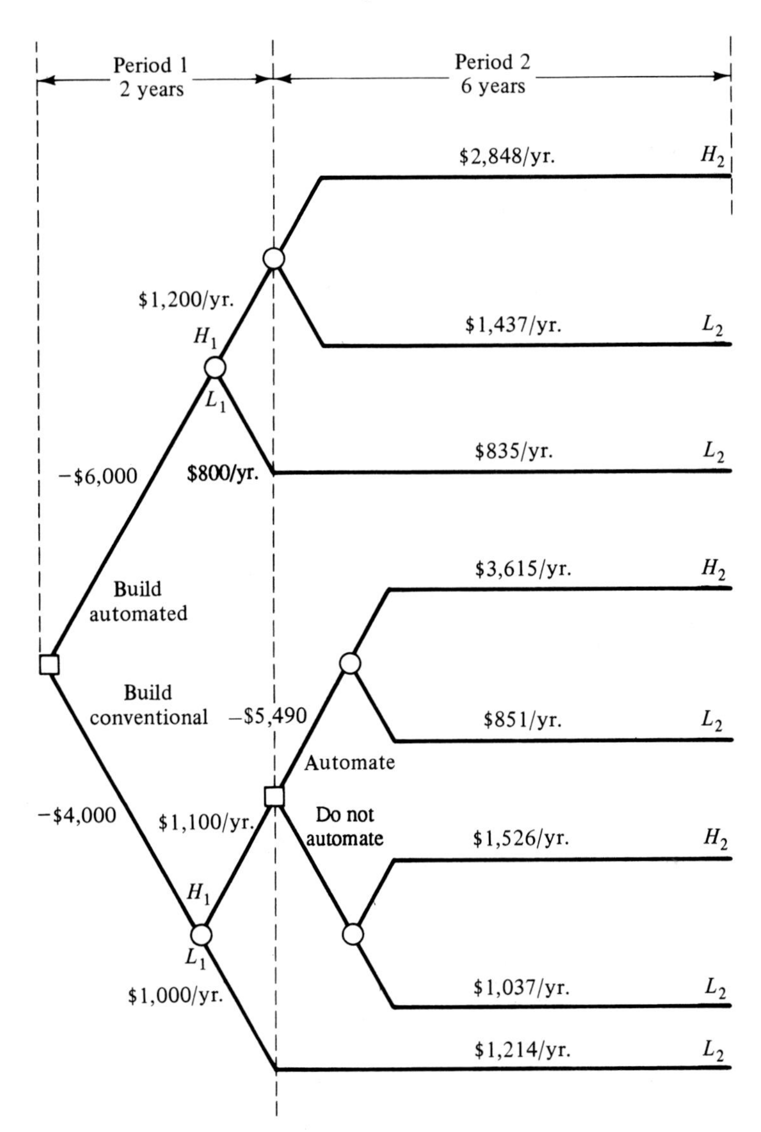

FIGURE 15.4. The decision tree with the forecasts of receipts and disbursements (in thousands).

automation is $5,490,000 if this is done. These costs are shown as negative values on the decision tree, while the revenues are positive amounts. The amounts that are shown on a per-year basis represent the annual net profit received during the two years of Period 1 or the six years of Period 2.

Since the costs and revenues occur at different points in time over the 10-year study period, it is appropriate to convert the various amuounts on the tree's branches to their equivalent amounts. For the problem being considered, the MARR is 13%, and Figure 15.5 shows the receipts and disbursements on the branches transformed to their present-worth equivalents. To illustrate, if the automated plant is built, the present worth of the net revenues that result from a high demand for the first 2 years is

$$PW(13) = \$1{,}200{,}000(\overset{P/A,13,2}{1.6681}) = \$2{,}000{,}000.$$

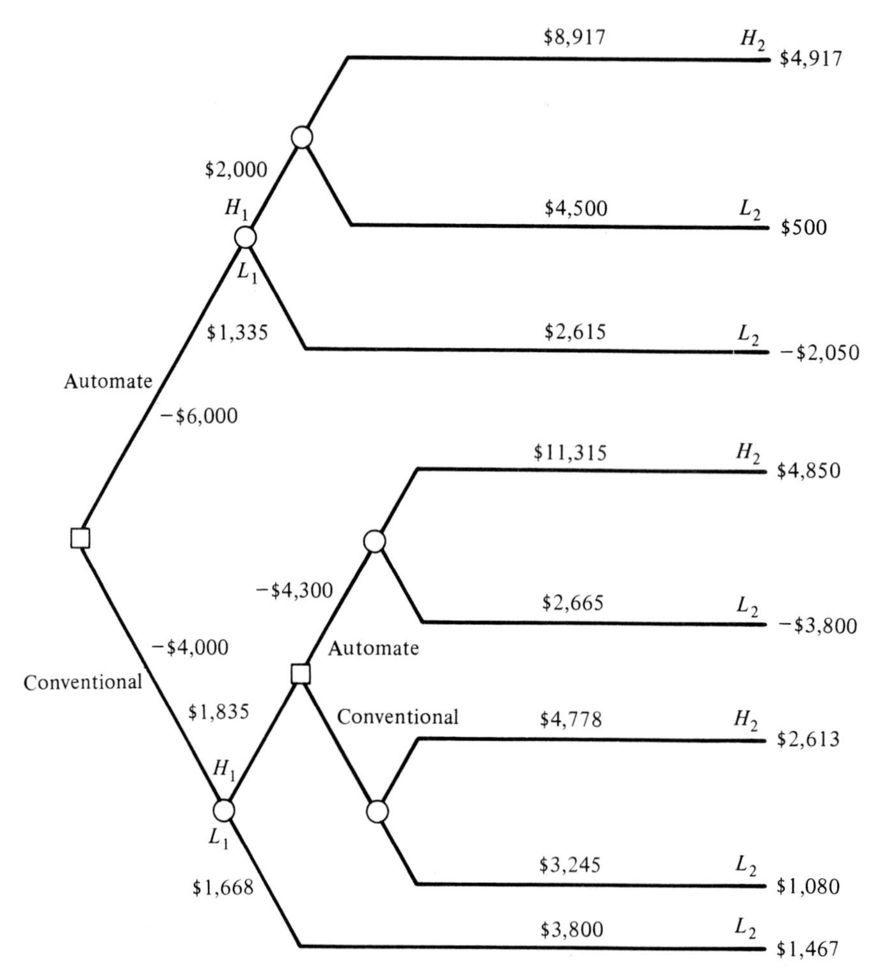

FIGURE 15.5. Present-worth amounts of receipts and disbursements for each income (in thousands).

If the automated plant is built, and a high demand is experienced for the remaining 6 years, the present worth of revenues is

$$PW(13) = \$2{,}848{,}000(\overset{P/A,13,6}{3.9975})(\overset{P/F,13,2}{0.7831})$$

$$= \$8{,}917{,}000$$

Once the present equivalent costs and revenues have been determined, one can sum these figures along each possible sequence of branches, starting each time at the leftmost node. Thus, if the automated plant is built, and the demand is high in Period 1 and Period 2, the total present worth is

$$-\$6{,}000{,}000 + \$2{,}000{,}000 + \$8{,}917{,}000 = \$4{,}917{,}000.$$

This amount is placed at the end of the rightmost branch, representing such an outcome. This procedure is repeated for each possible sequence of branches, and the results are shown in Figure 15.5.

If the conventional plant is built, demand is high for Period 1, the plant is expanded, and demand is low in Period 2, the net present worth is

$$-\$4{,}000{,}000 + \$1{,}835{,}000 - \$4{,}300{,}000 + \$2{,}665{,}000 = -\$3{,}800{,}000.$$

15.5.2 Nature's Tree

Once the cost and revenue information is in the form shown in Figure 15.5, it is necessary to place the probabilities of occurrence at the chance nodes on the decision tree. Suppose that the probabilities of product being popular, initially popular, or not popular are estimated by the marketing department as

$$P(\text{product popular}) = P(H_1 H_2) = \frac{3}{8};$$

$$P(\text{product initially popular}) = P(H_1 L_2) = \frac{1}{4};$$

$$P(\text{product not popular}) = P(L_1 L_2) = \frac{3}{8}.$$

Whenever it is necessary to compute the probabilities for a decision tree, it is usually very helpful first to construct Nature's tree—that is, construct a tree that indicates Nature's options, as shown in Figure 15.6. In Nature's tree all the nodes are chance nodes. At the beginning of each branch is placed the probability of following that branch, given that you are at the node preceding that

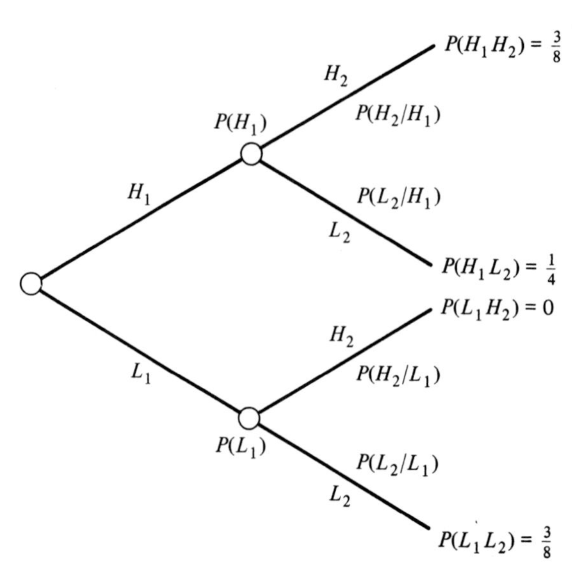

FIGURE 15.6. Nature's tree.

branch. The probability that branch H_2 is selected, given that there has been high demand in Period 1 (H_1), is the conditional probability $P[H_2 \mid H_1]$, and it is placed at the beginning of the H_2 branch that radiates from the H_1 branch. At each tip of Nature's tree is placed the probability that the sequence of events represented by that tip will occur. These probabilities are calculated by multiplying the probabilities on all the branches that lead from the initial node on Nature's tree to each of the tips. Thus, to find the probability that H_1 and L_2 occur, the probability on branch H_1 is multiplied by the probability on branch L_2 following H_1. This calculation gives $P(H_1 L_2) = P(H_1)P(L_2 \mid H_1)$.

The probabilities required for the decision tree are $P(H_1)$, $P(L_1)$, $P(H_2 \mid H_1)$, and $P(L_2 \mid H_1)$. By having Nature's tree and the probabilities for the ends of the tree, we then need to find $P(H_1)$ by adding each of the probabilities that contain an event H_1. Thus,

$$P(H_1) = P(H_1 H_2) + P(H_1 L_2) = \frac{3}{8} + \frac{1}{4} = \frac{5}{8}.$$

Similarly,

$$P(L_1) = P(L_1 H_2) + P(L_1 L_2) = 0 + \frac{3}{8} = \frac{3}{8}.$$

With the $P(H_1)$ now known, conditional probabilities are calculated as follows.

$$P(H_2 \mid H_1) = \frac{P(H_1 H_2)}{P(H_1)} = \frac{\left(\frac{3}{8}\right)}{\left(\frac{5}{8}\right)} = \frac{3}{5}.$$

$$P(L_2) \mid H_1) = \frac{P(H_1 L_2)}{P(H_1)} = \frac{\left(\frac{1}{4}\right)}{\left(\frac{5}{8}\right)} = \frac{2}{5}.$$

These probabilities are placed at the appropriate chance nodes on the decision tree, as shown in Figure 15.7. For example, the chance node following the decision to automate the conventional plant has two branches, H_2 and L_2. The probability of taking the H_2 branch is the probability that H_2 occurs *given* that H_1 has already occurred.

Therefore, $P(H_2 \mid H_1) = \frac{3}{5}$ is placed on the H_2 branch and $P(L_2 \mid H_1) = \frac{2}{5}$ is placed on the L_2 branch. For the case where the demand is low in Period 1, the probability $P(L_1) = P(L_1 L_2) = \frac{3}{8}$, since $P(L_2 \mid L_1) = 1$ and $P(H_2 \mid L_1) = 0$.

15.5.3 The Rollback Procedure

Next, the decision tree is solved in order to see which alternative should be undertaken. The solution technique is relatively simple and called the *rollback* procedure. Starting at the ends of the decision tree's branches and working back toward the initial node of the tree, we calculate according to the following rules:

1. If the node is a chance node, calculate the *expected* value of that node based on the rolled-back values on the adjacent nodes to the right of the node being considered.
2. If the node is a decision node, *select* the maximum profit or minimum cost from the adjacent nodes to the right of the node being considered.

As each node is considered, starting at the right of the decision tree, the values calculated for Rules 1 and 2 should be placed just to the right of the node and circled. By working backward through the decision tree, certain alternatives can be eliminated from further consideration. Thus the rollback procedure is efficient for large decision trees.

The rollback technique is illustrated in Figure 15.7. Starting at the top of the decision tree, note that the first node to the left is a chance node. Its ex-

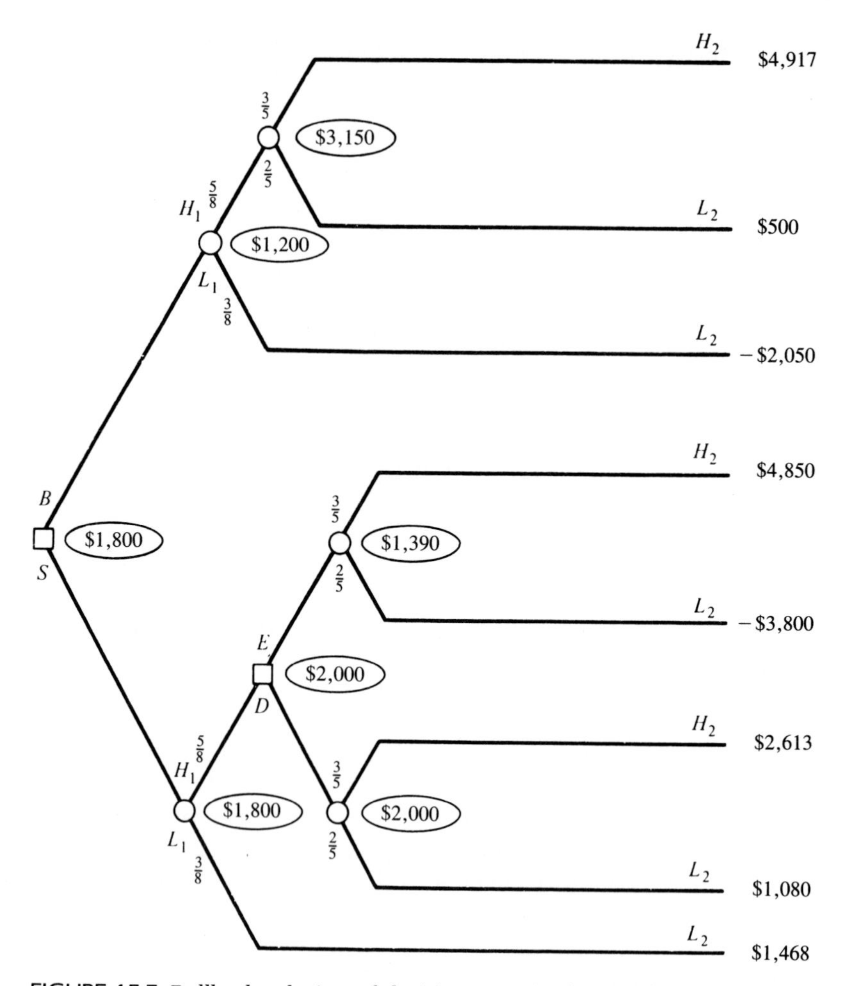

FIGURE 15.7. Rollback solution of decision tree (in thousands).

pected value is $\frac{3}{5}$(\$4,917,000) + $\frac{2}{5}$(\$500,000) = \$3,150,000. The next chance node that is farthest right is the node following the plant automation branch. The expected value for this node is $\frac{3}{5}$(\$4,850,000) + $\frac{2}{5}$(−\$3,800,000) = \$1,390,000. The other rightmost node is also a chance node, and its expected value is $\frac{3}{5}$(2,613,000) + $\frac{2}{3}$(1,080,000) = \$2,000,000. The rightmost node remaining to be evaluated is the decision node concerning the automation of the conventional plant. If the plant is automated, the expected future profits will be \$1,390,000, while if the plant is not automated, the expected future profits will be \$2,000,000. Therefore, the optimum policy at this node is not to automate the conventional plant. The $2,000,000 is written after the decision node, indicating the expected profit that will be realized if the optimum policy is followed at that node.

The next nodes to the left that must be evaluated by the rollback procedure are chance nodes. Expected values at those nodes are calculated from the values on the adjacent nodes to the right that were previously calculated. These expected values are

$$\frac{5}{8}(\$3{,}150{,}000) + \frac{3}{8}(-\$2{,}050{,}000) = \$1{,}200{,}000$$

and

$$\frac{5}{8}(\$2{,}000{,}000) + \frac{3}{8}(\$1{,}467{,}000) = \$1{,}800{,}000.$$

The last node to be evaluated is the initial decision node. If the automated plant is built, the expected profits will be $1,200,000, while if the conventional plant is built, the expected profit is $1,800,000. (Expected profit assumes that the optimum policy is followed in the future.) Therefore, the maximum is selected for the decision node, and the optimum policy is to build the conventional plant. If the demand is high in Period 1, the decision not to automate the conventional plant should be followed. If the demand is low in Period 1, not automating the conventional plant is the only course of action. If this policy is followed, the expected profit is $1,800,000. Since the present worth amount is positive, this policy assures the firm an expected return of better than 13%, besides being the best of the alternatives under consideration.

15.5.4 The Expected Value of Perfect Information

With the information that is available to the firm in this example, and if an optimal policy is followed, the expected present worth of the profit is $1,800,000. If the firm could obtain additional information about the occurrence of future demand levels, the profit amount might be improved. Usually such additional information costs money, since the firm must allocate resources for further research, or it must purchase the information from external sources such as market research consultants.

It is foolish to pay for additional information if it will not provide additional profits that exceed the cost of the information. In order to evaluate the possible benefits that can be derived from additional information, it is necessary to calculate the *expected value of perfect information* (*EVPI*).

TABLE 15.14.
MAXIMUM PAYOFF

State of Nature	Maximum Payoff
Good sales (H_1, H_2)	\$4,917,000
Initially good sales (H_1, L_2)	\$1,080,000
Poor sales (L_1, L_2)	\$1,467,000

Suppose a market research group does have perfect information about the future demand levels for the product being considered. How much would the firm be willing to pay for this additional information? Assume the firm knew with certainty that there would be high demand in both periods (H_1, H_2). Then the optimum policy is to build the automated plant, since that payoff (\$4,917,000) exceeds the payoffs possible from building the conventional plant and automating (\$4,850,000) or building the conventional plant and not automating (\$2,613,000). If the demand is known to be (H_1, L_2), then the best strategy is to build the conventional plant and not automate, since that payoff (\$1,080,000) exceeds the returns received if the automated plant is built (\$500,000) or the conventional plant is constructed and then automated (−\$3,800,000). Thus, for each possible state of nature, it is necessary to determine the strategy that will maximize the payoff. For this example, the three states of nature are (H_1, H_2), (H_1, L_2), and (L_1, L_2) and the maximum payoffs for each state of nature are given in Table 15.14.

Before receiving the perfect information from the market research group, the firm can calculate the *expected profit with perfect information* (*EPPI*). To determine this, for each possible state of nature, one multiplies the probability that a

TABLE 15.15.
CALCULATING THE EXPECTED PROFIT FOR PERFECT INFORMATION

State of Nature *A*	Best Strategy *B*	Maximum Payoff *C*	Probability the State of Nature Occurs *D*	Expected Payoff for Each State *C* × *D*
H_1, H_2	Automated plant	\$4,917,000	$\left(\frac{5}{8}\right)\left(\frac{3}{5}\right) = \frac{3}{8}$	\$1,843,875
H_1, L_2	Conv. plant; no automation	\$1,080,000	$\left(\frac{5}{8}\right)\left(\frac{2}{6}\right) = \frac{1}{4}$	\$ 270,000
L_1, L_2	Conv. plant	\$1,467,000	$\left(\frac{3}{8}\right)(1) = \frac{3}{8}$	\$ 550,125

Expected Value of Perfect Information (EVPI) = \$2,664,000

particular state will occur by the maximum payoff achievable for that state of nature. These calculations are shown in Table 15.15. The expected profit for perfect information is $2,664,000. Since the firm can achieve expected profits of $1,800,000 without the additional information, the expected value of perfect information is $2,664,000 − $1,800,000 = $864,000. The firm should never pay over $864,000 for additional information, even if that information predicts the future with certainty. Thus, the EVPI gives an upper limit reflecting the firm's expected improvement in profits if a source of perfect information were available. Information that is less than perfect can only provide a profit improvement that is less than this expected value.

KEY POINTS

- The economic elements upon which a course of action depends may vary from their estimated values due to chance causes.
- To formally incorporate risk and uncertainty about future events into a decision process, it is appropriate to utilize probability theory.
- When probability distributions are used to describe economic elements, the expected value of the economic outcome may be used to compare alternatives.
- Expectation variance is an extension of decision making based on expected outcomes. It brings variability in each alternative's outcome to the decision process.
- Multiple sources of variation in decision making require the use of simulated sampling known as Monte Carlo analysis to fully capture the impact of estimating errors.
- Decision-tree analysis is appropriate to use when a sequence of decisions is to be made; when actions taken in the present affect future decision options.
- It is not wise to pay for additional information if it will not provide additional benefit that exceeds the cost of the information.

PROBLEMS

1. Because the geological structure along a certain highway route is unknown, the cost per mile of construction is a random variable described as follows.

Cost (X)	Probability That Cost Is X
$ 8,000,000	0.1
10,000,000	0.3
12,000,000	0.3
14,000,000	0.2
16,000,000	0.1

What is the expected value of the cost of the highway per mile? Is the expected-cost approach appropriate in this situation? What bid should be submitted if the contractor wishes to be 90% sure that the cost will not exceed the income? Answers: $11,800,000; no; $14,000,000

2. Suppose that an asset costing $75,000 will result in an annual saving of $30,000 for as long as the asset remains serviceable. The probabilities that the asset will remain serviceable for a certain number of years are as follows.

Number of Years Asset Functions (n)	Probability That Asset Is Productive Exactly n Years
1	0.1
2	0.2
3	0.2
4	0.3
5	0.1
6	0.1

What is the expected net present worth of the venture if the interest rate is 16%? Should the investment be made?

3. Uncertainty as to the rate of technological innovation means that a proposed computer system will have a service life that is unknown. If the initial cost of developing a computer system is $10 million and each year of service life produces net revenues of $4.5 million, what is the expected present worth for an interest rate of 10%, if the lifetime of the system is described by the following probabilities?

Number of Years of Life (n)	Probability That Life Is Exactly n Years
1	0.1
2	0.2
3	0.4
4	0.2
5	0.1

4. A company is considering the purchase of a concrete plant for $4,000,000. The success of the plant depends on the amount of highway construction undertaken over the next 4 years. It is known that there are three possible levels (*A*, *B*, and *C*) of federal support for the construction of new highways. Shown below are the receipts and disbursements (in millions of dollars) the company expects from the concrete plant for each level of government support.

	End of Year				
Support Level	0	1	2	3	4
A	−$4.0	$2.0	$2.0	$2.0	$2.0
B	− 4.0	1.0	1.0	1.0	1.0
C	− 4.0	0.5	0.5	1.0	2.5

The probabilities that support levels *A*, *B*, and *C* are realized over the next 4 years are 0.3, 0.1, 0.6, respectively. If the minimum attractive rate of return is 12%, what is the expected present worth of this investment opportunity? Answer: $13,540

5. A retail firm has experienced three basic responses to its previous advertising programs. Presently the firm has two programs under consideration. Because each has a different emphasis, it is expected that the percentage of people responding in a particular manner will vary according to the program undertaken. Shown below are the cash flows that are expected for each of the three possible customer responses. Each cash flow shown assumes 100% response.

PROGRAM A

		End of Year			
Response	Percentage Responding	0	1	2	3
1	20%	−$500,000	$300,000	$200,000	$100,000
2	60%	− 500,000	200,000	200,000	300,000
3	20%	− 500,000	200,000	200,000	200,000

PROGRAM B

		End of Year			
Response	Percentage Responding	0	1	2	3
1	30%	−$600,000	300,000	200,000	200,000
2	40%	− 600,000	200,000	300,000	300,000
3	30%	− 600,000	300,000	300,000	300,000

For an interest rate of 10% which advertising compaign has the largest expected present worth?

6. A company is considering the introduction of a "new" umbrella, although its future success is uncertain. The product will be sold for only 5 years, and if the 5 years are "wet" years, the following cash flows (in hundred thousands) are anticipated.

		Year					
Probability	*Cash Flow*	0	1	2	3	4	5
0.6	W1	−$4	$2	$3	$2	$1	$0
0.4	W2	− 4	1	2	4	3	2

On the other hand, if the next 5 years are "dry" years, the cash flows are likely to be as follows:

		Year					
Probability	*Cash Flow*	0	1	2	3	4	5
0.5	D1	−$4	$0	$1	$1	$1	$1
0.5	D2	− 4	1	1	2	1	1

The interest rate is 10%.

a. What is the expected present worth if the next 5 years are going to be "wet"?

b. What is the expected present worth if the next 5 years are going to be "dry" ?

c. If the best available inofrmation indicates that the next 5 years will be "wet" with probability 0.7,
(i) What is the probability that the next 5 years will be "dry"?
(ii) What is the expected present woth of the proposal?

7. A software firm must decide whether to market a new package or to update a package that is currently being marketed. The firm has the following information to use in making its decision.

			Probability of Duration of Sales (in years)				
	Initial Cost	*Net Return per Year*	1	2	3	4	5
New package	$50,000	$25,000	0.1	0.2	0.3	0.2	0.2
Updated old package	20,000	12,000	0.4	0.3	0.2	0.1	0.0

Using the expected present worth, which alternative should the firm choose if the interest rate is 15%?

8. A plant is to be built to produce blasting devices for construction jobs, and the decision must be made as to the extent of automation in the plant. Ad-

ditional automatic equipment increases the investment costs but lowers the probability of shipping a defective device to the field, which must then be shipped back to the factory and dismantled at a cost of $10. The operating costs are identical for the different levels of automation. It is estimated that the plant will operate 10 years, the interest rate is 20%, and the rate of production is 100,000 devices per year for all levels of automation. Find the level of automation that will minimize the expected annual cost for the investment costs and probabilities given below. Answer: Level 3, $67,700

Level of Automation	*Probability of Producing a Defective*	*Investment*
1	0.100	$100,000
2	0.050	150.000
3	0.020	200.000
4	0.010	275,000
5	0.005	325,000
6	0.002	350,000

9. A dam is being planned for a certain stream of erratic flow. It has been determined by past experience that a dam of sufficient capacity to withstand various flow rates, where the probability of these rates' being exceeded in any one year is 0.10, 0.05, 0.025, 0.0125, and 0.00625, will cost $142,000, $154,000, $170,000, $196,000, and $220,000, respectively; will require annual maintenance amounting to $4,600, $4,900, $5,400, $6,500, and $7,200, respectively; and will suffer damage of $122,000, $133,000, $145,000, $170,000, and $190,000, respectively, if subjected to flows exceeding its capacity. The life of the dam will be 40 years with no salvage value. For an interest rate of 10% calculate the annual cost of the dam including probable damage for each of the five proposed plans and determine the dam size that will result in a minimum cost.

10. A company has developed probability distributions representing the probabilities that various annual equivalent amounts will be realized from the three mutually exclusive projects that are under consideration. These distributions are given below.

Annual Equivalent Profit	*Alternative*		
	*A*1	*A*2	*A*3
$ 5,000	0.10	0.00	0.00
10,000	0.10	0.20	0.00
15,000	0.20	0.20	0.20
20,000	0.30	0.20	0.50
25,000	0.30	0.20	0.20
30,000	0.00	0.20	0.10

Calculate the expected annual equivalent profit and the variance for each probability distribution. Which alternative would you consider the most attractive?

11. An advertising agency has developed four alternative advertising compaigns for one of its clients. The client has studied the alternatives and has developed probability distributions describing the present worth of the net profits expected if it invests in a particular advertising program. The probability distributions for the four ad programs are as follows:

	Net Present Worth of Profits				
Ad Program	−\$50,000	−\$10,000	\$10,000	\$50,000	\$100,000
A	0.10	0.20	0.30	0.30	0.10
B	0.05	0.15	0.40	0.40	0.00
C	0.40	0.00	0.00	0.00	0.60
D	0.00	0.10	0.40	0.50	0.00

Calculate the mean, the variance, and the probability that the net present worth of profits is less than zero. Which advertising program would you undertake? Explain your choice.

12. The salvage value of a certain asset depends upon its service life as follows:

Service Life	*Salvage Value*
3 years	\$18,000
4 years	\$12,000
5 years	\$ 6,000
6 years	\$ 2,000

If the asset has a first cost of \$120,000 and the interest rate is 12%, find its expected equivalent annual cost if each service life is equally likely to occur.

13. To buy a computer-controlled machine a manufacturer must pay \$200,000. If the machine is purchased, five different manufacturing processes can utilize it. By using the machine exclusively in Process *A, B, C, D, E*, respectively, the annual income that will be realized is \$120,000, \$130,000, \$150,000, \$160,000, or \$200,000. The life of the machine is expected to vary according to where the machine is used. The probability that the machine will provide service for exactly 1, 2, 3, or 4 years is shown below for each process.

Process	Life of Machine (Years)			
	1	2	3	4
A	0.25	0.25	0.25	0.25
B	0.30	0.30	0.30	0.10
C	0.30	0.40	0.25	0.05
D	0.20	0.60	0.20	0.00
E	0.50	0.50	0.00	0.00

Calculate the mean and variance of the present-worth amounts for each of these five processes. The interest rate is 10%. Which process could use this machine most effectively? Answer: *D*

14. Apply the Monte Carlo method with 100 trials to the situation described in Problem 1.

a. Plot the average cost after 10, 20, 30, . . . , 100 trials to illustrate convergence.

b. Plot the statistical distribution of cost and find the ratio of costs above $14,000,000 to the total number of cost elements generated.

c. Discuss the findings of this exercise in comparison with the theoretical results found in the solution to Problem 1.

15. Two equipment investment alternatives are under consideration. Alternative *A* requires an initial investment of $15,000 with an annual operating cost of $2,000. The service life–salvage value possibilities are

Service Life	*Salvage Value*
3 years	$4,000
4 years	$2,500

Each service life is equally likely to occur. Alternative *B* requires an initial investment of $30,000 with a negligible annual operating cost. The salvage value as a function of the service life is $8,000 − $500($n$), where n is the service life with the following probability distribution:

Service Life	*Probability*
7 years	0.4
8 years	0.3
9 years	0.3

If the interest rate is 20%, find the average annual equivalent cost for each alternative by 10 Monte Carlo trials. Discuss your result as a basis for choosing one alternative over the other.

16. Develop theoretical distributions from the information given in Section 15.4 that may be compared with the frequency distributions developed by

Monte Carlo and exhibited in Tables 15.11 and 15.12. What is the expected cost for each alternative? Why does it differ from that found after 100 Monte Carlo trials?

17. An inventor has developed an electronic device to monitor impurities in a chemical process. A company that makes this type of equipment is considering the purchase of the patent rights from the inventor for $40,000. The company feels that there is 1 chance in 5 of the device becoming a success in the market. It is estimated that if the device is successful, it will produce net revenues of $150,000 a year for the next 5 years. If the device is not a success, no revenues will be received. The company's interest rate is 15%.

a. Draw a decision tree describing the decision options, and determine the best decision policy. Answer: Market the product.

b. If a market research group can provide perfect information about the success of this product, what would be the most the company would be willing to pay for their service? Answer: $32,000

c. Suppose the market research group can make a market survey that, with probability 0.7, will give a favorable result if the device will be a success, and, with probability 0.9, will give an unfavorable result if the device will not sell. How much would this survey be worth to the company? Answer: $990

18. An electronics firm is trying to decide whether or not to manufacture a new communications device. The decision to produce the device means an investment of $5 million, and the demand for such a device is not known. If demand is *high*, the company expects a return of $2.0 million each year for 5 years. If the demand is *moderate*, the return will be $1.6 million each year for 4 years, and a *light* demand means a return of $0.8 million each year for 4 years. It is estimated that the probability of a light demand is 0.1 and of a high demand is 0.5. Interest is 10%.

a. On the basis of expected present worth, should the company make the investment?

b. It is proposed that a survey be taken to establish the actual demand to be experienced. What is the maximum value of such a survey?

c. A survey is available at a cost of $75,000 that has the following characteristics:

Survey results say favorable or unfavorable with following probabilities:

If demand is	*Favorable*	*Unfavorable*
High	0.9	0.1
Moderate	0.5	0.5
Low	0.0	1.0

Should the company make the survey? If so, should the company produce the device if the survey says unfavorable? What is the expected profit in this case?

19. A wholesaler is studying his warehouse needs for the next 8 years. At present three alternatives are under consideration: A new warehouse can be built to replace the existing facility, the existing facility can be expanded, or the decision to expand can be postponed. If the decision is postponed, the wholesaler will wait 4 years and then decide whether to expand the existing facility or to leave it as is. To expand the present facility now will cost \$400,000, while to build a new facility will cost \$700,000. Consumer demand for the wholesaler's products is expected to be either high for all 8 years (H_1, H_2), high for 4 years and low for the remaining 4 years (H_1, L_2), low for the first 4 years and high for the last 4 years (L_1, H_2), or low for all 8 years (L_1, L_2). Depending upon these demand levels, the following annual receipts are expected to be received from the two alternatives that require an investment now.

	Demand			
	$H_1 H_2$	$H_1 L_2$	$L_1 H_2$	$L_1 L_2$
Alternatives				
Build new warehouse now	\$320,000	\$160,000	\$110,000	\$80,000
Expand existing warehouse now	200,000	150,000	100,000	50,000

If the decision to construct additional warehouse capacity is postponed, the cost of expanding the warehouse 4 years from now is expected to be \$600,000. In this case it is anticipated that the annual revenues for high and low demand during the first 4 years will be \$50,000 and \$20,000, respectively. The annual revenues for the remaining 4 years are

	Demand	
	H_2	L_2
Decision		
Expand after 4 years	\$400,000	\$100,000
Do not expand after 4 years	80,000	40,000

The interest rate is 12% and the probabilities that demand will be at a particular level through the 8-year period are

	Demand			
	$H_1 H_2$	$H_1 L_2$	$L_1 H_2$	$L_1 L_2$
Probability	0.3	0.2	0.1	0.4

Using decision-tree analysis, determine the decision policy that should be followed so that the wholesaler's expected profits are maximized.

20. Urn 1 contains 6 white balls and 4 black balls, while Urn 2 contains 7 white balls and 3 black balls. A sample of 2 balls is going to be taken, and on this basis you must decide from which urn the sample was selected. Although you do not see the drawing of the sample, it is known that the sample is equally likely to come from either urn. If you make the correct decision you will win $10, and if you make an incorrect decision you win nothing. Using decision-tree analysis, find the decision policy that will maximize your expected winnings and indicate what the expected winnings will be.

a. Assume that you may choose whether the sample is drawn with or without replacement. (With replacement, the first ball drawn is returned to the urn before the second ball is selected.)

b. Assume that you may choose whether the second ball shall be drawn with or without replacement after you have seen the result of drawing the first ball.

16

Decision Making Under Uncertainty

It may be impossible to assign probabilities to the occurence of future events associated with some decision situations. Often no meaningful data are available from which probabilities may be derived. In other instances the decision maker may be unwilling to assign a probability, as is sometimes the case when the event is unpleasant or has severe consequences.

When probabilities are not available for the assignment to future events, the situation is called *decision making under uncertainty.* As compared with decision making under certainty and under risk, this decision situation is more abstract. In this chapter decisions under uncertainty are structured in a formal manner and some useful decision rules are applied.

16.1 THE PAYOFF MATRIX

A particular decision can result in one of several outcomes, depending upon which of several future events takes place. For example, a decision to go on a picnic can result in a high degree of satisfaction if the day turns out to be sunny or in a low degree of satisfaction if it rains. These levels of satisfaction would be reversed if the decision were made to stay home. Thus, for the two states of nature, sun and rain, there are different payoffs depending upon the alternative chosen.

A *payoff matrix* is a formal way of exhibiting the interaction of decision alternatives and the states of nature. In this usage alternatives have the same meaning as before—that is, courses of action between which choice is contemplated. The states of nature need not be natural events such as sun and rain. This phrase is used to describe a wide variety of future events over which the decision maker has no control. The payoff matrix gives a qualitative or quantitative payoff for each possible future state and for each alternative under consideration.

As an example of the structuring of a payoff matrix, consider the following situation. An engineering and construction firm has the opportunity to bid on two contracts. The first pertains to the design and construction of a plant to convert solid waste into steam for heating purposes in a city. The second pertains to the design and construction of a steam distribution system within the city. The firm may be awarded either contract X or contract Y or both. Thus, there are three possible outcomes or "states of nature."

In considering the opportunities afforded by these contracts, the firm identifies five alternatives. Alternative $A1$ is for the firm to serve as project manager, with all of the work to be subcontracted. Alternative $A2$ is for the firm to subcontract the design but to do the construction. Alternative $A3$ is for the firm to subcontract the construction but to do the design. Alternative $A4$ is for the firm to do both the design and construction. Alternative $A5$ calls for the firm to bid jointly with another organization that has the capability to undertake an innovative project of this type.

Once the states of nature and the alternatives are identified, the next step is to derive payoff values. In this example, 15 payoff values must be developed. By listing anticipated disburesements and receipts over time identified with each alternative, for each state of nature, the present value of profit is found. Suppose that these present values are in the thousands of dollars, as exhibited in Table 16.1.

From the payoff matrix it can be seen that the firm can incur a present loss of $4 million if Alternative $A1$ is chosen and contract X is awarded. If contract Y is awarded, the present profit will be $1 million. The present profit will be $2

TABLE 16.1.
PAYOFF MATRIX FOR PROFIT IN THOUSANDS OF DOLLARS

		States of Nature		
		X	Y	X and Y
	$A1$	−4,000	1,000	2,000
	$A2$	1,000	1,000	4,000
Alternatives	$A3$	−2,000	1,500	6,000
	$A4$	0	2,000	5,000
	$A5$	1,000	3,000	2,000

TABLE 16.2.
REDUCED PAYOFF MATRIX IN THOUSANDS OF DOLLARS

		States of Nature		
		X	*Y*	*X* and *Y*
Alternatives	*A*2	1,000	1,000	4,000
	*A*3	−2,000	1,500	6,000
	*A*4	0	2,000	5,000
	*A*5	1,000	3,000	2,000

million if both contracts are awarded. Thus, each row of the payoff matrix represents the outcomes expected for each state of nature (column) for a particular alternative (row).

Individual payoff values in a payoff matrix need not be monetary in character. They may be qualitative or quantitative expressions of the utility expected from each of the several alternatives. It is essential, however, that the payoff values be expressed in some common and directly comparable measure such as present worth or annual equivalent amount. In Table 16.1 the payoff values are present-worth amounts.

Before proceeding, we should examine the payoff matrix for dominance. If for two alternatives one will always be preferred no matter which future occurs, the preferred alternative dominates and the other alternative may be discarded.

In Table 16.1, Alternative *A*1 may be discarded, since it is dominated by other alternatives. Therefore, the payoff matrix can be reduced to the form shown in Table 16.2. This reduced payoff matrix completely rules out the alternative of the firm's serving as project manager with all the design and construction work to be subcontracted. The rules presented in the following sections will assist in the selection of one of the four remaining alternatives.

16.2 THE LAPLACE RULE

If the firm were willing to assign probabilities to the states of nature in Table 16.2, the decision situation would be classified as decision making involving risk. The techniques of the previous chapter could then be applied, and the best alternative would be chosen by applying the proper criteria.

Suppose, however, that the firm is unwilling to assess the states of nature in terms of their probabilities of occurrence. One might then reason that each possible state of nature is as likely to occur as any other. The rationale is that

TABLE 16.3.
COMPUTATION OF AVERAGE PAYOFF IN MILLIONS OF DOLLARS

Alternative	Average Payoff
$A2$	(\$1,000 + \$1,000 + \$4,000) ÷ 3 = \$2,000
$A3$	(−\$2,000 + \$1,500 + \$6,000) ÷ 3 = \$1,833
$A4$	(\$0 + \$2,000 + \$5,000) ÷ 3 = \$2,333
$A5$	(\$1,000 + \$3,000 + \$2,000) ÷ 3 = \$2,000

there is no stated basis for one state of nature to be more likely than any other. This is called the *Laplace principle* or the *principle of insufficient reason,* based upon the philosophy that nature is assumed to be indifferent.

Under the Laplace principle the probability of the occurrence of each future state of nature is assumed to be $1/n$, where n is the number of possible future states. To select the best alternative one would compute the arithmetic average for each. For the payoff matrix of Table 16.2 this is accomplished as shown in Table 16.3. Alternative $A4$ results in a maximum profit of \$2,333,000 and would be selected by this procedure.

16.3 MAXIMIN AND MAXIMAX RULES

Two very simply decision rules are available for dealing with decisions under uncertainty. The first, the *maximin* rule, is based on an extremely pessimistic view of the outcome of nature. The use of this rule is justified if it is judged that nature will do her worst. The second, the *maximax* rule, is based on an extremely optimistic view of the outcome of nature. Use of this rule is justified if it is judged that nature will do her best.

Because of the pessimism embraced by the maximin rule, its application will choose the alternative that assures the best of the worst possible outcomes. If P_{ij} is used to represent the payoff for the ith alternative and the jth state of nature, then the required computation is

$$\max_i[\min_j P_{ij}]. \tag{16.1}$$

Consider the decision situation described by the payoff matrix of Table 16.2. The application of the maximin rule requires that the minimum value in each row be selected. Then the maximum value is identified from these and associated with the alternative that would produce it. This procedure is illustrated in Table 16.4. Selection of either Alternative $A2$ or $A5$ assures the firm of a payoff of at least \$1,000,000 regardless of the outcome of nature.

TABLE 16.4.
PAYOFF IN THOUSANDS OF DOLLARS BY THE MAXIMIN RULE

Alternative	$\min_j P_{ij}$
$A2$	\$1,000
$A3$	−\$2,000
$A4$	\$0
$A5$	\$1,000

The optimism of the maximax rule contrasts sharply with the pessimism of the maximin rule. Its application will choose the alternative that assures the best of the best possible outcomes. As before, if P_{ij} represents the payoff for the ith alternative and the jth state of nature, then the required computation is

$$\max_i [\max_j P_{ij}]. \tag{16.2}$$

Consider the decision situation of Table 16.2. The application of the maximax rule requires that the maximum value in each row be selected. Then the maximum value is identified from these and associated with the alternative that would produce it. This procedure is illustrated in Table 16.5. Selection of Alternative $A3$ is indicated. Thus, the decision maker may receive a payoff of \$6,000,000 if nature is benevolent.

A decision maker who chooses the maximin rule considers only the worst possible occurrence for each alternative and selects that alternative which promises the best of the worst possible outcomes. In the example where $A2$ was chosen, the firm would be assured of a payoff of at least \$1,000,000, but it could

TABLE 16.5.
PAYOFF IN THOUSANDS OF DOLLARS BY THE MAXIMAX RULE

Alternative	$\max_j P_{ij}$
$A2$	\$4,000
$A3$	\$6,000
$A4$	\$5,000
$A5$	\$3,000

not receive a payoff any greater than \$4,000,000. Or, if $A5$ were chosen, the firm could not receive a payoff any greater than \$3,000,000. Conversely, the firm that chooses the maximax rule is an optimistic one that decides solely on the basis of the higher payoff offered for each alternative. Accordingly, in the example in which $A3$ was chosen, the firm faces the possibility of a loss of \$2,000,000 in the quest of a payoff of \$6,000,000.

16.4 THE HURWICZ RULE

Because of the extreme nature of the decision rules presented in the previous section, they are alien to many decision makers. Most people possess a degree of optimism or pessimism somewhere between the extremes. A third approach to decision making under uncertainty involves an index of relative optimism and pessimism. It is called the *Hurwicz rule.*

A compromise between optimism and pessimism is embraced in the Hurwicz rule by allowing the decision maker to select an index of optimism, α, such that $0 \le \alpha \le 1$. When $\alpha = 0$, the decision maker is pessimistic about the outcome of nature, while an $\alpha = 1$ indicates optimism about nature. Once α is selected, the Hurwicz rule requires the computation of

$$\boxed{\max_i \{\alpha [\max_j P_{ij}] + [1 - \alpha][\min_j P_{ij}]\}} \tag{16.3}$$

where P_{ij} is the payoff for the ith alternative and the jth state of nature.

As an example of the Hurwicz rule, consider the payoff matrix of Table 16.2 with $\alpha = 0.2$. The required computations are shown in Table 16.6, and Alternative $A2$ would be chosen by the firm.

Additional insight into the Hurwicz rule can be obtained by graphing each alternative for all values of α between zero and one. This makes it possible to

TABLE 16.6.
PAYOFF IN THOUSANDS OF DOLLARS BY THE HURWICZ RULE WITH $\alpha = 0.2$

Alternative	$\alpha[\max_j P_{ij}] + (1 - \alpha)[\min_j P_{ij}]$	
$A2$	0.2(\$4,000) + 0.8(\$1,000)	= \$1,600
$A3$	0.2(\$6,000) + 0.8(−\$2,000)	= −\$400
$A4$	0.2(\$5,000) + 0.8(0)	= \$1,000
$A5$	0.2(\$3,000) + 0.8(\$1,000)	= \$1,400

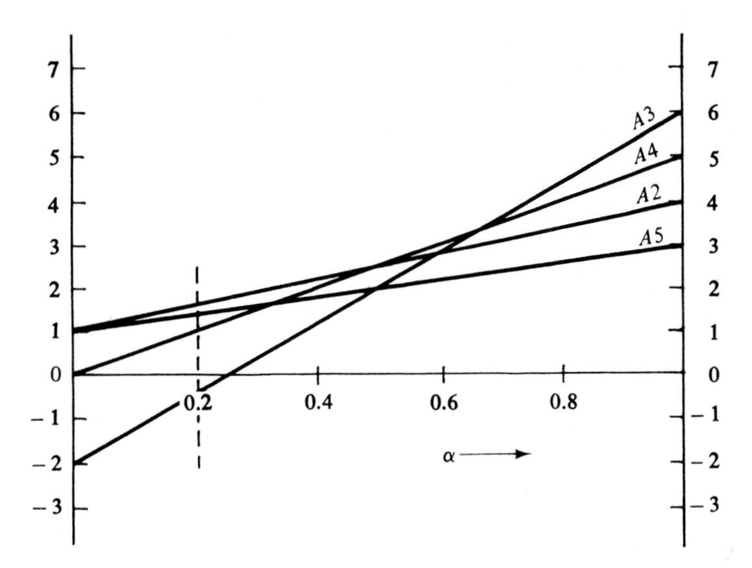

FIGURE 16.1. Payoff values for the Hurwicz rule representing four alternatives (in millions).

identify the values of α for which each alternative would be favored. Such a graph is shown in Figure 16.1. It may be observed that Alternative $A2$ yields a maximum expected payoff for all values of $\alpha \leq \frac{1}{2}$. Alternative $A4$ exhibits a maximum for $\frac{1}{2} \leq \alpha \leq \frac{2}{3}$, and Alternative $A3$ gives a maximum for $\frac{2}{3} \leq \alpha \leq 1$. There is no value of α for which Alternative $A5$ would be best except at $\alpha = 0$, where it is as good an alternative as $A2$.

When $\alpha = 0$, the Hurwicz rule gives the same result as the maximin rule, and when $\alpha = 1$, it is the same as the maximax rule. This may be shown for the case where $\alpha = 0$ as

$$\max_i \{0[\max_j P_{ij}] + (1 - 0)[\min_j P_{ij}]\} = \max_i [\min_j P_{ij}].$$

And, for the case where $\alpha = 1$

$$\max_i \{1[\max_j P_{ij}] + (1 - 1)[\min_j P_{ij}]\} = \max_i[\max_j P_{ij}].$$

Thus, the maximin rule and the maximax rule are special cases of the Hurwicz rule.

The philosophy behind the Hurwicz rule is that many people focus on the most extreme outcomes or consequences in arriving at a decision. By use of this rule, the decision maker may weight the extremes in such a manner as to reflect the relative importance attached to them.

16.5 THE MINIMAX REGRET RULE

If a decision maker selects an alternative and a state of nature occurs such that he could have done better by selecting another alternative, he "regrets" his original selection. This regret is the difference between the payoff that could have been achieved with perfect knowledge of the future and the payoff that was actually received from the alternative chosen. The *minimax regret rule* is based on the premise that a decision maker wishes to avoid any regret, or at least to minimize his maximum regret about a decision.

Application of the minimax regret rule requires the formulation of a regret matrix. This is accomplished by identifying the maximum payoff for each state (column). Next, each payoff in the column is subtracted from the maximum payoff identified, and this is repeated for each column. For the payoff matrix of Table 16.2 the maximum payoffs are \$1,000, \$3,000, and \$6,000 for X, Y, and X and Y, respectively. Thus, the regrets for X, applicable to Alternatives $A2$ through $A5$, are \$1,000 − \$1,000 = 0; \$1,000 − (−\$2,000) = \$3,000; \$1,000 − \$0 = \$1,000; \$1,000 − \$1,000 = \$0. Repeating this computation for each state results in the regret matrix shown in Table 16.7.

If the regret values are designated R_{ij} for the ith alternative and the jth state, then the minimax regret rule requires the computation of

$$\min_i \left[\max_j R_{ij}\right]. \tag{16.4}$$

This computation is shown in Table 16.8. Selection of Alternative $A4$ assures the firm of a maximum regret of \$1,000,000.

A decision maker who uses the minimax regret rule as a decision criterion will make that decision which will result in the least possible opportunity loss. Individuals who have a strong aversion to criticism would be tempted to apply this rule because it puts them in a relatively safe position with respect to the fu-

TABLE 16.7.
REGRET MATRIX IN THOUSANDS OF DOLLARS

		States of Nature		
		X	Y	X and Y
Alternative	$A2$	0	2,000	2,000
	$A3$	3,000	1,500	0
	$A4$	1,000	1,000	1,000
	$A5$	0	0	4,000

TABLE 16.8.
PAYOFF IN THOUSANDS OF DOLLARS BY THE MINIMAX REGRET RULE

Alternative	$\max_j P_{ij}$
$A2$	\$2,000
$A3$	\$3,000
$A4$	\$1,000
$A5$	\$4,000

ture states of nature that might occur. In this regard this criterion has a conservative underlying philosophy.

16.6 SUMMARY OF DECISION RULES

The alternatives selected by the decision rules presented in this chapter are summarized in Table 16.9. It will be noted that the rules do not give consistent results. They are developed to give insight into those decision situations in which probabilities are not or cannot be assigned to the occurrence of future events.

Examination of the courses of action recommended by the five decision rules indicates that each has its own merit. Several factors may influence a decision maker's choice of a rule in a given decision situation. The decision maker's attitude toward uncertainty (pessimistic or optimistic) and his personal utility function are important influences. Thus, the choice of a particular decision rule for a given decision situation must be based upon the subjective judgment of the decision maker. This is what one would expect, for in the absence of probabilities concerning future events it is not possible to derive a completely objective decision procedure.

TABLE 16.9.
COMPARISON OF RULES

Decision Rule	Alternative Selected
Laplace	$A4$
Maximin	$A2$ or $A5$
Maximax	$A3$
Hurwicz ($\alpha = 0.2$)	$A2$
Minimax regret	$A4$

KEY POINTS

- Decision making under uncertainty arises when it is impossible, or when individuals are unwilling, to assign probabilities to future events.
- As contrasted with decision making under risk, decision making under uncertainty is more abstract and more dependent upon predetermined decision rules.
- A payoff matrix is the centerpiece of decision making under uncertainty, in that it structures the interaction of alternatives and various states of nature.
- The preferred alternative in decision making under uncertainty depends directly upon the decision rule chosen by the decision maker.

PROBLEMS

1. The following matrix gives the payoffs in utiles for three alternatives and three possible states of nature.

		States of Nature		
		*S*1	*S*2	*S*3
Alternatives	*A*1	50	80	80
	*A*2	60	70	20
	*A*3	90	30	60

Which alternative would be chosen under the Laplace principle? The maximin rule? The maximax rule? The Hurwicz rule with $\alpha = 0.75$? The minimax regret rule? Answers: *A*1; *A*1; *A*3; *A*3; *A*1

2. The following matrix gives the payoff in utiles for four alternatives and four possible states of nature.

		States of Nature			
		*S*1	*S*2	*S*3	*S*4
Alternatives	*A*1	2	6	0	0
	*A*2	2	2	2	2
	*A*3	0	8	0	0
	*A*4	4	4	0	2

Which alternative would be chosen under the Laplace principle? The maximin rule? The maximax rule? The Hurwicz rule with $\alpha = 0.60$? The minimax regret rule?

3. Graph the Hurwicz rule for all values of α using the payoff matrix of Problem 2.

4. The following matrix gives the dollar profits expected for five investments and four different levels of sales.

		Levels of Sales			
		*L*1	*L*2	*L*3	*L*4
	*I*1	15	11	12	9
	*I*2	7	9	12	20
Investments	*I*3	8	8	14	17
	*I*4	17	5	5	5
	*I*5	6	14	8	19

Which investment would be chosen under the maximin rule? The maximax rule? The Hurwicz rule with $\alpha = 0.7$? The minimax regret rule? Answrs: *I*1; *I*2; *I*2; *I*3

5. The following matrix gives the expected profit in thousands of dollars for five marketing strategies and five potential levels of sales.

		Levels of Sales				
		*L*1	*L*2	*L*3	*L*4	*L*5
	*M*1	10	20	30	40	50
	*M*2	20	25	25	30	35
Strategies	*M*3	50	40	5	15	20
	*M*4	40	35	30	25	25
	*M*5	10	20	25	30	20

Which marketing strategy would be chosen under the maximin rule? The maximax rule? The Hurwicz rule with $\alpha = 0.4$? The minimax regret rule?

6. Graph the Hurwicz rule for all values of α using the payoff matrix of Problem 5.

7. The following payoff matrix indicates the costs associated with three mutually exclusive decision options and four states of nature.

		State of Nature			
		*S*1	*S*2	*S*3	*S*4
	*D*1	20	25	30	35
Option	*D*2	40	30	40	20
	*D*3	10	60	30	25

Which option should be selected by (a) maximin rule, (b) the minimax regret rule, and (c) the Hurwicz rule with $\alpha = 0.2$?

8. A construction firm is considering the purchase of a number of different pieces of equipment. The firm knows that, depending upon future projects won by bidding, certain types of equipment will have varying costs. Shown below is a cost matrix indicating the equivalent annual costs associated with a particular piece of equipment and its use on a particular project. The equipment alternatives are mutually exclusive, and the firm anticipates that they will win the contract on only one of the projects. What piece of equipment should be purchased based on the (a) minimax rule, (b) minimin rule, (c) Hurwicz rule for $\alpha = .03$, (d) minimax regret rule, and (e) Laplace rule? Graph the Hurwicz rule for each alternative for all values of α. Answers: *E*2; *E*4; *E*2; *E*2; *E*3

		Projects		
		A	*B*	*C*
Equipment	*E*1	\$100	\$90	\$ 60
	*E*2	70	80	90
	*E*3	30	30	140
	*E*4	100	20	120

9. Assume that an engineer has a sum of money to invest. After some thought and preliminary analysis, she has narrowed the possibilities to the following:

*A*1 = Invest in speculative stocks.

*A*2 = Invest in blue chip stocks.

*A*3 = Invest in government bonds.

The following three future states of the economy have also been considered:

*S*1 = International conflict.

*S*2 = Peace without economic recession.

*S*3 = Peace with economic recession.

Her preliminary analysis yields the following payoff matrix in terms of rate of return.

	S1	S2	S3
A1	20	1	−6
A2	9	8	0
A3	4	4	4

a. What course of action do the following decision criteria indicate: Laplace, maximin, maximax, and Hurwicz with $\alpha = 0.5$?

b. Which investment is preferred? Why?

10. Consider the example of an engineering and construction firm used in this chapter. Suppose that the firm restructures the states of nature as follows:

$S1$ = The firm receives no contract.

$S2$ = The firm gets at least one contract.

$S3$ = The firm gets both contracts.

The following payoff matrix applies, in which the values are present profits in thousands of dollars, with the values of $S2$ being the averages of the payoffs in columns X and Y of Table 16.1.

		States of Nature		
		S1	S2	S3
	A1	−8,000	−1,500	2,000
	A2	−4,000	1,000	4,000
Alternatives	A3	−5,000	−250	6,000
	A4	−3,000	1,000	5,000
	A5	0	2,000	2,000

a. Will the firm make a different decision under the restructured payoff matrix if the Laplace rule is applied?

b. Does the decision depend upon the order in which the states of nature are listed if the Laplace rule is applied? The maximin rule? The minimax rule?

Part Six

ECONOMIC ANALYSIS OF OPERATIONS

In this final part of the text, attention is turned exclusively to operations economy: the economic analysis of operational alternatives. This is in contrast to the economic evaluation of investment opportunities for bringing new products and systems into being. Ongoing operations may be evaluated and improved, or even optimized, by the proper application of the methods and techniques presented in previous chapters. At the heart of this thinking are the life-cycle ideas first presented in Section 2.5. The analysis procedures and models in this part of the text are an essential component of engineering economy studies which consider the entire system life cycle.

17 Analysis of Construction and Production Operations

Engineering proposals cannot meet human needs if they remain in the form of plans and specifications. Once economic feasibility is assured, design proposals may be converted into products, structures, or systems through construction and/or production. The physical environment is altered through these operations, resulting in the creation of utility and the satisfaction of human needs.

Construction and production operations require a sequence of activities and the employment of producer goods of various types. Producer goods in the form of construction and production equipment, coupled with organized activity, are instruments used to alter the physical environment. In this chapter some economic aspects of construction and production operations are examined from the standpoint of the design activities, equipments, and personnel utilized.

17.1 DESIGN FOR ECONOMY OF OPERATIONS

The results of design effort, such as in plans and specifications, crystallize the form of the product to be produced or the structure to be constructed, its physical attributes, material, and production or construction requirements. For this reason design affords many opportunities for economy. Design, no matter how poorly done, is predicated on the thought that the effort devoted to it will be outweighed by the results.

One important aspect of design is that it makes it possible to try out a number of ideas more or less conclusively. The cost of these trials may be negligible when compared with the actual accomplishment. A decision to accept a design is a decision to assume the advantages and disadvantages associated with it and to discard other designs that may have been considered.

This section begins with an explanation of the importance of life-cycle thinking in design for economy. It then presents selected examples of situations in which design plays a key role in the economy of production and operations.

17.1.1 Life-Cycle Considerations in Design

Traditionally, engineers have focused mainly on the acquisition phase of the product life-cycle as depicted in Figure 2.2. Experience shows conclusively that an economically competitive product, system, or structure cannot be realized if too much emphasis is placed on design for the product's primary function. Other functions such as reliability, maintainability, overall quality, and economic feasibility must be addressed during the design phase of the life cycle.

Concern for the entire life cycle is very strong within the Department of Defense. This may be attributed to the fact that acquired systems are owned, operated, and maintained by DOD. This is unlike the situation most often encountered in the private sector, where the consumer is usually not the producer.

Private firms serving as defense contractors are obliged to design and develop in accordance with DOD directives. Since DOD is the customer and also the user of the ultimate product or system, considerable intervention by DOD takes place during the acquisition phase.

Many firms that produce for private-sector markets have chosen to design with the life cycle in mind. For example, design for energy efficiency is now quite common in appliances such as water heaters and air conditioners. Fuel efficiency is a required design characteristic of automobiles. Some truck manufacturers promise that life-cycle maintenance requirements will be within stated limits. These developments are good, but they do not go far enough. When the producer is not the consumer, it is less likely that potential operational problems will be addressed during development. Undesirable outcomes too often end up with the user of the product instead of with the producer.

Well managed, high-quality life-cycle design offers many benefits. It can create distinctiveness and create a personality for a new product or system. It can be used to reinvigorate product interest in the mature stage of its life. It can communicate value to the consumer and lead to greater consumer satisfaction. Finally, it can reduce operating and maintenance costs.

17.1.2 Elimination of Overdesign

In the design of a certain building requiring 180 rafters, 2-inch by 8-inch by 16-foot pieces, 16 inches center-to-center, were found to be just sufficient to meet the anticipated load. It was also found that the sheathing to span rafters placed 24 inches center-to-center was just adequate for strength and stiffness. Consequently, the sheathing will be stronger and stiffer than necessary for the 16-inch spacing. The excess is called *overdesign*, since it has no functional value and may be disadvantageous because of its extra weight.

An alternative design is contemplated using 2-inch by 10-inch cross-section rafters. On the principle that the bending moment resisted by a rafter is in accordance with $bd^2/6$, the ratio of the load-carrying capacity of the 2-inch by 10-inch rafter to the 2-inch by 8-inch rafter is 1.56 to 1.00.

As far as load-carrying capacity is concerned, the 2-inch by 10-inch rafters may be spaced 16 × 1.56 inches, or 25 inches. In accordance with building standards, the rafters will be spaced 24 inches center-to-center. The analysis required for 2-inch by 8-inch by 16-foot rafters, 16-inches center-to-center, is

Number of rafters required — 180

Overdesign of rafters — 0%

Overdesign of sheathing, $\frac{24 \text{ in.} - 16 \text{ in.}}{24 \text{ in.}}$ — 33%

Cost of rafters @ $0.38 per board foot,

$$\frac{180 \text{ rafters} \times 2 \text{ in.} \times 8 \text{ in.} \times 16 \text{ ft}}{12 \text{ in}} \times \$0.38 \qquad \$1{,}459$$

For 2-inch by 10-inch by 16-foot rafters, 24 inches center-to-center, the required analysis is

Number of rafters required, $\frac{180 \times 16 \text{ in.}}{24 \text{ in.}}$ — 120

Overdesign of rafters, $\frac{25 \text{ in.} - 24 \text{ in.}}{24 \text{ in.}}$ — 4%

Overdesign of sheathing — 0%

Cost of rafters @ $0.38 per board foot,

$$\frac{120 \text{ rafters} \times 2 \text{ in.} \times 10 \text{ in.} \times 16 \text{ ft}}{12 \text{ in.}} \times \$0.38 \qquad \$1{,}216$$

In this example, the design with 2-inch by 10-inch rafters not only is adequate, but it is less expensive by $243. Additional savings will result from having to handle fewer pieces during the construction operation.

17.1.3 Material Selection and Design

When two or more materials may serve equally well from a functional standpoint, the relationship of their cost, availability, and processing cost should be considered in determining which is best. For example, brass is often found to be less costly for parts than cold rolled steel in spite of its greater weight per unit volume and greater cost per pound, because it can be machined at a higher rate. Aluminum, which is easily machinable and in addition has a low specific weight, is being used in increasing amounts as a replacement for steel, cast iron, and other metals whose cost per pound is considerably less. Because of the ease with which they can be processed, plastics have proved to be more economical in many applications as a replacement for materials of less cost per pound.

In some cases the decision to substitute one material for another will result in an entirely different sequence of processing. For instance, a change from grey iron to zinc alloy castings will require marked change in equipment. To determine the comparative economic desirability of two materials, it is necessary to make a detailed study of the costs that arise when each is used. For example, suppose that a designer has a choice of specifying either grey iron or aluminum for an instrument mounting in a power plant. It has been determined that either material will serve equally well. The aluminium casting will weigh 1.3 pounds and the grey iron casting 2.4 pounds. An analysis of the unit cost of providing each type of casting follows:

	Grey Iron	*Aluminum*
Cost of casting delivered to factory	$ 7.60	$ 9.20
Cost of machining	6.30	5.52
	$13.90	$14.72

On the basis that either material will serve equally well and provide an identical service, the grey iron casting will be selected because its immediate or present cost is the lower of the two.

In the above example, where the instrument is to be mounted in a power plant, differences in weight were not considered to be a factor. However, for an aircraft instrument the difference in weight might be the deciding factor. In aircraft service the lighter instrument casting would have an economic value in lessened fuel consumption or greater payload and would have greater utility.

17.1.4 Economy of Interchangeable Design

Interchangeability in design is an extension of the mathematical axiom that things equal to the same thing are equal to each other. Components of an assembly that are to be interchangeable with components of other assemblies must possess specific tolerance limits.

As an example of the economy inherent in interchangeability, consider an engine remanufacturing process. If the parts were not originally designed to be interchangeable with like parts of other engines of the same model, it would be necessary to keep all parts of each engine together during remanufacturing. The costs of coordination, tagging, and location of stray parts are eliminated through initial design that allows interchangeability. In this example interchangeability contributes to maintainability.

17.1.5 Designing for Economy of Production

A product or system may exhibit excellent functional design but may be poorly designed from the standpoint of producibility. The drawings and specifications establishing the design also establish a minimum manufacturing cost regardless of efforts to reduce labor, material, and other costs. Product first cost depends upon costs incurred during production.

The effect of design on production cost is recognized by designers in many fields. Plastics have replaced metal in children's toys, reducing production costs and increasing safety. In fastening, quick attachments have replaced nuts and bolts in some applications, making assembly easier while increasing the ease of disassembly for inspection and repair. Specifying certain quick-setting finishes makes possible the production of consumer goods with a minimum need for curing time.

Designers concerned with economy must take into consideration the availability of production or construction equipment as a prerequisite to the choice of methods. For example, there is little point in specifying a granular finish for the ceiling of a certain public building if the equipment to apply such a finish cannot be made available economically. In many instances designs must be adapted to account for the availability of both machines and the skills of construction or production workers.

17.1.6 Design for Economy of Shipping

The shipping costs of products can often be reduced by proper design. Some products are designed so they can be easily assembled upon receipt by the customer. Others are designed so they may be easily packaged for shipment. Weight reduction is often considered in design so that shipping costs will be minimized.

Often it may be advantageous to design so that the product can be nested to save space in shipment. In the shipment of pipe to Saudi Arabian oil fields this principle was used. The design called for pipe approximately 30 inches in diameter. To facilitate shipment by water, where bulk freight rates were in effect, half of the pipe was made 30 inches in diameter and the other half 31 inches in diameter to make nesting possible. As a result, shipping costs were reduced by millions of dollars.

17.1.7 Design for Economy of Maintenance

Engineering drawings and specifications establishing the design of a product or system should exhibit consideration for maintainability. A commercial product or a public system is not available to serve its intended purpose if it is down for maintenance. Because of the high cost of maintenance skills and the associated costs of test equipment, spare parts, transportation, and other logistic elements, engineers must consider maintainability during design to a greater degree than at present.

The challenge is simple. Engineers have the opportunity to reduce the need for maintenance by striving for a product or system that will fail less frequently in use. High reliability will assure system availability when needed and, in addition, avoid the costs associated with and resulting from maintenance. Of course, high reliability is not achieved without cost.

The ultimate goal of maintainability design is the economic reduction of support requirements for a system after it goes into operation, together with the facilitation of whatever remaining maintenance will be required. This goal is logically the concern of the design engineer, for it is not likely that the disruption of service and the burden of maintenance will be reduced unless engineers assume responsibility for this important area. The need is critical, for it appears that both commercial and military products, structures, and systems now consume more economic resources for maintenance and related services than their first cost.

17.2 ECONOMIC ASPECTS OF LOCATION

In many situations the geographical location of a construction project or a production facility is an important economic consideration. Contractors will normally not be interested in projects that are located too far from their equipment and labor pools. Corporate managers normally seek to locate new manufacturing plants in close proximity to markets, materials, and labor.

17.2.1 Locating Temporary Facilities

It is not uncommon for temporary service facilities to be set up in the area where temporary activities are to be undertaken. The economy of setting up such facilities and of selecting among various locations within the area must be determined.

Consider an example of a contractor who is to build a dam requiring 300,000 cubic yards of gravel. Two feasible locations have been identified, and their characteristics are given in Table 17.1.

To make a selection on the basis of economy, the cost of securing the required gravel from either source should be determined. The cost of 300,000 cubic yards of gravel from Source *A* is

Purchase price of pit	\$ 7,200
Road construction	2,400
Hauling cost, 300,000 yd^3 × 1.8 mi × \$0.08/yd^3/mi	43,200
	\$52,800

TABLE 17.1.
COMPARISON OF TWO SOURCES OF MATERIAL

Location and Cost Data	Source *A*	Source *B*
Distance, pit to dam site in miles	1.8	0.6
Cost of gravel per cubic yard at pit	—	$0.08
Purchase price of pit	$7,200	—
Road construction necessary	$2,400	None
Overburden to be removed at $0.24 per cubic yard	—	60,000
Hauling cost per cubic yard per mile	$0.08	$0.10

The cost of 300,000 cubic yards of gravel from Source *B* is

Cost of gravel at pit, 300,000 yd @ $0.08	$24,000
Removal of overburden, 60,000 yd @ $0.24	14,400
Hauling cost, 300,000 $yd^3 \times 0.6$ mi $\times$ \$0.10/yd^3/mi	18,000
	$56,400

17.2.2 Locating Permanent Facilities

The selection of a location for a factory or plant is a long-term commitment. Once a plant has been constructed, the expense and disruption necessary to move it to a more favorable location are generally so great that they are impractical, even though failure may result from the unfavorable characteristics of the original location. Therefore, the search for and evaluation of plant sites deserves very careful consideration.

When profit is the measure of success, the best location is the one in which production and marketing effort will result in the greatest profit. Location may affect the cost at which raw materials are gathered, the cost of production, the cost of marketing, and the volume for product that can be sold.

Evaluation begins with research to determine the volume of sales and income anticipated at the given location. Then research is directed along the lines suggested by the factors above to gain data with which to calculate the cost of gathering raw material, processing it, and delivering finished products to the consumer in the volume that can be sold. Evaluation of plant locations consists essentially of operating the enterprise under consideration "on paper" at each location studied.

The results of an evaluation of three locations for a small glassware plant requiring an investment of approximately $180,000 are given in Table 17.2. Consider the competitive advantage of a plant at Location *C* over one at Location *B*: an advantage in net income of more than one-third by virtue of its lower production costs to offset price reductions and operating and selling efficiency of a competing plant at Location *B*.

TABLE 17.2.
COMPARATIVE EVALUATION OF THREE PLANT LOCATIONS

Factor		Location *A*	Location *B*	Location *C*
Market data:				
Annual sales	(*A*)	$260,000	$260,000	$260,000
Selling expense......................	(*B*)	44,000	43,000	46,000
Net income from sales		$216,000	$217,000	$214,000
Production costs:				
Supplies and raw materials		$ 69,000	$ 71,000	$ 62,000
Transportation—in and out.......		27,000	26,000	23,000
Fuel, power and water............		13,000	17,000	18,000
Wages and salaries		64,000	63,000	61,000
Miscellaneous items................		8,000	8,000	10,000
Fixed costs other than interest...		11,000	11,000	11,000
Net production costs................	(*C*)	$192,000	$196,000	$185,000
Selling and production costs,				
$B + C$	(*D*)	236,000	239,000	231,000
Annual profit, $A - D$..................	(*E*)	24,000	21,000	29,000
Rate of return, $E \div \$180,000$		13.3%	11.7%	16.1%

17.3 ECONOMIC OPERATION OF EQUIPMENT

Most production facilities may be operated at rates below, equal to, or above their normal capacity. Facilities are usually inefficient and costly to operate when utilized at a rate of output below their normal capacity. For example, if a production department is alloted more space than is needed for its efficient operation, a number of losses may be expected. The fixed costs such as maintenance, taxes, insurance, and interest will be higher than necessary. Heat, light, and janitor service will be wasted in the unused space, and the cost of supervision and material handling may be higher than necessary.

The load that is considered proper to impose on equipment is usually indicated by its manufacturer. Such indicated normal loads are often not the optimum for economy. Operation of equipment above its normal capacity will result in increased production at the expense of increased power consumption and shortened life. In many cases, however, the overall result is a reduced cost per unit produced. If such is the case, the equipment in question should be overloaded so that overall economy may be realized.

17.3.1 The Economy of Loading to Normal Capacity

A mining company that is expanding its operations is in need of direct current for an electrolytic process. During the next year, an average of 12 kW of direct-current power will be needed. Planned expansion of the existing operations is expected to increase by 2 kW per year the need for power until 30 kW are needed. At this point, the demand will remain constant. Power will be needed 2,000 hours per year regardless of how much is used.

Investigation reveals that the need for direct current can best be provided by an ac-to-dc motor-generator set. These sets can be purchased in a variety of capacities. Alternative *A* involves the purchase of a 20-kW set now for 5 years of use, at which time the demand will have reached 20 kW. At that time, a 30-kW set will be purchased.

Efficiency-load curves provided by that vendor show that the 20-kW set has efficiencies of 48, 69, 78, and 76% at $\frac{1}{4}$, $\frac{1}{2}$, $\frac{3}{4}$, and full load respectively. The 30-kW set is slightly more efficient at each of these loads. The purchase rate of ac power is $0.06 per kWh. Interest is taken at 10%. Taxes, insurance, maintenance, and operating costs will be neglected. The 20-kW set will cost $3,690 installed. It is estimated that $1,400 will be allowed for the 20-kW set on the installed purchase price of $4,800 of the 30-kW set 5 years hence. An analysis of the present equivalent cost of the power consumed for each of the 5 years is given in Table 17.3.

The present equivalent cost of providing 5 years of service under Alternative *A* is calculated as follows:

Present worth of power for 5 years (ΣE in table)	$ 9,275
Present cost of 20-kW set installed	3,690
Present worth of 20-kW set trade-in receipt, $1,400$\overset{P/F,10,5}{(0.6209)}$	−869
Present worth of 30-kW set, $4,800$\overset{P/F,10,5}{(0.6209)}$	2,980
	$15,076

TABLE 17.3.
PRESENT EQUIVALENT COST OF CONSUMED POWER FOR ALTERNATIVE *A*

Year	Output Rate dc Power, in kW (*A*)	Efficiency at dc Output Rate (*B*)	Input Rate ac Power, *A* ÷ *B* (*C*)	Annual Power Bill C × $0.06 × 2,000h (*D*)	Present Equivalent of Annual Power, $D \times \overset{P/F,10,n}{(\quad)}$ (*E*)
1	12	0.74	16.2	$1,944	$1,767
2	14	0.77	18.2	2,184	1,805
3	16	0.78	20.5	2,460	1,850
4	18	0.78	23.1	2,772	1,893
5	20	0.76	26.3	3,156	1,960

TABLE 17.4.
PRESENT EQUIVALENT COST OF CONSUMED POWER FOR ALTERNATIVE *B*

Year	Output Rate dc Power, in kW (*A*)	Efficiency at dc Output Rate (*B*)	Input Rate ac Power, $A \div B$ (*C*)	Annual Power Bill C × \$0.06 × 2,000h (*D*)	Present Equivalent of Annual Power, $D \times (P/F, 10, n)$ (*E*)
1	12	0.65	18.5	\$2,220	\$2,018
2	14	0.70	20.0	2,400	1,983
3	16	0.74	21.6	2,592	1,947
4	18	0.77	23.4	2,808	1,918
5	20	0.79	25.3	3,036	1,885

Alternative *B* involves the purchase of the 30-kW set now so that it will be available to meet the anticipated demand. The present equivalent cost of the power consumed for each of the first 5 years is given in Table 17.4.

The present equivalent cost of providing 5 years of service under Alternative *B* is calculated as follows:

Present worth of power for 5 years (ΣE in table)	\$ 9,751
Present cost of 30-kW set installed	4,800
	\$14,551

Power costs beyond the first 5 years have not been taken into account in the analysis because they will be equal in subsequent years, since a 30-kW set will be used regardless of whether Alternative *A* or *B* is adopted. The assumption that a 30-kW unit 5 years old is equivalent to a new set was made to simplify the analyses and because electrical equipment ordinarily has a long service life.

17.3.2 The Economy of Loading Above Normal Capacity

The life of a piece of equipment is usually inversely proportional to the load imposed upon it. However, the output is directly proportional to the load imposed. When this is the case, there exists an optimum load that will determine the level of operation for maximum economy.

In some cases, particularly with short-lived assets, the economic life may be determined by experiment. If the effects of various loadings upon the life, maintenance, and operating costs of a piece of equipment are known or can be estimated with reasonable accuracy, it is easy to determine the load that will result in the greatest economy.

Consider the following example: A manufacturer has an automatic plastic molding machine that produces one piece for each revolution. The machine is now being operated at the rate of 150 r/min and produces 18.6 pounds of product per hour. This machine costs \$61,500 and is estimated to have an operating life of 10,000 hours. Direct labor plus overhead applicable to the operation amount to \$13.70 per hour. Average maintenance and power costs are \$0.72 and \$0.64 per hour, respectively. Output of product and the cost of power used per hour are estimated to be directly proportional to the r/min of the machine. Since centrifugal forces on machine parts increase with the square of the speed, it is estimated that maintenance costs will increase in proportion to the square of the speed, and the useful life of the machine will be inversely proportional to the square of the speed. Table 17.5 is developed under these assumptions.

On the basis of the estimated use, it would be desirable to increase the speed of the machine to 200 r/min. The expected saving per pound of product would be (\$1.14 − \$1.08) ÷ \$1.14 = 5.26%. This is a worthwhile saving, particularly since it shortens the capital-recovery period and so lessens the possibilities of losses from obsolescence and inadequacy.

17.3.3 Economic Load Distribution Between Machines

In many cases two or more machines are available for the same kind of production. Thus, two or more boilers may be available to produce process steam, or two or more turbines may be available to produce power. If the total load is less than the combined capacity of the available machines, the economy of operation will usually depend upon the portion of the total load that is carried by each machine.

As an example method of determining the load distribution between two machines, suppose that two diesel-engine-driven dc generators, one of 1,000-kW and the other of 500-kW capacity, are available. The efficiencies for different outputs and the corresponding inputs for each are given in Table 17.6.

Suppose that the total load to be met at a certain point in time is 1,200 kW. This load may be distributed in several ways, as shown in Table 17.7. It is observed that the desired output of 1,200 kW can be produced with a minimum input when the larger unit carries 800 kW and the smaller unit 400 kW.

In this example, input can readily be converted to units other than kilowatts. Suppose that it is desired to express input in terms of dollars. Assume that fuel oil having an energy content of 18,800 Btu weighs 7.48 pounds per gallon and costs \$1.18 per gallon delivered. One kWh is equivalent to 3,410 Btu. From these data the cost per kilowatthour is calculated to be

$$\frac{\$1.18 \times 3{,}410}{7.48 \times 18{,}800} = \$0.0286.$$

TABLE 17.5.
RELATIONSHIP OF COST TO OPERATING SPEED

Operating Speed in r/min	Estimated Life in hours, $10{,}000 \times (150)^2/A^2$	Average Output in pounds per hour, $18.6 \times A/150$	Labor Cost per hour,	Average Maintenance Cost per hour, $\$0.72 \times A^2/(150)^2$	Power Cost per hour, $\$0.64 \times A/150$	Average Depreciation Cost per hour, $\$61{,}500 \div B$	Total cost of Operation per hour $D + E + F + G$	Cost per pound, $H \div C$
A	*B*	*C*	*D*	*E*	*F*	*G*	*H*	*I*
150	10,000	18.6	$13.70	$0.72	$0.64	$ 6.15	$21.21	$1.14
200	5,630	24.8	13.70	1.28	0.85	10.92	26.75	1.08
250	3,600	31.0	13.70	2.00	1.07	17.08	33.85	1.09

TABLE 17.6.
EFFICIENCIES FOR DIFFERENT OUTPUTS

	Machine *A*, 1,000 kW Output		Machine *B*, 500 kW Output	
Output, in kW	Efficiency in percent	Input in kW	Efficiency in percent	Input in kW
100	15.1	663	20.7	483
200	22.0	909	27.9	719
300	26.3	1,141	31.8	943
400	29.5	1,356	32.5	1,231
500	31.2	1,603	32.0	1,563
600	32.3	1,858		
700	32.5	2,154		
800	32.6	2,454		
900	32.4	2,778		
1,000	32.3	3,096		

At this rate of cost for fuel, 1,200 kWh output of energy can be produced for $3{,}685 \times \$0.0286 = \$105.39$ if the loads on the two machines are 800 kW and 400 kW, respectively. If the load distribution had been 1,000 kW and 200 kW, the cost would have been $109.11, an increase over the better method of operation of 3.5%. This is a significant rate of saving, considering the ease with which it is determined. The saving is a result of a correct decision based on knowledge of load distribution.

TABLE 17.7.
DISTRIBUTION OF LOAD BETWEEN TWO MACHINES

Output in kW.			Input in kW.		
Machine *A*	Machine *B*	Total	Machine *A*	Machine *B*	Total
1,000	200	1,200	3,096	719	3,815
900	300	1,200	2,778	943	3,721
800	400	1,200	2,454	1,231	3,685
700	500	1,200	2,154	1,563	3,717

17.4 SELECTION OF PERSONNEL

Of the several factors used in production, construction, and operations, none has such a variety of characteristics as personnel. Characteristics of both body and mind are of concern. One individual may accomplish several times as much of a task requiring physical strength, dexterity, or acuteness of vision as another. The range of mental proficiencies for given tasks is even wider.

The physical and mental endowments of people appear to be subject only to limited change by training and leadership. For this reason it is important that personnel be selected who have inherent characteristics as nearly compatible as possible with the work they are to perform and the position they are to occupy. Contrasted with the meticulous care with which equipment and materials are selected, the selection of personnel in most organizations is given only superficial attention.

17.4.1 The Range of Human Capacities

The ratio of the poorest to the best performance time is often as much as 1 to 2 in common operations. It is apparent that the cost of labor input per unit of output might be reduced by selecting only employees capable of better-than-average performance. In many cases superior employees receive little or no greater compensation than less capable employees. This is particularly true when performance is hard to measure. Superior managerial ability, for example, may go unrecognized and therefore unrewarded for long periods because it cannot be measured. But even when superior ability is proportionally compensated for, savings may result.

Superior ability that remains unused is not an advantage and may often be a disdvantage because of the resulting frustration and dissatisfaction of the employee. Generally speaking, the work assignment should require a person's highest skill. For many tasks, tests can be devised for selection of prospective employees with superior ability. The economic desirability of having most employees above average in ability for the task under consideration (even though compensation be in proportion to ability, which it rarely is) can hardly be overestimated.

In this relatively new field, results of the installation of a program of selection practices are most difficult to estimate. Since the input cost of a program can usually be estimated with reasonable accuracy, it may be helpful in arriving at a decision to calculate the benefits necessary to justify a contemplated selection program.

Suppose that a certain plant employs 40 operators. The average output is 28 pieces per hour, the average spoilage is 1.7%, and each rejected piece results in a loss of $1.50. Records reveal that average spoilage for the best 20 operators is 1.1%, whereas that for the poorest 20 operators is 2.3%. A consultant agrees to prepare a set of tests and administer them for a year for $5,000. Administration subsequent to the first year is expected to cost $8 per operator hired or $350 per year based upon the estimated turnover.

Let R equal the percentage reduction in spoilage during the first year to justify expenditure of $5,000. Then $R \times 40$ operators $\times$ 28 pieces per hour $\times$ 2,000 hours per year $\times$ $1.50 per reject = $5,000, from which $R = 0.15\%$.

This percentage reduction in spoilage seems attainable, since a program that will reject applicants of less-than-average ability in relation to spoilage will result in an average reduction of $1.7 - 1.1$ or 0.6%. Moreover, it is reasonable to expect other benefits from the program.

By tending to place people in work for which they are best fitted, sound selection is generally beneficial. Even those who are rejected will generally be benefited; for, if continually rejected for work for which they are unfitted, they must eventually find a job that will permit them to work at their highest skill.

17.4.2 The Economy of Proficiency

The value of proficiency is not necessarily directly proportional to degree of proficiency. Thus, a baseball player whose batting average is 0.346 will ordinarily command more than twice as much salary as one whose average is 0.173, if they are equal in other respects. In some activities the compensation of those with unusually high proficiencies is extremely high, but those with ordinary abilities can find no market for their services. Consider acting; the motion picture industries pay fabulous salaries for those whose box office appeal is high, when they undoubtedly could cast all their productions with volunteer actors of fair ability without cost.

The cost of human effort is a considerable portion of the total cost of carrying on nearly all production and construction activities. People are employed with the idea of earning a profit on the skills they possess. In general, the higher a person's proficiency in any skill, be it manual dexterity, creativeness, leadership, inventiveness, or physical ability, the greater his value to his employer.

Consider a machine operation for which a worker is paid \$8 per hour for producing 10 pieces per hour on a machine whose rate is \$8 per hour. In this example the cost per piece is

$$\frac{\$8 + \$8}{10} = \$1.60.$$

If a second worker of less proficiency completes 9 pieces per hour, the relative worth to his employer (as compared with that of the first worker) is calculated as follows: Let W equal the hourly pay of the second worker to result in a cost per piece of \$1.60. Thus,

$$\frac{W + \$8}{9} = \$1.60$$

and

$$W = \$6.40.$$

Here a 10% reduction in the ability of the worker to turn out pieces makes the services worth 20% less per hour to his employer, all other things being equal.

If the second worker also receives \$8 per hour, the cost per piece of the 9 pieces he turns out in an hour will be

$$\frac{\$8 + \$8}{9} = \$1.78.$$

It is conceivable that the resulting cost per piece may be so high that the employer cannot profit if he must pay the second worker \$8 per hour.

17.4.3 The Economy of Specialization

To specialize is to restrict a person or thing to a particular activity, place, time, or situation. Specialization in this sense results when there is a divison of labor, as for example, when operations in a process performed by a single person are divided among a number of persons. It is clear that machines can also be specialized in this sense; that is, specialized as to activity.

Persons or things can also be specialized as to use, place, or time. Thus, a woman whose assignments require that she report to work in Atlanta is specialized as to place. A passenger train that operates on a schedule is specialized as to time and place; so is a person who works in one place from nine to five o'clock, five days a week.

Specialization is of interest in economy studies because it is often a means whereby the cost of accomplishing a given result can be reduced. In any design or planning effort the desirability of specialization should be considered as a routine matter. It is widely recognized that specialization, particularly as it relates to people, may result in improved performance.

Often, specialization requires preplanning and specialized arrangements, the cost and maintenance of which may more than offset the advantage gained. The gains of a specialization of one kind may require specialization of another that is too costly. For instance, it might be desirable to have two persons perform one each of two activities that must be performed one after the other, unless the new arrangement would also require the people to be specialized as to time. The latter specialization might easily result in one person being idle while the other works.

Specialization is subject to the law of diminishing returns. It is apparent that it would not be economical to specialize to such an extent, for example, that one truck driver made only right turns and another made only left turns.

Generalization, the reverse of specialization, may also be a factor in economy. Thus, in machine design it is often desirable to have one part serve several purposes; the crankcase of an engine, for instance, serves as a base for the engine, a container for oil, and a support for the crankshaft and cylinder. The term "all-around man" is evidence of generalization with respect to personnel. Generalization is gaining in popularity recently as more responsibility is being passed to workers for quality and productivity.

17.4.4 The Economy of Dependability

There is an economy in each of the innumerable personal qualities, but few are more important than dependability. Dependability in relation to organized action means behaving as agreed upon, or as expected. For example, the chairman of a planning committee has called a meeting of 12 persons but arrives 15 minutes after the specified time. If the average salary rate of committee members is $20 per hour, the chairman's lack of dependability may have resulted in a loss of time of 12 × 0.25 × $20, or $60. This direct loss points out the value of dependability, but it probably is not as great as the frustration and loss of confidence that the lack of dependability engenders in the persons affected by it.

Dependability rests upon ability, consideration for others, willingness to cooperate, and honesty. Lack of dependability in one person, regardless of the reason for it, invariably results in the need for increased vigilance and effort on the part of others. People in supervisory positions are keenly aware of the costs of deviation from dependability, and they thus prize dependability highly.

Willful undependability and dishonesty are particularly difficult to cope with. Even infrequent dishonesty requires that a supervisor be continually prepared against the entire range of defections a dishonest person may cause. Thus, for many positions, even extremely limited dishonesty may outweigh the benefits of honest services a person may render.

KEY POINTS

- Engineering proposals cannot meet human needs if they remain in the form of plans and specifications; they must be converted to products through construction or production operations.
- Design offers many opportunities for economy: in production or construction, in utilization, in maintenance, and in disposal.
- Experience has shown that an economically competitive product system or structure cannot be realized if attention is focused entirely on the design's primary function; design for economy of operations is essential.
- Production or construction is carried out at a specific location or set of locations, making location an important consideration in operations economy.
- Operations economy often depends upon the loading or rate of use of equipment, making loading and load distribution an important consideration.
- Personnel have the greatest variety of characteristics of any element in an operational system, making the selection of personnel a matter for close study.

PROBLEMS

1. The chief engineer in charge of refinery operations was not satisfied with the preliminary design for storage tanks to be used as part of a plant expansion program. The engineer who submitted the design was called in and asked to reconsider the overall dimensions in the light of an article in *The Chemical Engineer*, titled "How to Size Future Process Vessels." The original design submitted called for 4 tanks 5.2 meters in diameter and 7.0 meters high. From a graph in the article, the engineer found that the present ratio of height to diameter of 1.35 was 111% of the minimum cost and that the minimum cost for a tank occurred when the ratio of height to diameter was 4 to 1. The cost for the tank design as originally submitted was estimated to be $28,000. What are the optimum tank dimensions if the volume remains the same as for the original design? What total saving may be expected through redesign?
2. Two alternate designs are under consideration for a tapered fastening pin. Either design will serve equally well and will involve the same material and manufacturing cost except for the lathe and grinder operations. Design *A* will require 16 hours of lathe time and 4.5 hours of grinder time per

1,000 units. Design B will require 7 hours of lathe time and 12 hours of grinder time per 1,000 units. The operating cost of the lathe including labor is $22 per hour. The operating cost of the grinder including labor is $16 per hour. Which design should be adopted if 95,000 units are required per year, and what is the saving over the alternate design? Answer: B, $7,410

3. It has been decided by a previous design that 4-inch by 4-inch rough pine shores will be used to support certain concrete forms. The contractor will need 175 shores, each 10 feet long and each capable of supporting a load of 5,000 pounds induced by the forms and concrete. The maximum safe load on a shore is given by the expression $P = 1{,}000\ (1 - g/80b)bh$, where P = maximum safe load in pounds; g = height of shore in inches; b = width of shore in inches; h = depth of shore in inches. It has been decided that the 4-inch width of the shores will be maintained in any redesign consideration. If rough pine lumber costs $190 per 1,000 board feet, what total amount can be saved by the elimination of overdesign?

4. In the design of buildings to be constructed in Northern Greenland, the designer is considering the type of window frame to specify. Either steel or aluminum window frames will satisfy design criteria. Because of the remote location of the building site and the lack of building materials in Greenland, the window frames will be purchased in the United States and transported a distance of 3,500 miles to the site. The price of window frames of the type required is $84 each for steel frames and $112 each for aluminum frames. The weight of steel window frames is 150 pounds each and the weight of aluminum window frames 58 pounds each. The transportation shipping rate is $0.03 per pound per 100 miles. Which design should be specified, and what is the economic advantage of the selection?

5. In the design of a jet engine part, the designer has a choice of specifying either an aluminum alloy casting or a steel casting. Either material will provide equal service, but the aluminum casting will weigh 1.20 pounds as compared with 1.35 pounds for the steel casting. The aluminum can be cast for $4.90 per pound and the steel for $2.80 per pound. The cost of machining per unit is $9.60 for the aluminum and $10.10 for the steel. Every pound of excess weight is assessed a penalty of $90 due to increased fuel consumption. Which material should be specified, and what is the economic advantage of the selection per unit? Answer: Aluminum, $11.90

6. A contractor has been awarded the contract to construct an earth-fill dam requiring 500,000 cubic yards of fill. Two possible sources of fill have been found: Source A, above the dam site at a distance of 4 miles, at which the pit will be operated and maintained by the contractor; Source B, below the dam site at a distance of 2.6 miles, from which the contractor must purchase the fill by the cubic yard. The haul cost per cubic yard will be different, because in one instance the loaded trucks will travel down slope and in the other they will travel up slope. The following estimates are available:

	A	*B*
Purchase price of pit	\$10,000	—
Cost of fill per yd^3 at pit	—	\$0.22
Construction of haul road	6,200	—
Maintenance of haul road	2,400	—
Hauling cost per yd^3 per mile	0.18	0.24

Which source should the contractor choose? What savings will result from choosing this source?

7. Two locations are under consideration for a new brick plant. A cost analysis applicable to each location is given below:

	Location A	*Location B*
Capital cost per 1,000 bricks	\$16.50	\$14.70
Material cost per 1,000 bricks	8.75	10.10
Labor cost per 1,000 bricks	30.20	26.90
Distribution cost per 1,000 bricks	18.00	19.25
Total cost per 1,000 bricks	\$73.45	\$70.95

a. Calculate the ratio of the profit per 1,000 bricks produced at Location *A* to the profit per 1,000 bricks produced at Location *B* for selling prices of \$72, \$76, \$80, and \$84 per 1,000 bricks.

b. What is the minimum selling price per 1,000 bricks at each location if the profit must be at least 10%?

8. Electric lamps rates at 200 watts and 110, 115, or 120 volts can be purchased for \$0.82 each. When the lamps are placed in a 115-volt circuit, the following data apply:

	110-v. Lamp	*115-v. Lamp*	*120-v. Lamp*
Average watts input	209.1	195.2	182.8
Average lumens output per watt of input	18.30	16.84	15.48
Average life of lamps in hours	420	750	1,340

Energy costs \$0.055 per 1,000 watthours. Determine the cost per million lumen-hours for each lamp under the three conditions. Answers: \$3.37, \$3.60, \$3.77.

9. The normal operating speed of wire weaving looms is 300 r/min. At this speed the looms have a life expectancy of 5 years, an annual maintenance cost of \$860, and an annual power cost of \$540. The machines cost \$12,000 each and have negligible salvage value. Annual space and overhead charges are estimated at \$1,000 per machine. The operator's rate is \$6.10 per hour and each operator runs three machines regardless of their speed. The interest rate is 13%. Experimental work on the operation of these machines has resulted in the following data:

Speed, r/min	*Life in Years*	*Annual Maintenance Cost*	*Annual Power Cost*
250	7.2	\$ 680	\$500
300	5.0	860	540
350	4.2	1,100	600
400	2.8	1,420	675
450	2.2	1,800	770

Summarize the items of cost in a table and determine the most economical speed at which to run the machines. Assume that a loom operating at a normal operating speed of 300 r/min will produce 300 units of product per year (each year consists of 40 hours a week for 50 weeks) and that production is directly proportional to the speed of the loom. Answer: 350 r/min.

10. An order for 8 tons of a certain insulating material has been received. The material is packaged in 1-ton lots for shipment. Two trucks are available, one with a capacity of 7 tons and the other with a capacity of 4 tons. The material must be hauled 25 miles to a job site. Find the economic load distribution between the trucks if the following costs apply:

	Fuel Cost per Mile	*Labor Cost per Ton*	*Other Costs per Ton-Mile*
7-ton truck	\$0.16	\$10	\$1.90
4-ton truck	0.11	14	1.30

11. In the operation of a certain type of equipment, repairs average \$60 each and the cost of a shutdown for emergency repairs entails an additional loss of \$140. In the past, there has been no routine periodic inspection of the equipment. It is believed that periodic inspections would result in a worthwhile saving. In order to get data on which to determine the frequency of periodic inspections for greatest economy, five groups of an equal number of machines *A, B, C, D,* and *E* were inspected 0, 1, 2, 3, and 5 times per week, respectively. The cost per inspection was found to be \$12.50. The plant operates 5 days per week, 50 weeks per year. The results of a year's run are shown below:

Machine Group	*Inspections per Week*	*Emergency Repairs*	*Non-emergency Repairs*
A	0	31	27
B	1	20	36
C	2	13	41
D	3	9	43
E	5	4	46

How many inspections should be made per week and what will be the total yearly cost for inspection, loss due to emergency repairs, and cost of repairs for this many inspections per week?

12. A firm is considering the expansion of its operations to include a plating department that will require 5,000,000 kWh of electrical power per year. The maximum rate of consumption is estimated to be 1,400 kW. The power can be purchased from a power company for $0.051 per kWh, providing the firm will construct the required 25 miles of transmission line and furnish necessary transformers. The cost of the high-tension line will be $30,000 per mile and the transformers required will cost $42,000. The life of these assets will be 30 years with a salvage value of $18,500.

As an alternative, the firm can construct and operate its own power plant. The plant can be located on a river that will furnish adequate cooling water for condensing purposes if a steam plant is selected or for cooling if a diesel plant is installed. A good location is also available for a hydrogenerating plant 36 miles from the firm's plant site. Costs involved in the construction and maintenance of the three types of power plants are as follows.

	Steam	*Diesel*	*Hydro*
Total investment per kW of capacity	$440	$510	$640
Transmission equipment cost	0	0	$710,000
Fuel required in pounds per kWh	1.85	0.10	0
Cost of fuel per pund	$0.018	$0.042	—
Annual cost of maintenance	$84,000	$55,000	$48,000
Taxes and insurance as a % of investment	2	2	2
Life of plant and equipment in years	20	16	30
Salvage value as a % of original cost	10	12	5

Determine the cost of each of the four methods of supplying the required power if the interest rate is 8% compounded continuously.

13. In a certain manufacturing activity, 40 employees are engaged in identical activities. The average output of the group as a whole is 46.4 units per hour. The average output of the less productive half is 40.2 satisfactory and 1.4 unsatisfactory units per hour and the average for the more productive half is 52.6 satisfactory and 0.8 unsatisfactory units per hour. The employees work on a straight piecework plan and receive $3.30 per hundred satisfactory units. The firm sustains a loss of $0.07 for each unsatisfactory unit. One machine is required for each employee. Each machine has an annual fixed cost of $320 and a variable cost of $0.18 per hour. Supervision and other overhead costs are estimated at $720 per employee per year. The average employee works 1,900 hours per year. How much could be paid annually for a selection, training, and transfer program that would result in raising the average productivity of the entire group to 52.6 satisfactory and 0.8 unsatisfactory units per hour?

14. A foreman supervises the work of Mr. *B* and 8 other men. The foreman states that "Mr. *A* requires twice as much and Mr. *B* requires half as much of my time as the average of my men." Mr. *A*'s output is 8 units per day and Mr. *B*'s output is 7 units per day. On the basis of equal cost per unit, what monthly salary is justified for Mr. *B* if the foreman receives $1,450 per month and Mr. *A* receives $1,100 per month?

15. A foreman is in charge of a construction crew of 8 men and takes great pride in the amount of work he personally performs on the job. Observation shows that the accomplishment of the men is impaired by lack of direction as a result of the foreman's active participation in the work to be done. The foreman receives $12 and his men $9.40 per hour. If the foreman does one-third as much work as the average of his men would do if they were properly directed, what loss in the effectiveness of the crew will just be compensated for by the actual work performed by the foreman? Answer: 5.32%.

16. A machine-tool operator produces 46 units per hour, of which an average of 6 are defective. He is paid on a straight piecework basis at the rate of $5 per 100 satisfactory units. The firm sustains a loss of $0.02 per defective unit produced. The machine used in the process has a total operating cost of $3 per hour, which includes depreciation, return on investment, and overhead. The operator works 1,800 hours per year. If a training program is initiated, it is anticipated that the production rate will increase to 52 units per hour, only 2 of which will be defective.

a. What is the maximum amount that the firm can spend on the training program per year?

b. What would be the effect of the training program on the hourly wage of the operator?

17. A manufacturing concern has sales offices in 5 states in addition to the main plant and offices. Each branch sales manager is paid $32,000 per year, and the vice-president for sales receives $48,000 per annum. At the present time, average annual sales amount to $5,400,000. An annual sales conference is being considered, which will meet at the main plant for 2 days each year.

a. If each branch sales manager is paid $560 for travel and other expenses, what is the total cost of the conference on the basis of a 240-day work year? Answer: $4,533

b. What percentage increase in sales attributable to the conference is necessary to justify the annual meeting if the gross profits are 12% of sales? Answer: 0.70%

18. Foreman *A* directs the work of 12 men. He plans his work "as he goes" during the 8-hour work period. Foreman *B* also directs the work of 12 men, but spends 2 hours per day in planning the work to be done, thus reducing the time available for active supervision. What percentage more effective must his supervison be per unit of time so that his supervisory effectiveness equals that of Foreman *A*?

19. An engineer can do certain required computations in 3 hours, or she can delegate the work to an engineering aide. If the work is delegated, it will take 0.75 hour to explain the computational procedure and 0.50 hour to check the results. The actual calculations will take 4 hours to do if done by the aide. If the engineer receives a salary of $34,000 per year and the aide $17,200 per year, what are the comparative costs for each of the methods for a working year of 2,080 hours? Answer: $26.83 by engineer; $59.71 if delegated

18

Profit-Volume Analysis of Production Operations

With the accumulation of knowledge about physical phenomena, complex operational systems have been developed to produce goods and services. Many methods and techniques concerned with the economic acquisition of assets to provide these goods and services have been discussed. However, once the system is in place the organization must make many operating decisions to accommodate unanticipated variations in the level of activity. These decisions will determine whether the firm is to be operated economically.

This chapter presents an analytical framework for understanding the profit contribution of any productive activity. By examining the relationships among the cost of inputs, the price obtained for the output, and the levels of output, it is possible to determine optimum operating conditions.

18.1 CHARACTERISTICS OF PRODUCTION ACTIVITIES

Any production activity, whether it represents a single machine or a complete manufacturing facility, can be modeled in terms of its cost and revenue-generating characteristics. Use of these models to assist in making numerous operating decisions is critical to the ultimate profitability of the firm. Therefore, the simplicity of the mathematical form of these models belies their importance to those confronted with the complex operating decisions ordinarily encountered.

For any production activity there is an input-output relationship, whether it be a gasoline engine, a grocery check-out function, an electric generating plant, or a manufacturing facility. The output is usually expressed in terms of the number of units produced by the system, such as horsepower, number of persons serviced, kilowatts, or the number of items produced. The inputs are generally measured as the costs of resources consumed in the productive activity. These costs include the productive costs of direct labor, direct material, and overhead, as described in Section 11.2 and shown in Figure 11.4.

18.1.1 Capacity

For every production activity there is a potential maximum level of output that defines the capacity of that activity. This measure of maximum output reflects the upper limit of production achievable, assuming that there is no limit on the availability of inputs. A gasoline engine's capacity is measured by the maximum horsepower achievable, while a manufacturing facility is measured by the maximum items per unit time that could be produced if the optimum combination of inputs were possible. The level of output of a production activity is often stated as a percentage of the maximum capacity of the activity. Thus, an engine producing an output of 100 hp when it has a capacity of 200 hp is is said to be operating at 50% of capacity.

The **capacity factor** is defined as the ratio of the average output to the maximum capacity over some time interval.

Figure 18.1 shows the essential elements which specify a capacity factor, where

$$\text{Capacity factor} = \frac{\text{Average actual output}}{\text{Maximum capacity}}. \tag{18.1}$$

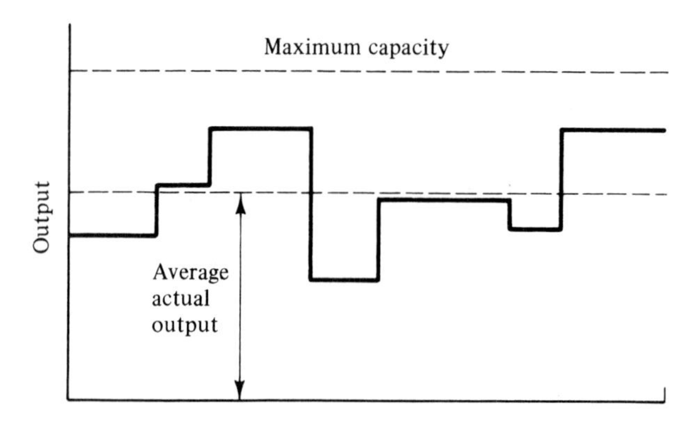

FIGURE 18.1. Output as related to capacity.

18.1.2 Physical Efficiency

The physical efficiency of a productive entity measures the output as a percentage of the inputs, as described in Section 1.3. Plotting the physical efficiency of an engine might lead to a curve as shown in Figure 18.2. The shape of the curve indicates that as the level of output increases beyond 80% of capacity, more units of fuel (input) are required to realize a unit increase of horsepower (output). This effect is a basic concept referred to as the *law of diminishing return.* Although this law does not apply to every productive activity, it does describe the behavior of a wide range of these activities.

As described in Section 1.3, the utilization of physical processes to produce goods and services can create desirable economic efficiencies. Even though the *physical* efficiency can never exceed 100%, successful activities have *economic* efficiencies that surpass 100%. The models presented in subsequent sections are concerned with improvement of economic efficiency.

18.1.3 Cost Characteristics

For most production activities there are two types of costs. These are the fixed and variable costs defined in Section 2.4. As the level of activity within the production unit changes, fixed costs will remain relatively constant.

Variable costs are costs which fluctuate according to changes in the number of units produced. The ability to identify which costs are variable and which ones are fixed is crucial to the analytical framework to be presented.

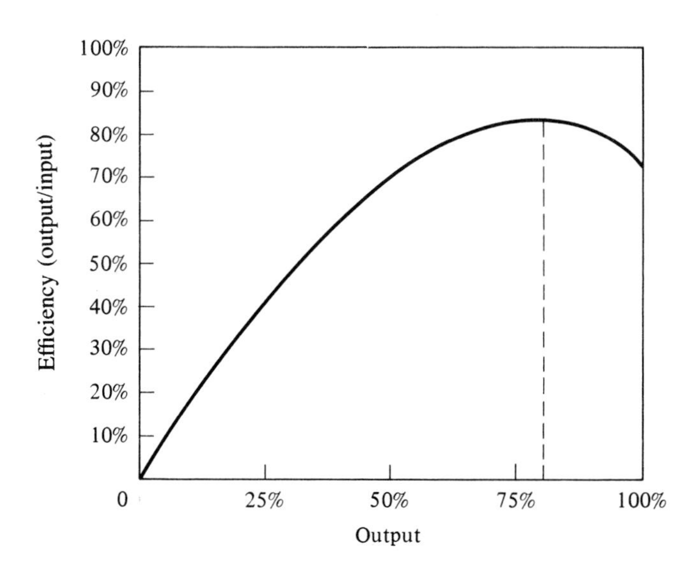

FIGURE 18.2. Physical efficiency of production activity.

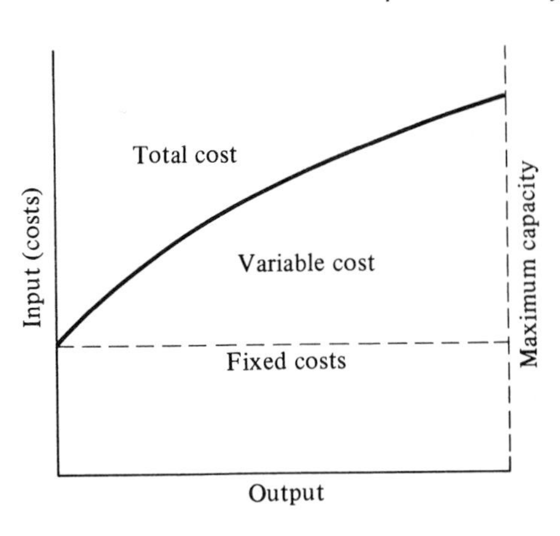

FIGURE 18.3. Fixed and variable costs as inputs.

For any production activity the relationship between the inputs and the outputs must be understood. Figure 18.3 illustrates the fixed and variable costs that represent the input and the output to be realized. The total cost of providing a particular level of output is seen to be the sum of the fixed costs and the variable costs incurred for that output level. The total cost (fixed cost and variable cost) over the entire range of possible outputs forms the basis for the analysis presented in the following sections.

> **Incremental cost** represents the change in total cost that occurs for a change in the level of output.

The decisions to be considered are those represented by the various levels of output. To make decisions about the appropriate output level for a production activity it is essential to have knowledge of the incremental costs as a function of the output.

In terms of total cost shown in Figure 18.3, the incremental costs are described as the slope of the total-cost curve. Since fixed costs do not change for various levels of output, it is only the variable costs that lead to cost differences between output levels. Therefore, the incremental costs can also be described as the slope of the variable cost curve. Either way, the incremental cost measures the *change* or *difference* in cost that occurs for an additional level of output.

> **Average unit cost** is the total input cost divided by the level of output.

Therefore, average unit cost is an average cost per unit and is a direct function of the amount of output achieved. For productive equipment such as a vehicle, output in miles might be averaged over the input of fuel cost, yielding an average unit cost stated in terms of miles per dollar spent on fuel. In manufacturing one usually measures input as total cost and output as the number of units produced. In this case, the average unit cost is the total cost divided by the number of units produced. It is stated in terms of dollars per unit.

Because average unit costs consider *all* the inputs, both fixed and variable costs are included. A caution is appropriate, since average unit costs do not reflect the true cost differences between output levels. Therefore, they should not be used for decisions concerning levels of output where fixed cost must be ignored.

Average unit costs do provide insight regarding the average cost per unit for various levels of output. In addition, these costs indicate how the overall cost of a unit of output is affected as the level of production increases and the fixed costs are spread over more units of output. The phenomenon of average unit costs decreasing with increasing levels of output is referred to as "economy of scale."

18.2 BREAK-EVEN ANALYSIS

There are two aspects to a production activity. One consists of assembling facilities, labor, and material for production. The other consists of marketing the goods or services that have been produced. The success of the activity depends upon a positive net difference between receipts for goods and services sold and the input necessary to produce and market them. Break-even analysis can be used to evaluate the situation.

18.2.1 Linear Break-Even Models

If receipts and costs are assumed to be linear functions of the quantity of product made and sold, an analysis of their relationship to profit is greatly simplified. The patterns of income and costs will appear as in Figure 18.4. Fixed production costs are represented by the line HL. The sum of variable production costs and marketing costs is represented by the line HK. Income from sales is represented by the line OJ. This linear representation is an approximation of actual operation conditions and is only a model.

Analysis of existing or proposed operations represented with a break-even model can be made mathematically or graphically. For an illustration of the mathematical method, let

N = number of units of product made and sold per year;

R = the amount received per unit of product in dollars; R is equal to the slope of OJ;

I = RN, the annual income from sales in dollars; $I = RN$ is the equation of line OJ;

F = fixed cost in dollars per year, represented by OH and HL;

V = variable cost per unit of product; V is equal to the slope of HK;

TC = the sum of fixed and variable cost of N units of product, $F + VN$; $TC = F + VN$ is the equation of line HK;

P = annual profit in dollars per year; $P = I - TC$; negative values of P represent loss;

N^* = break-even point; at this point $P = 0$;

Q = capacity of plant in units per year.

The break-even point occurs when income is equal to cost. In Figure 18.4 this occurs where lines OJ and HK interesect. At this point $I = TC$ and $RN = F + VN$. Solving for N^*, the break-even point, where

$$N^* = \frac{F}{R - V} \tag{18.2}$$

yields the level of output designated N^* in Figure 18.4.

The quantity $(R - V)$ is commonly referred as the *marginal contribution.* In this model the price, R, is constant, as is the variable cost, V. Since each item produced incurs the variable cost amount, V, and each sale generates a revenue amount, R, the difference represents a residual amount per unit that can be used to cover the fixed costs. Once the fixed costs are covered, the marginal contribu-

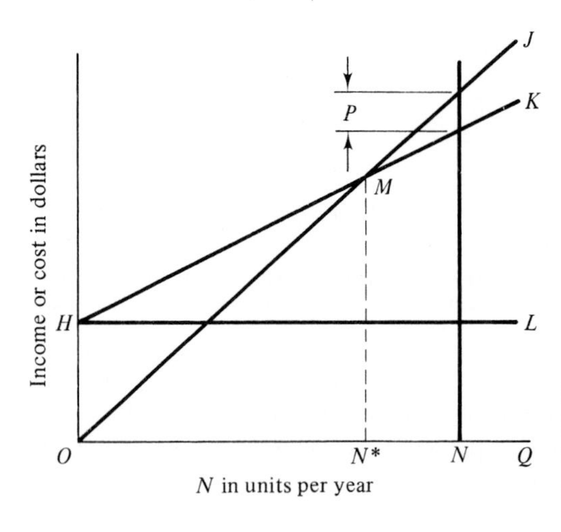

FIGURE 18.4. General graphic representation for income, cost, units of output, and profit.

tion represents the amount of profit resulting from the sale of each additional unit of production.

If $F/(R - V)$ is substituted for N in $I = RN$ or $TC = F + VN$, the ordinate at the break-even point may be found. This value, in terms of dollars of income or cost, will be

$$I = R\left(\frac{F}{R - V}\right) \tag{18.3}$$

and

$$TC = F + \frac{VF}{R - V}. \tag{18.4}$$

As an example, suppose it is desired to find the break-even point when $R = \$11$, $F = \$4{,}000$, and $V = \$5$. Then

$$N^* = \frac{F}{R - V} = \frac{\$4{,}000}{\$11 - \$5} = 667 \text{ units per year}$$

and

$$I = TC = RN = \$11(667) = \$7{,}337.$$

The marginal contribution is

$$R - V = -\$11 - \$6 = \$5 \text{ per unit.}$$

Since P is the annual profit, it is often desirable to have a relationship that expresses P as a function of the number of units made and sold, N. This relationship may be derived as follows:

$$\begin{aligned} P &= I - TC \\ &= RN - (F + VN) \\ &= (R - V)N - F. \end{aligned} \tag{18.5}$$

The profit amount P represents the difference between line OJ and line HK in Figure 18.4. The profit-volume function in Figure 18.5 is a means for displaying this relationship. Note that the slope of the profit line is equal to the marginal contribution, $R - V$. For example, to find the annual profit when $R = \$11$, $F = \$4{,}000$, $V = \$5$, and $N = 800$, calculate

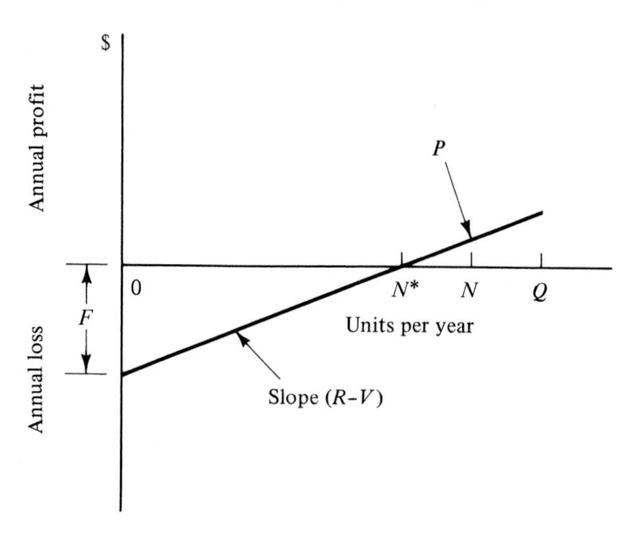

FIGURE 18.5. Profit-volume graph.

$$P = (R - V)N - F$$

$$= (\$11 - \$5)800 - \$4{,}000$$

$$= \$6(800) - \$4{,}000 = \$800.$$

All important elements contributing to the profitability of a production activity are represented on the profit-volume function. This simple graphical framework assists in the understanding of the relationship among price, fixed costs, variable costs, and profit. For example, a decrease in fixed cost by the amount $F - A$, with no other changes as shown in Figure 18.6 defines a profit line AB parallel to the original line FC. This reduction in fixed costs shifts the break-even point from N^* to N_1.

A change in either the price per unit, R, or the variable cost per unit, V, or both, affects only the marginal contribution amount, $R - V$. Any change in this amount with no change in the fixed cost could shift the line FC to the line FD as illustrated in Figure 18.6. A change in the amount $R - V$ alters the slope of FC, forcing the original profit line to rotate up or down while pivoting around the point F. In Figure 18.6, reducing the marginal contribution, $R - V$, decreases the slope of the profit line with the accompanying shift of the break-even point from N^* to N_2.

The profit-volume function allows for the analysis of simultaneous changes in price, fixed cost, and variable cost, so that the resulting change in profit and the shift in the break-even point can be easily assessed. Within this framework

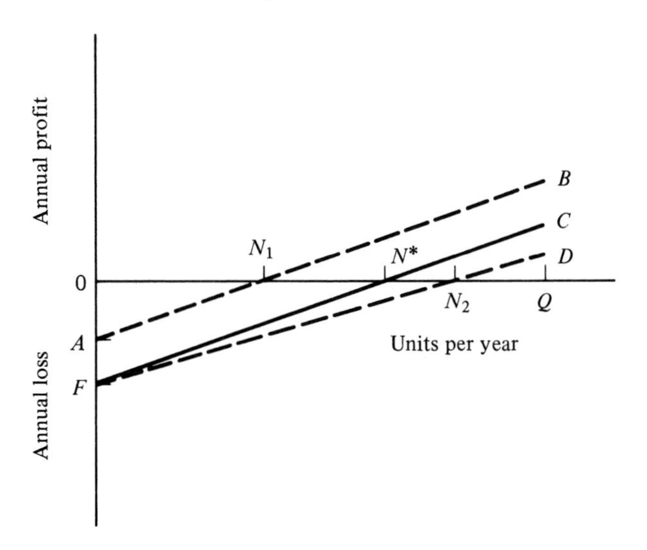

FIGURE 18.6. Changing parameters for the profit-volume graph.

the advantage to a firm for investing resources in marketing programs or cost-reduction programs can be evaluated. Also, the firm can study profit trade-offs connected with the change in the number of items sold due to the raising and lowering of prices.

Even though the analysis presented was based on the assumption that fixed cost, variable cost, and income are linear functions of the quantity made and sold, it is useful in evaluating the effect on profit of proposals for new operations not yet implemented and for which no data exist. Consider a proposed activity consisting of the manufacturing and marketing of a certain plastic template for which the sale price per unit is estimated to be \$3. The machine required in the operation will cost \$14,000 and will have an estimated life of 8 years. It is estimated that the cost of production including power, labor, space, and selling expense will be \$1.10 per unit sold. Material will cost \$0.95 per unit. An interest rate of 8% is considered necessary to justify the required investment. The costs associated with this activity are assumed to be

	Fixed Costs (Annual)	*Variable Costs (Per Unit)*
Capital recovery with return = \$14,000($\overset{A/P,8,8}{0.1740}$)	\$2,436	
Insurance and taxes	344	
Repairs and maintenance	220	\$0.05
Material		0.95
Labor, electricity, and space		1.10
Total	\$3,000	\$2.10

The difficulty of making a clear-cut separation between fixed and variable costs becomes apparent when attention is focused on the item for repair and maintenance in the above classification. In practice, it is very difficult to distinguish between repairs that are a result of deterioration that takes place with the passage of time and those that result from the wear and tear of use. However, in theory, the separation can be made as shown in this example and is in accord with fact, with the exception, perhaps, of the assumption that repairs from wear and tear will be in direct proportion to the number of units manufactured. To be in accord with actualities, capital recovery also undoubtedly should have been separated so that part would appear as variable cost.

In this example, $F = \$3{,}000$, $TC = \$3{,}000 + \$2.10N$, and $I = \$3N$. If F, TC, and I are plotted as N varies from 0 to 6,000 units, the result will be as shown in Figure 18.7.

The cost of producing templates will vary with the number made per year. The production cost per unit is given by $F/N + V$. If production cost per unit, variable cost per unit, and income per unit are plotted as N varies from 0 to 6,000, the results will be as shown in Figure 18.8. It is noted that the fixed cost per unit may be infinite. Thus, in determining average unit costs, fixed cost has little meaning unless the number of units to which it applies is known.

Income for most enterprises is directly proportional to the number of units sold. However, the income per unit may easily be exceeded by the sum of the

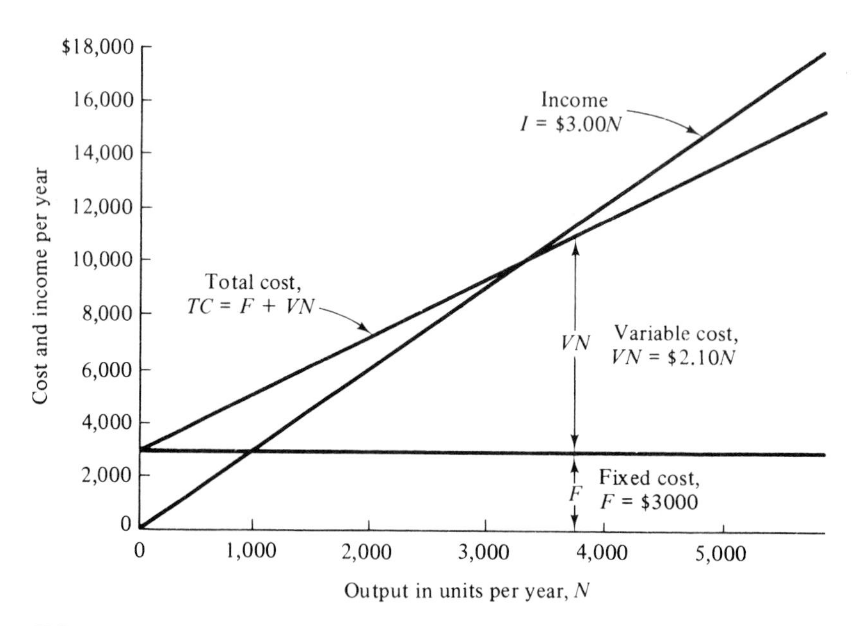

FIGURE 18.7. Fixed cost, variable cost, and income per year.

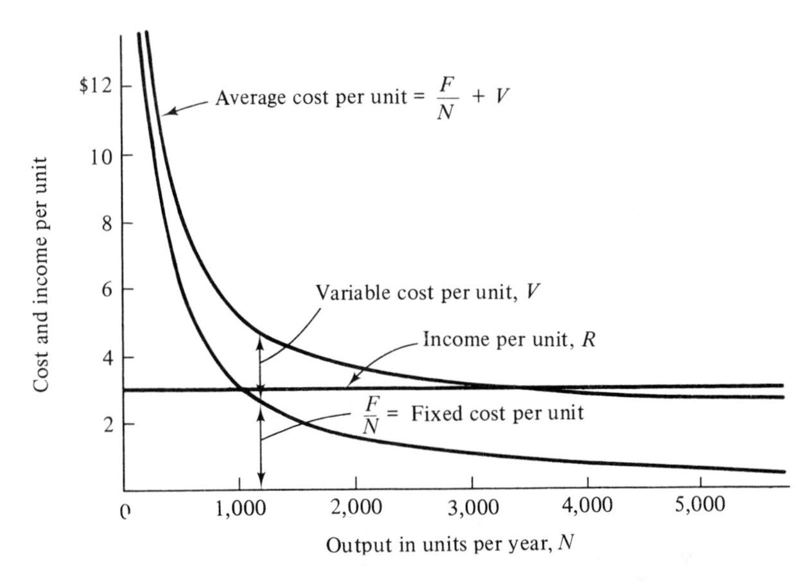

FIGURE 18.8. Fixed cost, variable cost, and income per unit.

fixed and the variable cost per unit for low production volumes. This is shown by comparing the average-cost-per-unit curve and the income-per-unit curve in Figure 18.8.

The *incremental cost* for this particular example is represented by the variable cost per unit, V. The variable cost per unit is constant. (If the variable cost is not constant, the incremental cost is the *change* in variable costs.) Thus, the incremental cost is constant and equal to \$2.10 per unit. This cost is shown in Figure 18.8 as the difference between the curves for the average cost per unit and the fixed cost per unit.

Although the break-even point occurs where the average-cost-per-unit (average-unit-cost) curve intersects the income-per-unit (price) curve, decisions about increasing or decreasing production output should be based on incremental costs. For this example, price per unit (\$3.00) always exceeds the incremental cost per unit (\$2.10). Therefore, the more units produced the better. However, suppose the present level of demand for the template is 5,000 units per year with the maximum capacity at 6,000 units per year. The costs associated with these levels of output are

1. Average unit cost for 5,000 units per year = (\$3,000/5,000 units) + \$2.10/unit = \$2.70 per unit
2. Average unit cost for 6,000 units per year = (\$3,000/6,000 units) + \$2.10/unit = \$2.60 per unit
3. Incremental cost = \$2.10 per unit.

Now suppose a buyer is willing to pay $2.50 per unit for an additional 1,000 units. Since fixed costs are the same for *any* level of production (unless the production facility is disposed of), it is only the differences in costs between the alternative production levels that should affect the decision. This difference is represented by the incremental cost per unit ($2.10). Since the price per unit ($2.50) exceeds this amount, the sale of the additional units is desirable. Note that this action ignores the fact that the average unit cost over the additional 1,000 units ranges from $2.70 per unit to $2.60 per unit. Since average costs include a fixed-cost component, they *do not* reflect the actual differences between the alternatives. Therefore, their use may result in misleading conclusions.

18.2.2 Nonlinear Break-Even Models

Table 18.1 gives cost data for a hypothetical plant which has a maximum capacity of 10 units of product per year and rates of production ranging from 0 to 10 units per year. These data and those presented in Tables 18.2 and 18.3 will be used as a framework for an analysis of relationships among production costs, marketing cost, income, and profit where the functions are nonlinear.

The annual total production cost of the hypothetical plant as its rate of production varies from 0 to 10 units per year is given in Column 2 of Table 18.1. It should be specifically noted that marketing costs are not embraced in these data.

Annual fixed cost is given as $200 per year in Column 3. The annual fixed cost in this example should be considered to be the annual cost of maintaining the plant in operating condition at a production rate of zero units per year. (This is the cost of the readiness to produce.) The capacity of the plant is 10 units per year.

The difference between the annual total production cost in Column 2 and the fixed production cost in Column 3 constitutes the annual variable cost and appears in Column 4. Average unit production, fixed cost, and variable cost appear in Columns 5, 6, and 7, respectively.

The incremental production costs per unit given in Column 8 were obtained by dividing the differences between successive values of annual total production cost as given in Column 2 by the corresponding difference between successive values of annual output given in Column 1. In this case the divisor will be equal to unity, but it should be understood that an increment of production of several units may be convenient in some analyses.

The values in all columns of Table 18.1, except those in Column 6, and the values of all columns of Table 18.2 and 18.3, have been plotted with respect to the annual output in number of units in Figures 18.9, 18.10, and 18.11 and have been keyed for identification. Consider curve *A* in Figure 18.9. Total-production-cost curves of actual operations take a great variety of forms. Ordinarily, however, a producing unit will produce at minimum average cost at a

TABLE 18.1.
RELATIONSHIP OF PRODUCTION COST, NET INCOME FROM SALES, PROFIT, AND THE NUMBER OF UNITS MADE AND SOLD PER YEAR

(1) Annual Output, Number of Units	(2) Annual Total Production Cost, *A*, Fig. 18.9	(3) Annual Fixed Production Cost, *B*, Fig. 18.9	(4) Annual Variable Production Cost, *C*, Fig. 18.9	(5) Average Production Cost per Unit, *D*, Fig. 18.9	(6) Average Fixed Cost per Unit	(7) Average Variable Cost per Unit, *E*, Fig. 18.9	(8) Incremental Production Cost per Unit, *F*, Fig. 18.9
0	$200	$200	$ 0	$ ∞	$ ∞	$ 0	—
1	300	200	100	300.00	200.00	100.00	$100
2	381	200	181	190.50	100.00	90.50	81
3	450	200	250	150.00	66.67	83.30	69
4	511	200	311	127.75	50.00	77.75	61
5	568	200	368	113.60	40.00	73.60	57
6	623	200	423	103.83	33.33	70.50	55
7	679	200	479	97.00	28.57	68.48	56
8	740	200	540	92.50	25.00	67.50	61
9	814	200	614	90.44	22.22	68.22	74
10	924	200	724	92.40	20.00	72.40	110

level of production between zero output and its maximum rate of output. Thus, the average unit cost of production will decrease with an increase in the rate of production from zero until a minimum average cost of production is reached. Then the average cost per unit will rise.

The average production cost per unit is given by curve *D* in Figure 18.9. It reaches a minimum at a rate of nine units per year. The average variable production cost is given in curve *E*; its minimum occurs at eight units per year. Curves *D* and *E* should be considered in relation to Column 6 of Table 18.1. Fixed cost per unit is inversely proportional to the number of units per year. Reduction of fixed cost per unit is an important factor leading toward lower average cost with increases in rates of production.

Incremental production costs are given in curve *F*; their minimum value is reached at six units per year. The form of this curve shows that the incremental

TABLE 18.2.
RELATIONSHIP OF GROSS INCOME FROM SALES, NET INCOME FROM SALES, MARKETING COST, AND NUMBER OF UNITS SOLD PER YEAR

(1) Annual Output, Number of Units	(2) Annual Gross Income from Sales, *J*, Fig. 18.10	(3) Annual Marketing Cost, *K*, Fig. 18.10	(4) Annual Net Income from Sales, *G*, Fig. 18.10	(5) Incremental Marketing Cost per Unit, *L*, Fig. 18.10	(6) Incremental Annual Net Income from Sales per Unit, *H*, Fig. 18.10
0	$ 0	$ 0	$ 0	—	—
1	140	38	102	$ 38	$102
2	280	72	208	34	106
3	420	104	316	32	108
4	560	136	424	32	108
5	700	170	530	34	106
6	840	211	629	41	99
7	980	261	719	50	90
8	1,120	325	795	64	76
9	1,260	405	855	80	60
10	1,400	505	895	100	40

TABLE 18.3.
RELATIONSHIP OF GROSS INCOME, PRODUCTION COST, MARKETING COST, PROFIT, AND NUMBER OF UNITS MADE AND SOLD PER YEAR

Annual Output, Number of Units	Annual Total Production Cost, *A*, Fig. 18.11	Annual Marketing Cost, *K*, Fig 18.11	Annual Total of Production and Marketing Cost, *M*, Fig. 18.11	Annual Gross Income from Sales, *J*, Fig. 18.11	Annual Profit, *N*, Figs. 18.9 and 18.11	Incremental Total Production and Marketing Costs, *O*, Fig 18.11
0	$200	$ 0	$ 200	$ 0	−$200	—
1	300	38	338	140	−198	$138
2	381	72	453	280	−173	115
3	450	104	554	420	−134	101
4	511	136	647	560	−87	93
5	568	170	738	700	−38	91
6	623	211	834	840	6	96
7	679	261	940	980	40	106
8	740	325	1,065	1,120	55	125
9	814	405	1,219	1,260	41	154
10	924	505	1,429	1,400	−29	210

cost of production per unit declines until six units per year are produced. Then it rises.

Observe that the incremental-cost curve F intersects the average-unit-production-cost curve D and the average-variable-production-cost curve E at their minimum points. This phenomenon is the consequence of each additional unit being produced at an incremental cost less than the average cost, lowering that average cost. Once the incremental cost for the next unit produced exceeds the average cost, the average cost will increase. The incremental-cost curve F is a measure of the slope of curve A. It may be noted that the slope of curve A decreases until six units per year is reached, and then it increases. The same could have been said of a variable-cost curve for C if one had been plotted.

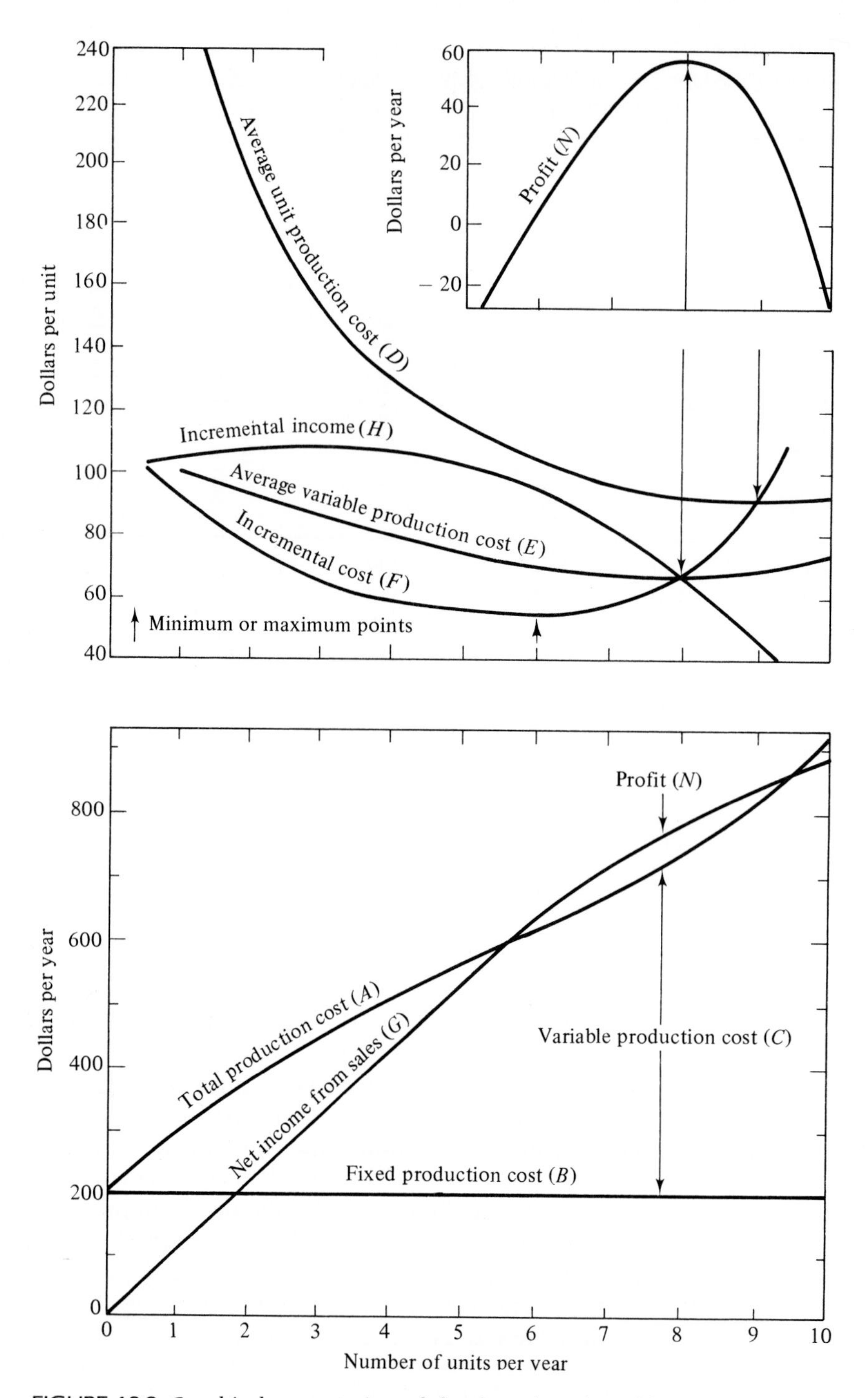

FIGURE 18.9. Graphical presentation of the data given in Table 18.1.

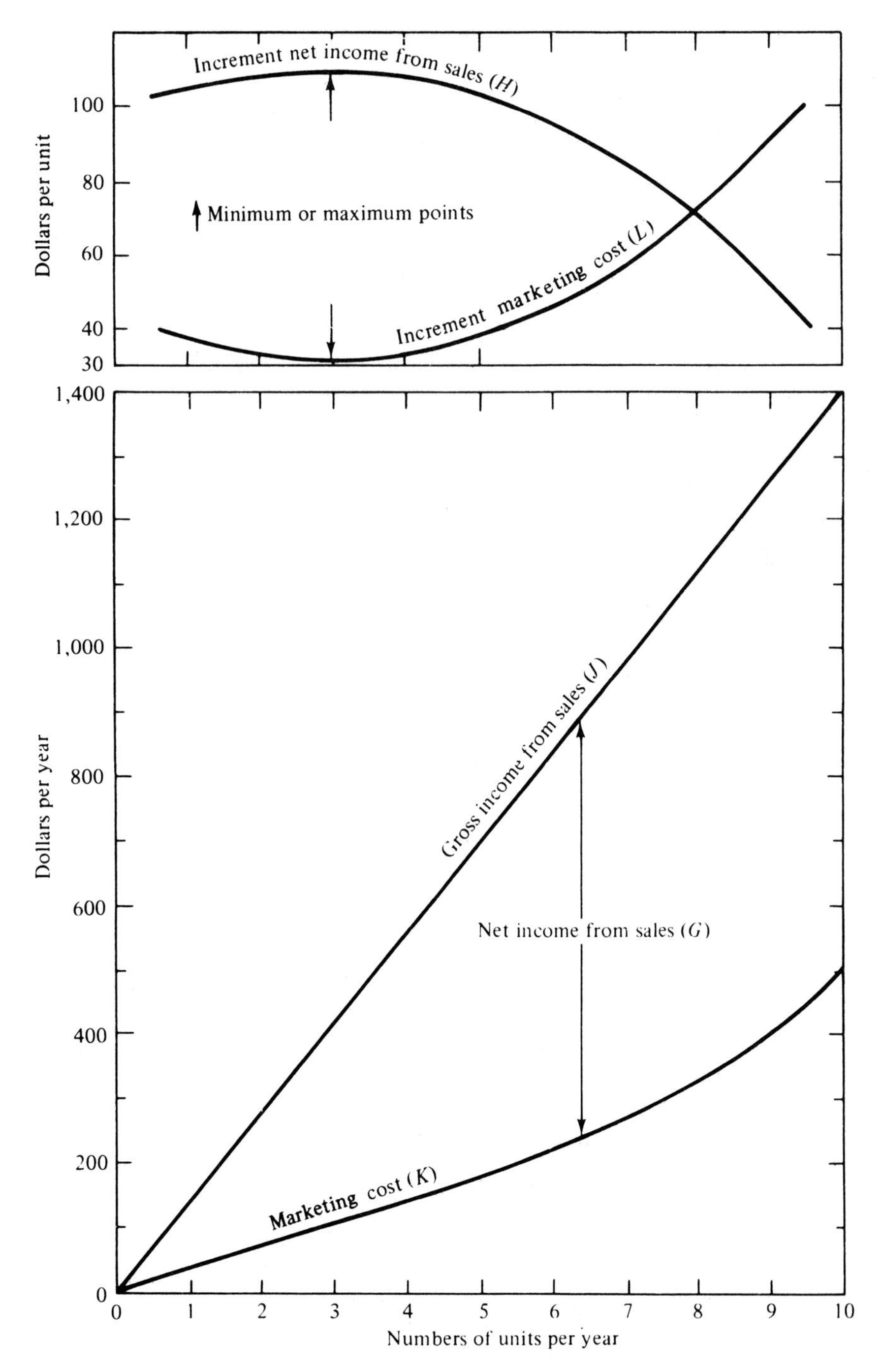

FIGURE 18.10. Graphical presentation of the data given in Table 18.2.

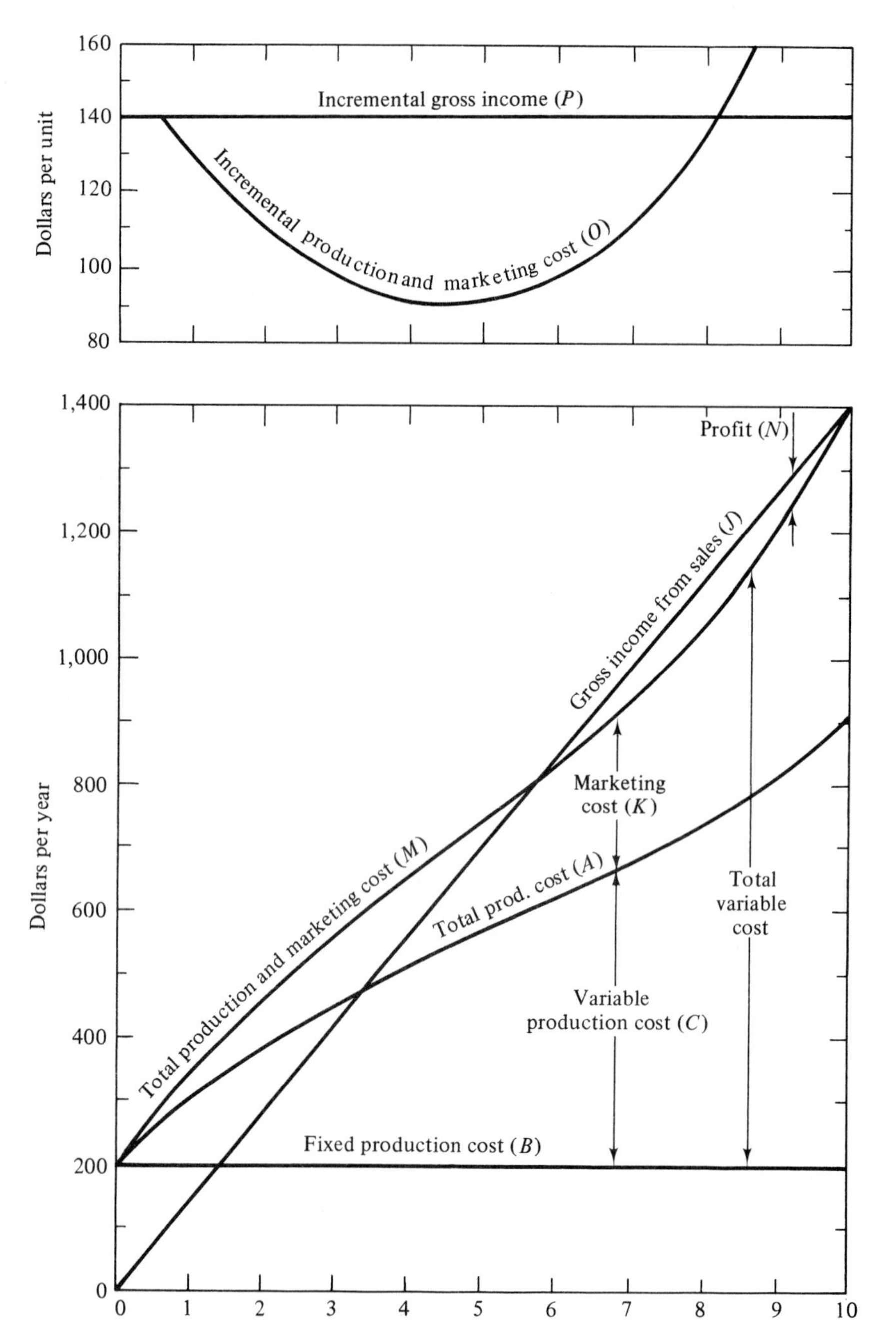

FIGURE 18.11. Graphical presentation of the data given in Table 18.3.

18.2.3 Pattern of Income and Cost of Marketing

Further consideration will be given to the curves of Figure 18.9 after examining income from sales of product and the cost of marketing. For simplicity, the cost of marketing will be considered to be a summation of an enterprise's expenditures to influence the sales of its products and services. Such items as advertising, sales administration, sales salaries, and expenditures for packaging and decoration of products done primarily for sales appeal will be included in the cost of marketing. Gross annual income from sales is the total income received from customers as payment for products. In columns 2 and 3 of Table 18.2 annual gross income from sales and the annual cost of marketing are given for various rates of product sales of the firm for which cost-of-production data were given in Table 18.1 and Figure 18.9.

The difference between the gross annual income from sales and the annual cost of marketing is equal to the annual net income from sales given in Column 4. Incremental marketing costs per unit are given in Column 5, and incremental net income from sales per unit appears in Column 6. The values of each of these columns have been plotted with respect to annual output in Figure 18.10. The gross annual income from sales, curve *J* in Figure 18.10, is a straight line and is typical of situations in which products are sold at a fixed price. The annual incremental marketing cost, curve *L*, first falls slightly until a minimum is reached, and then it rises. This is typical of situations in which sales effort is relatively inefficient at low levels and in which sales resistance increases with increased number of units sold.

A representation of the net income from sales is given as *G* in Figure 18.10. The positive difference between net income from sales and the total cost of production, curve *A* in Figure 18.9, represents profit. Profit is shown as *N* in Figure 18.9. Note that the maximum profit, or minimum loss if no profit is made, occurs at a rate of production at which the incremental-income curve *H* intersects the incremental-cost curve *F* (at eight units per year). From Column 8, Table 18.1, and Column 6, Table 18.2, it may be observed that the incremental production cost and incremental net income for a ninth unit are $74 and $60, respectively. Thus, a loss of $14 would be sustained if the activity were increased to nine units.

When the profit motive governs, there is no point in producing beyond the point where the incremental cost of the next unit will exceed the incremental income from it.

If the sales effort in the example is resulting in sales of nine or ten units per year, a number of steps might be taken. The sales effort could be reduced, causing sales to drop. The price could be increased, causing sales to decrease and the income per unit to increase. The plant could be expanded, or other changes could be made to alter the pattern of production costs. Consideration of any of these steps would require a new analysis embracing the altered factors.

18.2.4 Consolidation of Production and Marketing Cost

It is common practice to consolidate production and marketing cost to aid in the analysis of operations. To illustrate this practice, an analysis of the prior example will be made. The method is illustrated in Table 18.3, where values have been plotted as several curves in Figure 18.11.

Profit in this case will be equal to the difference between gross income J and total production and marketing cost M. The point of maximum profit will occur when the incremental gross income curve P is intersected by a rising incremental production and marketing cost curve O.

18.3 OPERATION AND PRODUCTION DECISIONS

Because there are unforeseen variations in the demand for a product and/or its cost of production, the firm must continually readjust its mode of operation. Although these variations may affect only one production unit or the entire production facility, the analysis methods are generally applicable to any production activity, whether it be large or small. An electric motor with its inputs and outputs can be analyzed in the same manner as a highly automated manufacturing plant. In both cases, there are prices to be charged for the output and costs associated with the inputs. It is the balance of these elements that ultimately determines the profitability of the activity.

There are usually a variety of actions that can be undertaken by management that will adjust prices and costs, the goal being the proper management of the elements to improve profits. This goal may include the minimization of losses. This section presents a framework for representing the interactions among price, fixed cost, variable cost, incremental cost, and profit. By using this framework, analyses can be made that will provide great assistance in deciding the appropriate management actions.

18.3.1 Break-Even for Profit Maximization

For every productive activity, it is vital that the cost characteristics of that activity be well understood. It is observed from the example in Figure 18.8 that the average unit cost decreases with each additional unit of output. The rate of decrease of this amount is important in determining cost competitiveness at various levels of output. In this example, a fixed price of \$3.00 per unit breaks even at an average unit cost of \$3.00 or at

$$N^* = \frac{\$3{,}000}{\$3.00 - \$2.10} = 3{,}333 \text{ units}$$

while a fixed price of \$4.00 per unit breaks even at

$$N^* = \frac{\$3{,}000}{\$4.00 - \$2.10} = 1{,}579 \text{ units.}$$

The average-unit-cost curve defines the break-even point for a fixed price equal to any point on the curve.

Also of interest is the incremental cost over the range of output. For this example the incremental cost is a constant \$2.10 per unit. As long as the incremental income received for each additional unit exceeds the incremental cost, additional production should be undertaken. Since the price of \$3.00 is constant for each unit, the strategy is to produce and sell as many units as possible. The profit-volume graph of Figure 18.12 confirms this. Profit is maximized when the operation is at the full capacity of 6,000 units per year. For the example in Figure 18.9 profits are increased, as long as the incremental cost of an item, curve *F*, is less than its incremental income, curve *H*. Thus, profit is maximized at 8 units per year, as indicated by curve *N*.

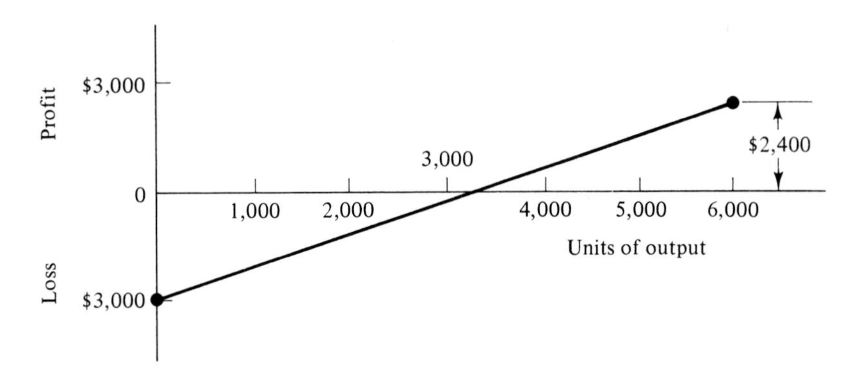

FIGURE 18.12. Profit-volume graph for Figure 18.7.

18.3.2 Making Pricing Decisions

There are numerous situations where it may be appropriate to modify prices for various levels of output. These situations can arise when there are either temporary reductions or increases in the demand for output. When there are reductions, the goal should be to increase demand so that the production operation can remain relatively stable. In other cases, a production activity operating at 100% capacity may be required to increase output through overtime and extra shifts. The prices set should help compensate for these variations in demand and yet assure a profitable outcome.

An example of the use of differential pricing to accomplish these goals follows: a firm produces a single product in a manufacturing facility having its full capacity at 100,000 items per year. Each unit of output requires one unit of material. The costs associated with this product are

Direct-Material Costs

1st 20,000 units	\$4/item
2nd 50,000 units	\$6/item
3rd 30,000 or more units	\$9/item

Direct-Labor Costs

Normal	\$8/item
Overtime	\$12/item

Overhead or Fixed Costs

At 100% capacity	\$1/item

If the facility is worked overtime using an extra shift, an additional 20,000 units can be produced. By examining the incremental costs attributed to this production facility, insights regarding product pricing are possible. Figure 18.13 presents the incremental costs for the various operating levels.

For a constant price of \$16 per item, the break-even point occurs at an output level of 30,000 items per year where profit is zero as

$$\text{Profit} = \$0 = (\$16 - \$4 - \$8)\,(20{,}000 \text{ items}) + (\$16 - \$6 - \$8)\,(10{,}000 \text{ items}) - \$10{,}000.$$

Profit for this price will be maximized by producing and selling items as long as the incremental price, \$16 per item, is greater than the incremental cost. This condition exists for the first 70,000 items produced, as shown in Figure 18.13.

If the production facility is producing 60,000 items per year and a new cus-

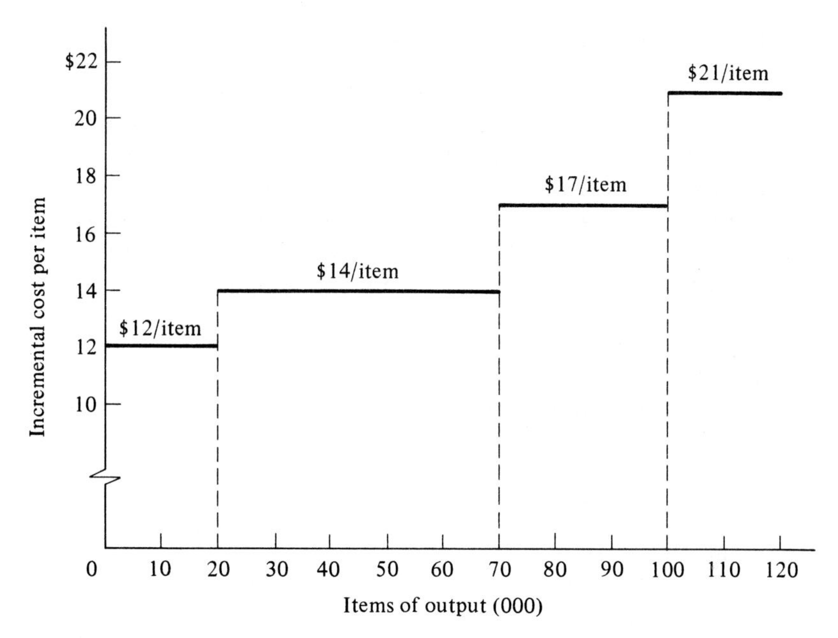

FIGURE 18.13. Incremental costs for a manufacturing facility.

tomer is willing to buy 30,000 additional items at $16.50, should the firm sell? Since the items should never be sold for less than their incremental costs, the answer is that only 10,000 more items should be sold for that price. Those are the items produced at an incremental cost of $14 per item between the output levels of 60,000 and 70,000 items. If the price offered is over $17 per item, it is profitable to produce an additional 40,000 items. Certainly no overtime production will be justified, unless a price exceeding $21 per item could be received for those items produced on the extra shift.

Changes in prices will often affect the demand for an item. Suppose the firm is charging a fixed price of $16 per item and it is presently producing at 60% of capacity (60,000 items per year). The marketing department believes reducing the price to $15 will increase sales by 16%. Marketing also estimates that an increase in the price by $1 will cut the demand in half. Examination of the incremental costs indicates that for either pricing change the incremental income will exceed the incremental cost for the range of output being considered (30,000 items to 69,600 items). No pricing strategy should be adopted that indicates otherwise. Calculating the profit for each option gives

$1 decrease in price per item for a 16% increase in output

Output = (1.16)(60,000 items) = 69,600 items

$$\text{Profit} = (\$15 - \$4 - \$8)\ (20{,}000\text{ items})$$

$$+ (\$15 - \$6 - \$8)\ (49{,}600\text{ items})$$

$$- \$100{,}000 = \$9{,}600$$

\$1 increase in price per item for a 50% decrease in output

$$\text{Output} = (0.50)(60{,}000\text{ items}) = 30{,}000\text{ items}$$

$$\text{Profit} = (\$17 - \$4 - \$8)\ (20{,}000\text{ items})$$

$$+ (\$17 - \$6 - \$8)\ (10{,}000\text{ items})$$

$$- \$100{,}000 = \$30{,}000.$$

In this example a price increase will prove to be more profitable, even though the number of items being sold is expected to decrease. This conclusion depends on the accuracy of the marketing department's information regarding the elasticity of demand for this product.

18.3.3 Stopping Production

In certain situations the demand for, or the price of the output of, a production activity may be so low that substantial losses are being recorded. At some point, the temporary closing down of production becomes a viable alternative to be considered. Suppose a manufacturing cell has the following cost characteristics:

100% capacity	10,000 units per year
Fixed cost (operating cell)	\$100,000
Fixed cost (cell closed)	\$75,000
Variable cost	\$40 per unit
Price per unit	\$60 per unit

If the cell is temporarily closed, certain fixed costs amounting to \$25,000 can be eliminated. These fixed costs include some utilities that can be shut off and a supervisory person who would be temporarily released. Presently the line is operating at 20% of capacity.

Analyzing the profit that will be realized for each option provides the basis for determining the appropriate course of action.

$$\text{Profit (20\% capacity)} = \$60(2{,}000\text{ units}) - \$100{,}000$$

$$- \$40(2{,}000\text{ units}) = -\$60{,}000$$

$$\text{Profit (closed temporarily)} = -\$75{,}000$$

From these results, the preferred choice is to minimize losses by operating at

20% of capacity. Operating the line at this reduced level results in a $15,000-per-year saving when compared to closing down. Because the incremental income of $60 per unit is greater than the incremental cost of $40 per unit, $40,000 extra can be generated at 2,000 units of output to cover a portion of the $100,000 fixed costs.

If the fixed costs had been the same whether the line is running or not running, the line should be operated at *any* level of output. In this case, the fixed costs are the same for both options, and the only difference between them is the extra income generated when the line is producing. This extra income on a per-unit basis is the marginal contribution, $R - V$.

When the option being considered is the complete disposal of the production activity rather than its temporary closure, the fixed costs related to operations are then zero. (This does not include the capital costs associated with the decision to completely eliminate the activity. A more detailed discussion of retirement decisions is given in Section 8.6.) For this example, if permanent disposal is an option and it appears that the line would continue to produce at 20% capacity for the foreseeable future, the line should be eliminated. This action would save $60,000 annually.

18.3.4 Operating Multiple Production Units

Some production operations consist of multiple production facilities or units that all produce the same product. An electric generating system may have a combination of nuclear, fossil, and hydro generating units. Other examples include manufacturing facilities that contain multiple and distinct production lines and pumping operations having engines or motors that can be used in combination. Since all the output of each unit is identical when viewed by the system, a common problem is how to most economically assign the various levels of output to each of the separate units within the system. The task is to determine for each level of the total system output the various operating levels of the individual units.

Incremental costs are the basis on which these type of operating decisions are made. Assuming that the price for the output is sufficient so it does not affect the assignment of the output levels, the objective is to minimize the total cost of production. In many situations incremental costs can facilitate the determination of how this minimization is to be achieved. There are three possible situations that require different decision rules for the assignment of multiple-unit output. These situations are described in the following discussion.

When incremental costs are *constant* for different output levels, the principle is to

Assign the unit with the lowest incremental cost and load it to full capacity before bringing the next productive unit into operation. Continue this procedure until the system output is achieved.

Figure 18.14 indicates how four separate production units, *A, B, C,* and *D,* have been placed in increasing order of constant incremental costs. By assiging output at the lowest possible incremental cost until the total desired output is achieved, the most efficient assignment of output to the individual units will be achieved.

When incremental costs are *always increasing* for different output levels, the principle is to

Assign the initial unit of output to the production unit having the smallest incremental cost. Then assign each additional unit of output to the production unit having the lowest incremental cost, given its present level of output.

This situation is depicted in Figure 18.15 for three production units, *A, B,* and *C.* In this example, Unit *A* initially produces all the output until the lowest incremental cost of Unit *C* is equal to that of Unit *A*. The production of each additional unit of output is then switched between unit *A* and *C* until the incremental costs of *A* and *C* are equal to the lowest incremental cost of Unit *B*. (This situation occurs between incremental cost 1 and 2.) The production of each additional unit of output is then switched among units *A, B,* and *C* (between incremental cost 2 and 3) so that each unit is produced at its lowest possible incremental cost. For incremental costs between 3 and 4, each additional unit of output switches between *A* and *B*. This switching process dictates that at any total-output level for the system, the marginal costs for the individual units assigned will essentially be equal.

Suppose there exists an electric generating system with a maximum total capacity of 1,400 megawatts (1 megawatt = 1,000 kilowatts) consisting of two separate generating plants. Plant *A* is a 600-MW unit and Plant *B* produces

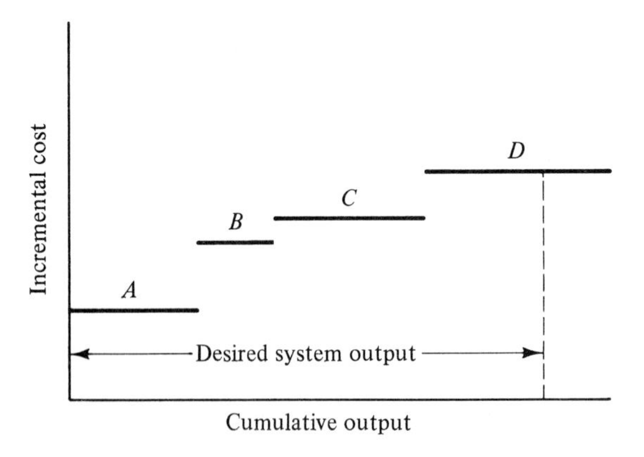

FIGURE 18.14. Constant incremental costs.

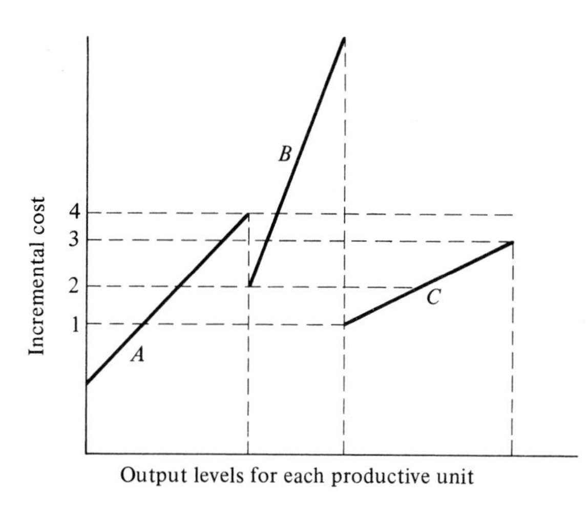

FIGURE 18.15. Increasing incremental costs for three production activities.

power at 800 MW at 100% capacity. The incremental-cost curves for each of these plants is linear and increasing. The following equations represent the incremental costs per megawatt hour (MWh):

$$IC_A = \$20 + (\$10/600 \text{ MWh}) (\text{Number of MWh produced by Plant } A)$$

$$IC_B = \$25 + (\$20/800 \text{ MWh}) (\text{Number of MWh produced by Plant } B)$$

Table 18.4 gives the incremental-cost values for particular levels of output for each plant. Because the output from each of these plants can range at any level between the values shown in Table 18.4, it is assumed that the incremental costs are continuous over all levels of output as defined by the equations.

Now suppose it is necessary to determine how these plants should be operated for various levels of required output for the entire system. Examination of Table 18.4 indicates the following schedule:

System Output Required	*Plants Assigned*
0 – 300 MWh	*A* (lowest possible incremental cost)
300 – 800 MWh	*A* and *B* (keeping incremental costs equal)
800 – 1,400 MWh	*A* (maximum capacity) and *B* (lowest possible incremental cost)

If 500 MWh is to be produced by this electric generating system, the most economic method is to produce 420 MWh at Plant *A* and 80 MWh at Plant *B*. This result is determined by the fact that for the range of outputs for each plant

TABLE 18.4.
INCREASING INCREMENTAL COSTS FOR TWO GENERATING PLANTS

	Plant *A*		Plant *B*	
Capacity	Output	Incremental Cost	Output	Incremental Cost
0	0	$20.00/MWh	0	$25.00/MWh
25%	150 MWh	22.50	200 MWh	30.00
50%	300	25.00	400	35.00
75%	450	27.50	600	40.00
100%	600	30.00	800	45.00

where the incremental costs can be the same, these costs must be equal. Let

$$x_1 = \text{output of Plant } A;$$

$$x_2 = \text{output of Plant } B.$$

Therefore,

$$\$20 + (\$10/600\ \text{MWh})x_1 = \$25 + (\$20/800\ \text{MWh})x_2$$

and

$$x_1 + x_2 = 500\ \text{MWh}.$$

Solving the above equation gives $x_1 = 420$ MWh and $x_2 = 80$ MWh.

If there are more than two units in the system, trial-and-error calculations are required. Beginning with the lowest incremental cost, continue to increase this cost until the desired total output is reached. For each incremental-cost value tried, the total output is the sum of the individual outputs determined by the value being tested.

When incremental costs are not constant or not always increasing for different output levels, the principle is to

Compute the total incremental cost for a given level of output from the total production activity (the entire system).

Since the total incremental cost is defined as the total variable costs divided by the units of output, the effect of this decision rule is to select the combination that minimizes total variable costs. Fixed costs continue to be ignored.

Consider a 100-horsepower electric pump motor that has an efficiency curve similar to the one in Figure 18.2 except that the maximum efficiency is reached at 50% of capacity. Table 18.5 presents the input and output data associated with this type of motor. Note that the incremental costs are not constant or increasing with increased output.

It is assumed that the energy output has the same value or price per horsepower-hour (hp-h) and that the total energy required by the activity using this type motor is 300,000 hp-hr per year. This requirement is based on a pumping operation that requires a fixed amount of wastewater to be pumped to a holding pond over a year's time.

The problem is to determine whether a single motor should be operated at full capacity or other identical motors available for this activity should be operated at partial capacity. Suppose the only options in this case include operating a single motor at 100%, two motors at 50% and 50% or 75% and 25%, or 4 motors at 25% each.

The total incremental costs for the combination of motors and operating capacities to produce a total output of 300,000 hp-h are calculated as follows:

1 motor (100% capacity)

$$\text{Total incremental cost} = \frac{\$34{,}286}{300{,}000 \text{ hp-h}} = \$0.1143 \text{ per hp-hr}$$

2 motors (1 motor at 25% capacity and 1 motor at 75% capacity)

$$\text{Total incremental cost} = \frac{\$10{,}000 + \$22{,}500}{300{,}000 \text{ hp-h}} = \$0.1083 \text{ per hp-hr}$$

TABLE 18.5.
INPUT AND OUTPUT COSTS FOR 100 HP ELECTRIC MOTOR

Capacity	Annual Fixed Cost	Annual Variable Cost	Total Cost (Input)	Energy (Output)	Average Incremental Cost
0%	$500	$ 0	$ 500	0 hp-hr	
25%	500	10,000	10,500	75,000	$0.1333/hp-hr
50%	500	14,118	14,618	150,000	0.0549
75%	500	22,500	23,000	225,000	0.1118
100%	500	34,286	34,786	300,000	0.1571

2 motors (1 motor at 50% capacity and 1 motor at 50% capacity)

$$\text{Total incremental cost} = \frac{\$14{,}118 + \$14{,}118}{300{,}000 \text{ hp-h}} = \$0.0941 \text{ per hp-hr}$$

4 motors (each motor at 25% capacity)

$$\text{Total incremental cost} = \frac{(4)(\$10{,}000)}{300{,}000 \text{ hp-h}} = \$0.1333 \text{ per hp-hr}$$

To provide the required output of 300,000 hp-h, two motors should be used at 50% of capacity. This result is confirmed by the stated fact that the maximum efficiency of this type of motor occurred at 50% of capacity. A similar analysis for total output levels of 75,000 hp-h, 150,000 hp-h, and 225,000 hp-h indicates that one motor at 25%, one motor at 50%, and one motor at 75%, respectively, will provide the desired output at least cost.

Note that in all of these analyses, fixed costs have been ignored. The methods presented are based on incremental costs, which by definition omit the fixed-cost component. Operating decisions usually do not affect fixed costs, and so those costs are identical for the various options under consideration. When the differences among the alternatives are calculated, the fixed costs cancel each other, eliminating the need to consider them.

Key Points

- By examining the relationships among the cost of inputs, the price obtained for the output, and the levels of output, it is possible to determine optimum operating conditions.
- Any production activity, whether it represents a single machine or a complete manufacturing facility, can be modeled in terms of its cost and revenue-generating characteristics.
- For every production activity there is a potential maximum level of output that defines its capacity; its efficiency then derives from the output as a percentage of the input.
- Fixed and variable costs are the primary considerations in production operations, and the ability to distinguish these cost types is essential in economy studies of production systems.
- Break-even analysis is a very useful tool in the analysis of production operations and may be applied under the assumption of linearity or nonlinearity.

- Economy of scale in production operations occurs when fixed cost is amortized over a larger and larger number of output units.
- The firm must continually readjust its mode of operation owing to unforeseen variations in demand, changes in the price obtained, and increases or decreases in the cost of input factors.

PROBLEMS

1. The fixed cost of a machine (depreciation, interest, space charges, maintenance, indirect labor, supervision, insurance, and taxes) is f dollars per year. The variable cost of operating the machine (power, supplies, and similar items, excluding direct labor) is V dollars per hour of operation. N is the number of hours the machine is operated per year, TC the annual total cost of operating the machine, TC_h the hourly cost of operating the machine, t the time in hours to process one unit of product, M the machine cost of processing a unit of product, and n the number of units of product processed per year. In terms of these symbols, write expressions for (a) TC, (b) TC_h, (c) M.

2. In Problem 1, $F = \$600$ per year, $t = 0.2$ hour, $V = \$0.50$ per hour, and n varies from 0 to 10,000 in increments of 1,000.

a. Plot values of M as a function of n.

b. Write the expression for TC_u, the total cost of direct labor and machine cost per unit, using the symbols given in Problem 1 and letting W equal the hourly cost of direct labor.

3. A machine that is used in an industrial process to make a certain product has an initial cost of \$50,000 and it is expected to have a zero salvage value. The maximum capacity of this machine is 7,000 units per year, and its life is estimated to be 4 years. The unit variable cost (raw material, energy, etc.) is \$12.40, the total fixed annual cost of supervision and taxes is \$20,000. The company uses 15% interest compounded continuously. If the machine works at (1) 50%, (2) 60%, (3) 70%, (4) 85% of maximum capacity

a. Calculate the average unit cost and annual total cost.

b. Calculate the incremental production cost.

4. A semiautomatic machine that is used for a certain joining process costs \$10,000. The machine has a life of 5 years and a salvage value of \$1,000. Maintenance, taxes, interest, and other fixed costs amount to \$500 per year. The cost of power and supplies is \$5 per hour, and the operator receives \$7.60 per hour. The cycle time per unit of product is 1.2 minutes. If interest is 17%, calculate the cost per unit of processing the product if (a) 200, Answer: \$0.587 (b) 600, (c) 1,200, (d) 2,500 units of product are made per week. Assume the plant operates 52 weeks per year.

5. A manufacturer uses a new turret lathe to produce motor shafts having a selling price of $27.50 per unit. He estimates the life of the new machine at 5 years and its salvage value at 15% of its installed cost of $38,000. The production time per 10 units is 45 minutes, and the machine operator is paid $15 per hour. The raw material and energy cost is $11.86 per unit if the number of units produced monthly is 100 or less, and this cost increases at the rate of $0.09 per unit as the number of items produced increases beyond 100. Interest is to be taken at 12% per year compounded monthly. Determine the average unit cost and the monthly profit if the number of units made per month is (a) 50, (b) 105, (d) 150, (e) 175. Calculate the amount of output which produces the break-even point.

6. A manufacturing plant produces a commodity having a selling-price function $R = (\$500 - \$0.0025x)$ per unit, where x is the number of units manufactured. This price function is the result of a policy of lowering the price in order to achieve a higher plant utilization. Total fixed costs are considered to be $26,335 per year, and the variable cost of production is $V = (\$440 + \$0.005x)$ per unit. The plant is designed to manufacture 5,000 units per year. Determine the level of output that produces the maximum annual profit, the minimum average unit cost, and the break-even point. On a graph, show the curves for average variable production cost per unit, incremental production cost per unit, and incremental annual profit from sales per unit.

7. An analysis of the current operations of a firm results in the conclusion that the production cost is $1,500,000 + $20,000$x$, where x is the number of units produced. The gross income will be $300,000$x^{0.6}$. Plot the production cost, the gross income from sales, and profit for production levels ranging from 0 to 800 units per year. Determine the level of production for maximum profit mathematically. Answer: 243 units

8. An analysis of an enterprise and the market in which its product is sold results in the following data.

Level of Operation Units of Product	*Net Income from Sales*	*Production Cost*	*Level of Operation Units of Product*	*Net Income from Sales*	*Production Cost*
0	$ 0	$2,180	70	$ 8,600	$ 8,340
10	1,600	3,520	80	9,360	8,960
20	3,080	4,600	90	10,000	9,600
30	4,400	5,500	100	10,540	10,260
40	5,600	6,300	110	10,960	10,960
50	6,700	7,020	120	11,260	11,700
60	7,700	7,700			

a. Determine the profit for each level of operation.

b. On a graph, show, for each level of operation, the curves for net income from sales, production cost, profit, average incremental production cost per unit, and average incremental net profit from sales per unit.

9. A manufacturing concern estimates that its expenses per year in thousands of dollars for different levels of operation would be as follows.

Output in units of product (millions)	0	10	20	30	40	50
Administrative and sales	$4,900	$ 5,700	$ 6,200	$ 6,700	$ 7,100	$ 7,500
Direct labor and materials	0	2,500	4,600	6,400	8,100	9,800
Overhead expense	4,120	4,190	4,270	4,350	4,400	4,550
Total	$9,020	$12,390	$15,070	$17,450	$19,640	$21,850

a. What is the incremental cost of maintaining the plant ready to operate (the incremental cost of making zero units of product)?

b. What is the average incremental cost per unit of manufacturing the first increment of 10 million units of product per year? Answer: $0.337

c. What is the average incremental cost per unit of manufacturing the increment of 31 to 40 million units per year?

d. What is the average total cost per unit when manufacturing at the rate of 30 million units per year? Answer: $0.581

e. At a time when the rate of manufacture is 20 million units per year, a salesman reports that he can sell 10 million additional units at $0.31 per unit without disturbing the market in which the company sells. Would it be profitable for the company to undertake the production of the additional units?

10. The product of an enterprise has a fixed selling price of $81. An analysis of production and sales costs and the market in which the product is sold has produced the following results.

Level of Operation *Units of Product*	*Total Production and Selling Cost, Dollars*	*Level of Operation* *Units of Product*	*Total Production and Selling Cost, Dollars*
0	$14,300	600	$36,100
100	19,000	700	39,700
200	22,500	800	48,200
300	25,700	900	56,700
400	28,300	1,000	66,500
500	32,600		

a. Determine the profit for each level of operation.

b. Plot production and selling cost, income from sales, profit, average incremental production and selling cost per unit, and average incremental profit per unit for each level of operation.

11. A plant can produce a maximum of 2,500 lamps per year, and the price of the lamp is a function of the number of units sold as follows.

Level of Sales	*Price for Lamp*
first 1,250 units	$100/unit
second 1,250 units	$ 90/unit

The annual operation costs are given below.

Utilization of Capacity	*20%*	*40%*	*60%*	*80%*	*100%*
Total fixed costs	$31,250	$31,250	$ 31,250	$ 38,250	$ 40,250
Total variable costs	33,415	66,830	100,245	143,660	192,075

a. For these prices and costs, what level of output will maximize profits?

b. At what level of output does the firm break even?

c. What is the lowest price per lamp for sales ranging from 2,000 to 2,500 lamps that will make operating at 100% capacity the most profitable? Assume the first 2,000 lamps are sold at the prices listed above.

d. While operating at full capacity under the price schedule listed above, a buyer desires to purchase 500 additional lamps that have a new feature. This order will raise plant utilization to 120% at an additional variable cost of $48,032. If these additional lamps can be sold for $120 apiece, what level of output provides the greatest profit? Fixed costs for the additional 500 lamps will remain the same as those at 100% capacity.

12. A manufacturer of receivers has an annual fixed cost of $300,000 and a variable cost of $13.20 per unit produced. An engineering proposal under consideration involves redesign of the present AM unit to include FM. The production engineering staff estimates that the redesigned product will increase the annual fixed cost by $200,000 and the variable cost by $5.20 per unit. A market survey produced the following information.

Unit Selling Price	*Sales with AM Only*	*Sales with AM and FM*
$70.00	0	6,200
60.00	3,000	21,000
50.00	11,000	39,500
40.00	25,500	64,000
30.00	48,000	100,000
25.00	74,000	—
22.00	98,000	—

a. Plot the total fixed cost of each plan, and plot the total income from each plan for the various selling prices against output as the abscissa.

b. Determine the selling price for maximum annual profit if only AM receivers are manufactured; if AM and FM receivers are manufactured.

c. Should the design change be adopted?

13. Annual sales of a given product are 8,000 units; the selling price is $8 per unit. Annual fixed production costs are $6,000 and the current annual profit is $18,000. The company is planning to spend funds to influence the sale of its product. The marketing department estimates that this new advertising campaign will increase total fixed costs by 15% and the unit variable cost by 25%. Sales are expected to be 9,600 units per year. Determine the new selling price in order to keep the same annual profit. Answer: $8.84 per unit

14. A factory producing electric power drills operates at 65% of its capacity and produces 22,800 drills per year. The unit manufacturing cost is computed as follows.

Direct-labor cost	$21.25
Direct-material cost	15.75
Overhead	11.00
	$48.00

The drills are marketed through a factory distributor for $56.75 each. It is anticipated that the volume of production can be increased to 30,000 units per year if the price is lowered to $46.50 per unit. This action would not increase the present total overhead cost. Compute the present profit per year and the profit per year if the volume of production is increased.

15. A certain firm has the capacity to produce 840,000 units of product per year. At present, it is operating at 60% of capacity. The firm's annual income is $504,000. Annual fixed costs are $212,000, and the variable costs are equal to $0.392 per unit of product.

a. What is the firm's annual profit or loss? Answer: $94,432 profit

b. At what volume of sales does the firm break even?

c. What will be the profit or loss at 75%, 85%, and 90% of capacity on the basis of constant income per unit and constant variable cost per unit?

16. A publishing company has issued a textbook that it distributes through college bookstores. It is estimated that 13,000 books per year will be sold, and the total fixed cost of production is $70,125. The average cost of materials and the direct labor cost are $32.50 and $10.25 per book, respectively.

a. If the company sells the book for $51, what is the annual profit, the break-even level in books produced, and the break-even level in terms of dollars of sales? Answer: $37,125; 8,500; $433,500

b. Repeat part (a) if the selling price is $55.

c. Repeat part (a) if the selling price is $47.

17. A company has priced its product at $1.50 per pound and is operating at a loss. Sales at this price total 910,000 pounds per year, which is equivalent to 65% of the total production capacity. The company's fixed cost of manufacture and selling is $510,000 per year and the variable cost is $0.95 per pound. It appears, from information obtained by a market survey, that price reductions of $0.05, $0.10, $0.15, and $0.16 per pound from the present selling price will result in total annual sales of 68%, 75%, 85%, and 96% of the total production capacity per year, respectively.

a. Calculate the annual profit or loss that will result from each of the selling prices given, assuming that variable cost per unit will be the same for all production levels.

b. Determine graphically the annual profit or loss that will result from each of the selling prices by the use of a break-even type chart.

c. Develop a general formula for the profit or loss that will result from each of the selling prices.

18. A company is operating at full capacity for one shift per day. Annual sales are 4,650 units per year and the income per unit is $215. Fixed costs are $316,400 and total variable costs are $553,700 per year at the present rate of operation. If output can be increased, 450 additional units can be sold at $226 per unit during the coming year. These additional units can be produced through overtime operation at the expense of a 20% increase in the unit variable cost of the additional units. The elasticity of demand for the product in question is believed to be such that an increase in selling price to $235 will result in curtailing demand to 3,800 units. For greatest profit in the coming year, should output be increased or should selling price be increased?

19. A manufacturing company owns two Plants, *A* and *B*, that produce an identical product. The capacity of Plant *A* is 65,000 units annually while that of Plant *B* is 85,000 units. The annual fixed cost of Plant *A* is $260,000 per year and the variable cost is $3.20 per unit. The corresponding values for Plant *B* are $280,000 and $3.90 per unit. At present, Plant *A* is being operated at 35% of capacity and Plant *B* at 40% of capacity.

a. What are the average unit costs of production of Plant *A* and Plant *B*? Answer: $14.63; $12.14

b. What are the total cost and the average cost of the total output of both plants?

c. What would be the total cost to the company and the average unit cost if all production were transferred to Plant *A*?

d. What would be the total cost to the company and the average unit cost if all production were transferred to Plant *B*?

20. A municipal electric power system consists of three generating units, Plant *A*, Plant *B* and Plant *C*. The total capacity of the system is 240 MW (megawatts), with maximum capacities of Plant A, Plant B, and Plant C at 100 MW, 50 MW, and 90 MW, respectively. The incremental costs for each plant are linearly increasing, and they are a function of the energy or MWh (megawatt-hours) produced. Let

x_j = the MWh produced per hour by plant j;

IC_j = the incremental cost per MWh of capacity for plant j;

$IC_A = \$26 + (\$0.020/\text{MWh})x_A \qquad 0 \leq x_A \leq 100 \text{ MWh};$

$IC_B = \$25 + (\$0.080/\text{MWh})x_B \qquad 0 \leq x_B \leq 50 \text{ MWh};$

$IC_C = \$27 + (\$0.025/\text{MWh})x_C \qquad 0 \leq x_C \leq 120 \text{ MWh}.$

Answer the following questions about assigning the production of electric energy within this system. Assume that the plants are assigned their loads in the most economic manner.

a. If there was no electric load on the system, what plant would be assigned the initial MWh to be produced? How much energy could be produced per hour before the next plant would be brought on line? What plant would be assigned at this new level of output? Answer: Plant *B*, 12.5 MWh, Plant *A*

b. At what minimum level of output for the system will all three plants be operating?

c. If the demand on the system is 100 MWh, how will the production of this energy be assigned to the individual plants?

d. Assuming that the owners of this system desire to increase their profit on every unit of energy produced, what is the lowest fixed price they could charge per MWh? Answer: \$29.25/MWh

21. A chemical processing plant is adding two production lines that produce identical products. Line 1 has a capacity of 100,000 lb per month while Line 2 can produce a maximum of 80,000 lb per month. The investments required to develop lines 1 and 2 are \$2,500,000 and \$2,000,000, and it is estimated that they will last 10 years. Once the two lines are operational, Line 1 has a variable cost of \$1.00/lb, increasing by \$0.0001 for each additional pound produced. Line 2 has a variable cost of \$2.00/lb, increasing by \$0.00005 for each additional pound of output. For the levels of output below find the most economic assignment of the production to each of the lines.

a. 1 lb per month.
b. 5000 lb per month.
c. 10,000 lb per month.
d. 12,000 lb per month.
e. 50,000 lb per month.
f. 100,000 lb per month.
g. 120,000 lb per month.
h. 160,000 lb per month.
i. 180,000 lb per month.

22. A company produces solid-state devices for desktop computers. One of the items can be produced at two separate plants, one within the United States and the other in another country. Because wage rates are lower for labor in the plant outside the United States, their direct labor costs are less. However, the plant in the United States is automated and the efficiency of production for this item is high. Both plants together can produce a maximum of 500,000 items per year, with 300,000 items produced within the United States and 200,000 outside the United States. The cost data for each plant are given below

	Within United States	Outside United States
Average unit cost at full capacity	\$30.25/item	\$26.00/item
Fixed costs	\$2,400,000	\$1,000,000
Variable costs		
Direct materials	\$16/item	\$15/item
Direct labor	1st 150,000 items, \$6.00/item 2nd 150,000 items, \$6.50/item	1st 80,000 items, \$3.00/item 2nd 120,000 items, \$8.00/item

Assign the output to each of the plants if the company has to deliver its customers the following annual quantities.

a. 60,000 items.
b. 200,000 items.
c. 250,000 items.
d. 300,000 items.
e. 400,000 items.
f. 450,000 items.
g. For a fixed price of \$30 find the break-even point.
h. For a fixed price of \$35 find the break-even point.
i. For a total price of \$40 find the break-even point.
j. Would you sell an additional 10,000 items for \$21/item if current production is 90,000 items?
k. Would you sell an additional 20,000 items for \$24/item if current production is 400,000 items?

l. Would you sell an additional 50,000 items for $23/item if current production is 250,00 items?

m. Would you sell an additional 10,000 items for $21/item if current production is 60,000 items?

n. Would you sell an additional 50,000 items for $21/item if current production is 60,000 items?

23. Machine *J* and Machine *K* produce the same product, and they originally cost $50,000 and $100,000, respectively. Machine *J* has a maximum capacity of 6,000 units per year, which it produces in lots of 2,000, 4,000 or 6,000 units per year. The capacity of Machine *K* is 8,000 units per year, having production lots of 2,000, 4,000, 6,000 and 8,000 units per year. No partial lots are produced for either machine. Given below are the fixed and variable costs associated with each machine's level of output

	Fixed Cost	*Variable Cost Per Unit for Ranges of Output*			
		0 units to 2,000 units	*2,000 units to 4,000 units*	*4,000 units to 6,000 units*	*6,000 units to 8,000 units*
Machine *J*	$8,000/year	$ 7/unit	$9/unit	$3/unit	—
Machine *K*	$20,000/year	10/unit	8/unit	5/unit	$4/unit

a. For the levels of total output (i) 2,000 units, (ii) 4,000 units, (iii) 6,000 units, (iv) 8,000 units, (v) 10,000 units, and (vi) 12,000 units indicate how the production should be assigned to each machine. Use total incremental costs.

b. If 6,000 units are produced, plot the profit-volume graph and find the break-even point for a price of $14.00 per unit.

c. If 10,000 units are produced, plot the profit-volume graph and find the break-even point for a price of $10.00 per unit.

d. If 12,000 units are produced, plot the profit-volume graph and find the break-even point for a price of $9.40 per unit.

24. A loading facility where ships are filled with bulk material is comprised of three material handling systems, *R*, *S*, and *T*. Each of these systems has the same maximum capacity and they can be operated at only three speeds, one-third of capacity, two-thirds of capacity, and full capacity. When the facility must operate at full capacity, all three material handling systems must be operated at full capacity. At less than full capacity, *R*, *S*, and *T* can be operated in various combinations to provide the total output desired. For example, at two-thirds of the total facility output there are 10 possible combinations among *R*, *S*, and *T* that will provide that output.

(One possible combination, one-third capacity, *R*; full capacity, *S*; two-thirds capacity, *T*.)

The monthly variable costs for operating at the various levels of capacity for each material handling system are as follows:

	Material Handling System		
Capacity	*R*	*S*	*T*
$\frac{1}{3}$	\$30,000/mo.	\$25,000/mo.	\$20,000/mo.
$\frac{2}{3}$	45,000	39,000	39,000
Full	55,000	52,000	57,000

Using total incremental costs, find the most economical assignments of output to the three material handling systems if the desired total output for the facility is two-thirds of capacity.

19

Mathematical Models for Optimizing Operations

Mathematical models are now routinely used to optimize many operational systems. Such models are quantitative abstractions of those aspects of the system important to its description and control. By formulating a mathematical model it is possible to determine values for policy variables that lead to a least-cost or maximum-profit operation.

The delay in developing models for operational systems may be attributed to an early preoccupation with the physical environment. During much of history the limiting factors were predominantly physical. But, with the accumulation of knowledge about physical phenomena, complex operational systems have been developed to produce goods and services. Decision makers are becoming increasingly aware that experience, intuition, and judgment must be augmented with analyses based on mathematical models if such systems are to be operated most efficiently.

19.1 MODELS FOR PROCUREMENT OPERATIONS

When the decision has been made to procure a certain item, it becomes necessary to determine the procurement quantity that will result in a minimum cost. The demand for the item may be met by procuring a year's supply at the beginning of each year or by procuring a day's supply at the beginning of each day. Neither of these extremes may be the most economical in terms of the sum of costs associated with procurement and holding the item in inventory. Models for inventory operations are used to determine the most economical procurement quantity.

19.1.1 The Economic Purchase Quantity

If the demand for an item is met by purchasing once per year, the cost incident to purchasing will occur once, but the large quantity received will result in a relatively high inventory holding cost for the year. Conversely, if orders are placed several times per year, the cost incident to purchasing will be incurred several times per year, but since small quantities will be received, the cost of holding the item in inventory will be relatively small. If the decision is to be based on economy of the total operation, the purchase quantity that will result in a minimum annual cost must be determined. Let

TC = total yearly cost of providing the item;
D = yearly demand for the item;
N = number of purchases per year;
t = time between purchases;
Q = purchase quantity;
C_i = item cost per unit (purchase price);
C_p = purchase cost per purchase order;
C_h = holding cost per unit per year made up of such items as interest, insurance, taxes, storage space, and handling.

If it is assumed that the demand for the item is constant throughout the year, the purchase lead time is zero, and no shortages are allowed, the resulting inventory flow may be represented graphically as in Figure 19.1.

The total yearly cost will be the sum of the item cost for the year, the purchase cost for the year, and the holding cost for the year. That is,

$$TC = IC + PC + HC. \tag{19.1}$$

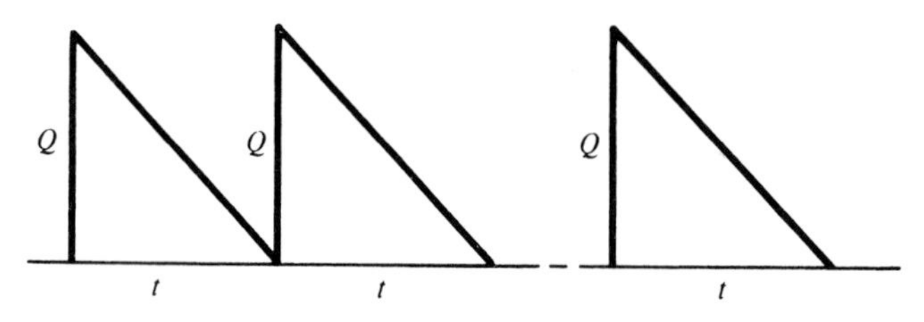

FIGURE 19.1. Graphical representation of inventory flow for purchasing.

The item cost for the year will be the item cost per unit times the yearly demand in units, or

$$IC = C_i(D). \tag{19.2}$$

The purchase cost for the year will be the cost per purchase times the number of purchases per year, or

$$PC = C_p(N).$$

But since N is the yearly demand divided by the purchase quantity,

$$PC = \frac{C_p(D)}{Q}. \tag{19.3}$$

Since the interval t begins with Q units in stock and ends with none, the average inventory during the cycle will be $Q/2$. Therefore, the holding cost for the year will be the holding cost per unit times the average number of units in stock for the year, or

$$HC = \frac{C_h(Q)}{2}. \tag{19.4}$$

The total yearly cost of providing the required item is the sum of the item cost, purchase cost, and holding cost, or

$$TC = C_i(D) + \frac{C_p(D)}{Q} + \frac{C_h(Q)}{2}. \tag{19.5}$$

The purchase quantity resulting in a minimum yearly cost may be found by differentiating with respect to Q, setting the result equal to zero, and solving for Q as follows:

$$\frac{dTC}{dQ} = -\frac{C_p(D)}{Q^2} + \frac{C_h}{2} = 0$$

$$Q^2 = \frac{2C_p(D)}{C_h}$$

$$\boxed{Q = \sqrt{\frac{2C_p(D)}{C_h}}.} \tag{19.6}$$

TABLE 19.1.
TOTAL COST AS A FUNCTION OF PURCHASE QUANTITY

Purchase Quantity	Total Cost
50	\$6,233
100	\$6,166
123	\$6,162
150	\$6,165
200	\$6,182
300	\$6,231
400	\$6,289
600	\$6,413

As an example of the use of this model, assume that the annual demand for a certain item is 1,000 units. The cost per unit is \$6 delivered. Purchasing cost per purchase order is \$10, and the cost of holding one unit in inventory for one year is estimated to be \$1.32.

The economic purchase quantity may be found by substituting the appropriate values in the derived relationship as follows:

$$Q = \sqrt{\frac{2(\$10)(1{,}000)}{\$1.32}}$$

$$= 123 \text{ units.}$$

Total cost may be expressed as a function of Q by substituting the costs and various values of Q into the total-cost equation. The result is shown in Table 19.1. The tabulated total-cost value for $Q = 123$ is the minimum-cost purchase quantity for the condition specified. Total cost as a function of Q is illustrated in Figure 19.2.

19.1.2 The Economic Production Quantity

When the decision has been made to produce a certain item, it becomes necessary to determine the optimum production quantity. This may be done in a manner similar to determining economic purchase quantities. The difference in analysis occurs because a purchased lot is received at one time whereas a production lot accumulates as it is made. Let

TC = total yearly cost of providing the item;
D = yearly demand for the item;
N = number of production runs per year;
t = time between production runs;

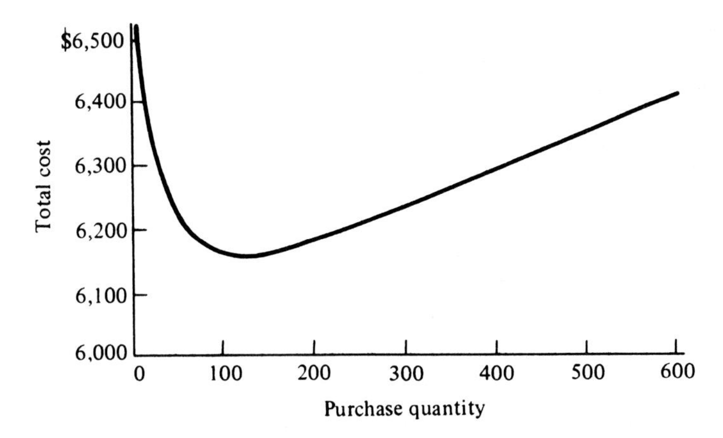

FIGURE 19.2. Total cost as a function of purchase quantity.

Q = production quantity;
C_i = item cost per unit (production cost);
C_s = set-up cost per production run;
C_h = holding cost per unit per year made up of such items as interest, insurance, taxes, storage space, and handling;
R = production rate.

If it is assumed that the demand for the item is constant, the production rate is constant during the production period, the production lead time is zero, and no shortages are allowed, then the resulting inventory flow may be represented graphically as in Figure 19.3.

The total yearly cost will be the sum of the item cost for the year, the set-up cost for the year, and the holding cost for the year. That is,

$$TC = IC + SC + HC. \tag{19.7}$$

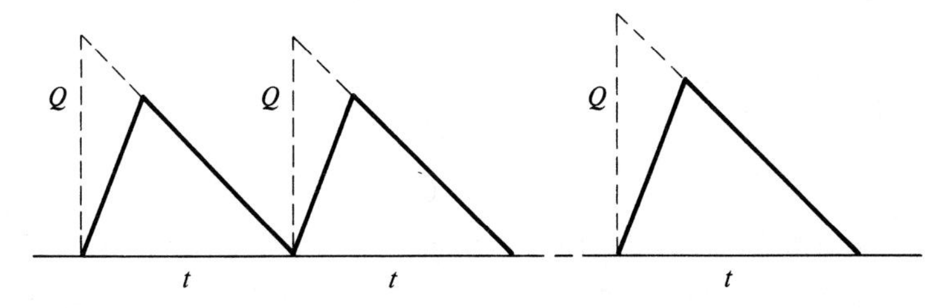

FIGURE 19.3. Graphical representation of inventory flow for production.

The item cost for the year will be the item cost per unit times the yearly demand in units, or

$$IC = C_i(D). \tag{19.8}$$

The set-up cost for the year will be the cost per set-up times the number of set-ups per year, or

$$SC = C_s(N).$$

But since N is the yearly demand divided by the production quantity,

$$SC = \frac{C_S(D)}{Q}. \tag{19.9}$$

When items are added to inventory at the rate of R units per year and are taken from inventory at a rate D units per year (where R is greater than D), the net rate of accumulation is $(R - D)$ units per year. The time required to produce D units at the rate R units per year is D/R years. If D units are made in a single lot, the maximum accumulation in inventory will be $(R - D)(D/R)$. Since no units will be in storage at the end of the year, the average number in inventory will be

$$\frac{(R - D)\dfrac{D}{R} + 0}{2} = (R - D)\frac{D}{2R}.$$

If N lots are produced per year, the average number of units in storage will be

$$(R - D)\frac{D}{2RN}.$$

But, since $N = D/Q$, the average number of units in storage may be expressed as

$$(R - D)\frac{Q}{2R}.$$

The holding cost for the year will be the holding cost per unit times the average number of units in storage for the year, or

$$HC = C_h(R - D)\frac{Q}{2R}. \tag{19.10}$$

The total yearly cost of producing the required item is the sum of the item cost, set-up cost, and holding cost, or

$$TC = C_i(D) + \frac{C_s(D)}{Q} + C_h(R - D)\frac{Q}{2R}. \tag{19.11}$$

The production quantity resulting in a minimum yearly cost may be found by differentiating with respect to Q, setting the result equal to zero, and solving for Q as follows:

$$\frac{dTC}{dQ} = -\frac{C_s(D)}{Q^2} + \frac{C_h(R - D)}{2R} = 0$$

$$Q^2 = \frac{C_s(D)2R}{C_h(R - D)} = \frac{C_s(D)2}{C_h\left(1 - \frac{D}{R}\right)}$$

$$\boxed{Q = \sqrt{\frac{2C_s(D)}{C_h\left(1 - \frac{D}{R}\right)}}.} \tag{19.12}$$

As an example of the use of this model assume that the annual demand for a certain item is 1,000 units. The cost of production is \$5.90 per unit and includes the usual cost elements of direct labor, direct material, and factory overhead. The set-up cost per lot is \$50 and the item can be produced at the rate of 6,000 units per year. The cost of holding one unit in inventory for one year is estimated to be \$1.30.

The economic production quantity may be found from the derived relationship by substituting the appropriate values as follows:

$$Q = \sqrt{\frac{2(\$50)(1{,}000)}{\$1.30\left(1 - \frac{1{,}000}{6{,}000}\right)}}$$

$$= 302 \text{ units.}$$

Total cost may be expressed as a function of Q by substituting cost and various values of Q into the total-cost equation. The result is shown in Table 19.2. The tabulated total-cost value for $Q = 302$ is the minimum-cost production quantity for the condition specified. Total cost as a function of Q is illustrated in Figure 19.4.

TABLE 19.2.
TOTAL COST AS A FUNCTION
OF PRODUCTION QUANTITY

Production Quantity	Total Cost
100	\$6,454
150	\$6,314
200	\$6,258
300	\$6,229
302	\$6,228
400	\$6,241
500	\$6,270
600	\$6,307

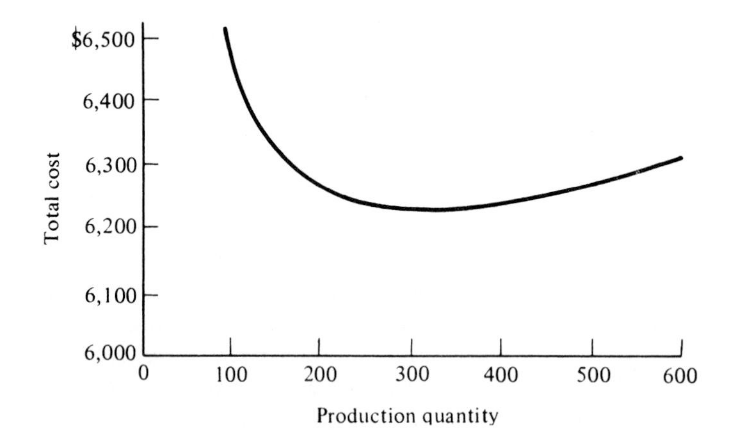

FIGURE 19.4. Total cost as a function of production quantity.

19.1.3 The "Make or Buy" Decision

The question of whether to manufacture or purchase a needed item may be resolved by the application of minimum-cost analysis for multiple alternatives. The alternative of producing may be compared with the alternative of purchasing if the minimum-cost procurement quantity for each is computed and used to find the respective total-cost values. Choice of the total-cost value that is a minimum identifies the best of the two alternatives.

For example, suppose that an item will have a yearly demand of 1,000 units and that the costs associated with purchasing and producing are the same as assumed in the previous sections and summarized in Table 19.3.

For the conditions assumed, total cost as a function of the purchase quantity was given in Table 19.1, and total cost as a function of production quantity was given in Table 19.2. Total cost as a function of purchase quantity was

TABLE 19.3.
COSTS ASSOCIATED WITH PURCHASING AND PRODUCING ALTERNATIVES

Category	Purchase	Produce
Item cost	\$ 6.00	\$ 5.90
Purchase cost	\$10.00	—
Set-up cost	—	\$50.00
Holding cost	\$ 1.32	\$ 1.30

graphed in Figure 19.2, and total cost as a function of production quantity was graphed in Figure 19.4. If Figures 19.2 and 19.4 are superimposed, the result is as shown in Figure 19.5.

The decision of whether to produce or purchase may be made by examining and comparing the minimum cost for each alternative. In this case, the decision to purchase will be the least-cost alternative and will result in a saving of \$6,228 less \$6,162, or \$66 per year. If the decision were made on the basis of item cost alone, the needed item would be supplied by producing, with a resultant loss of \$66 per year.

The optimal procurement policy may be formally stated by specifying that the item will be procured when the available stock falls to zero units on hand, for a procurement quantity of 123 units, from the purchasing source. In essence, the policy states when, how much, and from what source. The policy, if the decision has been made to purchase, states when and how much, the source being fixed by restriction. Therefore, the decision to "make or buy" is essentially a re-

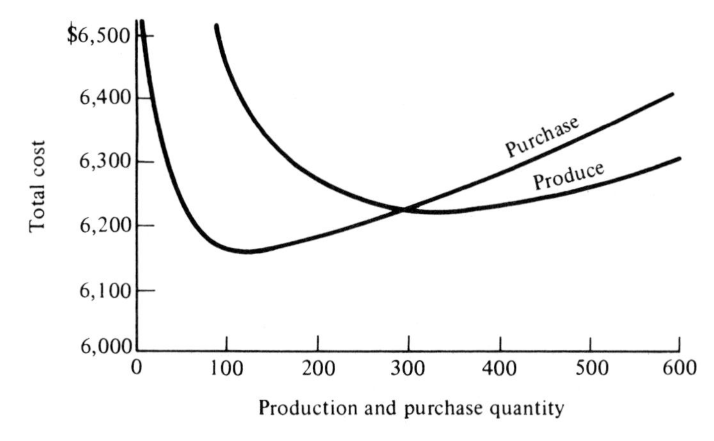

FIGURE 19.5. Total cost as a function of production and purchase quantities.

lease of the restriction that the source is fixed. This analysis may be extended to any number of sources; for example, it may be used to compare remanufacturing with purchasing, to compare alternate manufacturing facilities, or to evaluate alternate vendors.

19.2 PRODUCING TO A VARIABLE DEMAND

Demand for many manufactured goods is seasonal. Some seasonal variations in demand reflect changes in human wants caused by differences in temperature, rainfall, and hours of sunshine. Other items may vary in demand because of seasonality of social customs, as is the case with the sale of fireworks.

The seasonal goods manufacturer may produce product at a relatively low rate throughout the year and store it until it is needed. Or he may acquire sufficient facilities to manufacture the product at a rate equal to the demand during the period in which it is sold. The disadvantage of the first plan is that storage cost is relatively high; the disadvantage of the second method is that a rather large equipment investment will be required. The most desirable plan may be a compromise.

A method of solution of this and similar situations may be illustrated by an example. Assume that 36,000 units of a product are required during a 4-month period of each year as follows:

Month of Year	*Demand*
9	2,000
10	10,000
11	15,000
12	9,000

One machine can make 36,000 units of this product during a year. The fixed charges on the machine for such items as interest, taxes, insurance, space to house machine, and depreciation due to causes exclusive of usage amount to $2,000 per year. The cost for depreciation due exclusively to usage, power, supplies, maintenance, and so forth amounts to $0.25 per hour or $40 per month of operation on the basis of a 160-hour month.

Although one machine can meet the demand for the product, the accumulation of finished products throughout the year will result in considerable expense for storage. That expense can be reduced by using more machines to shorten the period required to make the year's needs.

The costs associated with plans of production based on using one, two, and three machines will be determined. For an output of 36,000 units per machine, 12, 6, and 4 months, respectively, will be required to produce the annual output of 36,000 units, with one, two, and three machines. The number sold each

TABLE 19.4.
STORAGE REQUIREMENTS FOR PRODUCING TO A SEASONAL DEMAND

		Number in Storage at End of Month for		
Month	Sales During Month	1 Machine	2 Machines	3 Machines
1		3,000		
2		6,000		
3		9,000		
4		12,000		
5		15,000		
6		18,000		
7		21,000	6,000	
8		24,000	12,000	
9	2,000	25,000	16,000	7,000
10	10,000	18,000	12,000	6,000
11	15,000	6,000	3,000	0
12	9,000	0	0	0

month and the number in storage at the end of each month during the year are given in Table 19.4 and are shown graphically in Figure 19.6.

The average number in storage for one, two, and three machines is 13,083, 4,083, and 1,083, respectively. These results are obtained by adding the average of the number of units of product in storage at the beginning and end of each month in the year for each plan and dividing the resulting sum by 12. The product is valued at $3 per unit. The sum of interest, taxes, and insurance is taken as 10% of the unit cost and storage costs as $0.20 per unit per year.

On the basis of the above data the cost with one, two, and three machines will be as follows:

One Machine

Fixed charges on machine, 1 × $2,000	$2,000
Variable charge on machine, 12 × $40	480
Interest, taxes, and insurance, 13,083 × $3 × 0.10	3,925
Storage cost, 13,083 × $0.20	2,617
Total cost with one machine	$9,022

Two Machines

Fixed charges on machines, 2 × $2,000	$4,000
Variable charge on machines, 6 × 2 × $40	480
Interest, taxes, and insurance, 4,083 × $3 × 0.10	1,225
Storage cost, 4,083 × $0.20	817
Total cost with two machines	$6,522

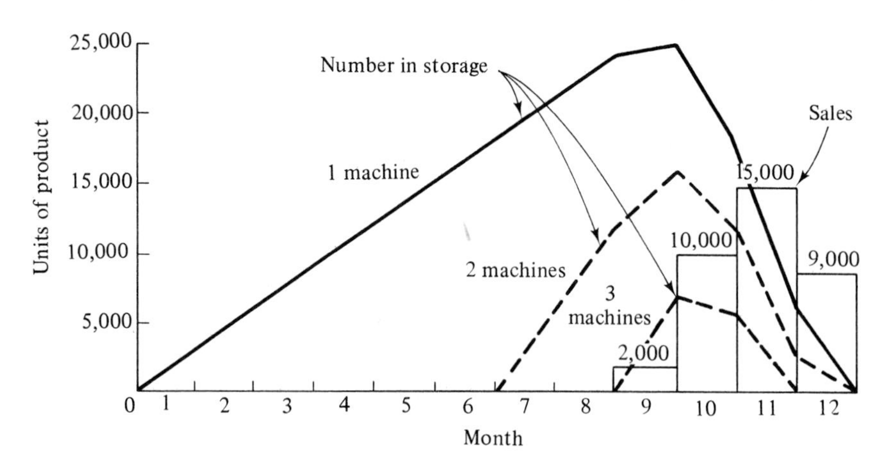

FIGURE 19.6. Graphical presentation of storage requirements.

Three Machines

Fixed charges on machines, 3 × \$2,000	\$6,000
Variable charge on machines, 4 × 3 × \$40	480
Interest, taxes and insurance, 1,083 × \$3 × 0.10	325
Storage cost, 1,083 × \$0.20	217
Total cost with three machines	\$7,022

On the basis of the analysis above, a saving of \$500 per year would result from using two machines in place of three, and a saving of \$2,500 would result from using two machines in place of one. It should be realized that there may be considerable hazard in the use of any of the three plans. If only one machine is used, there is a possibility that the product may become outmoded before the sales period or that sales may not be up to expectations. If three machines are used, there is a large investment that may not be recovered before the machines are rendered obsolete or inadequate.

19.3 LINEAR PROGRAMMING MODELS

Operational situations involving many activities that compete for scarce resources can often be optimized by use of the general linear programming model. Usually, any operation involving a linear effectiveness function and linear resource constraints can be expressed in the format of the linear programming model formally stated as:

Optimize the evaluation function

$$E = \sum_{j=1}^{n} e_j x_j$$

subject to the constraints

$$\sum_{j=1}^{n} a_{ij} x_j = b_i \qquad i = 1, 2, \ldots, m$$
$$x_j \geq 0 \qquad j = 1, 2, \ldots, n. \tag{19.13}$$

Optimization requires either maximization or minimization of E, depending upon the measure involved. The decision maker has control of the variables x_j. Not directly under his control are the coefficients e_j, the constants a_{ij}, and the constants b_j.

A good introduction to the structure and nature of linear programming problems can be obtained by utilizing graphical optimization methods. Although a graphical approach is not feasible for operations with more than three activities competing for scarce resources, these simple cases provide very good insight. The sections that follow present linear programming examples involving two and three variables.

19.3.1 Graphical Maximization for Two Activities

When two activities compete for scarce resources, a two-dimensional space is defined. Each restriction corresponds to a line on this surface. These restrictions identify a region of feasible solutions. The effectiveness function is also a line, its orthogonal distance from the origin being proportional to its value. The optimum value for the effectiveness function occurs when it is located so that it just touches an extreme point of the region.

Consider the following production example. Two products are to be produced. A single unit of product A requires 2.4 minutes of molding time and 5.0 minutes of assembly time. The profit for product A is \$0.60 per unit. A single unit of product B requires 3.0 minutes of molding time and 2.5 minutes of finishing time. The profit for product B is \$0.70 per unit. The molding department has time available to allocate to these products of 1,200 minutes per week. The finishing department has idle capacity of 600 minutes per week, and the assembly department can supply 1,500 minutes of capacity per week. The production and marketing data for this production situation are summarized in Table 19.5.

In this example, two products compete for the available production time. The objective is to determine the quantity of product A and the quantity of

TABLE 19.5.
PRODUCTION AND MARKETING DATA FOR TWO PRODUCTS

Process	Product A	Product B	Capacity
Molding	2.4	3.0	1,200
Finishing	0.0	2.5	600
Assembly	5.0	0.0	1,500
Profit	\$0.60	\$0.70	

product B to produce per week so that total profit will be maximized. This will require maximizing

$$TP = \$0.60A + \$0.70B \tag{19.14}$$

subject to

$$2.4A + 3.0B \leq 1{,}200 \tag{19.15}$$

$$0.0A + 2.5B \leq 600 \tag{19.16}$$

$$5.0A + 0.0B \leq 1{,}500 \tag{19.17}$$

$$A \geq 0 \text{ and } B \geq 0.$$

The graphical equivalent of the algebraic statement of this two-product problem is shown in Figure 19.7. The set of linear constraints define a region of feasible solutions. This region lies below $2.4A + 3.0B = 1{,}200$ and is constrained further by the requirements that $B \leq 240$, $A \leq 300$, and both A and B be non-negative. Thus, the scarce resources determine which combinations of the activities are feasible and which are not.

Feasible production-quantity combinations of products A and B may be found at the extreme points by working with the constraints as equalities using linear algebra. First, it is evident that a feasible combination exists at point 0 in Figure 19.7 when $A = 0$ and $B = 0$. From Equation (19.16) it is evident that point 1 in Figure 19.7 exhibits a feasible combination of $A = 0$ and $B = 240$. From Equation (19.17) it is also evident that point 4 exhibits a feasible combination of $A = 300$ and $B = 0$. Next, by utilizing Equation (19.15), the feasible combination at point 2 in Figure 19.7 is found to be

$$2.4A + 3.0B = 1{,}200$$

$$2.4A + 3.0(240) = 1{,}200$$

$$2.4A = 1{,}200 - 720$$

$$A = 200$$

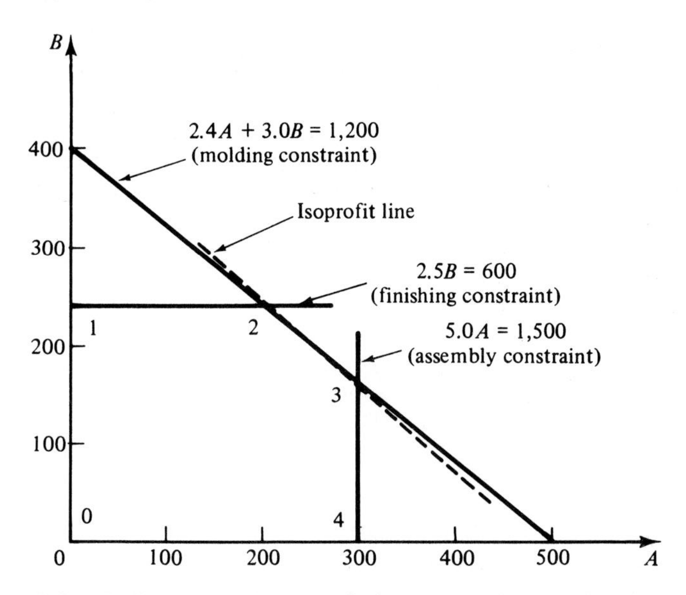

FIGURE 19.7. Maximizing profit for two-product production.

since $B = 240$ from Equation (19.16). Finally, by utilizing Equation (19.15), the feasible combination at point 3 in Figure 19.7 is found to be

$$2.4A + 3.0B = 1{,}200$$

$$2.4(300) + 3.0B = 1{,}200$$

$$3.0B = 1{,}200 - 720$$

$$B = 160$$

since $A = 300$ from Equation (19.17).

Each of these production-quantity combinations may now be evaluated for total profit, *TP*, by the use of Equation (19.14). For example, when $A = 300$ and $B = 160$ (point 3), total profit is $0.60(300) + $0.70(160) = $292. The results are exhibited in Table 19.6. Inspection of the total profit indicates that profit is maximized at point 3, which has the coordinates 300 and 160. No other production quantity combination of products *A* and *B* will result in a higher profit than the $292.00 shown.

A more formal approach to finding the maximum-profit combination of products *A* and *B* may be utilized by noting that the relationship between *A* and *B* is $A = 1.167B$, based on the relative profit contribution of each product. The total profit realized will depend on the production-quantity combination chosen. Thus, there is a family of isoprofit lines, one of which will have at least one point in the region of feasible production-quantity combinations and be a maximum

TABLE 19.6.
TOTAL-PROFIT COMBINATIONS AT EXTREME POINTS OF FIGURE 19.7

	Coordinate		
Point	*A*	*B*	*Total Profit*
0	0	0	0
1	0	240	$168.00
2	200	240	288.00
3	300	160	292.00
4	300	0	180.00

orthogonal distance from the origin. The member that satisfies this condition intersects the region of feasible solutions at the extreme point $A = 300$, $B = 160$. This is shown as a dashed line in Figure 19.6 and represents a total profit of $292, as was shown algebraically.

19.3.2 Graphical Maximization for Three Activities

When three activities compete for scarce resources, a three-dimensional space is involved. Each constraint is a plane in this space, and all constraints taken together identify a volume of feasible solutions. The effectiveness function is also a plane, its orthogonal distance from the origin being proportional to its value. The optimum value for the effectiveness function occurs when this plane is located so that it is at the extreme point of the volume of feasible solutions.

As an example, suppose that the production operations in the previous example are to be expanded to include a third product, designated *C*. A single unit of product *C* will require 2.0 minutes of molding time, 1.5 minutes of finishing time, and 2.5 minutes of assembly time. The profit associated with product *C* is $0.50 per unit. Production and marketing data for this revised situation are summarized in Table 19.7.

In this example, three products compete for the available production time. The objective is to determine the quantity of product *A*, the quantity of product

TABLE 19.7.
PRODUCTION AND MARKETING DATA FOR THREE PRODUCTS

Process	Product A	Product B	Product C	Capacity
Molding	2.4	3.0	2.0	1,200
Finishing	0.0	2.5	1.5	600
Assembly	5.0	0.0	2.5	1,500
Profit	$0.60	$0.70	$0.50	

B, and the quantity of product C to produce so that total profit will be maximized. This will require maximizing

$$TP = \$0.60A + \$0.70B + \$0.50C \tag{19.18}$$

subject to

$$2.4A + 3.0B + 2.0C \leq 1{,}200 \tag{19.19}$$

$$0.0A + 2.5B + 1.5C \leq 600 \tag{19.20}$$

$$5.0A + 0.0B + 2.5C \leq 1{,}500 \tag{19.21}$$

$$A \geq 0,\ B \geq 0, \text{ and } C \geq 0.$$

The graphical equivalent of the algebraic statement of this three-product production situation is shown in Figure 19.8. The set of constraining planes defines a volume of feasible solutions. The region lies below $2.4A + 3.0B + 2.0C = 1{,}200$ and is restricted further by the requirement that $2.5B + 1.5C \leq 600$, $5.0A + 2.5C \leq 1{,}500$, and that A, B, and C be nonnegative. Thus, the scarce resources determine which combinations of the activities are feasible and which are not feasible.

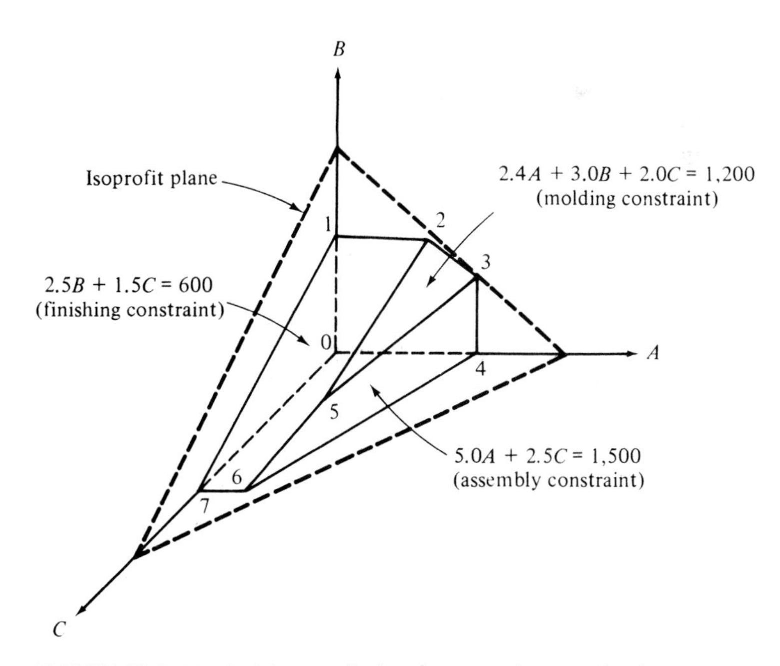

FIGURE 19.8. Maximizing profit for three-product production.

The production-quantity combinations of products A, B, and C that fall within the volume of feasible solutions constitute feasible production programs. The combination or combinations of products A, B, and C that maximizes total profit is sought. The expression $0.60A + 0.70B + 0.50C$ gives the relationship among products A, B, and C based on the relative profit of each. The total profit realized will depend on the production-quantity combination chosen. Thus, there exists a family of isoprofit planes, one for each value of total profit. One of these planes will have at least one point in the volume of feasible solutions and will be a maximum orthogonal distance from the origin. The plane that maximizes profit will intersect the volume at an extreme point.

Feasible production-quantity combinations may be found at the extreme points by the use of the elimination substitution technique of linear algebra. For example, the values of products A, B, and C at extreme point 5 in Figure 19.8 may be found as follows:

1. From Equation (19.20), $B = 240 - 0.6C$.
2. Substitute $B = 240 - 0.6C$ into Equation (19.19), giving $2.4A + 0.2C = 480$.
3. Multiply $2.4A + 0.2C = 480$ by 12.50, giving $30.0A + 2.5C = 6{,}000$.
4. Subtract $30.0A + 2.5C = 6{,}000$ from Equation (19.21) and solve for A, giving $A = 180$.
5. Substitute $A = 180$ into Equation (19.21) and solve for C, giving $C = 240$.
6. Substitute $C = 240$ into Equation (19.20) and solve for B, giving $B = 96$.

The total profit resulting from $A = 180$, $B = 96$, and $C = 240$ (point 5) is found from Equation (19.18) as \$0.60(180) + \$0.70(96) + \$0.50(240) = \$295.20. This is recorded in Table 19.8 together with the total profit at all other extreme points in Figure 19.8.

TABLE 19.8.
TOTAL PROFIT VALUES AT EXTREME POINTS OF FIGURE 19.8

		Coordinate		
Point	A	B	C	Total
0	0	0	0	$ 0
1	0	240	0	$168.00
2	200	240	0	$288.00
3	300	160	0	$292.00
4	300	0	0	$180.00
5	180	96	240	$295.20
6	100	0	400	$260.00
7	0	0	400	$200.00

Inspection of the total-profit values in Table 19.8 indicates that profit is maximized at point 5, which has the coordinates 180, 96, and 240. No other production-quantity combination of products *A*, *B*, and *C* will result in a higher profit. Also, no alternative optimum solutions exist, since the total-profit plane intersects the volume of feasible solutions at only a single point.

Addition of a third product to the two-product production operation increased total profit from $292.00 per week to $295.20 per week. This increase in profit results from reallocation of the available production time and from greater utilization of the idle capacity in the finishing department. The number of units of product *A* was reduced from 200 to 180, and the number of units of *B* was reduced from 240 to 96. This made it possible to add 240 units of product *C* to the production program with the resulting increase in profit.

19.4 ANALYSIS OF WAITING-LINE OPERATIONS

A general waiting-line or queuing system is shown in Figure 19.9. The system exists because the population shown demands service. To satisfy the demand, a decision maker must specify the level of service capacity at the service facility. This is a problem in operations economy, which will involve altering the service rate at existing channels, or adding or deleting channels. One approach to the problem is illustrated by the example of this section.

Failures of equipment do not occur with regularity. In addition, the amount of repair work needed for any given failure may be expected to vary above and below the average amount needed over a long period of time. Each of these events—the occurrence of a failure and the amount of repair needed—is a function of many unpredictable chance causes.

If the number of repair crews provided is just sufficient to take care of the average amount of repair work needed, there will be a considerable backlog of work waiting to be done, and the cost of equipment downtime will be high. When downtime is costly, it may be wise to maintain repair-crew capacity in ex-

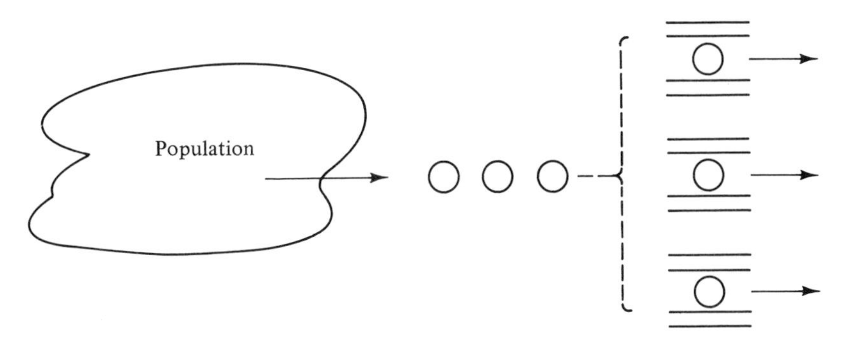

FIGURE 19.9. A multiple-channel waiting-line system.

cess of that needed for the average repair load, even though this will result in some idleness on the part of the repair crews.

As an example of the analysis required in finding the economical number of repair crews, consider an illustration from the petroleum industry. In petroleum production, heavy equipment is used to pump oil to the surface. When this equipment fails, it is necessary to remove it from the well for repairs. The required repairs are made by crews of three to five men who are equipped with heavy, portable machinery to pull the pumping equipment from the well.

Between the time the pumping equipment fails and the time it is repaired, the well is idle. This results in a loss, known as *lost production,* equal to the amount of oil that would have been produced if there had been no failure of equipment. In some cases lost production may be only production that is deferred to a later date. In other cases the loss may be due partially to drainage to competitors' wells. This loss may be quite large. Two days' downtime of a well producing 200 barrels of oil per day at $30 per barrel, for instance, when drainage to a competitor's well is judged to be half of lost production, results in a loss of $3,000.

The downtime of a well is made up of the time that the well is idle before the repair crew gets to it, and the time it is idle while repair is in process. Since the rate at which repairs can be made is substantially controlled by the repair equipment, reduction in loss is brought about by reducing the time a well is idle awaiting repair crews. This is done at the expense of having excess repair crews. Thus, the problem is to balance the number of repair crews with the lost production associated with delays of repairs to wells. Consider the following data, adapted from an actual situation in which the wells are operated 24 hours a day, 7 days a week.

N = number of oil wells in field = 30;

U = average interval at which individual wells fail = 15 days;

T = average actual time for company crew to repair well failure = 12 hours;

C_r = hourly cost for company-operated repair unit and repair crew = $180 per hour;

C_d = loss production per well when "down" = $3,000 per day.

Under the present situation, analysis has revealed that operation of the equivalent of one repair unit 24 hours per day, 7 days per week, can keep up with the repair of failures in the long run. This means that one channel is sufficient to service the 30 wells. However, serious losses are arising from delays in getting to wells after notification of their failure. Delay is caused by the chance bunching of well failures. For instance, a period during which no wells fail may be followed by a period of an above-average number of failures. Since repairs cannot be made before failures occur, it is clear that crews sufficient to take care of the average number of failures will have a backlog of failures awaiting them most of the time.

The first step in an analysis to determine the most economical number of repair crews is to determine the number of wells that may be expected to fail during each and every day of a period in the future. The number of failures to expect in the future may be estimated on the basis of past records or by mathematical analysis. The former method is used in the example because it is more revealing.

In the solution of this example the pattern of well failures of a previous 30-day period selected at random will be considered to be representative of future periods. A 30-day period will be taken in the interest of simplicity, even though experience has shown that longer periods are advantageous.

Since the company's unit can repair two wells per day when operating three 8-hour shifts per day, unrepaired wells will be carried over one day each day that the number of failures plus the carry-over from the previous day exceeds two.

Line *A* of Figure 19.10 shows the number of wells that failed during the 30-day period selected at random and considered to be typical. The number of wells failing during any one day ranged between 0 and 5, and their total was 57.

Line *B* gives the number of wells repaired each day with the company's unit and crews. During the 30-day period, 57 wells failed and 55 were repaired. Thus, in spite of the periodic backlogs, there was idle time of the unit and crews on the days numbered 3, 6, 7, and 8—sufficient to repair 5 wells.

The carry-over of unrepaired wells is given in line *C*. The total carry-over for the 30-day period is 52 well-days.

On the present basis of operation, employing the equivalent of one unit 24 hours per day, the total cost incident to well failures and repair during the 30-day period is calculated as follows:

$$TC = C_r(24)(30) + C_d(52)$$

$$= \$180(24)(30) + \$3{,}000(52)$$

$$= \$285{,}600.$$

Analysis of the pattern of occurrence of unrepaired wells carried over reveals that, once a backlog has accumulated, wells may remain unrepaired and unproductive for a long period. As a remedy the supervisor in charge considers the feasibility of hiring an additional unit and a crew (operating with two channels) whenever a backlog of unrepaired wells has accumulated. He finds that a repair unit and crew can be hired on short notice when needed for \$240 per hour. Because of greater travel distance and other reasons, a hired unit and crew has been found to require an average of 16 hours to repair a well.

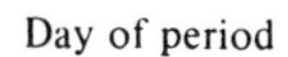

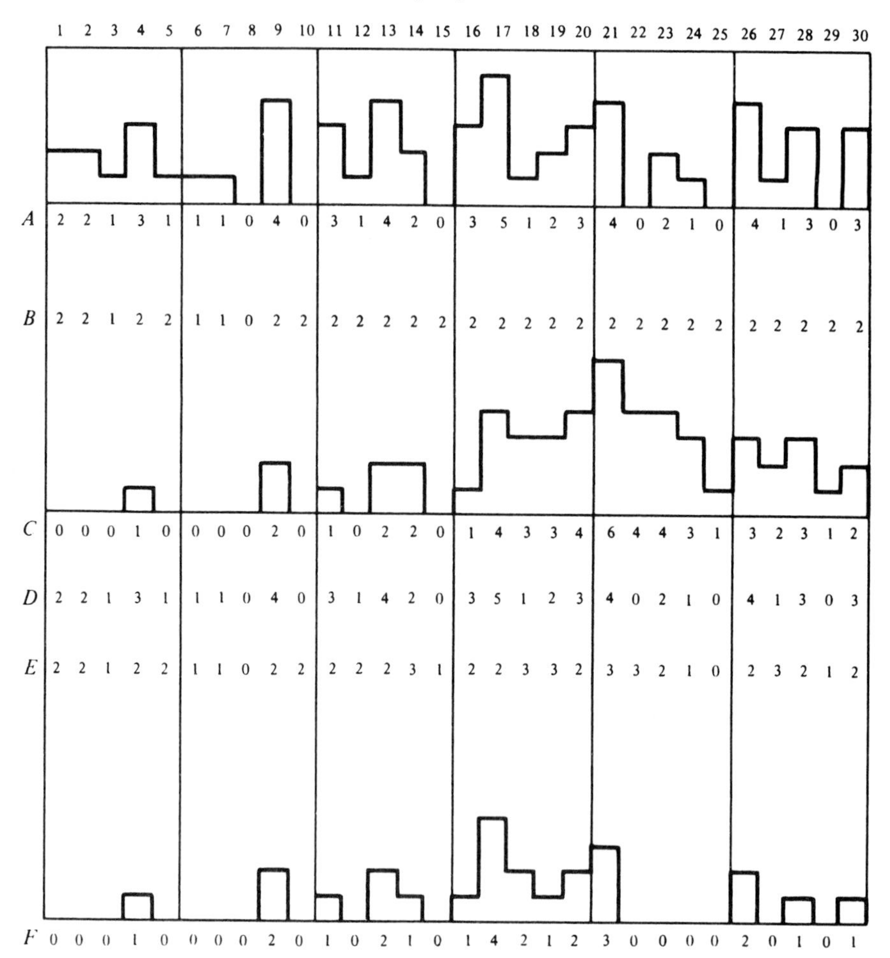

FIGURE 19.10. Pattern of oil-well failure and two repair methods.

The supervisor wishes to determine the effect of a policy of hiring an additional unit and crew for a 16-hour period on days when there is a carry-over of two or more unrepaired wells from the previous day. He assumes that the hired unit and crew will be paid for a minimum of 16 hours each time it is asked to report, whether or not it is used.

If this policy had been in effect during the 30-day period under consideration, the additional unit and repair crew would have been hired for 16 hours on the days numbered 10, 14, 18, 19, 21, 22, and 27; the total number of wells repaired during each day would have been as given in Line *E* of Figure 19.10.

Under this plan of operation, the total carry-over would have been 24 well-days, as shown in line *F*.

Line E indicates that there was idle time of units and crew during the days numbered 3, 6, 7, 8, 10, 15, 24, 25, and 29—sufficient for the repair of 11 wells.

On the basis of the analysis above, the total cost incident to well failures and repair for the 30-day period is calculated as follows:

$$TC = \$180(24)(30) + \$240(16)(7) + \$3{,}000(24)$$

$$= \$228{,}480.$$

The decision to hire an additional unit and crew on days when there is a carry-over of two or more unrepaired wells from a previous day would result in a reduction of the total cost of operations by $285,600 less $228,480, or $57,120 per month if future months experience the same pattern of failures. Actually, this is an average value, which will be achieved in the long run if the data used exhibit an average pattern.

KEY POINTS

- Progress in the physical sciences and engineering has pushed back limiting factors of a physical nature and made possible the development of complex operational systems to meet human needs.
- Mathematical models are quantitative abstractions of those aspects of a system important to its description and control and are now routinely used to optimize operational systems.
- Decision makers are becoming increasingly aware that experience, intuition, and judgment must be augmented with analyses based on mathematical models if systems are to be operated more economically.
- Models for procurement operations make it possible to determine the economic purchase quantity and the economic production quantity as well as the minimum-cost procurement source for a needed item.
- Producing goods to a variable demand requires a mathematical trade-off between the cost of production that patterns the variation in demand and the cost of uniform production with storage.
- Operational situations involving activities that compete for scarce resources can often be resolved by use of linear programming models, where a linear evaluation function is optimized in the face of linear constraints.
- In waiting-line operations, overall economy may be achieved by a mathematical model that trades off the cost of waiting against the cost of providing service capacity.

PROBLEMS

1. A contractor has a requirement for cement in the amount of 45,000 bags per year. Cement costs $2.85 per bag, and it costs $40 to process a purchase order. Holding cost is $4 per bag per year because of the high rate of spoilage. Find the minimum-cost purchase quantity. Answer: 949
2. A foundry uses 3,600 tons of pig iron per year at a constant rate. The cost per ton delivered to the foundry is $145. It costs $92 to place an order and $18 per ton per year for storage. Find the minimum-cost purchase quantity.
3. The demand for 1,000 units of a part to be used at a uniform rate throughout the year may be met by manufacturing. The part can be produced at the rate of 3 per hour in a department that works 1,880 hours per year. The set-up cost per lot is estimated to be $40, and the manufacturing cost has been established at $5.20 per unit. Interest, insurance, taxes, space, and other holding costs are $3.10 per unit per year. Calculate the economic manufacturing quantity. Answer: 178
4. An aircraft manufacturer uses 1,000,000 special rivets per year at a uniform rate. The rivets are made on a single-spindle automatic screw machine at the rate of 3,000 per hour. The manufacturing cost is $0.0052 per rivet and the storage cost is estimated to be $0.0008 per rivet per year. The set-up cost is $40 per set-up. Calculate the economic production quantity and the allowable variation in this quantity for a maximum deviation from the minimum cost of 5%. Assume the machine operates 1,700 hours per year.
5. Rederive the economic-purchase-quantity model to reflect holding cost for interest, insurance, taxes, and handling based on the average inventory level and the holding cost for space based on the maximum level.
6. Show that the economic-production-quantity model reduces to the economic-purchase-quantity model as R approaches infinity.
7. The annual demand for an item that can be either purchased or produced is 3,600 units. Costs associated with each alternative are as follows:

	Purchase	*Produce*
Item cost	$ 3.90	$ 3.75
Purchase cost per purchase	15.00	—
Set-up cost per set-up	—	175.00
Holding cost per item per year	1.25	1.25
Production rate per year	—	15,000

Should the item be manufactured or purchased? What is the economic lot size for the least-cost alternative? Answer: Purchase; 294

8. The annual demand for a certain item is 1,000 units per year. Holding cost is \$0.85 per unit per year. Demand can be met by either purchasing or producing the item, as described by the following data:

	Purchase	*Produce*
Item cost	\$ 8.00	\$ 7.40
Procurement cost	20.00	80.00
Production rate	—	2,500 per year

Find the minimum-cost procurement source and calculate the economic advantage over the alternate source. What is the minimum-cost procurement quantity?

9. A manufacturer of a seasonal item finds that her sales are at the rate of 10,000 units per month, except in September and October when the rate of sales is 40,000 units per month. This product is made on single-purpose machines that have an initial cost of \$24,000 each, an estimated life of 6 years, and no salvage value. Fixed charges other than interest and depreciation on each of these machines amount to \$810, and the variable cost amounts to \$0.80 per hour of operation. Machine operators are paid \$7.20 per hour. Each machine normally produces 25 units of product per hour, and the plant operates 160 hours per month. The material entering the product is received as used and costs \$5.10 per unit of product. Charges for interest, taxes, and insurance on the average inventory of finished goods are estimated to be 9% of the variable cost of manufacture and materials. Charges for storage of inventory amount to \$0.52 per unit per year of average inventory. How many machines should be employed in the production of the item for minimum cost if the interest rate is 14%? Answer: 5

10. The expected demand for a certain item during the months of January through June is 10,000 per month. It is expected that the demand will be 30,000 per month during the months of July through September and 20,000 per month during the months of October through December. The item can be ordered for a January 1 delivery at \$3.95 per unit, for an April 1 delivery at \$4 per unit, and for an August 1 delivery at \$4.05 per unit. The item cannot be delivered at other times during the year. The cost to initiate and complete a purchase is \$150 and the storage cost is \$0.10 per unit per year. Interest, insurance, taxes, and other storage costs are estimated to be 9% of the investment in average inventory.

As an alternative, the item can be manufactured on single-purpose machines with the following characteristics:

Machine cost	\$18,000
Service life in years	8
Salvage value	0

Annual fixed cost excluding depreciation	\$800
Variable cost per hour of operation	\$0.42
Hours available per month	160
Production in units per hour	25
Interest rate	10%

The direct-labor cost for the operator is \$6.65 per hour. The direct-material cost per unit is \$5. Calculate the number of machines for minimum cost if the manufacturing alternative is chosen, and compare this with the least-cost purchase policy.

11. Solve graphically for the values of A and B that maximize total profit expressed as

$$TP = \$0.30A = \$0.42B$$

subject to the constraints

$$A \leq 42$$

$$B \leq 30$$

$$A + B \leq 62$$

$$A \geq 0 \text{ and } B \leq 0.$$

Answer: $A = 32$, $B = 30$, $TP = \$22.20$

12. A small machine shop has capability in turning, milling, drilling, and welding. The machine capacity is 16 hours per day in turning, 16 hours per day in milling, 8 hours per day in drilling, and 8 hours per day in welding. Two products, designated A and B, are under consideration. Each will yield a net profit of \$0.25 per unit and will require the following hours of machine time:

	Product A	*Product B*
Turning	0.064	0.106
Milling	0.106	0.053
Drilling	0.000	0.080

Solve graphically for the number of units of each product that should be produced to maximize profit.

13. The firm in Problem 12 makes certain changes to provide 16 hours per day in turning, 8 hours per day in milling, 16 hours per day in drilling, and 8 hours per day in welding. Two products, designated *A* and *B*, are under consideration. Each will yield a net profit of $0.35 per unit and will require the following amount of machine time:

	Product A	*Product B*
Turning	0.046	0.124
Milling	0.112	0.048
Drilling	0.040	0.000
Welding	0.000	0.120

Solve graphically for the number of units of each product that should be produced to maximize profit.

14. A refinery produces three grades of gasoline: premium, regular, and economy. Each grade requires straight gasoline, octane, and additives, available in the amount of 3,200,000, 2,400,000, and 1,100,000 gal per week, respectively. A gallon of premium requires 0.22 gal of straight gasoline, 0.50 gal of octane, and 0.28 gal of additives. One gal of regular requires 0.55 gal of straight gasoline, 0.32 gal of octane, and 0.13 gal of additives. A gallon of economy requires 0.72 gal of straight gasoline, 0.20 gal of octane, and 0.08 gal of additives. A profit of $0.048, $0.040, and $0.029 per gallon is received for premium, regular, and economy, respectively. Formulate the effectiveness function and constraints and find the number of gallons of each grade to produce for maximum profit. Answer: 177,000 gal premium, 7,219,000 gal regular, 0 gal economy; $TP = \$2,963,000$

15. For a certain group of machines, a repair crew requires one day to repair a machine that breaks down. A loss of $180 in incurred for each day that a machine is carried over unrepaired. Each crew costs $200 per day to equip and maintain. The pattern of machine failures on succeeding days of a 30-day period is 7, 5, 8, 0, 3, 4, 5, 0, 2, 3, 8, 4, 5, 6, 1, 3, 4, 0, 2, 8, 6, 5, 5, 3, 4, 6, 7, 1, 3, 2. Assume that all breakdowns occur at the beginning of each day.

a. Determine the idle time of crews when four, five, or six crews are employed.

b. What is the optimum number of crews to employ?

16. It has been proposed to build a driven-in car wash, and the decision must be made as to the number of individual facilities to be provided. A greater number of facilities means fewer customers turned away (and thus more customers serviced), but each additional facility requires additional investment. Assume the interest rate is 10%, investment life is 10 years, and each car serviced produces an operating surplus of $1. Using the following data, determine the number of facilities that will result in the highest expected profit. Answer: 6

No. of Facilities	*Total Invest-ment*	*Probability of Averaging n Cars per Year*								
		2,000	4,000	6,000	8,000	10,000	12,000	14,000	16,000	18,000
1	$10,000	0.6	0.4	—	—	—	—	—	—	—
2	18,000	0.2	0.6	0.2	—	—	—	—	—	—
3	25,000	0.1	0.3	0.4	0.2	—	—	—	—	—
4	32,000	0.1	0.1	0.2	0.3	0.2	0.1	—	—	—
5	38,000	0.05	0.05	0.1	0.2	0.4	0.1	0.1	—	—
6	43,500	0.05	0.05	0.1	0.1	0.2	0.3	0.1	0.1	—
7	50,000	0.05	0.05	0.1	0.1	0.2	0.3	0.1	0.05	0.05
8	55,000	0.05	0.05	0.1	0.1	0.2	0.3	0.1	0.05	0.05

17. A company has 24 identical machines that are operated 24 hours per day, 360 days per year. On the average, each machine breaks down at intervals of 6 working days. The time required to repair each machine is 8 hours. Repairmen receive $12 per hour and work 8 hours per day. A loss of $280 is sustained for each day that a machine is carried over unrepaired.

a. Determine the number of machines that may be expected by chance to go "down" during each day of a month by tossing a die 24 times to represent the 24 machines. Let the downs for each day be represented by the aces that come up.

b. Assuming that all downs as found in part (a) occur at 12 midnight, determine the most economical number of regular repairmen to employ, if no extra repairmen or overtime wages are to be allowed and regular repairmen are not used except to repair the machines in question. Assume that the sequence for the 30-day month will be repeated 12 times during the year.

A

Interest Factors for Discrete Compounding

TABLE A.1.
$\frac{1}{2}$% INTEREST FACTORS FOR DISCRETE COMPOUNDING

	Single Payment		Equal Payment Series				Uniform gradient-series factor
n	Compound-amount factor	Present-worth factor	Compound-amount factor	Sinking-fund factor	Present-worth factor	Capital-recovery factor	Uniform gradient-series factor
	To find F Given P $F/P, i, n$	To find P Given F $P/F, i, n$	To find F Given A $F/A, i, n$	To find A Given F $A/F, i, n$	To find P Given A $P/A, i, n$	To find A Given P $A/P, i, n$	To find A Given g $A/G, i, n$
1	1.005	0.9950	1.000	1.0000	0.9950	1.0050	0.0000
2	1.010	0.9901	2.005	0.4988	1.9851	0.5038	0.4988
3	1.015	0.9852	3.015	0.3317	2.9703	0.3367	0.9967
4	1.020	0.9803	4.030	0.2481	3.9505	0.2531	1.4938
5	1.025	0.9754	5.050	0.1980	4.9259	0.2030	1.9900
6	1.030	0.9705	6.076	0.1646	5.8964	0.1696	2.4855
7	1.036	0.9657	7.106	0.1407	6.8621	0.1457	2.9801
8	1.041	0.9609	8.141	0.1228	7.8230	0.1278	3.4738
9	1.046	0.9561	9.182	0.1089	8.7791	0.1139	3.9668
10	1.051	0.9514	10.228	0.0978	9.7304	0.1028	4.4589
11	1.056	0.9466	11.279	0.0887	10.6770	0.0937	4.9501
12	1.062	0.9419	12.336	0.0811	11.6189	0.0861	5.4406
13	1.067	0.9372	13.397	0.0747	12.5562	0.0797	5.9302
14	1.072	0.9326	14.464	0.0691	13.4887	0.0741	6.4190
15	1.078	0.9279	15.537	0.0644	14.4166	0.0694	6.9069
16	1.083	0.9233	16.614	0.0602	15.3399	0.0652	7.3940
17	1.088	0.9187	17.697	0.0565	16.2586	0.0615	7.8803
18	1.094	0.9141	18.786	0.0532	17.1728	0.0582	8.3658
19	1.099	0.9096	19.880	0.0503	18.0824	0.0553	8.8504
20	1.105	0.9051	20.979	0.0477	18.9874	0.0527	9.3342
21	1.110	0.9006	22.084	0.0453	19.8880	0.0503	9.8172
22	1.116	0.8961	23.194	0.0431	20.7841	0.0481	10.2993
23	1.122	0.8916	24.310	0.0411	21.6757	0.0461	10.7806
24	1.127	0.8872	25.432	0.0393	22.5629	0.0443	11.2611
25	1.133	0.8828	26.559	0.0377	23.4456	0.0427	11.7407
26	1.138	0.8784	27.692	0.0361	24.3240	0.0411	12.2195
27	1.144	0.8740	28.830	0.0347	25.1980	0.0397	12.6975
28	1.150	0.8697	29.975	0.0334	26.0677	0.0384	13.1747
29	1.156	0.8653	31.124	0.0321	26.9330	0.0371	13.6510
30	1.161	0.8610	32.280	0.0310	27.7941	0.0360	14.1265
31	1.167	0.8568	33.441	0.0299	28.6508	0.0349	14.6012
32	1.173	0.8525	34.609	0.0289	29.5033	0.0339	15.0750
33	1.179	0.8483	35.782	0.0280	30.3515	0.0330	15.5480
34	1.185	0.8440	36.961	0.0271	31.1956	0.0321	16.0202
35	1.191	0.8398	38.145	0.0262	32.0354	0.0312	16.4915
40	1.221	0.8191	44.159	0.0227	36.1722	0.0277	18.8358
45	1.252	0.7990	50.324	0.0199	40.2072	0.0249	21.1595
50	1.283	0.7793	56.645	0.0177	44.1428	0.0227	23.4624
55	1.316	0.7601	63.126	0.0159	47.9815	0.0209	25.7447
60	1.349	0.7414	69.770	0.0143	51.7256	0.0193	28.0064
65	1.383	0.7231	76.582	0.0131	55.3775	0.0181	30.2475
70	1.418	0.7053	83.566	0.0120	58.9394	0.0170	32.4680
75	1.454	0.6879	90.727	0.0110	62.4137	0.0160	34.6679
80	1.490	0.6710	98.068	0.0102	65.8023	0.0152	36.8474
85	1.528	0.6545	105.594	0.0095	69.1075	0.0145	39.0065
90	1.567	0.6384	113.311	0.0088	72.3313	0.0138	41.1451
95	1.606	0.6226	121.222	0.0083	75.4757	0.0133	43.2633
100	1.647	0.6073	129.334	0.0077	78.5427	0.0127	45.3613

TABLE A.2.
$\frac{3}{4}$% INTEREST FACTORS FOR DISCRETE COMPOUNDING

	Single Payment		Equal Payment Series				Uniform gradient-series factor
n	Compound-amount factor	Present-worth factor	Compound-amount factor	Sinking-fund factor	Present-worth factor	Capital-recovery factor	
	To find F Given P $F/P, i, n$	To find P Given F $P/F, i, n$	To find F Given A $F/A, i, n$	To find A Given F $A/F, i, n$	To find P Given A $P/A, i, n$	To find A Given P $A/P, i, n$	To find A Given G $A/G, i, n$
1	1.008	0.9926	1.000	1.0000	0.9926	1.0075	0.0000
2	1.015	0.9852	2.008	0.4981	1.9777	0.5056	0.4981
3	1.023	0.9778	3.023	0.3309	2.9556	0.3384	0.9950
4	1.030	0.9706	4.045	0.2472	3.9261	0.2547	1.4907
5	1.038	0.9633	5.076	0.1970	4.8894	0.2045	1.9851
6	1.046	0.9562	6.114	0.1636	5.8456	0.1711	2.4782
7	1.054	0.9491	7.159	0.1397	6.7946	0.1472	2.9701
8	1.062	0.9420	8.213	0.1218	7.7366	0.1293	3.4608
9	1.070	0.9350	9.275	0.1078	8.6716	0.1153	3.9502
10	1.078	0.9280	10.344	0.0967	9.5996	0.1042	4.4384
11	1.086	0.9211	11.422	0.0876	10.5207	0.0951	4.9253
12	1.094	0.9142	12.508	0.0800	11.4349	0.0875	5.4110
13	1.102	0.9074	13.601	0.0735	12.3424	0.0810	5.8954
14	1.110	0.9007	14.703	0.0680	13.2430	0.0755	6.3786
15	1.119	0.8940	15.814	0.0632	14.1370	0.0707	6.8606
16	1.127	0.8873	16.932	0.0591	15.0243	0.0666	7.3413
17	1.135	0.8807	18.059	0.0554	15.9050	0.0629	7.8207
18	1.144	0.8742	19.195	0.0521	16.7792	0.0596	8.2989
19	1.153	0.8677	20.339	0.0492	17.6468	0.0567	8.7759
20	1.161	0.8612	21.491	0.0465	18.5080	0.0540	9.2517
21	1.170	0.8548	22.652	0.0442	19.3628	0.0517	9.7261
22	1.179	0.8484	23.822	0.0420	20.2112	0.0495	10.1994
23	1.188	0.8421	25.001	0.0400	21.0533	0.0475	10.6714
24	1.196	0.8358	26.188	0.0382	21.8892	0.0457	11.1422
25	1.205	0.8296	27.385	0.0365	22.7188	0.0440	11.6117
26	1.214	0.8234	28.590	0.0350	23.5422	0.0425	12.0800
27	1.224	0.8173	29.805	0.0336	24.3595	0.0411	12.5470
28	1.233	0.8112	31.028	0.0322	25.1707	0.0397	13.0128
29	1.242	0.8052	32.261	0.0310	25.9759	0.0385	13.4774
30	1.251	0.7992	33.503	0.0299	26.7751	0.0374	13.9407
31	1.261	0.7932	34.754	0.0288	27.5683	0.0363	14.4028
32	1.270	0.7873	36.015	0.0278	28.3557	0.0353	14.8636
33	1.280	0.7815	37.285	0.0268	29.1371	0.0343	15.3232
34	1.289	0.7757	38.565	0 0259	29.9128	0.0334	15.7816
35	1.299	0.7699	39.854	0.0251	30.6827	0.0326	16.2387
40	1.348	0.7417	46.446	0.0215	34.4469	0.0290	18.5058
45	1.400	0.7145	53.290	0.0188	38.0732	0.0263	20.7421
50	1.453	0.6883	60.394	0.0166	41.5665	0.0241	22.9476
55	1.508	0.6630	67.769	0.0148	44.9316	0.0223	25.1223
60	1.566	0.6387	75.424	0.0133	48.1734	0.0208	27.2665
65	1.625	0.6153	83.371	0.0120	51.2963	0.0195	29.3801
70	1.687	0.5927	91.620	0.0109	54.3046	0.0184	31.4634
75	1.751	0.5710	100.183	0.0100	57.2027	0.0175	33.5163
80	1.818	0.5501	109.073	0.0092	59.9945	0.0167	35.5391
85	1.887	0.5299	118.300	0.0085	62.6838	0.0160	37.5318
90	1.959	0.5105	127.879	0.0078	65.2746	0.0153	39.4946
95	2.034	0.4917	137.823	0.0073	67.7704	0.0148	41.4277
100	2.111	0.4737	148.145	0.0068	70.1746	0.0143	43.3311

TABLE A.3.
1% INTEREST FACTORS FOR DISCRETE COMPOUNDING

n	Single Payment		Equal Payment Series				Uniform gradient-series factor
	Compound-amount factor	Present-worth factor	Compound-amount factor	Sinking-fund factor	Present-worth factor	Capital-recovery factor	
	To find F Given P $F/P, i, n$	To find P Given F $P/F, i, n$	To find F Given A $F/A, i, n$	To find A Given F $A/F, i, n$	To find P Given A $P/A, i, n$	To find A Given P $A/P, i, n$	To find A Given G $A/G, i, n$
1	1.010	0.9901	1.000	1.0000	0.9901	1.0100	0.0000
2	1.020	0.9803	2.010	0.4975	1.9704	0.5075	0.4975
3	1.030	0.9706	3.030	0.3300	2.9410	0.3400	0.9934
4	1.041	0.9610	4.060	0.2463	3.9020	0.2563	1.4876
5	1.051	0.9515	5.101	0.1960	4.8534	0.2060	1.9801
6	1.062	0.9421	6.152	0.1626	5.7955	0.1726	2.4710
7	1.072	0.9327	7.214	0.1386	6.7282	0.1486	2.9602
8	1.083	0.9235	8.286	0.1207	7.6517	0.1307	3.4478
9	1.094	0.9143	9.369	0.1068	8.5660	0.1168	3.9337
10	1.105	0.9053	10.462	0.0956	9.4713	0.1056	4.4179
11	1.116	0.8963	11.567	0.0865	10.3676	0.0965	4.9005
12	1.127	0.8875	12.683	0.0789	11.2551	0.0889	5.3815
13	1.138	0.8787	13.809	0.0724	12.1338	0.0824	5.8607
14	1.149	0.8700	14.947	0.0669	13.0037	0.0769	6.3384
15	1.161	0.8614	16.097	0.0621	13.8651	0.0721	6.8143
16	1.173	0.8528	17.258	0.0580	14.7179	0.0680	7.2887
17	1.184	0.8444	18.430	0.0543	15.5623	0.0643	7.7613
18	1.196	0.8360	19.615	0.0510	16.3983	0.0610	8.2323
19	1.208	0.8277	20.811	0.0481	17.2260	0.0581	8.7017
20	1.220	0.8196	22.019	0.0454	18.0456	0.0554	9.1694
21	1.232	0.8114	23.239	0.0430	18.8570	0.0530	9.6354
22	1.245	0.8034	24.472	0.0409	19.6604	0.0509	10.0998
23	1.257	0.7955	25.716	0.0389	20.4558	0.0489	10.5626
24	1.270	0.7876	26.973	0.0371	21.2434	0.0471	11.0237
25	1.282	0.7798	28.243	0.0354	22.0232	0.0454	11.4831
26	1.295	0.7721	29.526	0.0339	22.7952	0.0439	11.9409
27	1.308	0.7644	30.821	0.0325	23.5596	0.0425	12.3971
28	1.321	0.7568	32.129	0.0311	24.3165	0.0411	12.8516
29	1.335	0.7494	33.450	0.0299	25.0658	0.0399	13.3045
30	1.348	0.7419	34.785	0.0288	25.8077	0.0388	13.7557
31	1.361	0.7346	36.133	0.0277	26.5423	0.0377	14.2052
32	1.375	0.7273	37.494	0.0267	27.2696	0.0367	14.6532
33	1.389	0.7201	38.869	0.0257	27.9897	0.0357	15.0995
34	1.403	0.7130	40.258	0.0248	28.7027	0.0348	15.5441
35	1.417	0.7059	41.660	0.0240	29.4086	0.0340	15.9871
40	1.489	0.6717	48.886	0.0205	32.8347	0.0305	18.1776
45	1.565	0.6391	56.481	0.0177	36.0945	0.0277	20.3273
50	1.645	0.6080	64.463	0.0155	39.1961	0.0255	22.4363
55	1.729	0.5785	72.852	0.0137	42.1472	0.0237	24.5049
60	1.817	0.5505	81.670	0.0123	44.9550	0.0223	26.5333
65	1.909	0.5237	90.937	0.0110	47.6266	0.0210	28.5217
70	2.007	0.4983	100.676	0.0099	50.1685	0.0199	30.4703
75	2.109	0.4741	110.913	0.0090	52.5871	0.0190	32.3793
80	2.217	0.4511	121.672	0.0082	54.8882	0.0182	34.2492
85	2.330	0.4292	132.979	0.0075	57.0777	0.0175	36.0801
90	2.449	0.4084	144.863	0.0069	59.1609	0.0169	37.8725
95	2.574	0.3886	157.354	0.0064	61.1430	0.0164	39.6265
100	2.705	0.3697	170.481	0.0059	63.0289	0.0159	41.3426

TABLE A.4.
$1\frac{1}{4}$% INTEREST FACTORS FOR DISCRETE COMPOUNDING

	Single Payment		Equal Payment Series				Uniform gradient-series factor
n	Compound-amount factor	Present-worth factor	Compound-amount factor	Sinking-fund factor	Present-worth factor	Capital-recovery factor	
	To find F Given P $F/P, i, n$	To find P Given F $P/F, i, n$	To find F Given A $F/A, i, n$	To find A Given F $A/F, i, n$	To find P Given A $P/A, i, n$	To find A Given P $A/P, i, n$	To find A Given G $A/G, i, n$
1	1.013	0.9877	1.000	1.0000	0.9877	1.0126	0.0000
2	1.025	0.9755	2.013	0.4970	1.9631	0.5095	0.4932
3	1.038	0.9635	3.038	0.3293	2.9265	0.3418	0.9895
4	1.051	0.9516	4.076	0.2454	3.8780	0.2579	1.4830
5	1.064	0.9398	5.127	0.1951	4.8177	0.2076	1.9729
6	1.077	0.9282	6.191	0.1616	5.7459	0.1741	2.4618
7	1.091	0.9168	7.268	0.1376	6.6627	0.1501	2.9491
8	1.105	0.9055	8.359	0.1197	7.5680	0.1322	3.4330
9	1.118	0.8943	9.463	0.1057	8.4623	0.1182	3.9158
10	1.132	0.8832	10.582	0.0946	9.3454	0.1071	4.3960
11	1.147	0.8723	11.714	0.0854	10.2177	0.0979	4.8744
12	1.161	0.8616	12.860	0.0778	11.0792	0.0903	5.3506
13	1.175	0.8509	14.021	0.0714	11.9300	0.0839	5.8248
14	1.190	0.8404	15.196	0.0659	12.7704	0.0784	6.2968
15	1.205	0.8300	16.386	0.0611	13.6004	0.0736	6.7669
16	1.220	0.8198	17.591	0.0569	14.4201	0.0694	7.2350
17	1.235	0.8097	18.811	0.0532	15.2298	0.0657	7.7009
18	1.251	0.7997	20.046	0.0499	16.0293	0.0624	8.1645
19	1.266	0.7898	21.296	0.0470	16.8191	0.0595	8.6264
20	1.282	0.7801	22.563	0.0444	17.5991	0.0569	9.0861
21	1.298	0.7704	23.845	0.0420	18.3695	0.0545	9.5439
22	1.314	0.7609	25.143	0.0398	19.1303	0.0523	9.9993
23	1.331	0.7515	26.457	0.0378	19.8818	0.0503	10.4528
24	1.347	0.7423	27.788	0.0360	20.6240	0.0485	10.9044
25	1.364	0.7331	29.135	0.0344	21.3570	0.0469	11.3539
26	1.381	0.7240	30.499	0.0328	22.0810	0.0453	11.8012
27	1.399	0.7151	31.880	0.0314	22.7960	0.0439	12.2465
28	1.416	0.7063	33.279	0.0301	23.5022	0.0426	12.6898
29	1.434	0.6976	34.695	0.0289	24.1998	0.0414	13.1311
30	1.452	0.6889	36.128	0.0277	24.8886	0.0402	13.5703
31	1.470	0.6804	37.580	0.0267	25.5690	0.0392	14.0074
32	1.488	0.6720	39.050	0.0257	26.2410	0.0382	14.4425
33	1.507	0.6637	40.538	0.0247	26.9047	0.0372	14.8756
34	1.526	0.6555	42.045	0.0238	27.5601	0.0363	15.3066
35	1.545	0.6475	43.570	0.0230	28.2075	0.0355	15.7357
40	1.644	0.6085	51.489	0.0195	31.3266	0.0320	17.8503
45	1.749	0.5718	59.915	0.0167	34.2578	0.0292	19.9144
50	1.861	0.5374	68.880	0.0146	37.0125	0.0271	21.9284
55	1.980	0.5050	78.421	0.0128	39.6013	0.0253	23.8925
60	2.107	0.4746	88.573	0.0113	42.0342	0.0238	25.8072
65	2.242	0.4460	99.375	0.0101	44.3206	0.0226	27.6730
70	2.386	0.4192	110.870	0.0091	46.4693	0.0216	29.4902
75	2.539	0.3939	123.101	0.0082	48.4886	0.0207	31.2594
80	2.702	0.3702	136.116	0.0074	50.3862	0.0199	32.9812
85	2.875	0.3479	149.965	0.0067	52.1696	0.0192	34.6560
90	3.059	0.3270	164.701	0.0061	53.8456	0.0186	36.2844
95	3.255	0.3073	180.382	0.0056	55.4207	0.0181	37.8671
100	3.463	0.2888	197.067	0.0051	56.9009	0.0176	39.4048

TABLE A.5.
$1\frac{1}{2}$% INTEREST FACTORS FOR DISCRETE COMPOUNDING

	Single Payment		Equal Payment Series				Uniform gradient-series factor
n	Compound-amount factor	Present-worth factor	Compound-amount factor	Sinking-fund factor	Present-worth factor	Capital-recovery factor	
	To find *F* Given *P* *F/P, i, n*	To find *P* Given *F* *P/F, i, n*	To find *F* Given *A* *F/A, i, n*	To find *A* Given *F* *A/F, i, n*	To find *P* Given *A* *P/A, i, n*	To find *A* Given *P* *A/P, i, n*	To find *A* Given *G* *A/G, i, n*
1	1.015	0.9852	1.000	1.0000	0.9852	1.0150	0.0000
2	1.030	0.9707	2.015	0.4963	1.9559	0.5113	0.4963
3	1.046	0.9563	3.045	0.3284	2.9122	0.3434	0.9901
4	1.061	0.9422	4.091	0.2445	3.8544	0.2595	1.4814
5	1.077	0.9283	5.152	0.1941	4.7827	0.2091	1.9702
6	1.093	0.9146	6.230	0.1605	5.6972	0.1755	2.4566
7	1.110	0.9010	7.323	0.1366	6.5982	0.1516	2.9405
8	1.127	0.8877	8.433	0.1186	7.4859	0.1336	3.4219
9	1.143	0.8746	9.559	0.1046	8.3605	0.1196	3.9008
10	1.161	0.8617	10.703	0.0934	9.2222	0.1084	4.3772
11	1.178	0.8489	11.863	0.0843	10.0711	0.0993	4.8512
12	1.196	0.8364	13.041	0.0767	10.9075	0.0917	5.3227
13	1.214	0.8240	14.237	0.0703	11.7315	0.0853	5.7917
14	1.232	0.8119	15.450	0.0647	12.5434	0.0797	6.2582
15	1.250	0.7999	16.682	0.0600	13.3432	0.0750	6.7223
16	1.269	0.7880	17.932	0.0558	14.1313	0.0708	7.1839
17	1.288	0.7764	19.201	0.0521	14.9077	0.0671	7.6431
18	1.307	0.7649	20.489	0.0488	15.6726	0.0638	8.0997
19	1.327	0.7536	21.797	0.0459	16.4262	0.0609	8.5539
20	1.347	0.7425	23.124	0.0433	17.1686	0.0583	9.0057
21	1.367	0.7315	24.471	0.0409	17.9001	0.0559	9.4550
22	1.388	0.7207	25.838	0.0387	18.6208	0.0537	9.9018
23	1.408	0.7100	27.225	0.0367	19.3309	0.0517	10.3462
24	1.430	0.6996	28.634	0.0349	20.0304	0.0499	10.7881
25	1.451	0.6892	30.063	0.0333	20.7196	0.0483	11.2276
26	1.473	0.6790	31.514	0.0317	21.3986	0.0467	11.6646
27	1.495	0.6690	32.987	0.0303	22.0676	0.0453	12.0992
28	1.517	0.6591	34.481	0.0290	22.7267	0.0440	12.5313
29	1.540	0.6494	35.999	0.0278	23.3761	0.0428	12.9610
30	1.563	0.6398	37.539	0.0266	24.0158	0.0416	13.3883
31	1.587	0.6303	39.102	0.0256	24.6462	0.0406	13.8131
32	1.610	0.6210	40.688	0.0246	25.2671	0.0396	14.2355
33	1.634	0.6118	42.299	0.0237	25.8790	0.0387	14.6555
34	1.659	0.6028	43.933	0.0228	26.4817	0.0378	15.0731
35	1.684	0.5939	45.592	0.0219	27.0756	0.0369	15.4882
40	1.814	0.5513	54.268	0.0184	29.9159	0.0334	17.5277
45	1.954	0.5117	63.614	0.0157	32.5523	0.0307	19.5074
50	2.105	0.4750	73.683	0.0136	34.9997	0.0286	21.4277
55	2.268	0.4409	84.530	0.0118	37.2715	0.0268	23.2894
60	2.443	0.4093	96.215	0.0104	39.3803	0.0254	25.0930
65	2.632	0.3799	108.803	0.0092	41.3378	0.0242	26.8392
70	2.835	0.3527	122.364	0.0082	43.1549	0.0232	28.5290
75	3.055	0.3274	136.973	0.0073	44.8416	0.0223	30.1631
80	3.291	0.3039	152.711	0.0066	46.4073	0.0216	31.7423
85	3.545	0.2821	169.665	0.0059	47.8607	0.0209	33.2676
90	3.819	0.2619	187.930	0.0053	49.2099	0.0203	34.7399
95	4.114	0.2431	207.606	0.0048	50.4622	0.0198	36.1602
100	4.432	0.2256	228.803	0.0044	51.6247	0.0194	37.5295

TABLE A.6.
2% INTEREST FACTORS FOR DISCRETE COMPOUNDING

n	Single Payment		Equal Payment Series				Uniform gradient-series factor
	Compound-amount factor	Present-worth factor	Compound-amount factor	Sinking-fund factor	Present-worth factor	Capital-recovery factor	
	To find F Given P $F/P, i, n$	To find P Given F $P/F, i, n$	To find F Given A $F/A, i, n$	To find A Given F $A/F, i, n$	To find P Given A $P/A, i, n$	To find A Given P $A/P, i, n$	To find A Given G $A/G, i, n$
1	1.020	0.9804	1.000	1.0000	0.9804	1.0200	0.0000
2	1.040	0.9612	2.020	0.4951	1.9416	0.5151	0.4951
3	1.061	0.9423	3.060	0.3268	2.8839	0.3468	0.9868
4	1.082	0.9239	4.122	0.2426	3.8077	0.2626	1.4753
5	1.104	0.9057	5.204	0.1922	4.7135	0.2122	1.9604
6	1.126	0.8880	6.308	0.1585	5.6014	0.1785	2.4423
7	1.149	0.8706	7.434	0.1345	6.4720	0.1545	2.9208
8	1.172	0.8535	8.583	0.1165	7.3255	0.1365	3.3961
9	1.195	0.8368	9.755	0.1025	8.1622	0.1225	3.8681
10	1.219	0.8204	10.950	0.0913	8.9826	0.1113	4.3367
11	1.243	0.8043	12.169	0.0822	9.7869	0.1022	4.8021
12	1.268	0.7885	13.412	0.0746	10.5754	0.0946	5.2643
13	1.294	0.7730	14.680	0.0681	11.3484	0.0881	5.7231
14	1.319	0.7579	15.974	0.0626	12.1063	0.0826	6.1786
15	1.346	0.7430	17.293	0.0578	12.8493	0.0778	6.6309
16	1.373	0.7285	18.639	0.0537	13.5777	0.0737	7.0799
17	1.400	0.7142	20.012	0.0500	14.2919	0.0700	7.5256
18	1.428	0.7002	21.412	0.0467	14.9920	0.0667	7.9681
19	1.457	0.6864	22.841	0.0438	15.6785	0.0638	8.4073
20	1.486	0.6730	24.297	0.0412	16.3514	0.0612	8.8433
21	1.516	0.6598	25.783	0.0388	17.0112	0.0588	9.2760
22	1.546	0.6468	27.299	0.0366	17.6581	0.0566	9.7055
23	1.577	0.6342	28.845	0.0347	18.2922	0.0547	10.1317
24	1.608	0.6217	30.422	0.0329	18.9139	0.0529	10.5547
25	1.641	0.6095	32.030	0.0312	19.5235	0.0512	10.9745
26	1.673	0.5976	33.671	0.0297	20.1210	0.0497	11.3910
27	1.707	0.5859	35.344	0.0283	20.7069	0.0483	11.8043
28	1.741	0.5744	37.051	0.0270	21.2813	0.0470	12.2145
29	1.776	0.5631	38.792	0.0258	21.8444	0.0458	12.6214
30	1.811	0.5521	40.568	0.0247	22.3965	0.0447	13.0251
31	1.848	0.5413	42.379	0.0236	22.9377	0.0436	13.4257
32	1.885	0.5306	44.227	0.0226	23.4683	0.0426	13.8230
33	1.922	0.5202	46.112	0.0217	23.9886	0.0417	14.2172
34	1.961	0.5100	48.034	0.0208	24.4986	0.0408	14.6083
35	2.000	0.5000	49.994	0.0200	24.9986	0.0400	14.9961
40	2.208	0.4529	60.402	0.0166	27.3555	0.0366	16.8885
45	2.438	0.4102	71.893	0.0139	29.4902	0.0339	18.7034
50	2.692	0.3715	84.579	0.0118	31.4236	0.0318	20.4420
55	2.972	0.3365	98.587	0.0102	33.1748	0.0302	22.1057
60	3.281	0.3048	114.052	0.0088	34.7609	0.0288	23.6961
65	3.623	0.2761	131.126	0.0076	36.1975	0.0276	25.2147
70	4.000	0.2500	149.978	0.0067	37.4986	0.0267	26.6632
75	4.416	0.2265	170.792	0.0059	38.6771	0.0259	28.0434
80	4.875	0.2051	193.772	0.0052	39.7445	0.0252	29.3572
85	5.383	0.1858	219.144	0.0046	40.7113	0.0246	30.6064
90	5.943	0.1683	247.157	0.0041	41.5869	0.0241	31.7929
95	6.562	0.1524	278.085	0.0036	42.3800	0.0236	32.9189
100	7.245	0.1380	312.232	0.0032	43.0984	0.0232	33.9863

TABLE A.7.
3% INTEREST FACTORS FOR DISCRETE COMPOUNDING

	Single Payment		Equal Payment Series				Uniform gradient-series factor
n	Compound-amount factor	Present-worth factor	Compound-amount factor	Sinking-fund factor	Present-worth factor	Capital-recovery factor	
	To find F Given P $F/P, i, n$	To find P Given F $P/F, i, n$	To find F Given A $F/A, i, n$	To find A Given F $A/F, i, n$	To find P Given A $P/A, i, n$	To find A Given P $A/P, i, n$	To find A Given G $A/G, i, n$
1	1.030	0.9709	1.000	1.0000	0.9709	1.0300	0.0000
2	1.061	0.9426	2.030	0.4926	1.9135	0.5226	0.4926
3	1.093	0.9152	3.091	0.3235	2.8286	0.3535	0.9803
4	1.126	0.8885	4.184	0.2390	3.7171	0.2690	1.4631
5	1.159	0.8626	5.309	0.1884	4.5797	0.2184	1.9409
6	1.194	0.8375	6.468	0.1546	5.4172	0.1846	2.4138
7	1.230	0.8131	7.662	0.1305	6.2303	0.1605	2.8819
8	1.267	0.7894	8.892	0.1125	7.0197	0.1425	3.3450
9	1.305	0.7664	10.159	0.0984	7.7861	0.1284	3.8032
10	1.344	0.7441	11.464	0.0872	8.5302	0.1172	4.2565
11	1.384	0.7224	12.808	0.0781	9.2526	0.1081	4.7049
12	1.426	0.7014	14.192	0.0705	9.9540	0.1005	5.1485
13	1.469	0.6810	15.618	0.0640	10.6350	0.0940	5.5872
14	1.513	0.6611	17.086	0.0585	11.2961	0.0885	6.0211
15	1.558	0.6419	18.599	0.0538	11.9379	0.0838	6.4501
16	1.605	0.6232	20.157	0.0496	12.5611	0.0796	6.8742
17	1.653	0.6050	21.762	0.0460	13.1661	0.0760	7.2936
18	1.702	0.5874	23.414	0.0427	13.7535	0.0727	7.7081
19	1.754	0.5703	25.117	0.0398	14.3238	0.0698	8.1179
20	1.806	0.5537	26.870	0.0372	14.8775	0.0672	8.5229
21	1.860	0.5376	28.676	0.0349	15.4150	0.0649	8.9231
22	1.916	0.5219	30.537	0.0328	15.9369	0.0628	9.3186
23	1.974	0.5067	32.453	0.0308	16.4436	0.0608	9.7094
24	2.033	0.4919	34.426	0.0291	16.9356	0.0591	10.0954
25	2.094	0.4776	36.459	0.0274	17.4132	0.0574	10.4768
26	2.157	0.4637	38.553	0.0259	17.8769	0.0559	10.8535
27	2.221	0.4502	40.710	0.0246	18.3270	0.0546	11.2256
28	2.288	0.4371	42.931	0.0233	18.7641	0.0533	11.5930
29	2.357	0.4244	45.219	0.0221	19.1885	0.0521	11.9558
30	2.427	0.4120	47.575	0.0210	19.6005	0.0510	12.3141
31	2.500	0.4000	50.003	0.0200	20.0004	0.0500	12.6678
32	2.575	0.3883	52.503	0.0191	20.3888	0.0491	13.0169
33	2.652	0.3770	55.078	0.0182	20.7658	0.0482	13.3616
34	2.732	0.3661	57.730	0.0173	21.1318	0.0473	13.7018
35	2.814	0.3554	60.462	0.0165	21.4872	0.0465	14.0375
40	3.262	0.3066	75.401	0.0133	23.1148	0.0433	15.6502
45	3.782	0.2644	92.720	0.0108	24.5187	0.0408	17.1556
50	4.384	0.2281	112.797	0.0089	25.7298	0.0389	18.5575
55	5.082	0.1968	136.072	0.0074	26.7744	0.0374	19.8600
60	5.892	0.1697	163.053	0.0061	27.6756	0.0361	21.0674
65	6.830	0.1464	194.333	0.0052	28.4529	0.0352	22.1841
70	7.918	0.1263	230.594	0.0043	29.1234	0.0343	23.2145
75	9.179	0.1090	272.631	0.0037	29.7018	0.0337	24.1634
80	10.641	0.0940	321.363	0.0031	30.2008	0.0331	25.0354
85	12.336	0.0811	377.857	0.0027	30.6312	0.0327	25.8349
90	14.300	0.0699	443.349	0.0023	31.0024	0.0323	26.5667
95	16.578	0.0603	519.272	0.0019	31.3227	0.0319	27.2351
100	19.219	0.0520	607.288	0.0017	31.5989	0.0317	27.8445

TABLE A.8.
4% INTEREST FACTORS FOR DISCRETE COMPOUNDING

	Single Payment		Equal Payment Series				Uniform gradient-series factor
n	Compound-amount factor	Present-worth factor	Compound-amount factor	Sinking-fund factor	Present-worth factor	Capital-recovery factor	
	To find *F* Given *P* *F/P, i, n*	To find *P* Given *F* *P/F, i, n*	To find *F* Given *A* *F/A, i, n*	To find *A* Given *F* *A/F, i, n*	To find *P* Given *A* *P/A, i, n*	To find *A* Given *P* *A/P, i, n*	To find *A* Given *G* *A/G, i, n*
1	1.040	0.9615	1.000	1.0000	0.9615	1.0400	0.0000
2	1.082	0.9246	2.040	0.4902	1.8861	0.5302	0.4902
3	1.125	0.8890	3.122	0.3204	2.7751	0.3604	0.9739
4	1.170	0.8548	4.246	0.2355	3.6299	0.2755	1.4510
5	1.217	0.8219	5.416	0.1846	4.4518	0.2246	1.9216
6	1.265	0.7903	6.633	0.1508	5.2421	0.1908	2.3857
7	1.316	0.7599	7.898	0.1266	6.0021	0.1666	2.8433
8	1.369	0.7307	9.214	0.1085	6.7328	0.1485	3.2944
9	1.423	0.7026	10.583	0.0945	7.4353	0.1345	3.7391
10	1.480	0.6756	12.006	0.0833	8.1109	0.1233	4.1773
11	1.539	0.6496	13.486	0.0742	8.7605	0.1142	4.6090
12	1.601	0.6246	15.026	0.0666	9.3851	0.1066	5.0344
13	1.665	0.6006	16.627	0.0602	9.9857	0.1002	5.4533
14	1.732	0.5775	18.292	0.0547	10.5631	0.0947	5.8659
15	1.801	0.5553	20.024	0.0500	11.1184	0.0900	6.2721
16	1.873	0.5339	21.825	0.0458	11.6523	0.0858	6.6720
17	1.948	0.5134	23.698	0.0422	12.1657	0.0822	7.0656
18	2.026	0.4936	25.645	0.0390	12.6593	0.0790	7.4530
19	2.107	0.4747	27.671	0.0361	13.1339	0.0761	7.8342
20	2.191	0.4564	29.778	0.0336	13.5903	0.0736	8.2091
21	2.279	0.4388	31.969	0.0313	14.0292	0.0713	8.5780
22	2.370	0.4220	34.248	0.0292	14.4511	0.0692	8.9407
23	2.465	0.4057	36.618	0.0273	14.8569	0.0673	9.2973
24	2.563	0.3901	39.083	0.0256	15.2470	0.0656	9.6479
25	2.666	0.3751	41.646	0.0240	15.6221	0.0640	9.9925
26	2.772	0.3607	44.312	0.0226	15.9828	0.0626	10.3312
27	2.883	0.3468	47.084	0.0212	16.3296	0.0612	10.6640
28	2.999	0.3335	49.968	0.0200	16.6631	0.0600	10.9909
29	3.119	0.3207	52.966	0.0189	16.9837	0.0589	11.3121
30	3.243	0.3083	56.085	0.0178	17.2920	0.0578	11.6274
31	3.373	0.2965	59.328	0.0169	17.5885	0.0569	11.9371
32	3.508	0.2851	62.701	0.0160	17.8736	0.0560	12.2411
33	3.648	0.2741	66.210	0.0151	18.1477	0.0551	12.5396
34	3.794	0.2636	69.858	0.0143	18.4112	0.0543	12.8325
35	3.946	0.2534	73.652	0.0136	18.6646	0.0536	13.1199
40	4.801	0.2083	95.026	0.0105	19.7928	0.0505	14.4765
45	5.841	0.1712	121.029	0.0083	20.7200	0.0483	15.7047
50	7.107	0.1407	152.667	0.0066	21.4822	0.0466	16.8123
55	8.646	0.1157	191.159	0.0052	22.1086	0.0452	17.8070
60	10.520	0.0951	237.991	0.0042	22.6235	0.0442	18.6972
65	12.799	0.0781	294.968	0.0034	23.0467	0.0434	19.4909
70	15.572	0.0642	364.290	0.0028	23.3945	0.0428	20.1961
75	18.945	0.0528	448.631	0.0022	23.6804	0.0422	20.8206
80	23.050	0.0434	551.245	0.0018	23.9154	0.0418	21.3719
85	28.044	0.0357	676.090	0.0015	24.1085	0.0415	21.8569
90	34.119	0.0293	817.983	0.0012	24.2673	0.0412	22.2826
95	41.511	0.0241	1012.785	0.0010	24.3978	0.0410	22.6550
100	50.505	0.0198	1237.624	0.0008	24.5050	0.0408	22.9800

TABLE A.9.
5% INTEREST FACTORS FOR DISCRETE COMPOUNDING

n	Single Payment		Equal Payment Series				Uniform gradient-series factor
	Compound-amount factor	Present-worth factor	Compound-amount factor	Sinking-fund factor	Present-worth factor	Capital-recovery factor	
	To find F Given P $F/P, i, n$	To find P Given F $P/F, i, n$	To find F Given A $F/A, i, n$	To find A Given F $A/F, i, n$	To find P Given A $P/A, i, n$	To find A Given P $A/P, i, n$	To find A Given G $A/G, i, n$
1	1.050	0.9524	1.000	1.0000	0.9524	1.0500	0.0000
2	1.103	0.9070	2.050	0.4878	1.8594	0.5378	0.4878
3	1.158	0.8638	3.153	0.3172	2.7233	0.3672	0.9675
4	1.216	0.8227	4.310	0.2320	3.5460	0.2820	1.4391
5	1.276	0.7835	5.526	0.1810	4.3295	0.2310	1.90[illegible]5
6	1.340	0.7462	6.802	0.1470	5.0757	0.1970	2.3579
7	1.407	0.7107	8.142	0.1228	5.7864	0.1728	2.8052
8	1.477	0.6768	9.549	0.1047	6.4632	0.1547	3.2445
9	1.551	0.6446	11.027	0.0907	7.1078	0.1407	3.6758
10	1.629	0.6139	12.587	0.0795	7.7217	0.1295	4.0991
11	1.710	0.5847	14.207	0.0704	8.3064	0.1204	4.5145
12	1.796	0.5568	15.917	0.0628	8.8633	0.1128	4.9219
13	1.886	0.5303	17.713	0.0565	9.3936	0.1065	5.3215
14	1.980	0.5051	19.599	0.0510	9.8987	0.1010	5.7133
15	2.079	0.4810	21.579	0.0464	10.3797	0.0964	6.0973
16	2.183	0.4581	23.658	0.0423	10.8378	0.0923	6.4736
17	2.292	0.4363	25.840	0.0387	11.2741	0.0887	6.8423
18	2.407	0.4155	28.132	0.0356	11.6896	0.0856	7.2034
19	2.527	0.3957	30.539	0.0328	12.0853	0.0828	7.5569
20	2.653	0.3769	33.066	0.0303	12.4622	0.0803	7.9030
21	2.786	0.3590	35.719	0.0280	12.8212	0.0780	8.2416
22	2.925	0.3419	38.505	0.0260	13.1630	0.0760	8.5730
23	3,072	0.3256	41.430	0.0241	13.4886	0.0741	8.8971
24	3.225	0.3101	44.502	0.0225	13.7987	0.0725	9.2140
25	3.386	0.2953	47.727	0.0210	14.0940	0.0710	9.5238
26	3.556	0.2813	51.113	0.0196	14.3752	0.0696	9.8266
27	3.733	0.2679	54.669	0.0183	14.6430	0.0683	10.1224
28	3.920	0.2551	58.403	0.0171	14.8981	0.0671	10.4114
29	4.116	0.2430	62.323	0.0161	15.1411	0.0661	10.6936
30	4.322	0.2314	66.439	0.0151	15.3725	0.0651	10.9691
31	4.538	0.2204	70.761	0.0141	15.5928	0.0641	11.2381
32	4.765	0.2099	75.299	0.0133	15.8027	0.0633	11.5005
33	5.003	0.1999	80.064	0.0125	16.0026	0.0625	11.7566
34	5.253	0.1904	85.067	0.0118	16.1929	0.0618	12.0[illegible]63
35	5.516	0.1813	90.320	0.0111	16.3742	0.0611	12.2498
40	7.040	0.1421	120.800	0.0083	17.1591	0.0583	13.3775
45	8.985	0.1113	159.700	0.0063	17.7741	0.0563	14.3644
50	11.467	0.0872	209.348	0.0048	18.2559	0.0548	15.2233
55	14.636	0.0683	272.713	0.0037	18.6335	0.0537	15.9665
60	18.679	0.0535	353.584	0.0028	18.9293	0.0528	16.6062
65	23.840	0.0420	456.798	0.0022	19.1611	0.0522	17.1541
70	30.426	0.0329	588.529	0.0017	19.3427	0.0517	17.6212
75	38.833	0.0258	756.654	0.0013	19.4850	0.0513	18.0176
80	49.561	0.0202	971.229	0.0010	19.5965	0.0510	18.3526
85	63.254	0.0158	1245.087	0.0008	19.6838	0.0508	18.6346
90	80.730	0.0124	1594.607	0.0006	19.7523	0.0506	18.8712
95	103.035	0.0097	2040.694	0.0005	19.8059	0.0505	19.0689
100	131.501	0.0076	2610.025	0.0004	19.8479	0.0504	19.2337

TABLE A.10.
6% INTEREST FACTORS FOR DISCRETE COMPOUNDING

	Single Payment		Equal Payment Series				Uniform gradient-series factor
n	Compound-amount factor	Present-worth factor	Compound-amount factor	Sinking-fund factor	Present-worth factor	Capital-recovery factor	
	To find F Given P $F/P, i, n$	To find P Given F $P/F, i, n$	To find F Given A $F/A, i, n$	To find A Given F $A/F, i, n$	To find P Given A $P/A, i, n$	To find A Given P $A/P, i, n$	To find A Given G $A/G, i, n$
1	1.060	0.9434	1.000	1.0000	0.9434	1.0600	0.0000
2	1.124	0.8900	2.060	0.4854	1.8334	0.5454	0.4854
3	1.191	0.8396	3.184	0.3141	2.6730	0.3741	0.9612
4	1.262	0.7921	4.375	0.2286	3.4651	0.2886	1.4272
5	1.338	0.7473	5.637	0.1774	4.2124	0.2374	1.8836
6	1.419	0.7050	6.975	0.1434	4.9173	0.2034	2.3304
7	1.504	0.6651	8.394	0.1191	5.5824	0.1791	2.7676
8	1.594	0.6274	9.897	0.1010	6.2098	0.1610	3.1952
9	1.689	0.5919	11.491	0.0870	6.8017	0.1470	3.6133
10	1.791	0.5584	13.181	0.0759	7.3601	0.1359	4.0220
11	1.898	0.5268	14.972	0.0668	7.8869	0.1268	4.4213
12	2.012	0.4970	16.870	0.0593	8.3839	0.1193	4.8113
13	2.133	0.4688	18.882	0.0530	8.8527	0.1130	5.1920
14	2.261	0.4423	21.015	0.0476	9.2950	0.1076	5.5635
15	2.397	0.4173	23.276	0.0430	9.7123	0.1030	5.9260
16	2.540	0.3937	25.673	0.0390	10.1059	0.0990	6.2794
17	2.693	0.3714	28.213	0.0355	10.4773	0.0955	6.6240
18	2.854	0.3504	30.906	0.0324	10.8276	0.0924	6.9597
19	3.026	0.3305	33.760	0.0296	11.1581	0.0896	7.2867
20	3.207	0.3118	36.786	0.0272	11.4699	0.0872	7.6052
21	3.400	0.2942	39.993	0.0250	11.7641	0.0850	7.9151
22	3.604	0.2775	43.392	0.0231	12.0416	0.0831	8.2166
23	3.820	0.2618	46.996	0.0213	12.3034	0.0813	8.5099
24	4.049	0.2470	50.816	0.0197	12.5504	0.0797	8.7951
25	4.292	0.2330	54.865	0.0182	12.7834	0.0782	9.0722
26	4.549	0.2198	59.156	0.0169	13.0032	0.0769	9.3415
27	4.822	0.2074	63.706	0.0157	13.2105	0.0757	9.6030
28	5.112	0.1956	68.528	0.0146	13.4062	0.0746	9.8568
29	5.418	0.1846	73.640	0.0136	13.5907	0.0736	10.1032
30	5.744	0.1741	79.058	0.0127	13.7648	0.0727	10.3422
31	6.088	0.1643	84.802	0.0118	13.9291	0.0718	10.5740
32	6.453	0.1550	90.890	0.0110	14.0841	0.0710	10.7988
33	6.841	0.1462	97.343	0.0103	14.2302	0.0703	11.0166
34	7.251	0.1379	104.184	0.0096	14.3682	0.0696	11.2276
35	7.686	0.1301	111.435	0.0090	14.4983	0.0690	11.4319
40	10.286	0.0972	154.762	0.0065	15.0463	0.0665	12.3590
45	13.765	0.0727	212.744	0.0047	15.4558	0.0647	13.1413
50	18.420	0.0543	290.336	0.0035	15.7619	0.0635	13.7964
55	24.650	0.0406	394.172	0.0025	15.9906	0.0625	14.3411
60	32.988	0.0303	533.128	0.0019	16.1614	0.0619	14.7910
65	44.145	0.0227	719.083	0.0014	16.2891	0.0614	15.1601
70	59.076	0.0169	967.932	0.0010	16.3846	0.0610	15.4614
75	79.057	0.0127	1300.949	0.0008	16.4559	0.0608	15.7058
80	105.796	0.0095	1746.600	0.0006	16.5091	0.0606	15.9033
85	141.579	0.0071	2342.982	0.0004	16.5490	0.0604	16.0620
90	189.465	0.0053	3141.075	0.0003	16.5787	0.0603	16.1891
95	253.546	0.0040	4209.104	0.0002	16.6009	0.0602	16.2905
100	339.302	0.0030	5638.368	0.0002	16.6176	0.0602	16.3711

TABLE A.11.
7% INTEREST FACTORS FOR DISCRETE COMPOUNDING

	Single Payment		Equal Payment Series				Uniform gradient-series factor
n	Compound-amount factor	Present-worth factor	Compound-amount factor	Sinking-fund factor	Present-worth factor	Capital-recovery factor	Uniform gradient-series factor
	To find *F* Given *P* *F*/*P*, *i*, *n*	To find *P* Given *F* *P*/*F*, *i*, *n*	To find *F* Given *A* *F*/*A*, *i*, *n*	To find *A* Given *F* *A*/*F*, *i*, *n*	To find *P* Given *A* *P*/*A*, *i*, *n*	To find *A* Given *P* *A*/*P*, *i*, *n*	To find *A* Given *G* *A*/*G*, *i*, *n*
1	1.070	0.9346	1.000	1.0000	0.9346	1.0700	0.0000
2	1.145	0.8734	2.070	0.4831	1.8080	0.5531	0.4831
3	1.225	0.8163	3.215	0.3111	2.6243	0.3811	0.9549
4	1.311	0.7629	4.440	0.2252	3.3872	0.2952	1.4155
5	1.403	0.7130	5.751	0.1739	4.1002	0.2439	1.8650
6	1.501	0.6664	7.153	0.1398	4.7665	0.2098	2.3032
7	1.606	0.6228	8.654	0.1156	5.3893	0.1856	2.7304
8	1.718	0.5820	10.260	0.0975	5.9713	0.1675	3.1466
9	1.838	0.5439	11.978	0.0835	6.5152	0.1535	3.5517
10	1.967	0.5084	13.816	0.0724	7.0236	0.1424	3.9461
11	2.105	0.4751	15.784	0.0634	7.4987	0.1334	4.3296
12	2.252	0.4440	17.888	0.0559	7.9427	0.1259	4.7025
13	2.410	0.4150	20.141	0.0497	8.3577	0.1197	5.0649
14	2.579	0.3878	22.550	0.0444	8.7455	0.1144	5.4167
15	2.759	0.3625	25.129	0.0398	9.1079	0.1098	5.7583
16	2.952	0.3387	27.888	0.0359	9.4467	0.1059	6.0897
17	3.159	0.3166	30.840	0.0324	9.7632	0.1024	6.4110
18	3.380	0.2959	33.999	0.0294	10.0591	0.0994	6.7225
19	3.617	0.2765	37.379	0.0268	10.3356	0.0968	7.0242
20	3.870	0.2584	40.996	0.0244	10.5940	0.0944	7.3163
21	4.141	0.2415	44.865	0.0223	10.8355	0.0923	7.5990
22	4.430	0.2257	49.006	0.0204	11.0613	0.0904	7.8725
23	4.741	0.2110	53.436	0.0187	11.2722	0.0887	8.1369
24	5.072	0.1972	58.177	0.0172	11.4693	0.0872	8.3923
25	5.427	0.1843	63.249	0.0158	11.6536	0.0858	8.6391
26	5.807	0.1722	68.676	0.0146	11.8258	0.0846	8.8773
27	6.214	0.1609	74.484	0.0134	11.9867	0.0834	9.1072
28	6.649	0.1504	80.698	0.0124	12.1371	0.0824	9.3290
29	7.114	0.1406	87.347	0.0115	12.2777	0.0815	9.5427
30	7.612	0.1314	94.461	0.0106	12.4091	0.0806	9.7487
31	8.145	0.1228	102.073	0.0098	12.5318	0.0798	9.9471
32	8.715	0.1148	110.218	0.0091	12.6466	0.0791	10.1381
33	9.325	0.1072	118.933	0.0084	12.7538	0.0784	10.3219
34	9.978	0.1002	128.259	0.0078	12.8540	0.0778	10.4987
35	10.677	0.0937	138.237	0.0072	12.9477	0.0772	10.6687
40	14.974	0.0668	199.635	0.0050	13.3317	0.0750	11.4234
45	21.002	0.0476	285.749	0.0035	13.6055	0.0735	12.0360
50	29.457	0.0340	406.529	0.0025	13.8008	0.0725	12.5287
55	41.315	0.0242	575.929	0.0017	13.9399	0.0717	12.9215
60	57.946	0.0173	813.520	0.0012	14.0392	0.0712	13.2321
65	81.273	0.0123	1146.755	0.0009	14.1099	0.0709	13.4760
70	113.989	0.0088	1614.134	0.0006	14.1604	0.0706	13.6662
75	159.876	0.0063	2269.657	0.0005	14.1964	0.0705	13.8137
80	224.234	0.0045	3189.063	0.0003	14.2220	0.0703	13.9274
85	314.500	0.0032	4478.576	0.0002	14.2403	0.0702	14.0146
90	441.103	0.0023	6287.185	0.0002	14.2533	0.0702	14.0812
95	618.670	0.0016	8823.854	0.0001	14.2626	0.0701	14.1319
100	867.716	0.0012	12381.662	0.0001	14.2693	0.0701	14.1703

TABLE A.12.
8% INTEREST FACTORS FOR DISCRETE COMPOUNDING

	Single Payment		Equal Payment Series				Uniform
n	Compound-amount factor	Present-worth factor	Compound-amount factor	Sinking-fund factor	Present-worth factor	Capital-recovery factor	gradient-series factor
	To find *F* Given *P* $F/P, i, n$	To find *P* Given *F* $P/F, i, n$	To find *F* Given *A* $F/A, i, n$	To find *A* Given *F* $A/F, i, n$	To find *P* Given *A* $P/A, i, n$	To find *A* Given *P* $A/P, i, n$	To find *A* Given *G* $A/G, i, n$
1	1.080	0.9259	1.000	1.0000	0.9259	1.0800	0.0000
2	1.166	0.8573	2.080	0.4808	1.7833	0.5608	0.4808
3	1.260	0.7938	3.246	0.3080	2.5771	0.3880	0.9488
4	1.360	0.7350	4.506	0.2219	3.3121	0.3019	1.4040
5	1.469	0.6806	5.867	0.1705	3.9927	0.2505	1.8465
6	1.587	0.6302	7.336	0.1363	4.6229	0.2163	2.2764
7	1.714	0.5835	8.923	0.1121	5.2064	0.1921	2.6937
8	1.851	0.5403	10.637	0.0940	5.7466	0.1740	3.0985
9	1.999	0.5003	12.488	0.0801	6.2469	0.1601	3.4910
10	2.159	0.4632	14.487	0.0690	6.7101	0.1490	3.8713
11	2.332	0.4289	16.645	0.0601	7.1390	0.1401	4.2395
12	2.518	0.3971	18.977	0.0527	7.5361	0.1327	4.5958
13	2.720	0.3677	21.495	0.0465	7.9038	0.1265	4.9402
14	2.937	0.3405	24.215	0.0413	8.2442	0.1213	5.2731
15	3.172	0.3153	27.152	0.0368	8.5595	0.1168	5.5945
16	3.426	0.2919	30.324	0.0330	8.8514	0.1130	5.9046
17	3.700	0.2703	33.750	0.0296	9.1216	0.1096	6.2038
18	3.996	0.2503	37.450	0.0267	9.3719	0.1067	6.4920
19	4.316	0.2317	41.446	0.0241	9.6036	0.1041	6.7697
20	4.661	0.2146	45.762	0.0219	9.8182	0.1019	7.0370
21	5.034	0.1987	50.423	0.0198	10.0168	0.0998	7.2940
22	5.437	0.1840	55.457	0.0180	10.2008	0.0980	7.5412
23	5.871	0.1703	60.893	0.0164	10.3711	0.0964	7.7786
24	6.341	0.1577	66.765	0.0150	10.5288	0.0950	8.0066
25	6.848	0.1460	73.106	0.0137	10.6748	0.0937	8.2254
26	7.396	0.1352	79.954	0.0125	10.8100	0.0925	8.4352
27	7.988	0.1252	87.351	0.0115	10.9352	0.0915	8.6363
28	8.627	0.1159	95.339	0.0105	11.0511	0.0905	8.8289
29	9.317	0.1073	103.966	0.0096	11.1584	0.0896	9.0133
30	10.063	0.0994	113.283	0.0088	11.2578	0.0888	9.1897
31	10.868	0.0920	123.346	0.0081	11.3498	0.0881	9.3584
32	11.737	0.0852	134.214	0.0075	11.4350	0.0875	9.5197
33	12.676	0.0789	145.951	0.0069	11.5139	0.0869	9.6737
34	13.690	0.0731	158.627	0.0063	11.5869	0.0863	9.8208
35	14.785	0.0676	172.317	0.0058	11.6546	0.0858	9.9611
40	21.725	0.0460	259.057	0.0039	11.9246	0.0839	10.5699
45	31.920	0.0313	386.506	0.0026	12.1084	0.0826	11.0447
50	46.902	0.0213	573.770	0.0018	12.2335	0.0818	11.4107
55	68.914	0.0145	848.923	0.0012	12.3186	0.0812	11.6902
60	101.257	0.0099	1253.213	0.0008	12.3766	0.0808	11.9015
65	148.780	0.0067	1847.248	0.0006	12.4160	0.0806	12.0602
70	218.606	0.0046	2720.080	0.0004	12.4428	0.0804	12.1783
75	321.205	0.0031	4002.557	0.0003	12.4611	0.0803	12.2658
80	471.955	0.0021	5886.935	0.0002	12.4735	0.0802	12.3301
85	693.456	0.0015	8655.706	0.0001	12.4820	0.0801	12.3773
90	1018.915	0.0010	12723.939	0.0001	12.4877	0.0801	12.4116
95	1497.121	0.0007	18701.507	0.0001	12.4917	0.0801	12.4365
100	2199.761	0.0005	27484.516	0.0001	12.4943	0.0800	12.4545

TABLE A.13.
9% INTEREST FACTORS FOR DISCRETE COMPOUNDING

	Single Payment		Equal Payment Series				Uniform gradient-series factor
n	Compound-amount factor	Present-worth factor	Compound-amount factor	Sinking-fund factor	Present-worth factor	Capital-recovery factor	Uniform gradient-series factor
	To find *F* Given *P* *F*/*P*, *i*, *n*	To find *P* Given *F* *P*/*F*, *i*, *n*	To find *F* Given *A* *F*/*A*, *i*, *n*	To find *A* Given *F* *A*/*F*, *i*, *n*	To find *P* Given *A* *P*/*A*, *i*, *n*	To find *A* Given *P* *A*/*P*, *i*, *n*	To find *A* Given *G* *A*/*G*, *i*, *n*
1	1.090	0.9174	1.000	1.0000	0.9174	1.0900	0.0000
2	1.188	0.8417	2.090	0.4785	1.7591	0.5685	0.4785
3	1.295	0.7722	3.278	0.3051	2.5313	0.3951	0.9426
4	1.412	0.7084	4.573	0.2187	3.2397	0.3087	1.3925
5	1.539	0.6499	5.985	0.1671	3.8897	0.2571	1.8282
6	1.677	0.5963	7.523	0.1329	4.4859	0.2229	2.2498
7	1.828	0.5470	9.200	0.1087	5.0330	0.1987	2.6574
8	1.993	0.5019	11.028	0.0907	5.5348	0.1807	3.0512
9	2.172	0.4604	13.021	0.0768	5.9953	0.1668	3.4312
10	2.367	0.4224	15.193	0.0658	6.4177	0.1558	3.7978
11	2.580	0.3875	17.560	0.0570	6.8052	0.1470	4.1510
12	2.813	0.3555	20.141	0.0497	7.1607	0.1397	4.4910
13	3.066	0.3262	22.953	0.0436	7.4869	0.1336	4.8182
14	3.342	0.2993	26.019	0.0384	7.7862	0.1284	5.1326
15	3.642	0.2745	29.361	0.0341	8.0607	0.1241	5.4346
16	3.970	0.2519	33.003	0.0303	8.3126	0.1203	5.7245
17	4.328	0.2311	36.974	0.0271	8.5436	0.1171	6.0024
18	4.717	0.2120	41.301	0.0242	8.7556	0.1142	6.2687
19	5.142	0.1945	46.018	0.0217	8.9501	0.1117	6.5236
20	5.604	0.1784	51.160	0.0196	9.1286	0.1096	6.7675
21	6.109	0.1637	56.765	0.0176	9.2923	0.1076	7.0006
22	6.659	0.1502	62.873	0.0159	9.4424	0.1059	7.2232
23	7.258	0.1378	69.532	0.0144	9.5802	0.1044	7.4358
24	7.911	0.1264	76.790	0.0130	9.7066	0.1030	7.6384
25	8.623	0.1160	84.701	0.0118	9.8226	0.1018	7.8316
26	9.399	0.1064	93.324	0.0107	9.9290	0.1007	8.0156
27	10.245	0.0976	102.723	0.0097	10.0266	0.0997	8.1906
28	11.167	0.0896	112.968	0.0089	10.1161	0.0989	8.3572
29	12.172	0.0822	124.135	0.0081	10.1983	0.0981	8.5154
30	13.268	0.0754	136.308	0.0073	10.2737	0.0973	8.6657
31	14.462	0.0692	149.575	0.0067	10.3428	0.0967	8.8083
32	15.763	0.0634	164.037	0.0061	10.4063	0.0961	8.9436
33	17.182	0.0582	179.800	0.0056	10.4645	0.0956	9.0718
34	18.728	0.0534	196.982	0.0051	10.5178	0.0951	9.1933
35	20.414	0.0490	215.711	0.0046	10.5668	0.0946	9.3083
40	31.409	0.0318	337.882	0.0030	10.7574	0.0930	9.7957
45	48.327	0.0207	525.859	0.0019	10.8812	0.0919	10.1603
50	74.358	0.0135	815.084	0.0012	10.9617	0.0912	10.4295
55	114.408	0.0088	1260.092	0.0008	11.0140	0.0908	10.6261
60	176.031	0.0057	1944.792	0.0005	11.0480	0.0905	10.7683
65	270.846	0.0037	2998.288	0.0003	11.0701	0.0903	10.8702
70	416.730	0.0024	4619.223	0.0002	11.0845	0.0902	10.9427
75	641.191	0.0016	7113.232	0.0002	11.0938	0.0902	10.9940
80	986.552	0.0010	10950.574	0.0001	11.0999	0.0901	11.0299
85	1517.932	0.0007	16854.800	0.0001	11.1038	0.0901	11.0551
90	2335.527	0.0004	25939.184	0.0001	11.1064	0.0900	11.0726
95	3593.497	0.0003	39916.635	0.0000	11.1080	0.0900	11.0847
100	5529.041	0.0002	61422.675	0.0000	11.1091	0.0900	11.0930

TABLE A.14.
10% INTEREST FACTORS FOR DISCRETE COMPOUNDING

	Single Payment		Equal Payment Series				Uniform gradient-series factor
n	Compound-amount factor	Present-worth factor	Compound-amount factor	Sinking-fund factor	Present-worth factor	Capital-recovery factor	
	To find F Given P $F/P, i, n$	To find P Given F $P/F, i, n$	To find F Given A $F/A, i, n$	To find A Given F $A/F, i, n$	To find P Given A $P/A, i, n$	To find A Given P $A/P, i, n$	To find A Given G $A/G, i, n$
1	1.100	0.9091	1.000	1.0000	0.9091	1.1000	0.0000
2	1.210	0.8265	2.100	0.4762	1.7355	0.5762	0.4762
3	1.331	0.7513	3.310	0.3021	2.4869	0.4021	0.9366
4	1.464	0.6830	4.641	0.2155	3.1699	0.3155	1.3812
5	1.611	0.6209	6.105	0.1638	3.7908	0.2638	1.8101
6	1.772	0.5645	7.716	0.1296	4.3553	0.2296	2.2236
7	1.949	0.5132	9.487	0.1054	4.8684	0.2054	2.6216
8	2.144	0.4665	11.436	0.0875	5.3349	0.1875	3.0045
9	2.358	0.4241	13.579	0.0737	5.7590	0.1737	3.3724
10	2.594	0.3856	15.937	0.0628	6.1446	0.1628	3.7255
11	2.853	0.3505	18.531	0.0540	6.4951	0.1540	4.0641
12	3.138	0.3186	21.384	0.0468	6.8137	0.1468	4.3884
13	3.452	0.2897	24.523	0.0408	7.1034	0.1408	4.6988
14	3.798	0.2633	27.975	0.0358	7.3667	0.1358	4.9955
15	4.177	0.2394	31.772	0.0315	7.6061	0.1315	5.2789
16	4.595	0.2176	35.950	0.0278	7.8237	0.1278	5.5493
17	5.054	0.1979	40.545	0.0247	8.0216	0.1247	5.8071
18	5.560	0.1799	45.599	0.0219	8.2014	0.1219	6.0526
19	6.116	0.1635	51.159	0.0196	8.3649	0.1196	6.2861
20	6.728	0.1487	57.275	0.0175	8.5136	0.1175	6.5081
21	7.400	0.1351	64.003	0.0156	8.6487	0.1156	6.7189
22	8.140	0.1229	71.403	0.0140	8.7716	0.1140	6.9189
23	8.954	0.1117	79.543	0.0126	8.8832	0.1126	7.1085
24	9.850	0.1015	88.497	0.0113	8.9848	0.1113	7.2881
25	10.835	0.0923	98.347	0.0102	9.0771	0.1102	7.4580
26	11.918	0.0839	109.182	0.0092	9.1610	0.1092	7.6187
27	13.110	0.0763	121.100	0.0083	9.2372	0.1083	7.7704
28	14.421	0.0694	134.210	0.0075	9.3066	0.1075	7.9137
29	15.863	0.0630	148.631	0.0067	9.3696	0.1067	8.0489
30	17.449	0.0573	164.494	0.0061	9.4269	0.1061	8.1762
31	19.194	0.0521	181.943	0.0055	9.4790	0.1055	8.2962
32	21.114	0.0474	201.138	0.0050	9.5264	0.1050	8.4091
33	23.225	0.0431	222.252	0.0045	9.5694	0.1045	8.5152
34	25.548	0.0392	245.477	0.0041	9.6086	0.1041	8.6149
35	28.102	0.0356	271.024	0.0037	9.6442	0.1037	8.7086
40	45.259	0.0221	442.593	0.0023	9.7791	0.1023	9.0962
45	72.890	0.0137	718.905	0.0014	9.8628	0.1014	9.3741
50	117.391	0.0085	1163.909	0.0009	9.9148	0.1009	9.5704
55	189.059	0.0053	1880.591	0.0005	9.9471	0.1005	9.7075
60	304.482	0.0033	3034.816	0.0003	9.9672	0.1003	9.8023
65	490.371	0.0020	4893.707	0.0002	9.9796	0.1002	9.8672
70	789.747	0.0013	7887.470	0.0001	9.9873	0.1001	9.9113
75	1271.895	0.0008	12708.954	0.0001	9.9921	0.1001	9.9410
80	2048.400	0.0005	20474.002	0.0001	9.9951	0.1001	9.9609
85	3298.969	0.0003	32979.690	0.0000	9.9970	0.1000	9.9742
90	5313.023	0.0002	53120.226	0.0000	9.9981	0.1000	9.9831
95	8556.676	0.0001	85556.760	0.0000	9.9988	0.1000	9.9889
100	13780.612	0.0001	137796.123	0.0000	9.9993	0.1000	9.9928

TABLE A.15.
11% INTEREST FACTORS FOR DISCRETE COMPOUNDING

n	Single Payment		Equal Payment Series				Uniform gradient-series factor
	Compound-amount factor	Present-worth factor	Compound-amount factor	Sinking-fund factor	Present-worth factor	Capital-recovery factor	
	To find *F* Given *P* *F/P, i, n*	To find *P* Given *F* *P/F, i, n*	To find *F* Given *A* *F/A, i, n*	To find *A* Given *F* *A/F, i, n*	To find *P* Given *A* *P/A, i, n*	To find *A* Given *P* *A/P, i, n*	To find *A* Given *G* *A/G, i, n*
1	1.110	0.9009	1.000	1.0000	0.9009	1.1100	0.0000
2	1.232	0.8116	2.110	0.4739	1.7125	0.5839	0.4739
3	1.368	0.7312	3.342	0.2992	2.4437	0.4092	0.9306
4	1.518	0.6587	4.710	0.2123	3.1024	0.3223	1.3700
5	1.685	0.5935	6.228	0.1606	3.6959	0.2706	1.7923
6	1.870	0.5346	7.913	0.1264	4.2305	0.2364	2.1976
7	2.076	0.4817	9.783	0.1022	4.7122	0.2122	2.5863
8	2.305	0.4339	11.859	0.0843	5.1461	0.1943	2.9585
9	2.558	0.3909	14.164	0.0706	5.5370	0.1806	3.3144
10	2.839	0.3522	16.722	0.0598	5.8892	0.1698	3.6544
11	3.152	0.3173	19.561	0.0511	6.2065	0.1611	3.9788
12	3.498	0.2858	22.713	0.0440	6.4924	0.1540	4.2879
13	3.883	0.2575	26.212	0.0382	6.7499	0.1482	4.5822
14	4.310	0.2320	30.095	0.0332	6.9819	0.1432	4.8619
15	4.785	0.2090	34.405	0.0291	7.1909	0.1391	5.1275
16	5.311	0.1883	39.190	0.0255	7.3972	0.1355	5.3794
17	5.895	0.1696	44.501	0.0225	7.5488	0.1325	5.6180
18	6.544	0.1528	50.396	0.0198	7.7016	0.1298	5.8439
19	7.263	0.1377	56.939	0.0176	7.8393	0.1276	6.0574
20	8.062	0.1240	64.203	0.0156	7.9633	0.1256	6.2590
21	8.949	0.1117	72.265	0.0138	8.0751	0.1238	6.4491
22	9.934	0.1007	81.214	0.0123	8.1757	0.1223	6.6283
23	11.026	0.0907	91.148	0.0110	8.2664	0.1210	6.7969
24	12.239	0.0817	102.174	0.0098	8.3481	0.1198	6.9555
25	13.585	0.0736	114.413	0.0087	8.4217	0.1187	7.1045
26	15.080	0.0663	127.999	0.0078	8.4881	0.1178	7.2443
27	16.739	0.0597	143.079	0.0070	8.5478	0.1170	7.3754
28	18.580	0.0538	159.817	0.0063	8.6016	0.1163	7.4982
29	20.624	0.0485	178.397	0.0056	8.6501	0.1156	7.6131
30	22.892	0.0437	199.021	0.0050	8.6938	0.1150	7.7206
31	25.410	0.0394	221.913	0.0045	8.7331	0.1145	7.8210
32	28.206	0.0355	247.324	0.0040	8.7686	0.1140	7.9147
33	31.308	0.0319	275.529	0.0036	8.8005	0.1136	8.0021
34	34.752	0.0288	306.837	0.0033	8.8293	0.1133	8.0836
35	38.575	0.0259	341.590	0.0029	8.8552	0.1129	8.1594
40	65.001	0.0154	581.826	0.0017	8.9511	0.1117	8.4659
45	109.530	0.0091	986.639	0.0010	9.0079	0.1110	8.6763
50	184.565	0.0054	1668.771	0.0006	9.0417	0.1106	8.8185

TABLE A.16.
12% INTEREST FACTORS FOR DISCRETE COMPOUNDING

n	Single Payment		Equal Payment Series				Uniform gradient-series factor
	Compound-amount factor	Present-worth factor	Compound-amount factor	Sinking-fund factor	Present-worth factor	Capital-recovery factor	
	To find F Given P $F/P, i, n$	To find P Given F $P/F, i, n$	To find F Given A $F/A, i, n$	To find A Given F $A/F, i, n$	To find P Given A $P/A, i, n$	To find A Given P $A/P, i, n$	To find A Given G $A/G, i, n$
1	1.120	0.8929	1.000	1.0000	0.8929	1.1200	0.0000
2	1.254	0.7972	2.120	0.4717	1.6901	0.5917	0.4717
3	1.405	0.7118	3.374	0.2964	2.4018	0.4164	0.9246
4	1.574	0.6355	4.779	0.2092	3.0374	0.3292	1.3589
5	1.762	0.5674	6.353	0.1574	3.6048	0.2774	1.7746
6	1.974	0.5066	8.115	0.1232	4.1114	0.2432	2.1721
7	2.211	0.4524	10.089	0.0991	4.5638	0.2191	2.5515
8	2.476	0.4039	12.300	0.0813	4.9676	0.2013	2.9132
9	2.773	0.3606	14.776	0.0677	5.3283	0.1877	3.2574
10	3.106	0.3220	17.549	0.0570	5.6502	0.1770	3.5847
11	3.479	0.2875	20.655	0.0484	5.9377	0.1684	3.8953
12	3.896	0.2567	24.133	0.0414	6.1944	0.1614	4.1897
13	4.364	0.2292	28.029	0.0357	6.4236	0.1557	4.4683
14	4.887	0.2046	32.393	0.0309	6.6282	0.1509	4.7317
15	5.474	0.1827	37.280	0.0268	6.8109	0.1468	4.9803
16	6.130	0.1631	42.753	0.0234	6.9740	0.1434	5.2147
17	6.866	0.1457	48.884	0.0205	7.1196	0.1405	5.4353
18	7.690	0.1300	55.750	0.0179	7.2497	0.1379	5.6427
19	8.613	0.1161	63.440	0.0158	7.3658	0.1358	5.8375
20	9.646	0.1037	72.052	0.0139	7.4695	0.1339	6.0202
21	10.804	0.0926	81.699	0.0123	7.5620	0.1323	6.1913
22	12.100	0.0827	92.503	0.0108	7.6447	0.1308	6.3514
23	13.552	0.0738	104.603	0.0096	7.7184	0.1296	6.5010
24	15.179	0.0659	118.155	0.0085	7.7843	0.1285	6.6407
25	17.000	0.0588	133.334	0.0075	7.8431	0.1275	6.7708
26	19.040	0.0525	150.334	0.0067	7.8957	0.1267	6.8921
27	21.325	0.0469	169.374	0.0059	7.9426	0.1259	7.0049
28	23.884	0.0419	190.699	0.0053	7.9844	0.1253	7.1098
29	26.750	0.0374	214.583	0.0047	8.0218	0.1247	7.2071
30	29.960	0.0334	241.333	0.0042	8.0552	0.1242	7.2974
31	33.555	0.0298	271.293	0.0037	8.0850	0.1237	7.3811
32	37.582	0.0266	304.848	0.0033	8.1116	0.1233	7.4586
33	42.092	0.0238	342.429	0.0029	8.1354	0.1229	7.5303
34	47.143	0.0212	384.521	0.0026	8.1566	0.1226	7.5965
35	52.800	0.0189	431.664	0.0023	8.1755	0.1223	7.6577
40	93.051	0.0108	767.091	0.0013	8.2438	0.1213	7.8988
45	163.988	0.0061	1358.230	0.0007	8.2825	0.1207	8.0572
50	289.002	0.0035	2400.018	0.0004	8.3045	0.1204	8.1597

TABLE A.17.
13% INTEREST FACTORS FOR DISCRETE COMPOUNDING

n	Single Payment		Equal Payment Series				Uniform gradient-series factor
	Compound-amount factor	Present-worth factor	Compound-amount factor	Sinking-fund factor	Present-worth factor	Capital-recovery factor	
	To find F Given P $F/P, i, n$	To find P Given F $P/F, i, n$	To find F Given A $F/A, i, n$	To find A Given F $A/F, i, n$	To find P Given A $P/A, i, n$	To find A Given P $A/P, i, n$	To find A Given G $A/G, i, n$
1	1.130	0.8850	1.000	1.0000	0.8850	1.1300	0.0000
2	1.277	0.7831	2.130	0.4695	1.6681	0.5995	0.4695
3	1.443	0.6931	3.407	0.2935	2.3612	0.4235	0.9187
4	1.630	0.6133	4.850	0.2062	2.9745	0.3362	1.3479
5	1.842	0.5428	6.480	0.1543	3.5172	0.2843	1.7571
6	2.082	0.4803	8.323	0.1202	3.9975	0.2502	2.1468
7	2.353	0.4251	10.405	0.0961	4.4226	0.2261	2.5171
8	2.658	0.3762	12.757	0.0784	4.7988	0.2084	2.8685
9	3.004	0.3329	15.416	0.0649	5.1317	0.1949	3.2014
10	3.395	0.2946	18.420	0.0543	5.4262	0.1843	3.5162
11	3.836	0.2607	21.814	0.0458	5.6869	0.1758	3.8134
12	4.335	0.2307	25.650	0.0390	5.9176	0.1690	4.0936
13	4.898	0.2042	29.985	0.0334	6.1218	0.1634	4.3573
14	5.535	0.1807	34.883	0.0287	6.3025	0.1587	4.6050
15	6.254	0.1599	40.417	0.0247	6.4624	0.1547	4.8375
16	7.067	0.1415	46.672	0.0214	6.6039	0.1514	5.0552
17	7.986	0.1252	53.739	0.0186	6.7291	0.1486	5.2589
18	9.024	0.1108	61.725	0.0162	6.8399	0.1462	5.4491
19	10.197	0.0981	70.749	0.0141	6.9380	0.1441	5.6265
20	11.523	0.0868	80.947	0.0124	7.0248	0.1424	5.7917
21	13.021	0.0768	92.470	0.0108	7.1016	0.1408	5.9454
22	14.714	0.0680	105.491	0.0095	7.1695	0.1395	6.0881
23	16.627	0.0601	120.205	0.0083	7.2297	0.1383	6.2205
24	18.788	0.0532	136.831	0.0073	7.2829	0.1373	6.3431
25	21.231	0.0471	155.620	0.0064	7.3300	0.1364	6.4566
26	23.991	0.0417	176.850	0.0057	7.3717	0.1357	6.5614
27	27.109	0.0369	200.841	0.0050	7.4086	0.1350	6.6582
28	30.633	0.0326	227.950	0.0044	7.4412	0.1344	6.7474
29	34.616	0.0289	258.583	0.0039	7.4701	0.1339	6.8296
30	39.116	0.0256	293.199	0.0034	7.4957	0.1334	6.9052
31	44.201	0.0226	332.315	0.0030	7.5183	0.1330	6.9747
32	49.947	0.0200	376.516	0.0027	7.5383	0.1327	7.0385
33	56.440	0.0177	426.463	0.0023	7.5560	0.1323	7.0971
34	63.777	0.0157	482.903	0.0021	7.5717	0.1321	7.1507
35	72.069	0.0139	546.681	0.0018	7.5856	0.1318	7.1998
40	132.782	0.0075	1013.704	0.0010	7.6344	0.1310	7.3888
45	244.641	0.0041	1874.165	0.0005	7.6609	0.1305	7.5076
50	450.736	0.0022	3459.507	0.0003	7.6752	0.1303	7.5811

TABLE A.18.
14% INTEREST FACTORS FOR DISCRETE COMPOUNDING

	Single Payment		Equal Payment Series				Uniform gradient-series factor
n	Compound-amount factor	Present-worth factor	Compound-amount factor	Sinking-fund factor	Present-worth factor	Capital-recovery factor	
	To find F Given P $F/P, i, n$	To find P Given F $P/F, i, n$	To find F Given A $F/A, i, n$	To find A Given F $A/F, i, n$	To find P Given A $P/A, i, n$	To find A Given P $A/P, i, n$	To find A Given G $A/G, i, n$
1	1.140	0.8772	1.000	1.0000	0.8772	1.1400	0.0000
2	1.300	0.7695	2.140	0.4673	1.6467	0.6073	0.4673
3	1.482	0.6750	3.440	0.2907	2.3216	0.4307	0.9129
4	1.689	0.5921	4.921	0.2032	2.9137	0.3432	1.3370
5	1.925	0.5194	6.610	0.1513	3.4331	0.2913	1.7399
6	2.195	0.4556	8.536	0.1172	3.8887	0.2572	2.1218
7	2.502	0.3996	10.730	0.0932	4.2883	0.2332	2.4832
8	2.853	0.3506	13.233	0.0756	4.6389	0.2156	2.8246
9	3.252	0.3075	16.085	0.0622	4.9464	0.2022	3.1463
10	3.707	0.2697	19.337	0.0517	5.2161	0.1917	3.4490
11	4.226	0.2366	23.045	0.0434	5.4527	0.1834	3.7333
12	4.818	0.2076	27.271	0.0367	5.6603	0.1767	3.9998
13	5.492	0.1821	32.089	0.0312	5.8424	0.1712	4.2491
14	6.261	0.1597	37.581	0.0266	6.0021	0.1666	4.4819
15	7.138	0.1401	43.842	0.0228	6.1422	0.1628	4.6990
16	8.137	0.1229	50.980	0.0196	6.2651	0.1596	4.9011
17	9.276	0.1078	59.118	0.0169	6.3729	0.1569	5.0888
18	10.575	0.0946	68.394	0.0146	6.4674	0.1546	5.2630
19	12.056	0.0829	78.969	0.0127	6.5504	0.1527	5.4243
20	13.743	0.0728	91.025	0.0110	6.6231	0.1510	5.5734
21	15.668	0.0638	104.768	0.0095	6.6870	0.1495	5.7111
22	17.861	0.0560	120.436	0.0083	6.7429	0.1483	5.8381
23	20.362	0.0491	138.297	0.0072	6.7921	0.1472	5.9549
24	23.212	0.0431	158.659	0.0063	6.8351	0.1463	6.0624
25	26.462	0.0378	181.871	0.0055	6.8729	0.1455	6.1610
26	30.167	0.0331	208.333	0.0048	6.9061	0.1448	6.2514
27	34.390	0.0291	238.499	0.0042	6.9352	0.1442	6.3342
28	39.204	0.0255	272.889	0.0037	6.9607	0.1437	6.4100
29	44.693	0.0224	312.094	0.0032	6.9830	0.1432	6.4791
30	50.950	0.0196	356.787	0.0028	7.0027	0.1428	6.5423
31	58.083	0.0172	407.737	0.0025	7.0199	0.1425	6.5998
32	66.215	0.0151	465.820	0.0021	7.0350	0.1421	6.6522
33	75.485	0.0132	532.035	0.0019	7.0482	0.1419	6.6998
34	86.053	0.0116	607.520	0.0016	7.0599	0.1416	6.7431
35	98.100	0.0102	693.573	0.0014	7.0700	0.1414	6.7824
40	188.884	0.0053	1342.025	0.0007	7.1050	0.1407	6.9300
45	363.679	0.0027	2590.565	0.0004	7.1232	0.1404	7.0188
50	700.233	0.0014	4994.521	0.0002	7.1327	0.1402	7.0714

TABLE A.19.
15% INTEREST FACTORS FOR DISCRETE COMPOUNDING

	Single Payment		Equal Payment Series				Uniform gradient-series factor
n	Compound-amount factor	Present-worth factor	Compound-amount factor	Sinking-fund factor	Present-worth factor	Capital-recovery factor	
	To find F Given P $F/P, i, n$	To find P Given F $P/F, i, n$	To find F Given A $F/A, i, n$	To find A Given F $A/F, i, n$	To find P Given A $P/A, i, n$	To find A Given P $A/P, i, n$	To find A Given G $A/G, i, n$
1	1.150	0.8696	1.000	1.0000	0.8696	1.1500	0.0000
2	1.323	0.7562	2.150	0.4651	1.6257	0.6151	0.4651
3	1.521	0.6575	3.473	0.2880	2.2832	0.4380	0.9071
4	1.749	0.5718	4.993	0.2003	2.8550	0.3503	1.3263
5	2.011	0.4972	6.742	0.1483	3.3522	0.2983	1.7228
6	2.313	0.4323	8.754	0.1142	3.7845	0.2642	2.0972
7	2.660	0.3759	11.067	0.0904	4.1604	0.2404	2.4499
8	3.059	0.3269	13.727	0.0729	4.4873	0.2229	2.7813
9	3.518	0.2843	16.786	0.0596	4.7716	0.2096	3.0922
10	4.046	0.2472	20.304	0.0493	5.0188	0.1993	3.3832
11	4.652	0.2150	24.349	0.0411	5.2337	0.1911	3.6550
12	5.350	0.1869	29.002	0.0345	5.4206	0.1845	3.9082
13	6.153	0.1625	34.352	0.0291	5.5832	0.1791	4.1438
14	7.076	0.1413	40.505	0.0247	5.7245	0.1747	4.3624
15	8.137	0.1229	47.580	0.0210	5.8474	0.1710	4.5650
16	9.358	0.1069	55.717	0.0180	5.9542	0.1680	4.7523
17	10.761	0.0929	65.075	0.0154	6.0472	0.1654	4.9251
18	12.375	0.0808	75.836	0.0132	6.1280	0.1632	5.0843
19	14.232	0.0703	88.212	0.0113	6.1982	0.1613	5.2307
20	16.367	0.0611	102.444	0.0098	6.2593	0.1598	5.3651
21	18.822	0.0531	118.810	0.0084	6.3125	0.1584	5.4883
22	21.645	0.0462	137.632	0.0073	6.3587	0.1573	5.6010
23	24.891	0.0402	159.276	0.0063	6.3988	0.1563	5.7040
24	28.625	0.0349	184.168	0.0054	6.4338	0.1554	5.7979
25	32.919	0.0304	212.793	0.0047	6.4642	0.1547	5.8834
26	37.857	0.0264	245.712	0.0041	6.4906	0.1541	5.9612
27	43.535	0.0230	283.569	0.0035	6.5135	0.1535	6.0319
28	50.066	0.0200	327.104	0.0031	6.5335	0.1531	6.0960
29	57.575	0.0174	377.170	0.0027	6.5509	0.1527	6.1541
30	66.212	0.0151	434.745	0.0023	6.5660	0.1523	6.2066
31	76.144	0.0131	500.957	0.0020	6.5791	0.1520	6.2541
32	87.565	0.0114	577.100	0.0017	6.5905	0.1517	6.2970
33	100.700	0.0099	664.666	0.0015	6.6005	0.1515	6.3357
34	115.805	0.0086	765.365	0.0013	6.6091	0.1513	6.3705
35	133.176	0.0075	881.170	0.0011	6.6166	0.1511	6.4019
40	267.864	0.0037	1779.090	0.0006	6.6418	0.1506	6.5168
45	538.769	0.0019	3585.128	0.0003	6.6543	0.1503	6.5830
50	1083.657	0.0009	7217.716	0.0002	6.6605	0.1501	6.6205

TABLE A.20.
16% INTEREST FACTORS FOR DISCRETE COMPOUNDING

	Single Payment		Equal Payment Series				Uniform gradient-series factor
n	Compound-amount factor	Present-worth factor	Compound-amount factor	Sinking-fund factor	Present-worth factor	Capital-recovery factor	
	To find *F* Given *P* *F*/*P*, *i*, *n*	To find *P* Given *F* *P*/*F*, *i*, *n*	To find *F* Given *A* *F*/*A*, *i*, *n*	To find *A* Given *F* *A*/*F*, *i*, *n*	To find *P* Given *A* *P*/*A*, *i*, *n*	To find *A* Given *P* *A*/*P*, *i*, *n*	To find *A* Given *G* *A*/*G*, *i*, *n*
1	1.160	0.8621	1.000	1.0000	0.8621	1.1600	0.0000
2	1.346	0.7432	2.160	0.4630	1.6052	0.6230	0.4630
3	1.561	0.6407	3.506	0.2853	2.2459	0.4453	0.9014
4	1.811	0.5523	5.066	0.1974	2.7982	0.3574	1.3156
5	2.100	0.4761	6.877	0.1454	3.2743	0.3054	1.7060
6	2.436	0.4104	8.977	0.1114	3.6847	0.2714	2.0729
7	2.826	0.3538	11.414	0.0876	4.0386	0.2476	2.4170
8	3.278	0.3050	14.240	0.0702	4.3436	0.2302	2.7388
9	3.803	0.2630	17.519	0.0571	4.6065	0.2171	3.0391
10	4.411	0.2267	21.321	0.0469	4.8332	0.2069	3.3187
11	5.117	0.1954	25.733	0.0389	5.0286	0.1989	3.5783
12	5.936	0.1685	30.850	0.0324	5.1971	0.1924	3.8189
13	6.886	0.1452	36.786	0.0272	5.3423	0.1872	4.0413
14	7.988	0.1252	43.672	0.0229	5.4675	0.1829	4.2464
15	9.268	0.1079	51.660	0.0194	5.5755	0.1794	4.4352
16	10.748	0.0930	60.925	0.0164	5.6685	0.1764	4.6086
17	12.468	0.0802	71.673	0.0140	5.7487	0.1740	4.7676
18	14.463	0.0691	84.141	0.0119	5.8178	0.1719	4.9130
19	16.777	0.0596	98.603	0.0101	5.8775	0.1701	5.0457
20	19.461	0.0514	115.380	0.0087	5.9288	0.1687	5.1666
21	22.574	0.0443	134.841	0.0074	5.9731	0.1674	5.2766
22	26.186	0.0382	157.415	0.0064	6.0113	0.1664	5.3765
23	30.376	0.0329	183.601	0.0054	6.0442	0.1654	5.4671
24	35.236	0.0284	213.978	0.0047	6.0726	0.1647	5.5490
25	40.874	0.0245	249.214	0.0040	6.0971	0.1640	5.6230
26	47.414	0.0211	290.088	0.0034	6.1182	0.1634	5.6898
27	55.000	0.0182	337.502	0.0030	6.1364	0.1630	5.7500
28	63.800	0.0157	392.503	0.0025	6.1520	0.1625	5.8041
29	74.009	0.0135	456.303	0.0022	6.1656	0.1622	5.8528
30	85.850	0.0116	530.312	0.0019	6.1772	0.1619	5.8964
31	99.586	0.0100	616.162	0.0016	6.1872	0.1616	5.9356
32	115.520	0.0087	715.747	0.0014	6.1959	0.1614	5.9706
33	134.003	0.0075	831.267	0.0012	6.2034	0.1612	6.0019
34	155.443	0.0064	965.270	0.0010	6.2098	0.1610	6.0299
35	180.314	0.0055	1120.713	0.0009	6.2153	0.1609	6.0548
40	378.721	0.0026	2360 757	0.0004	6.2335	0.1604	6.1441
45	795.444	0.0013	4965.274	0.0002	6.2421	0.1602	6.1934
50	1670.704	0.0006	10435.649	0.0001	6.2463	0.1601	6.2201

TABLE A.21.
17% INTEREST FACTORS FOR DISCRETE COMPOUNDING

n	Single Payment		Equal Payment Series				Uniform gradient-series factor
	Compound-amount factor	Present-worth factor	Compound-amount factor	Sinking-fund factor	Present-worth factor	Capital-recovery factor	
	To find F Given P $F/P, i, n$	To find P Given F $P/F, i, n$	To find F Given A $F/A, i, n$	To find A Given F $A/F, i, n$	To find P Given A $P/A, i, n$	To find A Given P $A/P, i, n$	To find A Given G $A/G, i, n$
1	1.170	0.8547	1.000	1.0000	0.8547	1.1700	0.0000
2	1.369	0.7305	2.170	0.4608	1.5852	0.6308	0.4608
3	1.602	0.6244	3.539	0.2826	2.2096	0.4526	0.8958
4	1.874	0.5337	5.141	0.1945	2.7432	0.3645	1.3051
5	2.192	0.4561	7.014	0.1426	3.1993	0.3126	1.6893
6	2.565	0.3898	9.207	0.1086	3.5892	0.2786	2.0489
7	3.001	0.3332	11.772	0.0849	3.9224	0.2549	2.3845
8	3.511	0.2848	14.773	0.0677	4.2072	0.2377	2.6969
9	4.108	0.2434	18.285	0.0547	4.4506	0.2247	2.9870
10	4.807	0.2080	22.393	0.0447	4.6586	0.2147	3.2555
11	5.624	0.1778	27.200	0.0368	4.8364	0.2068	3.5035
12	6.580	0.1520	32.824	0.0305	4.9884	0.2005	3.7318
13	7.699	0.1299	39.404	0.0254	5.1183	0.1954	3.9417
14	9.007	0.1110	47.103	0.0212	5.2293	0.1912	4.1340
15	10.539	0.0949	56.110	0.0178	5.3242	0.1878	4.3098
16	12.330	0.0811	66.649	0.0150	5.4053	0.1850	4.4702
17	14.426	0.0693	78.979	0.0127	5.4746	0.1827	4.6162
18	16.679	0.0592	93.406	0.0107	5.5339	0.1807	4.7488
19	19.748	0.0506	110.285	0.0091	5.5845	0.1791	4.8689
20	23.106	0.0433	130.033	0.0077	5.6278	0.1777	4.9776
21	27.034	0.0370	153.139	0.0065	5.6648	0.1765	5.0757
22	31.629	0.0316	180.172	0.0056	5.6964	0.1756	5.1641
23	37.006	0.0270	211.801	0.0047	5.7234	0.1747	5.2436
24	43.297	0.0231	248.808	0.0040	5.7465	0.1740	5.3149
25	50.658	0.0197	292.105	0.0034	5.7662	0.1734	5.3789
26	59.270	0.0169	342.763	0.0029	5.7831	0.1729	5.4362
27	69.345	0.0144	402.032	0.0025	5.7975	0.1725	5.4873
28	81.134	0.0123	471.378	0.0021	5.8099	0.1721	5.5329
29	94.927	0.0105	552.512	0.0018	5.8204	0.1718	5.5736
30	111.065	0.0090	647.439	0.0015	5.8294	0.1715	5.6098
31	129.946	0.0077	758.504	0.0013	5.8371	0.1713	5.6419
32	152.036	0.0066	888.449	0.0011	5.8437	0.1711	5.6705
33	177.883	0.0056	1040.486	0.0010	5.8493	0.1710	5.6958
34	208.123	0.0048	1218.368	0.0008	5.8541	0.1708	5.7182
35	243.503	0.0041	1426.491	0.0007	5.8582	0.1707	5.7380
40	533.869	0.0019	3134.522	0.0003	5.8713	0.1703	5.8073
45	1170.479	0.0009	6879.291	0.0001	5.8773	0.1701	5.8439
50	2566.215	0.0004	15089.502	0.0001	5.8801	0.1701	5.8629

TABLE A.22.
18% INTEREST FACTORS FOR DISCRETE COMPOUNDING

	Single Payment		Equal Payment Series				Uniform
n	Compound-amount factor	Present-worth factor	Compound-amount factor	Sinking-fund factor	Present-worth factor	Capital-recovery factor	Uniform gradient-series factor
	To find *F* Given *P* *F/P, i, n*	To find *P* Given *F* *P/F, i, n*	To find *F* Given *A* *F/A, i, n*	To find *A* Given *F* *A/F, i, n*	To find *P* Given *A* *P/A, i, n*	To find *A* Given *P* *A/P, i, n*	To find *A* Given *G* *A/G, i, n*
1	1.180	0.8475	1.000	1.0000	0.8475	1.1800	0.0000
2	1.392	0.7182	2.180	0.4587	1.5656	0.6387	0.4587
3	1.643	0.6086	3.572	0.2799	2.1743	0.4599	0.8902
4	1.939	0.5158	5.215	0.1917	2.6901	0.3717	1.2947
5	2.288	0.4371	7.154	0.1398	3.1272	0.3198	1.6728
6	2.700	0.3704	9.442	0.1059	3.4976	0.2859	2.0252
7	3.185	0.3139	12.142	0.0824	3.8115	0.2624	2.3526
8	3.759	0.2660	15.327	0.0652	4.0776	0.2452	2.6558
9	4.435	0.2255	19.086	0.0524	4.3030	0.2324	2.9358
10	5.234	0.1911	23.521	0.0425	4.4941	0.2225	3.1936
11	6.176	0.1619	28.755	0.0348	4.6560	0.2148	3.4303
12	7.288	0.1372	34.931	0.0286	4.7932	0.2086	3.6470
13	8.599	0.1163	42.219	0.0237	4.9095	0.2037	3.8449
14	10.147	0.0985	50.818	0.0197	5.0081	0.1997	4.0250
15	11.974	0.0835	60.965	0.0164	5.0916	0.1964	4.1887
16	14.129	0.0708	72.939	0.0137	5.1624	0.1937	4.3369
17	16.672	0.0600	87.068	0.0115	5.2223	0.1915	4.4708
18	19.673	0.0508	103.740	0.0096	5 2732	0.1896	4.5916
19	23.214	0.0431	123.414	0.0081	5.3162	0.1881	4.7003
20	27.393	0.0365	146.628	0.0068	5.3527	0.1868	4.7978
21	32.324	0.0309	174.021	0.0057	5.3837	0.1857	4.8851
22	38.142	0.0262	206.345	0.0048	5.4099	0.1848	4.9632
23	45.008	0.0222	244.487	0.0041	5.4321	0.1841	5.0329
24	53.109	0.0188	289.494	0.0035	5.4509	0.1835	5.0950
25	62.669	0.0160	342.603	0.0029	5.4669	0.1829	5.1502
26	73.949	0.0135	405.272	0.0025	5.4804	0.1825	5.1991
27	87.260	0.0115	479.221	0.0021	5.4919	0.1821	5.2425
28	102.967	0.0097	566.481	0.0018	5.5016	0.1818	5.2810
29	121.501	0.0082	669.447	0.0015	5.5098	0.1815	5.3149
30	143.371	0.0070	790.948	0.0013	5.5168	0.1813	5.3448
31	169.177	0.0059	934.319	0.0011	5.5227	0.1811	5.3712
32	199.629	0.0050	1103.496	0.0009	5.5277	0.1809	5.3945
33	235.563	0.0042	1303.125	0.0008	5.5320	0.1808	5.4149
34	277.964	0.0036	1538.688	0.0006	5.5356	0.1806	5.4328
35	327.997	0.0030	1816.642	0.0006	5.5386	0.1806	5.4485
40	750.378	0 0013	4163.213	0.0002	5.5482	0.1802	5.5022
45	1716.684	0.0006	9531.577	0.0001	5.5523	0.1801	5.5293
50	3927.357	0.0003	21813.094	0.0001	5.5541	0.1801	5.5428

TABLE A.23.
19% INTEREST FACTORS FOR DISCRETE COMPOUNDING

n	Single Payment		Equal Payment Series				Uniform gradient-series factor
	Compound-amount factor	Present-worth factor	Compound-amount factor	Sinking-fund factor	Present-worth factor	Capital-recovery factor	
	To find F Given P $F/P, i, n$	To find P Given F $P/F, i, n$	To find F Given A $F/A, i, n$	To find A Given F $A/F, i, n$	To find P Given A $P/A, i, n$	To find A Given P $A/P, i, n$	To find A Given G $A/G, i, n$
1	1.190	0.8403	1.000	1.0000	0.8403	1.1900	0.0000
2	1.416	0.7062	2.190	0.4566	1.5465	0.6466	0.4566
3	1.685	0.5934	3.606	0.2773	2.1399	0.4673	0.8846
4	2.005	0.4987	5.291	0.1890	2.6386	0.3790	1.2844
5	2.386	0.4190	7.297	0.1371	3.0576	0.3271	1.6566
6	2.840	0.3521	9.683	0.1033	3.4098	0.2933	2.0019
7	3.379	0.2959	12.523	0.0799	3.7057	0.2699	2.3211
8	4.021	0.2487	15.902	0.0629	3.9544	0.2529	2.6154
9	4.785	0.2090	19.923	0.0502	4.1633	0.2402	2.8856
10	5.695	0.1756	24.709	0.0405	4.3389	0.2305	3.1331
11	6.777	0.1476	30.404	0.0329	4.4865	0.2229	3.3589
12	8.064	0.1240	37.180	0.0269	4.6105	0.2169	3.5645
13	9.596	0.1042	45.244	0.0221	4.7147	0.2121	3.7509
14	11.420	0.0876	54.841	0.0182	4.8023	0.2082	3.9196
15	13.590	0.0736	66.261	0.0151	4.8759	0.2051	4.0717
16	16.172	0.0618	79.850	0.0125	4.9377	0.2025	4.2086
17	19.244	0.0520	96.022	0.0104	4.9897	0.2004	4.3314
18	22.901	0.0437	115.266	0.0087	5.0333	0.1987	4.4413
19	27.252	0.0367	138.166	0.0072	5.0700	0.1972	4.5394
20	32.429	0.0308	165.418	0.0060	5.1009	0.1960	4.6268
21	38.591	0.0259	197.847	0.0051	5.1268	0.1951	4.7045
22	45.923	0.0218	236.438	0.0042	5.1486	0.1942	4.7734
23	54.649	0.0183	282.362	0.0035	5.1668	0.1935	4.8344
24	65.032	0.0154	337.011	0.0030	5.1822	0.1930	4.8883
25	77.388	0.0129	402.032	0.0025	5.1951	0.1925	4.9359
26	92.092	0.0109	479.431	0.0021	5.2060	0.1921	4.9777
27	109.589	0.0091	571.522	0.0017	5.2151	0.1917	5.0145
28	130.411	0.0077	681.112	0.0015	5.2228	0.1915	5.0468
29	155.189	0.0064	811.523	0.0012	5.2292	0.1912	5.0751
30	184.675	0.0054	966.712	0.0010	5.2347	0.1910	5.0998
31	219.764	0.0046	1151.388	0.0009	5.2392	0.1909	5.1215
32	261.519	0.0038	1371.151	0.0007	5.2430	0.1907	5.1403
33	311.207	0.0032	1632.670	0.0006	5.2462	0.1906	5.1568
34	370.337	0.0027	1943.877	0.0005	5.2489	0.1905	5.1711
35	440.701	0.0023	2314.214	0.0004	5.2512	0.1904	5.1836
40	1051.668	0.0010	5529.829	0.0002	5.2582	0.1902	5.2251
45	2509.651	0.0004	13203.424	0.0001	5.2611	0.1901	5.2452
50	5988.914	0.0002	31515.336	0.0000	5.2623	0.1900	5.2548

TABLE A.24.
20% INTEREST FACTORS FOR DISCRETE COMPOUNDING

n	Single Payment		Equal Payment Series				Uniform gradient-series factor
	Compound-amount factor	Present-worth factor	Compound-amount factor	Sinking-fund factor	Present-worth factor	Capital-recovery factor	
	To find F Given P $F/P, i, n$	To find P Given F $P/F, i, n$	To find F Given A $F/A, i, n$	To find A Given F $A/F, i, n$	To find P Given A $P/A, i, n$	To find A Given P $A/P, i, n$	To find A Given G $A/G, i, n$
1	1.200	0.8333	1.000	1.0000	0.8333	1.2000	0.0000
2	1.440	0.6945	2.200	0.4546	1.5278	0.6546	0.4546
3	1.728	0.5787	3.640	0.2747	2.1065	0.4747	0.8791
4	2.074	0.4823	5.368	0.1863	2.5887	0.3863	1.2742
5	2.488	0.4019	7.442	0.1344	2.9906	0.3344	1.6405
6	2.986	0.3349	9.930	0.1007	3.3255	0.3007	1.9788
7	3.583	0.2791	12.916	0.0774	3.6046	0.2774	2.2902
8	4.300	0.2326	16.499	0.0606	3.8372	0.2606	2.5756
9	5.160	0.1938	20.799	0.0481	4.0310	0.2481	2.8364
10	6.192	0.1615	25.959	0.0385	4.1925	0.2385	3.0739
11	7.430	0.1346	32.150	0.0311	4.3271	0.2311	3.2893
12	8.916	0.1122	39.581	0.0253	4.4392	0.2253	3.4841
13	10.699	0.0935	48.497	0.0206	4.5327	0.2206	3.6597
14	12.839	0.0779	59.196	0.0169	4.6106	0.2169	3.8175
15	15.407	0.0649	72.035	0.0139	4.6755	0.2139	3.9589
16	18.488	0.0541	87.442	0.0114	4.7296	0.2114	4.0851
17	22.186	0.0451	105.931	0.0095	4.7746	0.2095	4.1976
18	26.623	0.0376	128.117	0.0078	4.8122	0.2078	4.2975
19	31.948	0.0313	154.740	0.0065	4.8435	0.2065	4.3861
20	38.338	0.0261	186.688	0.0054	4.8696	0.2054	4.4644
21	46.005	0.0217	225.026	0.0045	4.8913	0.2045	4.5334
22	55.206	0.0181	271.031	0.0037	4.9094	0.2037	4.5942
23	66.247	0.0151	326.237	0.0031	4.9245	0.2031	4.6475
24	79.497	0.0126	392.484	0.0026	4.9371	0.2026	4.6943
25	95.396	0.0105	471.981	0.0021	4.9476	0.2021	4.7352
26	114.475	0.0087	567.377	0.0018	4.9563	0.2018	4.7709
27	137.371	0.0073	681.853	0.0015	4.9636	0.2015	4.8020
28	164.845	0.0061	819.223	0.0012	4.9697	0.2012	4.8291
29	197.814	0.0051	984.068	0.0010	4.9747	0.2010	4.8527
30	237.376	0.0042	1181.882	0.0009	4.9789	0.2009	4.8731
31	284.852	0.0035	1419.258	0.0007	4.9825	0.2007	4.8908
32	341.822	0.0029	1704.109	0.0006	4.9854	0.2006	4.9061
33	410.186	0.0024	2045.931	0.0005	4.9878	0.2005	4.9194
34	492.224	0.0020	2456.118	0.0004	4.9899	0.2004	4.9308
35	590.668	0.0017	2948.341	0.0003	4.9915	0.2003	4.9407
40	1469.772	0.0007	7343.858	0.0002	4.9966	0.2001	4.9728
45	3657.262	0.0003	18281.310	0.0001	4.9986	0.2001	4.9877
50	9100.438	0.0001	45497.191	0.0000	4.9995	0.2000	4.9945

TABLE A.25.
25% INTEREST FACTORS FOR DISCRETE COMPOUNDING

	Single Payment		Equal Payment Series				Uniform gradient-series factor
n	Compound-amount factor	Present-worth factor	Compound-amount factor	Sinking-fund factor	Present-worth factor	Capital-recovery factor	
	To find F Given P $F/P, i, n$	To find P Given F $P/F, i, n$	To find F Given A $F/A, i, n$	To find A Given F $A/F, i, n$	To find P Given A $P/A, i, n$	To find A Given P $A/P, i, n$	To find A Given G $A/G, i, n$
1	1.250	0.8000	1.000	1.0000	0.8000	1.2500	0.0000
2	1.563	0.6400	2.250	0.4445	1.4400	0.6945	0.4445
3	1.953	0.5120	3.813	0.2623	1.9520	0.5123	0.8525
4	2.441	0.4096	5.766	0.1735	2.3616	0.4235	1.2249
5	3.052	0.3277	8.207	0.1219	2.6893	0.3719	1.5631
6	3.815	0.2622	11.259	0.0888	2.9514	0.3388	1.8683
7	4.768	0.2097	15.073	0.0664	3.1611	0.3164	2.1424
8	5.960	0.1678	19.842	0.0504	3.3289	0.3004	2.3873
9	7.451	0.1342	25.802	0.0388	3.4631	0.2888	2.6048
10	9.313	0.1074	33.253	0.0301	3.5705	0.2801	2.7971
11	11.642	0.0859	42.566	0.0235	3.6564	0.2735	2.9663
12	14.552	0.0687	54.208	0.0185	3.7251	0.2685	3.1145
13	18.190	0.0550	68.760	0.0146	3.7801	0.2646	3.2438
14	22.737	0.0440	86.949	0.0115	3.8241	0.2615	3.3560
15	28.422	0.0352	109.687	0.0091	3.8593	0.2591	3.4530
16	35.527	0.0282	138.109	0.0073	3.8874	0.2573	3.5366
17	44.409	0.0225	173.636	0.0058	3.9099	0.2558	3.6084
18	55.511	0.0180	218.045	0.0046	3.9280	0.2546	3.6698
19	69.389	0.0144	273.556	0.0037	3.9424	0.2537	3.7222
20	86.736	0.0115	342.945	0.0029	3.9539	0.2529	3.7667
21	108.420	0.0092	429.681	0.0023	3.9631	0.2523	3.8045
22	135.525	0.0074	538.101	0.0019	3.9705	0.2519	3.8365
23	169.407	0.0059	673.626	0.0015	3.9764	0.2515	3.8634
24	211.758	0.0047	843.033	0.0012	3.9811	0.2512	3.8861
25	264.698	0.0038	1054.791	0.0010	3.9849	0.2510	3.9052
26	330.872	0.0030	1319.489	0.0008	3.9879	0.2508	3.9212
27	413.590	0.0024	1650.361	0.0006	3.9903	0.2506	3.9346
28	516.988	0.0019	2063.952	0.0005	3.9923	0.2505	3.9457
29	646.235	0.0016	2580.939	0.0004	3.9938	0.2504	3.9551
30	807.794	0.0012	3227.174	0.0003	3.9951	0.2503	3.9628
31	1009.742	0.0010	4034.968	0.0003	3.9960	0.2503	3.9693
32	1262.177	0.0008	5044.710	0.0002	3.9968	0.2502	3.9746
33	1577.722	0.0006	6306.887	0.0002	3.9975	0.2502	3.9791
34	1972.152	0.0005	7884.609	0.0001	3.9980	0.2501	3.9828
35	2465.190	0.0004	9856.761	0.0001	3.9984	0.2501	3.9858

TABLE A.26.
30% INTEREST FACTORS FOR DISCRETE COMPOUNDING

n	Single Payment		Equal Payment Series				Uniform gradient-series factor
	Compound-amount factor	Present-worth factor	Compound-amount factor	Sinking-fund factor	Present-worth factor	Capital-recovery factor	
	To find F Given P $F/P, i, n$	To find P Given F $P/F, i, n$	To find F Given A $F/A, i, n$	To find A Given F $A/F, i, n$	To find P Given A $P/A, i, n$	To find A Given P $A/P, i, n$	To find A Given G $A/G, i, n$
1	1.300	0.7692	1.000	1.0000	0.7692	1.3000	0.0000
2	1.690	0.5917	2.300	0.4348	1.3610	0.7348	0.4348
3	2.197	0.4552	3.990	0.2506	1.8161	0.5506	0.8271
4	2.856	0.3501	6.187	0.1616	2.1663	0.4616	1.1783
5	3.713	0.2693	9.043	0.1106	2.4356	0.4106	1.4903
6	4.827	0.2072	12.756	0.0784	2.6428	0.3784	1.7655
7	6.275	0.1594	17.583	0.0569	2.8021	0.3569	2.0063
8	8.157	0.1226	23.858	0.0419	2.9247	0.3419	2.2156
9	10.605	0.0943	32.015	0.0312	3.0190	0.3312	2.3963
10	13.786	0.0725	42.620	0.0235	3.0915	0.3235	2.5512
11	17.922	0.0558	56.405	0.0177	3.1473	0.3177	2.6833
12	23.298	0.0429	74.327	0.0135	3.1903	0.3135	2.7952
13	30.288	0.0330	97.625	0.0103	3.2233	0.3103	2.8895
14	39.374	0.0254	127.913	0.0078	3.2487	0.3078	2.9685
15	51.186	0.0195	167.286	0.0060	3.2682	0.3060	3.0345
16	66.542	0.0150	218.472	0.0046	3.2832	0.3046	3.0892
17	86.504	0.0116	285.014	0.0035	3.2948	0.3035	3.1345
18	112.455	0.0089	371.518	0.0027	3.3037	0.3027	3.1718
19	146.192	0.0069	483.973	0.0021	3.3105	0.3021	3.2025
20	190.050	0.0053	630.165	0.0016	3.3158	0.3016	3.2276
21	247.065	0.0041	820.215	0.0012	3.3199	0.3012	3.2480
22	321.184	0.0031	1067.280	0.0009	3.3230	0.3009	3.2646
23	417.539	0.0024	1388.464	0.0007	3 3254	0.3007	3.2781
24	542.801	0.0019	1806.003	0.0006	3.3272	0.3006	3.2890
25	705.641	0.0014	2348.803	0.0004	3.3286	0.3004	3.2979
26	917.333	0.0011	3054.444	0.0003	3.3297	0.3003	3.3050
27	1192.533	0.0008	3971.778	0.0003	3.3305	0.3003	3.3107
28	1550.293	0.0007	5164.311	0.0002	3.3312	0.3002	3.3153
29	2015.381	0.0005	6714.604	0.0002	3.3317	0.3002	3.3189
30	2619.996	0.0004	8729.985	0.0001	3.3321	0.3001	3.3219
31	3405.994	0.0003	11349.981	0.0001	3.3324	0.3001	3.3242
32	4427.793	0.0002	14755.975	0.0001	3.3326	0.3001	3.3261
33	5756.130	0.0002	19183.768	0.0001	3.3328	0.3001	3.3276
34	7482.970	0.0001	24939.899	0.0001	3.3329	0.3001	3.3288
35	9727.860	0.0001	32422.868	0.0000	3.3330	0.3000	3.3297

TABLE A.27.
40% INTEREST FACTORS FOR DISCRETE COMPOUNDING

n	Single Payment		Equal Payment Series				Uniform gradient-series factor
	Compound-amount factor	Present-worth factor	Compound-amount factor	Sinking-fund factor	Present-worth factor	Capital-recovery factor	
	To find F Given P $F/P, i, n$	To find P Given F $P/F, i, n$	To find F Given A $F/A, i, n$	To find A Given F $A/F, i, n$	To find P Given A $P/A, i, n$	To find A Given P $A/P, i, n$	To find A Given G $A/G, i, n$
1	1.400	0.7143	1.000	1.0000	0.7143	1.4001	0.0000
2	1.960	0.5103	2.400	0.4167	1.2245	0.8167	0.4167
3	2.744	0.3645	4.360	0.2294	1.5890	0.6294	0.7799
4	3.842	0.2604	7.104	0.1408	1.8493	0.5408	1.0924
5	5.378	0.1860	10.946	0.0914	2.0352	0.4914	1.3580
6	7.530	0.1329	16.324	0.0613	2.1680	0.4613	1.5811
7	10.541	0.0949	23.853	0.0420	2.2629	0.4420	1.7664
8	14.758	0.0678	34.395	0.0291	2.3306	0.4291	1.9186
9	20.661	0.0485	49.153	0.0204	2.3790	0.4204	2.0423
10	28.925	0.0346	69.814	0.0144	2.4136	0.4144	2.1420
11	40.496	0.0247	98.739	0.0102	2.4383	0.4102	2.2215
12	56.694	0.0177	139.234	0.0072	2.4560	0.4072	2.2846
13	79.371	0.0126	195.928	0.0052	2.4686	0.4052	2.3342
14	111.120	0.0090	275.299	0.0037	2.4775	0.4037	2.3729
15	155.568	0.0065	386.419	0.0026	2.4840	0.4026	2.4030
16	217.794	0.0046	541.986	0.0019	2.4886	0.4019	2.4262
17	304.912	0.0033	759.780	0.0014	2.4918	0.4014	2.4441
18	426.877	0.0024	1064.691	0.0010	2.4942	0.4010	2.4578
19	597.627	0.0017	1491.567	0.0007	2.4959	0.4007	2.4682
20	836.678	0.0012	2089.195	0.0005	2.4971	0.4005	2.4761
21	1171.348	0.0009	2925.871	0.0004	2.4979	0.4004	2.4821
22	1639.887	0.0007	4097.218	0.0003	2.4985	0.4003	2.4866
23	2295.842	0.0005	5737.105	0.0002	2.4990	0.4002	2.4900
24	3214.178	0.0004	8032.945	0.0002	2.4993	0.4002	2.4926
25	4499.847	0.0003	11247.110	0.0001	2.4995	0.4001	2.4945
26	6299.785	0.0002	15746.960	0.0001	2.4997	0.4001	2.4959
27	8819.695	0.0002	22046.730	0.0001	2.4998	0.4001	2.4970
28	12347.570	0.0001	30866.430	0.0001	2.4998	0.4001	2.4978
29	17286.590	0.0001	43213.990	0.0001	2.4999	0.4001	2.4984
30	24201.230	0.0001	60500.580	0.0001	2.4999	0.4001	2.4988

TABLE A.28.
50% INTEREST FACTORS FOR DISCRETE COMPOUNDING

	Single Payment		Equal Payment Series				
n	Compound-amount factor	Present-worth factor	Compound-amount factor	Sinking-fund factor	Present-worth factor	Capital-recovery factor	Uniform gradient-series factor
	To find F Given P F/P, i, n	To find P Given F P/F, i, n	To find F Given A F/A, i, n	To find A Given F A/F, i, n	To find P Given A P/A, i, n	To find A Given P A/P, i, n	To find A Given G A/G, i, n
1	1.500	0.6667	1.000	1.0000	0.6667	1.5000	0.0001
2	2.250	0.4445	2.500	0.4000	1.1112	0.9001	0.4001
3	3.375	0.2963	4.750	0.2106	1.4075	0.7106	0.7369
4	5.063	0.1976	8.125	0.1231	1.6050	0.6231	1.0154
5	7.594	0.1317	13.188	0.0759	1.7367	0.5759	1.2418
6	11.391	0.0878	20.781	0.0482	1.8245	0.5482	1.4226
7	17.086	0.0586	32.172	0.0311	1.8830	0.5311	1.5649
8	25.629	0.0391	49.258	0.0204	1.9220	0.5204	1.6752
9	38.443	0.0261	74.887	0.0134	1.9480	0.5134	1.7597
10	57.665	0.0174	113.330	0.0089	1.9654	0.5089	1.8236
11	86.498	0.0116	170.995	0.0059	1.9769	0.5059	1.8714
12	129.746	0.0078	257.493	0.0039	1.9846	0.5039	1.9068
13	194.620	0.0052	387.239	0.0026	1.9898	0.5026	1.9329
14	291.929	0.0035	581.858	0.0018	1.9932	0.5018	1.9519
15	437.894	0.0023	873.788	0.0012	1.9955	0.5012	1.9657
16	656.841	0.0016	1311.681	0.0008	1.9970	0.5008	1.9757
17	985.261	0.0011	1968.522	0.0006	1.9980	0.5006	1.9828
18	1477.891	0.0007	2953.783	0.0004	1.9987	0.5004	1.9879
19	2216.837	0.0005	4431.671	0.0003	1.9991	0.5003	1.9915
20	3325.256	0.0004	6648.511	0.0002	1.9994	0.5002	1.9940
21	4987.882	0.0003	9973.765	0.0002	1.9996	0.5002	1.9958
22	7481.824	0.0002	14961.640	0.0001	1.9998	0.5001	1.9971
23	11222.730	0.0001	22443.470	0.0001	1.9999	0.5001	1.9980
24	16834.100	0.0001	33666.210	0.0001	1.9999	0.5001	1.9986
25	25251.160	0.0001	50500.330	0.0001	2.0000	0.5001	1.9991

B

Effective Interest Rates Corresponding to Nominal Rate r

TABLE B.1.
EFFECTIVE INTEREST RATES CORRESPONDING TO NOMINAL RATE r

r	Compounding Frequency					
	Semi-annually	Quarterly	Monthly	Weekly	Daily	Continuously
	$\left(1+\frac{r}{2}\right)^2-1$	$\left(1+\frac{r}{4}\right)^4-1$	$\left(1+\frac{r}{12}\right)^{12}-1$	$\left(1+\frac{r}{52}\right)^{52}-1$	$\left(1+\frac{r}{365}\right)^{365}-1$	$\left(1+\frac{r}{\infty}\right)^{\infty}-1$
.01	.010025	.010038	.010046	.010049	.010050	.010050
.02	.020100	.020151	.020184	.020197	.020200	.020201
.03	.030225	.030339	.030416	.030444	.030451	.030455
.04	.040400	.040604	.040741	.040793	.040805	.040811
.05	.050625	.050945	.051161	.051244	.051261	.051271
.06	.060900	.061364	.061678	.061797	.061799	.061837
.07	.071225	.071859	.072290	.072455	.072469	.072508
.08	.081600	.082432	.082999	.083217	.083246	.083287
.09	.092025	.093083	.093807	.094085	.094132	.094174
.10	.102500	.103813	.104713	.105060	.105126	.105171
.11	.113025	.114621	.115718	.116144	.116231	.116278
.12	.123600	.125509	.126825	.127336	.127447	.127497
.13	.134225	.136476	.138032	.138644	.138775	.138828
.14	.144900	.147523	.149341	.150057	.150217	.150274
.15	.155625	.158650	.160755	.161582	.161773	.161834
.16	.166400	.169859	.172270	.173221	.173446	.173511
.17	.177225	.181148	.183891	.184974	.185235	.185305
.18	.188100	.192517	.195618	.196843	.197142	.197217
.19	.199025	.203971	.207451	.208828	.209169	.209250
.20	.210000	.215506	.219390	.220931	.221316	.221403
.21	.221025	.227124	.231439	.233153	.233584	.233678
.22	.232100	.238825	.243596	.245494	.245976	.246077
.23	.243225	.250609	.255863	.257957	.258492	.258600
.24	.254400	.262477	.268242	.270542	.271133	.271249
.25	.265625	.274429	.280731	.283250	.283901	.284025
.26	.276900	.286466	.293333	.296090	.296796	.296930
.27	.288225	.298588	.306050	.309049	.309821	.309964
.28	.299600	.310796	.318880	.322135	.322976	.323130
.29	.311025	.323089	.331826	.335350	.336264	.336428
.30	.322500	.335469	.344889	.348693	.349684	.349859
.31	.334025	.347936	.358068	.362168	.363238	.363425
.32	.345600	.360489	.371366	.375775	.376928	.377128
.33	.357225	.373130	.384784	.389515	.390756	.390968
.34	.368900	.385859	.398321	.403389	.404722	.404948
.35	.380625	.398676	.411979	.417399	.418827	.419068

C

Funds Flow Conversion Factors

TABLE C.1.
FUNDS FLOW
CONVERSION
FACTORS

r	$\frac{e^r - 1}{r}$ $[A/\bar{A}, r]$
1	1.005020
2	1.010065
3	1.015150
4	1.020270
5	1.025422
6	1.030608
7	1.035831
8	1.041088
9	1.046381
10	1.051709
11	1.057073
12	1.062474
13	1.067910
14	1.073384
15	1.078894
16	1.084443
17	1.090028
18	1.095652
19	1.101313
20	1.107014
21	1.112752
22	1.118530
23	1.124347
24	1.130204
25	1.136101
26	1.142038
27	1.148016
28	1.154035
29	1.160094
30	1.166196
31	1.172339
32	1.178524
33	1.184751
34	1.191022
35	1.197335
36	1.203692
37	1.210093
38	1.216538
39	1.223027
40	1.229561

Selected References

AMERICAN TELEPHONE & TELEGRAPH COMPANY, *Engineering Economy*, 3rd ed., New York: McGraw-Hill Book Company, 1980.

ANG, A. H. and W. H. TANG, *Probability Concepts in Engineering Planning and Design*, New York: John Wiley and Sons, Inc., 1984.

AU, T. and T. P. AU, *Engineering Economics for Capital Investment Analysis*, 2nd ed., Englewood Cliffs, NJ: Prentice-Hall, Inc., 1992.

BARISH, N. N. and S. KAPLAN, *Economic Analysis for Engineering and Managerial Decision Making*, 2nd ed., New York: McGraw-Hill Book Company, 1978.

BARNARD, C. I., *The Functions of the Executive*, Cambridge, MA: Harvard University Press, 1938.

BIERMAN, H. and S. SMIDT, *The Capital Budgeting Decision*, 6th ed., New York: Macmillan Publishing Co., Inc., 1984.

BLANCHARD, B. S. and W. J. FABRYCKY, *Systems Engineering and Analysis*, 2nd ed. Englewood Cliffs, NJ: Prentice-Hall, Inc., 1990.

BLANK, L. T. and A. J. TARQUIN, *Engineering Economy*, 2nd ed., New York: McGraw-Hill Book Company, 1983.

BROWN, R. J. and R. R. YANUCK, *Introduction to Life Cycle Costing*, Englewood Cliffs, NJ: Prentice-Hall, Inc., 1985.

BUSSEY, L. E., and T. E. ESCHENBACH, *The Economic Analysis of Industrial Projects*, Englewood Cliffs, NJ: Prentice-Hall, Inc., 1992.

CANADA, J. R. and W. G. Sullivan, *Economic and Multiattribute Evaluation of Advanced Manufacturing Systems*, Englewood Cliffs, NJ: Prentice-Hall, Inc., 1989.

CLARK, F. D. and A. B. LORENZONI, *Applied Cost Engineering*, 2nd ed., New York: Mercel Dekker, Inc., 1985.

COLLIER, C. A. and W. B. LEDBETTER, *Engineering Cost Analysis*, New York: Harper and Row, Publishers, Inc., 1982.

DASGUPTA, A. K. and D. W. PEARCE, *Cost-Benefit Analysis: Theory and Practice*, New York: Barnes and Noble, 1972.

DEGARMO, E. P., W. G. SULLIVAN, and J. A. BONTIDELLI, *Engineering Economy*, 9th ed., New York: Macmillan Publishing Co., Inc., 1993.

DE LA MARE, R. F., *Manufacturing Systems Economics*, East Sussex, England: Holt, Rinehart and Winston, 1982.

ENGLISH, J. M., *Cost Effectiveness*, New York: John Wiley & Sons, Inc., 1968.

ENGLISH, J. M., *Project Evaluation*, New York: Macmillan Publishing Co., Inc., 1984.

FABRYCKY, W. J., and B. S. BLANCHARD, *Life-Cycle Cost and Economic Analysis*, Englewood Cliffs, NJ: Prentice-Hall, Inc., 1991.

FABRYCKY, W. J. and G. J. THUESEN, *Economic Decision Analysis*, 2nd ed., Englewood Cliffs, NJ: Prentice-Hall, Inc., 1980.

FABRYCKY, W. J., P. M. GHARE, and P. E. TORGERSEN, *Applied Operations Research and Management Science*, Englewood Cliffs, NJ: Prentice-Hall, Inc., 1984.

FLEISCHER, G. A., *Engineering Economy*, Monterey, CA: Brooks/Cole, 1984.

GOICOECHEA, A., D. R. HANSEN, and L. DUCKSTEIN, *Multiobjective Decision Analysis with Engineering and Business Applications*, New York: John Wiley and Sons, Inc., 1982.

GRANT, E. L., W. G. IRESON, and R. S. LEAVENWORTH, *Principles of Engineering Economy*, 8th ed., New York: John Wiley and Sons, Inc., 1990.

GRIFFITHS, R. F., *Dealing with Risk—The Planning Management and Acceptability of Technological Risk*, New York: John Wiley and Sons, Inc., 1981.

HALPIN, D. W., *Financial and Cost Concepts for Construction Management*, New York: John Wiley and Sons, Inc., 1985.

HERTZ, D. B. and H. THOMAS, *Risk Analysis and Its Applications*, New York: John Wiley and Sons, Inc., 1983.

HIRSLEIFER, J., *Investment, Interest, and Capital*. Englewood Cliffs, NJ: Prentice-Hall, Inc., 1970.

JELEN, F. C. and J. H. BLACK, *Cost and Optimization Engineering*, 2nd ed., New York: McGraw-Hill Book Company, 1983.

JEYNES, P. H., *Profitability and Economic Choice*, Ames, lowa: The lowa State University Press, 1968.

JONES, B. W., *Inflation in Engineering Economic Analysis*, New York: John Wiley and Sons Inc., 1982.

KAPLAN, S., *Energy Economics, Quantitative Methods for Energy and Environmental Decisions*, New York: McGraw-Hill Book Company, 1983.

KEENEY, R. L. and H. RAIFFA, *Decisions with Multiple Objectives: Preferences and Value Tradeoffs*, New York: John Wiley and Sons, Inc., 1976.

KURTZ, M., *Handbook of Engineering Economics,* New York: McGraw-Hill Book Company, 1984.

LINDLEY, D. V., *Making Decisions,* 2nd ed., New York: John Wiley and Sons, Inc., 1985.

MARSH, W. D., *Economics of Electric Utility Power Generation,* New York: Oxford University Press, 1980.

MARSTON, A., R. WINFREY, and J. C. HEMPSTEAD, *Engineering Valuation and Depreciation,* Ames, Iowa: Iowa State University Press, 1963.

MCNAMEE, P. and J. CELONA, *Decision Analysis for the Professional with Supertree,* Redwood City, CA: The Scientific Press, 1987.

NEWNAN, D. G., *Engineering Economic Analysis,* 3rd ed., San Jose, CA: Engineering Press, 1987.

OAKFORD, R. V., *Capital Budgeting,* New York: The Ronald Press Company, 1970.

OSTWALD, P. F., *Cost Estimating,* 3rd ed., Englewood Cliffs, NJ: Prentice-Hall, Inc., 1992.

PARK, C. S. AND G. P. SHARP-BETTE, *Advanced Engineering Economics,* New York: John Wiley and Sons, Inc., 1990.

PEARCE, D. W. and C. A. NASH, *The Social Appraisal of Projects—A Text in Cost-Benefit Analysis,* New York: Halsted Press, 1981.

PETERS, M. S. and K. D. TIMMERHAUS, *Plant Design and Economics for Chemical Engineers,* 4th ed., New York: McGraw-Hill Book Company, 1980.

PEURIFOY, R. L., *Estimating Construction Costs,* New York: McGraw-Hill Book Company, 1975.

RAIFFA, H., *Decision Analysis: Introductory Lectures on Choice Under Uncertainty,* Reading, MA: Addison-Wesley Publishing Company, Inc., 1968.

RIGGS, J. L. AND T. M. WEST, *Engineering Economics,* 2nd ed., New York: McGraw-Hill Book Company, 1982.

SEPULVEDA, J. A., W. E. SOUDER, and B. S. GOTTFRIED, *Theory and Problems of Engineering Economics* (Schaum's Outline Series), New York: John Wiley and Sons, Inc., 1984.

SMITH, G. W., *Engineering Economy,* 4th ed., Ames, Iowa: Iowa State University Press 1987.

SPRAGUE, J. C. and J. D. WHITTAKER, *Economic Analysis for Engineers and Managers,* Englewood Cliffs: NJ: Prentice-Hall, Inc., 1986.

STEINER, H. M., *Public and Private Investments, Socioeconomic Analysis,* New York: John Wiley and Sons, Inc., 1980.

STERMOLE, F. J., *Economic Evaluation and Investment Decision Methods,* Golden, CO: Investment Evaluations Corp., 1974.

STEVENS, G. T., *The Economic Analysis of Capital Investments for Managers and Engineers,* Reston, Virginia: Reston Publishing Co., Inc., 1983.

STEWART, R. D., *Cost Estimating,* New York: John Wiley and Sons, Inc., 1982.

STEWART, R. D. and R. M. WYSKIDA, *Cost Estimator's Reference Manual,* New York: John Wiley and Sons, Inc., 1987.

TAYLOR, G. A., *Managerial and Engineering Economy*, 3rd ed., New York: D. Van Nostrand Company, 1980.

VAN HORNE, J. C., *Financial Management and Policy*, 5th ed., Englewood Cliffs, NJ: Prentice-Hall, Inc., 1980.

WHITE, J. A., M. H. AGEE, and K. E. CASE, *Principles of Engineering Economic Analysis*, 3rd ed., New York: John Wiley & Sons. Inc., 1989.

WINFREY, R., *Economic Analysis for Highways*, Scranton, PA.: International Textbook Company, 1969.

ZELENY, M., *Multiple Criteria Decision Making*, New York: McGraw-Hill Book Company, 1982.

Index

A

B

C

D

E

F

J

K

L

Q

R

S

Find	Given	Formula
F	P	$F = P(F/P, i, n)$
P	F	$P = F(P/F, i, n)$
F	A	$F = A(F/A, i, n)$
A	F	$A = F(A/F, i, n)$
P	A	$P = A(P/A, i, n)$
A	P	$A = P(A/P, i, n)$
A	G	$A = G(A/G, i, n)$
P	F_1	$P = \frac{F_1(P/A, g', n)}{1 + g}$

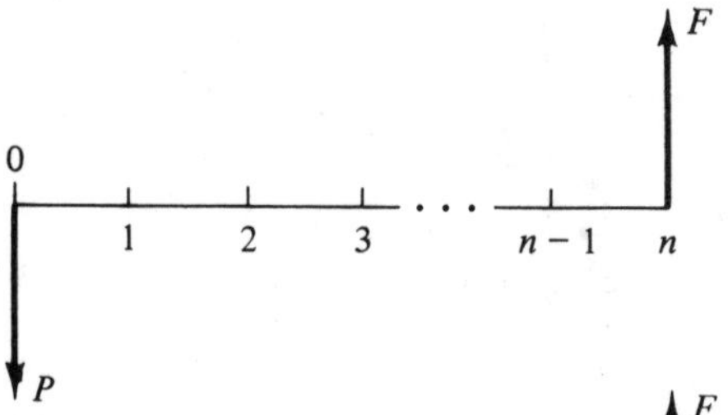

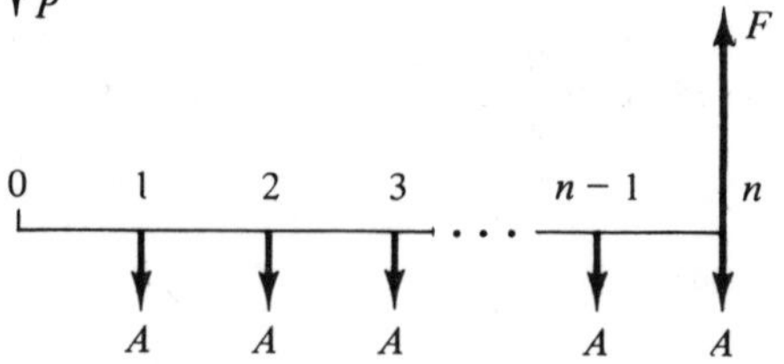

F
0
1 2 3 . . . $n - 1$ n
P

F
0 1 2 3 . . . $n - 1$ n
A A A A A

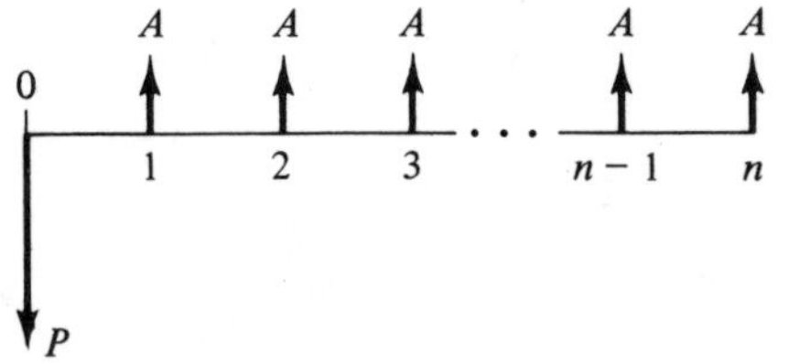

A A A A A
0
1 2 3 . . . $n - 1$ n
P

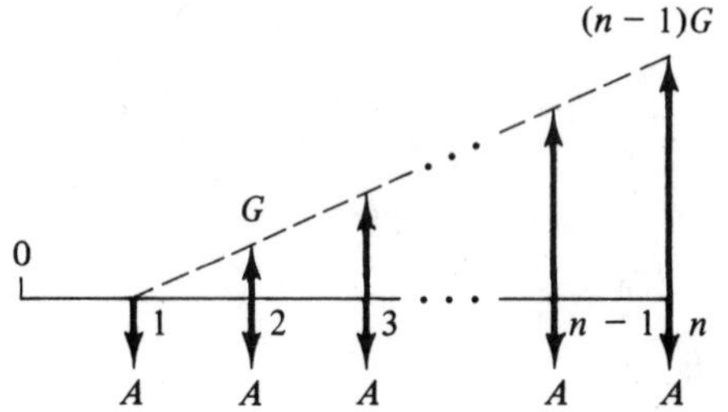

$(n - 1)G$
G
0
1 2 3 . . . $n - 1$ n
A A A A A

$F_1(1 + g)^{n-1}$
F_1
0
1 2 3 . . . $n - 1$ n
P

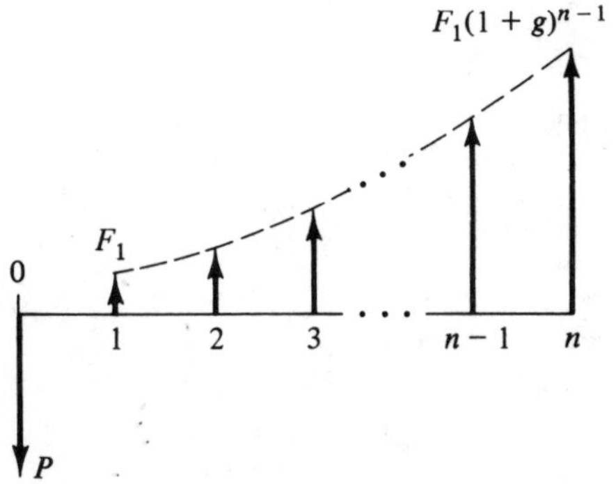